ADVANCES IN NATURAL GAS TECHNOLOGY

Edited by **Hamid A. Al-Megren**

INTECHOPEN.COM

Advances in Natural Gas Technology
Edited by Hamid A. Al-Megren

CBS, Edition **2017**

Published by InTech
Janeza Trdine 9, 51000 Rijeka, Croatia

Publishing Process Manager Ivana Zec

Technical Editor InTech DTP team

Cover Designer InTech Design Team

Additional hard copies can be obtained from orders@intechopen.com

Advances in Natural Gas Technology, Edited by Hamid A. Al-Megren
p. cm.
ISBN 978-953-51-0507-7

Contents

Contents

Contents

Preface

Natural gas is a vital component of the world's supply of energy and an important source of many bulk chemicals and speciality chemicals. It is one of the cleanest, safest, and most useful of all energy sources, and helps to meet the world's rising demand for cleaner energy into the future. However, exploring, producing and bringing gas to the user or converting gas into desired chemicals is a systematical engineering project, and every step requires thorough understanding of gas and the surrounding environment. Any advances in the process link could make a step change in gas industry. There have been increasing efforts in gas industry in recent years.

With state-of-the-art contributions by leading experts in the field, this book addressed the technology advances in natural gas industry. It consists of 21 chapters divided into five sections which cover Natural Gas Resources, Production, Marketing & Transportation, Utilization, and Combustion.

Section 1 consists of four chapters dealing with gas properties and reserves. It discusses the geological and geochemical setting in natural hydrocarbon emissions, and dynamics of natural gas hydrate system and shale gas development, in addition to geopolitics.

Section 2 consists of seven chapters on the topic of technology of gas production. This section deals with fluid-solid coupling numerical simulation on natural gas production from hydrate. Also, it presents a review of state-of-the-art and recent developments of unconventional gas production and technology. It provides an overview on basic shale gas technology; assessment was addressed on the fundamental technologies that are widely used for natural gas purification processes.

Section 3 has three chapters dealing with natural gas marketing and transportation. This section outlines the recoverable reserves, demand, supply, transportation of natural gas, and gas pricing and supply contracts. It also describes market options for developing countries. In addition, the section discuses the elements of gas pipeline network system, the variety of gas demand in a year, and mathematical model for gas network.

Section 4 summarizes the latest catalysis technology and gas utilization in three chapters. This section focuses on the development of novel channel monolithic reactors with meso-scale dimensions which can be coated with suitable metal oxide and used for steam reforming of methane to produce syngas and hydrogen. This section also describes the upgrading of existing and creation of new power production technologies, which increase the efficiency of natural gas utilization and decrease the harmful emissions to the environment. In addition, this section discusses the utilization of nature gas in reforming processes using both methane and carbon dioxide.

Section 5 describes the biggest consumption of gas - combustion for heating and powder generation or driving motors. This section deals with the application of natural gas in combustion engines and development of multi point injection system of natural gas engine. In addition, it describes the influence of modified atmosphere on natural gas combustion while defining the performance of gas turbine based on database model.

I hope one will be able to find the latest information for his / her needs in this book. Due to the constraint on time and very broad range of topics, there may be some improper descriptions or errors in the book. The editors would be grateful to anyone bringing them to their attention.

Dr. Hamid A. Al-Megren
Director of Petrochemical Research Institute,
Coordinator of Materials and Nanotechnology Sector,
King Abdulaziz City for Science and Technology, Riyadh
Kingdom of Saudi Arabia

Part 1

Natural Gas Resources

1

Shale Gas Development in the United States

Daniel J. Soeder
U.S. Department of Energy, National Energy Technology Laboratory, USA

1. Introduction

Although natural gas has been obtained from organic-rich shales in the United States since the first commercial gas well was produced in 1821 to provide gas light to four commercial establishments and a mill in the small town of Fredonia, New York, large-scale shale gas production is a recent phenomenon. Assessments of the geological and engineering challenges of shale gas resources were performed in the 1970s and 1980s, as new domestic energy sources were sought in response to an oil embargo imposed upon the United States, and the resulting "energy crisis" that followed. The amount of natural gas present in the shales was found to be significant, but commercial production had to await advances in drilling and completion technology that came about in the 1990s. The new technology allowed for the economic development of this resource in the 21st Century.

1.1 Basic shale geology

Shale is the name for a class of sedimentary rocks. The term shale refers to a rock that is composed primarily of tiny grains of clay minerals and quartz, the mineral components of mud. These materials were deposited as sediment in water, which was then buried, compacted by the weight of overlying sediment, and cemented together to form a rock through a process called lithification. Clay minerals are a type of sheet silicate related to mica that usually occurs in the form of thin plates or flakes. As the sediment was deposited, the flakes of clay tended to stack together flat, one on top of another like a deck of cards, and as a result, lithified shale often has the property of splitting into paper-thin sheets. This is called fissility, and it is an easy way to identify shale from other fine-grained rocks like limestone or siltstone.

Because the grains of material that make up shale are so small, pore spaces between these grains are equally small. Although shale can have porosity in the range of ten percent, the pores and flowpaths are so tiny that it is difficult for any fluids in the pores, like gas, oil or water, to flow out of the rock. Cracks or fractures are needed to for flowpaths.

Shale comes in two general varieties based on organic content: dark or light. Dark colored or black shales are organic-rich, whereas the lighter colored shales are organic-lean. Organic-rich shales were deposited under conditions of little or no oxygen in the water, which preserved the organic material from decay. The organic matter was mostly plant debris that had accumulated with the sediment. As these materials were buried deeply beneath younger sediments and subjected to intense heat and pressure over geologic time periods, they became hydrocarbons, or what we know as oil, gas and coal.

A location with a good exposure of rock that is representative of the formation as a whole is called the "type section," and formation names are assigned by geologists after a nearby geographic feature. Some well-known gas shales in the United States include the Marcellus Shale, named for the village of Marcellus, New York (figure 1), and the Barnett Shale, named for exposures in the valley of a creek called the Barnett Stream near Fort Worth, Texas (Stamm, 2011).

Fig. 1. Marcellus Shale type section near Marcellus, New York, showing natural fractures. Rock hammer is 13 inches (33 cm) in length, tip pointed north. Photo by D.J. Soeder.

1.2 Natural gas in shale

Shale gas resources are huge. Estimates tabulated by Bruner and Smosna (2011) from different authors on the size of the Marcellus Shale resource alone range from about 85 trillion cubic feet (TCF) to nearly 500 TCF of technically recoverable gas. (One TCF equals about 9.3 billion cubic meters). The Utica Shale below the Marcellus may have even greater reserves. Such numbers are of course built on many assumptions about the geology, gas generating potential, gas in place, and percentage of recoverable gas, resulting in a wide range of estimates. In nearly all cases, however, they are quite large when compared to conventional gas reservoirs. The amount of gas consumed annually in the United States is about 23 TCF, making the "hundreds" of TCF considered recoverable from domestic gas shales a significant resource, no matter what the exact figure might be.

To understand why the gas resources in these shale formations are so large, it is helpful to review the manner in which oil and gas are created over geologic time periods. Rocks that have the ability to produce hydrocarbons in commercial quantities with standard well drilling technology are known as conventional reservoirs. The hydrocarbons present in a conventional reservoir were usually created elsewhere, and migrated into the porous and permeable reservoir rock, where they were trapped. Creating a conventional oil and gas reservoir is a complicated process that requires a number of events to occur in a specific order. These are: 1) source rock, 2) thermal maturity, 3) reservoir rock, 4) trap and seal, and 5) migration pathway. The rarity of all these things happening with precise timing and in the proper sequence is the main reason oil and gas can often be quite difficult to find.

Source rock: Petroleum and natural gas were formed from decayed plant matter trapped and preserved in fine-grained sediments. Two common sources of organic material were algae or other water plants, and woody land plants. Some animals may have contributed as well, but most fossil fuel is derived from ancient plant material, not dead dinosaurs. Decay bacteria usually require oxygen, so if the dead plants settled to the bottom in water that contained low levels of dissolved oxygen, the organic matter was often preserved and buried under more sediment. These organic-rich sediments (later turned into rock) are known as source rocks. Until recently, source rocks were not considered to be much good for production, because they are generally made up of fine-grained, low permeability materials deposited in quiet water environments, like black shale.

Thermal maturity: In addition to containing a few percent of preserved organic matter, the source rock sediment had to be buried deeply, and subjected to heat and pressure within the Earth to become thermally mature. Temperatures within the Earth increase with depth. This is called the geothermal gradient and varies with location. In most places, the temperature generally increases by about 25 degrees C with every kilometer of depth (Blackwell and Richards, 2004). More deeply buried rocks were exposed to higher temperatures. Organic materials, exposed to high temperatures over geologic time periods slowly break down without oxidizing, turning organic carbohydrates into fossil fuel hydrocarbons such as coal, oil and natural gas.

The thermal maturity of a gas shale is related to its burial history. For example, Lash (2008) published an analysis for Devonian formations in western New York, determining that the Marcellus Shale was initially buried rapidly beneath a thick wedge of sediments, then uplifted and eroded by mountain-building before being buried again under more sediment derived from the new mountains. Parts of the Marcellus Shale that were deeply buried were exposed to temperatures above 175 degrees C for millions of years, thoroughly cooking everything in the rock. Most measurements of thermal maturity on the Marcellus Shale place it quite high, well beyond the liquid petroleum range. Dry methane gas is almost the only hydrocarbon remaining in this shale, although some ethane is present in the western part of the basin.

Other shales with lower thermal maturity ranges do produce liquids along with the gas. A shale in Texas called the Eagle Ford produces dry methane gas where it is deeply buried, and significant natural gas liquids in shallower areas. Natural gas liquids are known as condensate, which travels up the well as a vapor, but then condenses into liquid form under the lower pressures and temperatures at the surface. Condensates such as propane and butane are worth substantial money, and are eagerly sought by the petroleum industry.

Reservoir rock: Conventional oil and gas production comes from reservoir rocks, which consist of coarse-grained sandstones or limestones with significant porosity and high permeability. These formations are too open and sponge-like to be good source rocks, but if a source rock elsewhere is able to fill up the reservoir rock pore spaces with oil and gas, it is easy to produce with conventional wells.

Trap and seal: In order to contain the gas and oil in a conventional reservoir rock, there must be some kind of a trap, such as a fold or a fault, to create an underground structure that acts as container of sorts to hold the hydrocarbons in the reservoir rock. To be effective, the trap must also include an impermeable caprock to seal the reservoir and contain the hydrocarbons within.

Migration pathway: Because the source rocks and reservoir rocks are usually completely different formations, once the oil and gas have formed in the source rock, they need a migration path to get from the source rock to a reservoir rock. This can be a fracture like a fault that allows movement through the intervening rocks, or just tilted beds that will let hydrocarbons slowly flow updip. Timing is everything: if the migration pathway is in place before a reservoir rock is available, the oil and gas will be lost. On the other hand, if the reservoir rock is present but no migration path ever develops, the reservoir stays empty.

In summary, a driller will end up with a dry hole in a conventional oil or gas reservoir if any one of the five items described above is missing, or occurs out of sequence.

Gas shales are unconventional reservoirs. This means that they are significantly larger than a conventional reservoir, but they are also much more difficult to produce. The shale acts as both the source rock and the reservoir rock. The gas in the shale was created from organic material deposited with the sediment, and was not required to migrate anywhere to be trapped in a reservoir. However, because the gas has remained in the shale, it must be produced directly from this fine-grained, impermeable rock, and that is not a simple task.

If the quality of most natural resources is plotted against the quantity, a triangle or pyramid shape is typically produced, showing a small amount of high-quality resource, and significantly larger volumes as the quality goes down (figure 2). The lower quality resource is usually more difficult and expensive to produce, but if there is a technological or economic breakthrough that makes it competitive with the higher-graded resource, the production quantities can be enormous. This has happened with commodities like iron, coal, gold, and timber, to name a few. For example, high purity drinking water from protected springs or pristine mountain streams is in very limited supply. However, suppose a new technology allowed seawater to be turned into drinking water of the same quality at a similar cost, or better yet, cheaper. Supplies would suddenly expand greatly.

This is essentially what happened with shale gas. The application of new drilling and hydraulic fracturing technology has allowed drillers to extract this gas directly from the source rock at prices comparable to gas from conventional reservoirs.

The U.S. Geological Survey (USGS) has the responsibility for estimating oil and gas resources in the United States, including shale gas. USGS hydrocarbon assessments are based on mathematical models, which use an understanding of the geology of the rock unit and the production characteristics of existing wells. The production data used by the USGS

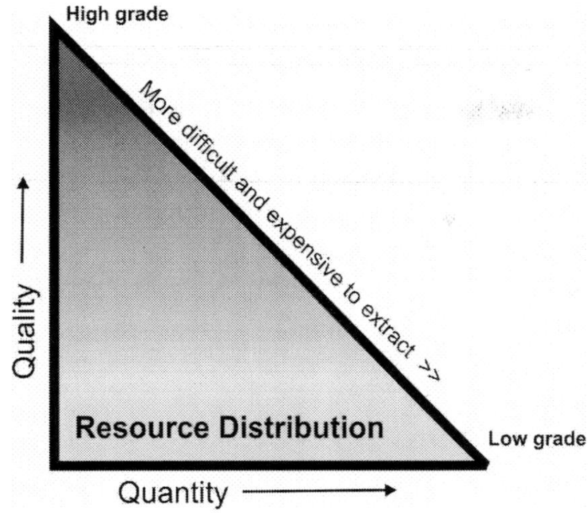

Fig. 2. The resource triangle illustrating the distribution of most natural resources, including natural gas, when quantity is plotted against quality.

are the decline curves, or the fall-off in production over time, from a large number of gas wells. Unfortunately, very little decline curve data have been made available, and production from gas shales is so new that most of the wells have not declined very much anyway. As such, calculations of the true size of the shale gas resource have been problematic at best (Coleman and others, 2011).

The wide range of estimates for U.S. domestic shale gas resources is a clear sign that a better understanding is needed of the processes that generate and store gas in the shale. Greater rigor may also need to be applied to some of the assumptions used in the various estimates. Discovering how the physical properties of the shale act to control the limits of gas content would help to constrain the numbers and provide more accurate and realistic estimates of gas in place, or GIP.

The amount of recoverable gas is always a fraction of the GIP, under the assumption that 100 percent of the gas will never be recovered, even under the best of circumstances. Hydraulic fractures don't contact every part of the formation, some pores may be blocked with water or oil, and others may not be connected to flowpaths. The value for this fraction varies from assessment to assessment, ranging from a low of about 10 percent up to 50 percent or even higher. In one of the classic publications on the resource, Engelder and Lash (2008) stated that the Marcellus Shale GIP exceeds 500 TCF over an area encompassing parts of New York, Pennsylvania, West Virginia, and Ohio. They assumed a technically recoverable gas fraction of 10%, leading to a reserve estimate of 50 TCF. This caused quite a stir at the time, because 50 TCF of producible gas from a single formation was more than double the annual consumption of natural gas in the entire United States.

It has only gone up from there. More refined calculations by Engelder (2009) came up with significantly higher estimates for GIP, and a 50 percent probability that the Marcellus Shale will ultimately yield 489 TCF of gas, assuming a power-law decline rate, 80-acre well spacing, and 50 year well life. U.S. government numbers for Marcellus Shale recoverable gas from the U.S. Energy Information Administration (EIA) are about 410 TCF (EIA, 2011). A recent Marcellus Shale assessment by the USGS (Coleman, 2011) has concluded that the median amount of technically recoverable gas from this formation is about 84 TCF, and it may go as high as 144 TCF, but this is still quite conservative compared to some of the other numbers out there. The EIA recently reduced their estimates to be more in line with the high end numbers from the USGS.

Why is there so much gas in shales like the Marcellus? Most geologists agree that the gas was derived from rich deposits of organic matter in the shale, formed from abundant marine algae that grew and died in a shallow inland sea during the time of Marcellus Shale deposition. Wrightstone (2011) suggests that the planktonic or floating marine plants were fertilized by mineral-laden dust blown into the basin by trade winds off arid highland areas to the east. Periodic dust storms from these deserts would have added fine particles of quartz to the water in the enclosed basin, along with a host of mineral nutrients, including iron and phosphorous. It has been known for some time that iron is an essential fertilizer for algae. The dust-blown minerals could have fertilized an explosion of plant growth in the water. Wrightstone cites documentation from a modern algae bloom that occurred in the Tasman Sea after an Australian dust storm of epic proportions in 2009, and similar blooms in the Atlantic Ocean from dust storms off the Sahara Desert. Under a microscope, a significant part of the mineral matter in the Marcellus Shale can be seen to be composed of small particles of quartz that are just the right size to have been carried by the wind.

Algal blooms concentrate a great deal of organic matter in the water column, and then transport this organic matter to the ocean bottom when the plants die and sink. When these blooms happen, bacteria trying to feed on large masses of dead algae rapidly remove any residual oxygen from the bottom waters and create anoxic sediments, which preserve the organic material. The anoxic muds were then buried, exposed to heat and pressure, and generated copious amounts of methane gas.

2. History of U.S. shale gas investigations

Although the first commercial American gas well was hand-dug into Devonian-age shale in Fredonia, NY by William Hart in 1821, serious shale gas studies did not begin in the United States until the 1970s, in response to oil shortages that led to an "energy crisis." This crisis was actually a series of oil shocks precipitated by a Middle East war in 1973, and the Iranian revolution in 1979. The energy shortages experienced during these episodes worried the American public, and influenced U.S. foreign policy.

The Middle East war known variously among historians as the Yom Kippur War, the Ramadan War, the 1973 Arab-Israeli War, or the Fourth Arab-Israeli War was fought between October 6 and October 25, 1973. Lasting less than a month, it involved armies from Egypt, Syria, Iraq and Jordan attacking Israel, followed by an Israeli counterattack, and ended with a U.N. brokered ceasefire (Rabinovich, 2004). Both the United States and Soviet Union enlisted the two sides as proxies, with the Soviets resupplying and supporting Egypt, and the Americans airlifting material and providing intelligence support to Israel. U.S. involvement led some members of the Organization of Petroleum Exporting Countries, an oil cartel better known by its acronym OPEC, to call for an embargo on oil exports to the United States. At a meeting of oil ministers in Kuwait on October 20, 1973, members of OPEC declared a total embargo on American oil deliveries (Yergin, 1991).

The oil embargo on the U.S. lasted until the spring of 1974. Although this was at a time when significantly less than half of the oil used in the United States was imported, and not all the member countries of OPEC had even joined in the embargo, the action still resulted in severe shortages, long lines at gasoline stations when there was fuel available, and consumer panic. The price of oil quadrupled almost overnight. The American driving public, who had not worried about gasoline supplies since the days of fuel rationing during the Second World War, were shocked and stunned.

The U.S. postwar housing boom had relocated many people into suburbs at long distances from city centers. Suburban life meant that automobiles were required for nearly all transportation needs. Fuel shortages and price hikes in the winter of 1973-74 raised the prospect of being stuck with an empty gas tank in a useless car, and unable to carry out the simplest tasks. In the rhetoric of the time, people demanded that something be done to prevent America from being held "hostage" to imported oil. Many people thought that if the United States could send men to the moon, we ought to be able to figure out how to fuel our automobiles. The public outcry forced the government to act. The United States Department of Energy (DOE) was formed from a number of smaller agencies as a cabinet-level entity of the U.S. federal government under President Jimmy Carter on August 4, 1977. Along with inherited responsibilities like running the national labs and maintaining the nation's nuclear weapons stockpile, a primary mission of the new DOE was to find technological solutions to the energy crisis.

The second oil shock hit in 1979, during the Iranian revolution. The student protests that eventually led to the fall of the Shah severely disrupted Iranian oil production, and essentially curtailed exports for several months. Although the United States received only a relatively small percentage of imported oil from Iran, the disruption to global supplies was enough to precipitate a second oil shortage, with the same gasoline station lines and panic as seen in 1973. The 1979 crisis was much shorter-lived, however, because Saudi Arabia and other exporting nations were able to make up for the Iranian oil shortages and return American imports to nearly steady levels (Yergin, 1991).

Schrider and Wise (1980) described some of the potential new domestic sources of fossil fuel, including natural gas, being investigated by DOE. These included unconventional gas resources such as coalbed methane, tight gas sands, gas dissolved in deep brines under high pressures, and shale gas. There was no doubt that the production of these would be a technical challenge, but if they could be exploited, the energy would help displace imported oil. A number of scientific and engineering investigations were begun on unconventional energy resources by the U.S. government, one of which was the Eastern Gas Shales Project.

2.1 U.S. Department of Energy Eastern Gas Shales Project

In 1975, the Energy Research and Development Administration, a predecessor agency to the U.S. Department of Energy, initiated the Eastern Gas Shales Project (EGSP) to assess the potential for a sequence of Devonian-age shales in the Appalachian Basin, as well as similar rock units in the Michigan and Illinois Basins to produce large amounts of natural gas under the proper conditions. The initial definition of proper conditions was to find organic-rich black shales that contained abundant natural fractures. The organic matter in the black shales would provide the gas, and the fractures would provide the flowpaths. Engineering experiments would seek to link the natural fractures through a series of man-made hydraulic fractures, creating a network of high-permeability flowpaths through the shales.

Under DOE, the EGSP had 3 major components: resource characterization, development of production technology, and the transfer of that technology to industry. The project was managed by the DOE Morgantown Energy Technology Center (METC) in West Virginia, which later became a campus of the present-day DOE National Energy Technology Laboratory (NETL). Over a period of about 6 years, from 1976 to 1982, the EGSP used cooperative agreements with drillers to collect oriented drill core from a variety of shale units in the Appalachian, Michigan and Illinois Basins. Directional or "oriented" core was necessary, because one of the major pieces of data being gathered was the strike and dip of the natural fractures. Most of the EGSP cores came out of the Appalachian Basin, and many of these were from the shallower, western side near the Ohio River. The shale sequence in the eastern part of the basin is considerably deeper, and therefore more expensive to drill. Only eight of the EGSP wells were drilled all the way down to the Marcellus Shale, and data from those are now in high demand.

Cores were collected from 34 different EGSP wells in the Appalachian Basin, in formations ranging from the Cleveland Shale to the Marcellus Shale. Three wells were also cored in the Devonian Antrim Shale of the Michigan Basin, and seven wells into the equivalent New Albany Shale in the Illinois Basin, for a total of 44. The locations of the EGSP wells in the Appalachian Basin are shown in figure 3 (Bolyard, 1981).

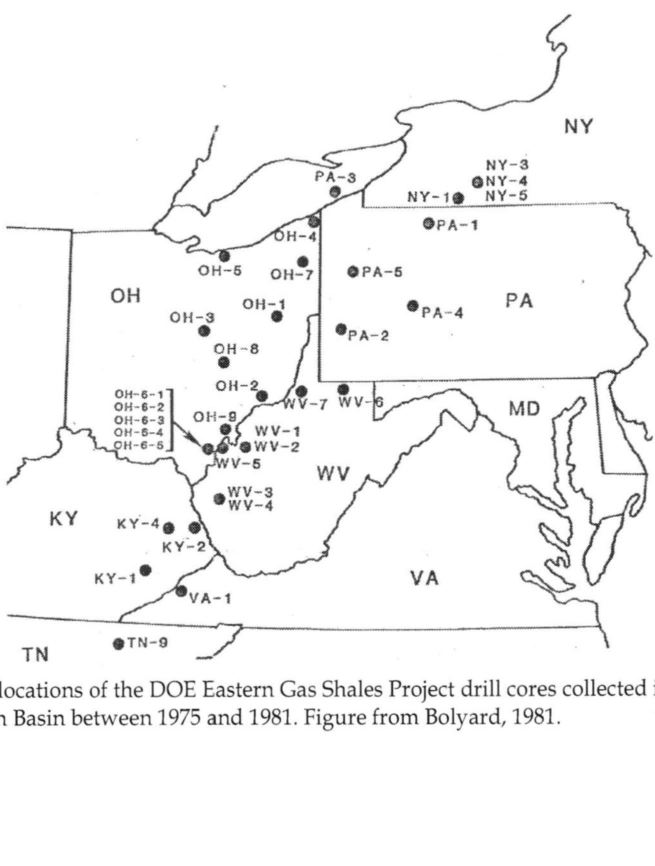

Fig. 3. Map locations of the DOE Eastern Gas Shales Project drill cores collected in the Appalachian Basin between 1975 and 1981. Figure from Bolyard, 1981.

As soon as the core was recovered, it was unloaded from the plastic sleeves used to line the core barrel, washed to remove drilling mud, assembled, aligned, measured for length and had depths marked on it. The cores were solid cylinders of rock, 3.5 inches (9 cm) in diameter that weighed about ten pounds per foot (~15 kg per meter). The field crew then set to work collecting and preserving samples, and creating a field description of the lithology, noting in particular any gas shows, natural fractures or other features. The most time-sensitive of the field samples were short segments of core designated for chemical gas analysis. As quickly as possible, but certainly within two hours of the core reaching the surface, these offgassing samples were hermetically sealed in steel cans. The cans were sent to a chemistry lab in Ohio, where the composition of the gas was analyzed as it came out of the rock. The final field task was to pack the core back up into the plastic liners and cap the ends for transport to the EGSP core lab in Morgantown, WV.

Once the core arrived at the lab, it was laid out on tables, carefully pieced together, cleaned, oriented and measured. Using the core orientation data and a circular plastic protractor, north lines were drawn on the rock cylinders in permanent marker, which allowed for the measurement of the orientation of any features or natural fractures encountered in the core (figure 4).

Fig. 4. Black Cleveland Shale above gray Chagrin Shale in an EGSP core from Ohio, with pyrite at the contact (2040.54 ft.). An orientation groove is visible to the right of the depth marks; the vertical line on the front of the core marks north. Photo by Daniel J. Soeder.

Cores were then processed for data collection. The lithology and color of the core were described from bottom to top. Wet core surfaces were compared to standardized color chips to determine color. Many so-called black shales are actually a yellowish black color similar in tint to ripe olives. Only the Marcellus Shale was found to be a true charcoal black. Correlations between organic carbon content, gas potential and "blackness" of the shale were impractical, because once the organic carbon content reached a few percent, the shale was black and didn't get any blacker with more carbon (Hosterman and Whitlow, 1980).

Fractures were identified as natural or coring-induced, based on criteria defined by Kulander and others (1977). Natural fractures were further classified as joints or faults. Joints are fractures where the two walls have simply pulled apart. Faults are fractures where the two walls have slid past each other; often leaving a polished, grooved surface behind called a slickenside. The orientation of the natural fractures was measured using the north directional line on the cores as a reference. The frequency of the coring- induced fractures was counted, but little else could be done with them. It was hoped that they could at least provide an indication of the brittleness of the rock, and a possible response to hydraulic fracturing. Rock samples were collected from the cores for the various labs, agencies and universities that had asked for them, and small wooden blocks were inserted into the core to mark where each sample was taken from, and who had it.

The cores were photographed with a specially made rig that could trundle a camera down the length of a core table. The photos were hand-pasted into albums and kept as a reference. Sadly, many of these photos suffered water damage years later when in storage, and have been lost. Gamma radiation readings were collected on the cores every foot (30 cm) using a scintillometer for comparison with gamma ray logs collected downhole on a wireline tool. Unfortunately, like the core photographs, the scintillometer readings have also been lost over the three decades that have passed since the data were first collected.

Funding for the Eastern Gas Shales Project formally ended in 1992, but the budget had been at relatively low levels since 1982. Despite the low funding levels, a number of cutting edge engineering experiments were run on shale. An air-drilled, horizontal test well was completed in the Huron Shale in December 1986 (Duda and others, 1991), which was drilled with the intent of intercepting existing fractures and improving the efficiency of natural gas recovery. Innovative logging techniques, directional drilling techniques, assessments of reservoir anisotropy, liquid CO2 fracturing, and other new technologies were tried out on gas shales during the last decade of the program. These studies greatly assisted industry in the commercial development of shale gas a decade later.

By 2007, many of the old EGSP reports, publications and data tables were being eagerly sought out by industry people interested in the potential of shale gas, and personnel at DOE-NETL started getting numerous requests for copies. Many of these had been packed away for years, or were quite rare. In response to increasing demands for information as unconventional gas drilling expanded, the NETL library assembled nearly every relevant document from the DOE unconventional gas program, some of which were literally down to the last copy in the known universe. The documents were transferred into an electronic format and placed on two DVDs, allowing hundreds of reports and scientific papers on western U.S. gas sands, secondary gas recovery, eastern U.S. gas shales, methane hydrates, deep source gas, and methane recovery from coalbeds to be carried in one's pocket.

There were many other aspects of the Eastern Gas Shales Project above and beyond the core recovery and analysis activity, but that part of the work is described here because it would later turn out to be an important factor in the decision to develop the Marcellus Shale gas resource. Many of the other EGSP efforts, including drilling engineering and well design, field tests on a wide variety of reservoir stimulation techniques, early attempts at computer modeling, development of an assessment procedure and nomenclature for rock fractures, a vast amount of gas chemistry and geochemistry analysis, and geological basin studies were equally pioneering in the development of shale gas and other unconventional gas resources.

2.2 Institute of Gas Technology shale gas research

The Institute of Gas Technology (IGT), located in Chicago, Illinois is now known as the Gas Technology Institute (GTI). It had been founded in 1949 as a research institute for the gas utility companies, who were moving away from the use of manufactured or town gas, made from coal and water, and replacing it with natural gas from oil wells that was being pipelined up from the Gulf Coast. Town gas was extremely hazardous, consisting of a mixture of carbon monoxide and hydrogen. Natural gas, on the other hand, is composed of non-toxic methane, and it was abundant in the oilfields of Texas and Louisiana. Interstate transmission companies were building thousand-mile pipelines from production areas on the Gulf Coast to market areas in the Northeast and Midwest. In order to recoup some of the costs of the pipelines and new distribution systems, the gas utility and transmission companies wanted to encourage the use of more natural gas. IGT was initially founded to conduct gas utilization research, developing new consumer appliances, and finding additional commercial and industrial applications for natural gas. It wasn't until the first energy crisis in 1973 that IGT began a natural gas supply research program.

In the early 1980s, IGT had a subcontract with Sandia National Lab to analyze the core from a DOE tight gas sand project called the Multiwell Experiment, or MWX. Tight gas sand is a sandstone formation with reasonable porosity and a fair amount of gas in the pores, but with permeability almost as low as that of a shale. It is an abundant gas resource in a number of southern and western states in the U.S., and in western Canada. The challenge is figuring out how to extract the gas economically. The MWX was a series of three wells drilled relatively close together into the Mesaverde Formation in the Piceance Basin of western Colorado. One of the wells was hydraulically fractured, and the other two were observed for effects. The final part of the experiment was to drill an angled borehole across the hydraulic fracture, and capture it in the core, which was done successfully.

The analysis of rock properties such as porosity, permeability, capillary entry pressure, pore volume compressibility, pore size distribution and flowpath width are collectively known as petrophysics. IGT developed a laboratory instrument to accurately measure the petrophysical properties of tight sandstone core samples under pressure conditions representative of the rocks at depth. The key was to maintain stable air temperatures inside the apparatus, so gas pressures would not fluctuate due to thermal changes. Under these steady temperatures, volume and flow measurements using gas were very accurate. The apparatus employed a reference pressure stable to about one part in half a million, and it could accurately measure gas flows lower than one millionth of a standard cubic centimeter (gas at room temperature and atmospheric pressure) per second. This is equivalent to a cubic centimeter of gas flowing from a rock over a time period of a million seconds, which is more than 11 ½ days. The reason for making this point is that the device was also used to measure gas flow through shale. Some people have questioned whether such low flow measurements could actually be made with this degree of precision, and the answer is yes. The flows were measured with electronic sensors, and the apparatus was controlled by a 1980s version of a desktop computer, which in those days used cassette tapes to transfer programs and record data. The device was named the Computer-Operated Rock Analysis Lab or CORAL. It was described in detail at a Society of Petroleum Engineers (SPE) meeting, and in a paper by Randolph (1983).

When the subcontract with Sandia National Lab expired, IGT received funding in 1983 directly from DOE to do additional core analysis and experimental work with the CORAL, including trying to measure simultaneous gas and water flow through tight sandstone cores. IGT suggested that the CORAL could also be used to try collecting some gas permeability data on EGSP shale core, because such permeability data were not in the literature. The DOE project manager agreed, and supplied IGT with a list of "zones of interest" in many of the original EGSP cores based on gas production or gas shows, correlation with gas-productive intervals in nearby wells, successful stimulation results, and indications of high organic content.

At the end of active core collection in 1982, the EGSP core lab had been shut down, and the shale cores were shipped to the state geological surveys in the state where they had been cut for storage and safekeeping. Twenty-eight zones of interest were sampled by IGT from thirteen cores in Ohio, Kentucky, New York, Pennsylvania and West Virginia.

In the end, IGT was only able to run two full loads of shale core in the CORAL. The device had four core holders, so this was a total of eight samples in all. Six of the samples were black Huron Shale, a member of the Ohio Shale, which was known to be gas productive in southwestern WV. One of the Huron Shale cores in the first batch had cracked in the coreholder, so core seven was a repeat run of another sample from this same well. Core number eight was a Marcellus Shale sample from the EGSP WV-6 well.

The CORAL had been upgraded in anticipation of the shale analyses. The upgrades included changing the flow directions of the air circulation system so temperatures of critical components were more stable and returned to equilibrium more quickly, improving the digitizing resolution with a better data logger, and rewriting the temperature control software so it could predict when temperatures were nearing a setpoint and reduce power to the heating coils beforehand, instead of overshooting and then having to correct. These changes, driven by the desire to make measurements on shale, led to an overall improvement in the performance of the apparatus and provided better data on all samples.

Small rock cylinders cut from the EGSP core samples for CORAL analysis were dried in a controlled relative humidity oven to remove water from the rock without dehydrating the clays. Proper drying under controlled relative humidity was important for obtaining useful measurements (Soeder, 1986). Harsh drying, under high temperatures and/or in a vacuum oven causes clay minerals to dry out and collapse, opening up pores and creating abnormal permeability. Samples dried at 60°C under 45% relative humidity retain a layer of bound water on clays and other hydrated minerals, although free water in the pores is removed. Many analyses on a variety of tight sandstones and other rocks had confirmed this.

IGT core analysis on the eight samples of EGSP shale revealed a number of important findings (Soeder, 1988). The first was that the Huron Shale samples contained small but significant amounts of petroleum in the pores, which blocked gas flow. Under low differential pressures, no measureable gas flowed through these cores at all. At high differential pressure, gas flow rates gradually increased over a period of hours before leveling out. The data appeared to show that a liquid phase was being pushed out of the pore system by the gas pressure, and gas permeability was slowly increasing as the flowpaths were cleared of liquid. Dozens of tight sandstone measurements made on samples dried under the same temperature and relative humidity conditions had never shown any evidence of a liquid water phase in the samples. Yet all of the Huron Shale cores showed some kind of liquid draining from the pores under pressure. The IGT analytical chemistry lab placed a sample of the Huron Shale core in a tagged solvent and ran the liquid through a gas chromatograph. The data revealed that the Huron Shale contained a light paraffinic petroleum, typical of Appalachian Basin oils. It is important to note that a similar analysis on the Marcellus Shale sample revealed that there was no oil present in this rock.

The discovery of oil blocking the pores of the Huron Shale helped explain some of the erratic results of the earlier EGSP reservoir stimulation experiments. Many different types of stimulations had been used, including hydraulic fracturing, explosives, and fracturing with cryogenic liquids, gas and foam, among others. The results had been hit or miss at best – some stimulation methods performed well on certain formations in certain locations, and poorly elsewhere. There were not enough data to explain why this should be, and a 1982 report concluded that reservoir stimulation alone was insufficient to achieve commercial shale gas production (Horton, 1982). This conclusion implied that the situation was a bit more complicated than the old EGSP idea that any black shale will produce gas if sufficiently fractured. With the finding that oil was blocking gas flow in the pores of at least some shales, a few of the stimulation failures became more understandable.

The Marcellus Shale sample was measured by IGT in August 1984. Gas flowed through this sample with remarkable ease, and excellent data were collected. The values for gas permeability (reported as K_∞) were 19.6 µd (microdarcies) at 3000 psi net confining pressure, and about 6 µd at 6000 psi net confining pressure (Soeder, 1988). The high sensitivity of permeability to net confining stress (i.e. doubling the net stress reduced permeability by more than two thirds) has implications with respect to production drawdown and the expected economics of shale gas wells. Loss of flow under higher net stress will be offset somewhat by increased gas slippage as pore pressures are lowered during production. Shale gas has not been produced long enough for many wells to have entered these stress and pressure regimes, but this is certainly a concern for the future. Obviously, a lot more data are needed to understand the petrophysics of shale.

The CORAL apparatus at IGT was also capable of measuring the pore volume of a core sample under representative net confining stress using a Boyle's Law (pressure-volume equivalence) technique. The device used a volume-calibrated, positive displacement pump, and a sensitive differential pressure transducer to measure pore volumes with an accuracy of 0.01 cm^3. Porosity of the Marcellus Shale core was measured using nitrogen gas at two pressures to check the validity of the data points. Instead of obtaining the same value for pore volume regardless of pressure, significantly more nitrogen gas went into the sample at lower pressure. This higher apparent porosity at low pressure is a sign that some of the gas was being adsorbed. The phenomenon of adsorption occurs when gas molecules attach themselves to electrostatic surfaces inside the pores. The nitrogen data prompted another, more thorough gas porosity measurement at a wider range of pressures using methane, the main component of natural gas. These data showed that the amount of methane the Marcellus Shale could hold was equal to 0.224 times the square root of the pressure. This is called an isotherm in adsorption chemistry and defines an adsorption function. When these data are plotted on a linear graph to show the volume of methane per volume of rock as a function of pressure, the curve shown in figure 5 emerges. The curved line is due to adsorption; if this was strictly a pressure-volume relationship, the line would be ruler-straight. When this curve is extended out to the value of 3500 psi reported for the initial gas pressure of the Marcellus Shale in the EGSP WV-6 well, where the core originated, the calculation shows a gas-in-place value for this shale of approximately 26.5 standard cubic feet of gas (scf) per cubic foot (ft^3) of rock. This was an important piece of data, because the National Petroleum Council (1980) had assessed the gas potential of Appalachian Basin shales at just 0.1 to 0.6 scf/ft^3. The IGT value of 26.5 scf/ft^3 in the Marcellus Shale core from EGSP WV-6 was an astounding 44 to 265 times greater than the NPC estimate. No one had ever reported this much gas in a black shale before. The results were published in a DOE report, and as an SPE journal article (Soeder, 1988).

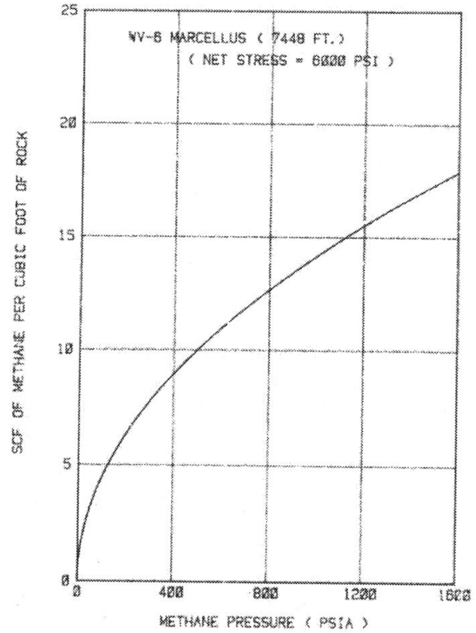

Fig. 5. Natural gas potential in the Marcellus Shale from methane porosity data measured as a function of pressure. From Soeder (1988).

3. Production of shale gas

The people bewildered by all the drill rigs setting up across the Pennsylvania countryside might be surprised to learn that the Marcellus Shale drilling did not simply come as a bolt from the blue. After the end of the Eastern Gas Shales Project, the drilling industry continued to innovate with the development of new drilling technology and new hydraulic fracturing techniques. The potential applications of these new techniques to the production of shale gas were carefully scrutinized by George P. Mitchell, the co-founder of Mitchell Energy, who had been involved with shale gas since the early days of the EGSP, drilling a number of cooperative wells with DOE in Ohio. Mitchell was interested in the gas potential of the Barnett Shale in Texas. Like the Marcellus and other shales, it was difficult to obtain economical amounts of gas from vertical wells in the Barnett. Mitchell Energy tried a lot of different drilling techniques and reservoir stimulation methods over a period of about 18 years, including massive hydraulic fracture stimulations, which did produce significant flows of gas, but at very high cost.

The key to producing economical quantities of shale gas has turned out to be the ability to drill and fracture horizontal boreholes through the rock, which contacts much more formation volume than a vertical well. The typical black shale thickness of only a few hundred feet (tens of meters) limits the amount of contact a single vertical well can have with the rock. Drilling horizontally, however, allows the wellbore to remain within the shale, and penetrate distances of thousands of feet (kilometers). The drilling is coupled with hydraulic fracturing to create high permeability flowpaths into the shale. Instead of the single hydrofracs done in vertical wells, the long horizontal boreholes allow for an entire series of hydraulic fractures to be spaced a few hundred feet apart (figure 6). There can be ten or more of these so-called "staged" hydrofracs in a horizontal borehole, resulting in large volumes of gas production. Drilling costs for a horizontal Marcellus well are approximately 2-3 times higher than for a vertical well, but the initial gas production can be 3-4 times greater (Engelder and Lash, 2008).

Advances in horizontal drilling, or more accurately, directional drilling, came about in the 1990s, driven by the needs of deepwater offshore oil production. As offshore rigs moved into deeper water, the engineering design of the platforms changed. Steel towers standing on the sea bed had worked fine in shallower water, but drilling in thousands of feet of water required the use of semi-submerged, floating platforms held firmly in place by tensioned steel cables anchored into the seafloor. These platforms and their associated seabed anchor facilities were expensive and complicated to rig up. The less frequently they needed to be moved, the better.

Directional drilling was the answer. If a driller could bore a well directionally into one reservoir pocket, and then drill another well in a different direction from the same location to intercept a second reservoir pocket, a great deal of oil could be recovered without having to move the rig. This need and the large sums of money behind it drove the development of directional drilling forward in the 1990s. Some deepwater platforms now routinely drill as many as 60 separate directional wells from a single location.

Directional drilling had been around for years, but there were two problems with it that needed to be overcome: steering the bit and knowing where it was located. The first technological advance in directional drilling was the downhole motor. Without the need to turn the entire drill string from the surface, the drill pipe is much more flexible and can turn relatively tight corners. The modern design uses hydraulic power, supplied by a slurry of drilling mud pumped down the drill string under high pressure and through an impeller on the downhole motor. The motor then turns the bit, which cuts the rock. The impeller, motor

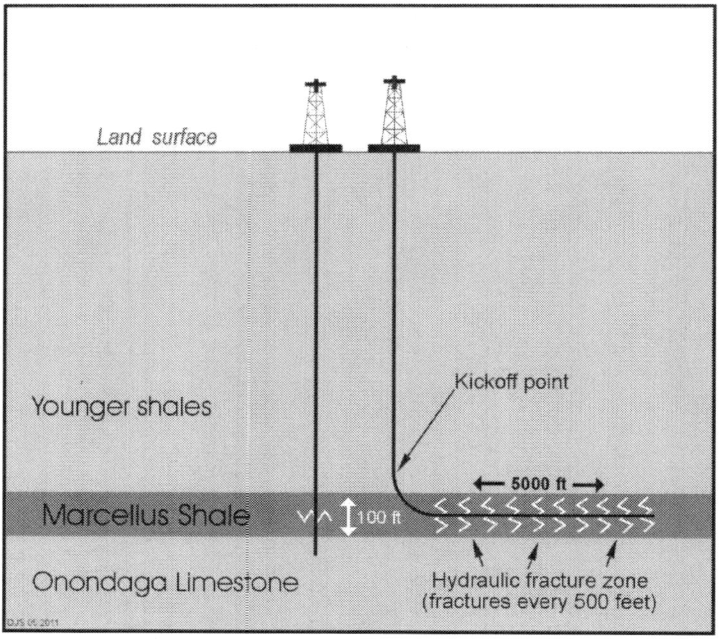

Fig. 6. Illustration of the combination of horizontal drilling and hydraulic fracturing technology used for shale gas. Not to scale. (Modified from Soeder and Kappel, 2009).

and bit together are known as the "bottomhole assembly," and it allows wells to be drilled in virtually any direction, including horizontally. The drill bit is steered by using a bent section of pipe near the bottomhole assembly to deviate the well away from vertical, by changing the pressure being applied against the cutting face, or by varying the rotational speed. Some advanced bits have thrust bearings controlled from the surface that change the angle of the cutting head to provide precise control.

Advances in downhole position measurement using a combination of inertial navigation with a gyroscopic compass and remote telemetry now allow drillers to more accurately monitor the downhole location of their drill bit and the configuration of the borehole. Data transmission methods vary, but usually involve digitally encoding the data and transmitting it to the surface as pressure pulses in the mud.

The depth at which a directional wellbore changes from vertical to some other orientation is called the kickoff point. The location in 3-dimensional space where the borehole is supposed to intercept the producing formation is called the target. The radius of the curve used to change the borehole direction from vertical to horizontal is called the build or the build rate. The horizontal stretch of the borehole is called the lateral. The path of the lateral through the target formation is called the trajectory.

Directional drilling in gas shale is laid out in patterns that look like the legs of a spider on a map. The body of the spider is the drill pad. Multiple wells will originate from a single drill pad, ranging from 6 to 10 or more in number. All of the wells start out vertical, and then from a kickoff point about 500 feet (150 m) above the target, build a curve. The trajectory takes them down into the target shale in a direction that is usually perpendicular to the trend of the most prominent natural fractures, or joints. Laterals can extend 5,000 feet (1.5 km) or more in length, and are often drilled parallel to one another at some optimal spacing for the most efficient recovery of natural gas from the formation.

Because of the need to protect aquifers, the finished boreholes are lined with casing, held in place by cement. Casing is made of heavy steel pipe, which screws together in approximately 30-foot (10 m) segments. Each length of casing that is made up of joined segments of a particular diameter is known as a string. There are several concentric strings of casing in a well, with each successive casing string being smaller in diameter and extending to a greater depth. As each string of casing is placed in the hole, cement is pumped down through the center, and is distributed by a shoe at the bottom of the string so that it oozes up into the annular space between the casing and the borehole wall. Enough cement is pumped in to completely fill this annular space, and then left to cure. A proper cement job is critical for sealing the casing and keeping it in place.

Different casing configurations are used in different climates, but in the Appalachian Basin of North America, the following design is typical: The conductor casing is installed from the surface to a depth of 30 to 60 feet (10-20 m) as a mechanical barrier to support the sides of the hole in unconsolidated soil. This is the largest diameter string of casing used in a gas well, usually about 24 inches (60 cm) in diameter. Inside the conductor casing, a second, narrower casing string known as the surface casing (sometimes called the water or coal casing) is run and cemented in place from the surface down to a depth of several hundred feet below the deepest freshwater. The surface casing is 14 to 20 inches (35-50 cm) in diameter and is used to isolate the gas well from the aquifers and coal seams. It is designed to protect the groundwater by preventing any gas or oil from entering the aquifer, while at the same time keeping groundwater from flooding the well. Regulations for fresh groundwater protection tend to be conservative, and may require surface casing to be set as deep as a thousand feet (300 m).

From the bottom of the surface casing, the main vertical, curved and lateral portions of the borehole are drilled. Another string of casing, called the intermediate casing, is installed through this part of the hole down to the kick-off point of the curved borehole, and cemented into place. This casing is 9 to 12 inches (23-30 cm) in diameter, and its purpose is to keep the borehole walls from collapsing and to prevent any gas and liquid in the rocks above the target formation from entering the annular space of the well. A final string of well casing, called the production casing, is installed in the finished hole. It is usually only about 5 inches (13 cm) in diameter, and extends from the surface down the vertical hole, through the curve and along the entire length of the lateral to the very bottom end or toe of the hole. It is cemented into place through the production zone to the base of the intermediate casing, and serves to channel all gas production directly to the surface, without any opportunities to go astray.

The completion process for the well begins by punching holes in a section of the lateral casing in the production zone using shaped explosive charges on a wireline carrier. This perforation of the casing creates contact with the reservoir. The holes and cement behind the casing are cleaned using a 15% solution of hydrochloric acid. The hydraulic fracturing process begins by pressure testing and calibrating all of the equipment. Water, chemicals and sand are mixed in a blender and pumped downhole. As the hydrofrac begins, the pump rate is brought up slowly while pressures at the wellhead, downhole, and in the annulus behind the production casing are carefully monitored. The frac fluid is pressurized until the breakdown pressure of the formation is exceeded and the rock cracks open. The initial part of the frac is just water, called the pad. It is followed by water mixed with sand pumped in as a proppant, to keep the fractures open after pressure is released. A flow meter on the blender measures the volume of fluid pumped downhole, and a densometer measures the amount of sand in the fluid. Engineers watch the wellhead, annulus and bottomhole pressures, pump rate, fluid density and material parameters throughout the frac. When the first stage of hydraulic fracturing is finished, the pressure is released and a seal called a bridge plug is set into the production casing to close off the perforated and fractured interval from the rest of the well. The hydraulic fracture treatment is repeated in a second stage, which is then also closed off with another bridge plug. The process continues until the last stage reaches the upper end of the lateral, called the heel, and begins to curve up out of the shale.

After all the stages of fracturing have been completed, the bridge plugs are removed. Gas pressure in the rocks is used to push the frac fluid out of the well during a procedure called "blowback," which is designed to remove as much of the liquid as possible. The expelled fluid is diverted into a holding tank. Because the well is not yet on production, this operation requires that the gas be burned-off or flared. The water being pushed out of the well by gas pressure is called flowback fluid, and the flow of liquid can persist intermittently for weeks. Once the gas production begins after the initial discharge of flowback fluid, a production wellhead is installed along with a gas meter and connector line, and the gas is sold to a transmission company. Most operators filter and recycle the flowback fluid into the next frac to avoid the costs of disposal.

3.1 Development of shale gas resources

George Mitchell truly believed in the gas potential of the Barnett Shale, and would not give up. Because of his determination, Mitchell Energy continued their field experiments in Texas, eventually developing something called a light sand frac, which was more effective on the shale at a lower cost than most other hydraulic fracture treatments. A rise in gas prices in the mid-1990s improved the economics. By 1997, Mitchell had perfected the light sand frac technique in vertical wells, and started trying it in horizontal wells. They began successfully producing commercial amounts of gas from the Barnett Shale using horizontal boreholes and staged hydrofracturing in the early 21st century.

In the summer of 2004, Southwestern Energy announced that the Fayetteville Shale in Arkansas had many of the same characteristics that made the Barnett Shale gas productive, which set off another gas drilling boom. Oil and gas producers familiar with the Barnett Shale rushed to northern Arkansas to get in on the Fayetteville Shale. Similar drilling booms followed soon afterward on the Haynesville Shale in the Arkansas-Louisiana-Texas border region known as the ArkLaTex, and the Marcellus Shale in Pennsylvania.

The Marcellus Shale was developed in the southwestern corner of Pennsylvania by a company called Range Resources, who remain a major producer in the area. In 2005, Range was drilling a vertical well called Rentz#1 in Washington County, PA to test oil and gas prospects in the Lockport Dolomite. This is a Silurian carbonate rock in the Appalachian Basin, older than the Marcellus Shale and located below it. The Lockport was originally deposited as a calcite-rich limestone, which was later altered into a different rock called dolomite (named after the Italian mountains where it is common) by magnesium-enriched groundwater. The alteration process causes calcium carbonate (calcite) to recrystallize into a magnesium/calcium carbonate mineral also called dolomite (the rock is often referred to as "dolostone" to distinguish it from the mineral). The mineral dolomite usually forms larger crystals than calcite, giving a sugary texture to the formerly fine-grained limestone, which creates porosity that may contain oil and gas. There is no guarantee that hydrocarbons will be present, however. A mantra of all oil and gas geologists is that despite all the geology and geophysics used for exploration, you never really know what's down there until you get down there, and the only way to get down there is to drill.

The Rentz well came back with poor gas shows from the target formation. Bill Zagorski, the Range Resources geologist in charge of the well, was left wondering what to do with this non-productive, dry hole. Zagorski found himself in Houston a few months later, trying to sell an interest in developing a shale gas prospect in Alabama using Mitchell Energy's production technology, when he recalled that he had seen evidence of gas in the Devonian shale section penetrated by the Rentz well above the Lockport Dolomite. Some of the historical data he had researched when selecting the well location reported gas shows and even a "blowout" in old drilling records, which he realized were at depths near the Marcellus Shale (Durham, 2010). Zagorski researched what was known about gas resources in the Marcellus Shale, which included finding many of the old DOE and SPE publications on the subject from the EGSP work two decades earlier, and got the go-ahead from his company to try a completion in the Marcellus.

Range Resources re-completed the vertical Rentz#1 well in the Marcellus Shale, and got a significant return of initial gas production. Thus encouraged, they drilled the first few horizontal Marcellus wells in 2006 with mixed results, but after some trial and error, Range eventually applied a modification to the Mitchell Energy light sand frac called the slickwater frac, which turned out to be an effective production technique that is now commonly used on Marcellus Shale wells throughout the basin. The first successful horizontal Marcellus well, Gulla#9, came online in 2007, returning an initial gas production rate of 4.9 million cubic feet per day, which is quite exceptional for any gas well, and until then practically unheard of for a gas shale. Zagorski considers Gulla#9 to be the "discovery" well for the Marcellus Shale, and the one that started the play. Between 2008 and 2011, nearly 8,000 gas wells have been drilled and hydraulically fractured in the Marcellus Shale in Pennsylvania and West Virginia.

In addition to the Marcellus Shale, gas production from the Barnett Shale in the Fort Worth Basin of Texas is still going strong, as are the Haynesville and Fayetteville shales in Arkansas. The Woodford Shale in the Anadarko Basin of Oklahoma is also being produced. A new shale play getting started is the Utica Shale, an Appalachian Basin black shale that is deeper and older than the Marcellus. The Utica actually covers more land area than the Marcellus, extending farther into the northern, western and eastern reaches of the basin. It is already being explored and produced in Canada along the St. Lawrence River in Quebec. In eastern New York, it fills fault bounded valleys called grabens, sometimes to thicknesses of several thousand feet. The Utica also extends farther westward than the Marcellus into central Ohio. One advantage of producing gas from the Utica Shale is that it underlies the Marcellus Shale in many parts of the Appalachian Basin, making "dual completion" wells possible: i.e. two production targets from a single borehole.

Other shales of interest in the Appalachian Basin include the Rhinestreet and Ohio shales above the Marcellus. In Utah, gas potentials of the Mancos, Manning Canyon, Paradox, and

Pierre-Niobrara shales are being investigated. Alabama is looking into possibilities with the Floyd Shale and the Conasauga Shale. Well-known, organic-rich shales like the Antrim in the Michigan Basin, the New Albany in the Illinois Basin, and others are now being reviewed for their gas potential. Even the black shales in some of the small, Triassic rift basins along the U.S. East Coast are being evaluated for shale gas.

The Eagle Ford Shale in Texas produces significant natural gas liquids, or condensate, along with the gas. The liquids are worth more money than gas, and hence are more attractive to the petroleum industry, making the Eagle Ford the current "hot prospect" for shale drilling in the U.S. The Eagle Ford ranges in depth from about 2,500 feet to over 15,000 feet (760 m to 4.6 km), which has taken the shale through a variety of thermal maturation windows, from dry gas at the deep end through crude oil to "wet gas" at the shallower end. The Eagle Ford has problems with what is called retrograde condensate, however. Under reservoir pressures and temperatures, the natural gas liquids exist as a vapor phase, and come up to the surface as such, condensing out in tanks at the well pad. When reservoir pressures drop because of production, however, the vapors condense into liquids downhole, plugging up pores in the shale just like the light oil in the Huron Shale did for IGT. Trying to figure out how to produce both gas and liquid from such a reservoir without losing permeability is a major engineering and technical challenge. The locations of major shale gas plays in the United States (excluding Alaska and Hawaii) are shown on the map in figure 7 from the U.S. DOE Energy Information Administration.

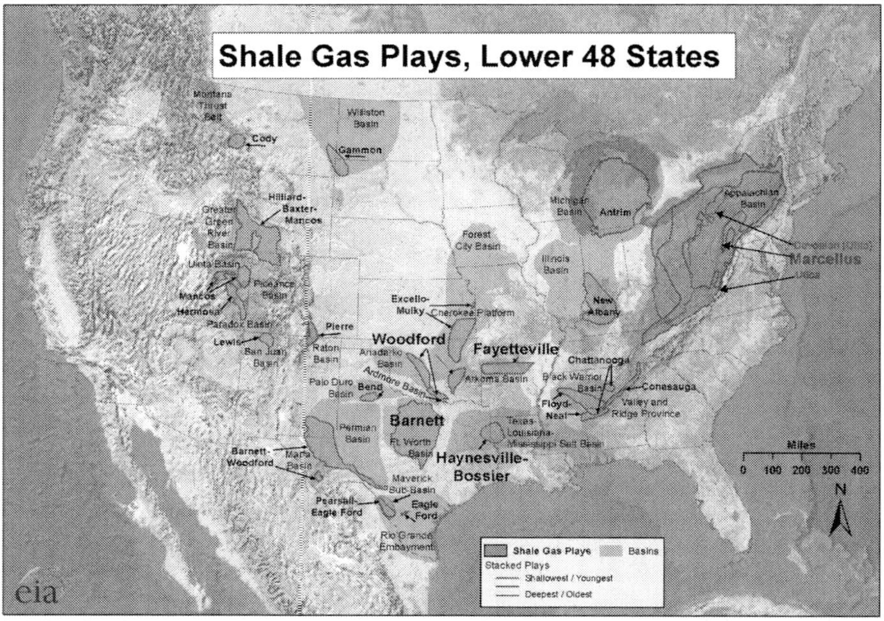

Source: Energy Information Administration based on data from various published studies.
Updated: March 10, 2010

Fig. 7. Location of shale gas plays in the contiguous United States. Source: U.S. Department of Energy, Energy Information Administration.

3.2 Shale gas world-wide

Gas and oil production from shales is of interest worldwide. Many countries who once thought they were limited on conventional hydrocarbon reservoirs are finding black shales and exploring them for gas and oil. Active drilling projects are underway or planned in Britain, Canada, Ukraine, South Africa, several North African countries, and Argentina. Mexico, Belarus, Poland, Germany, Brazil, Australia and China are also interested in shale gas development. Once George Mitchell's ideas about how to horizontally drill and hydraulically fracture these rocks became known, the exploration of shale energy resources took off nearly everywhere (figure 8).

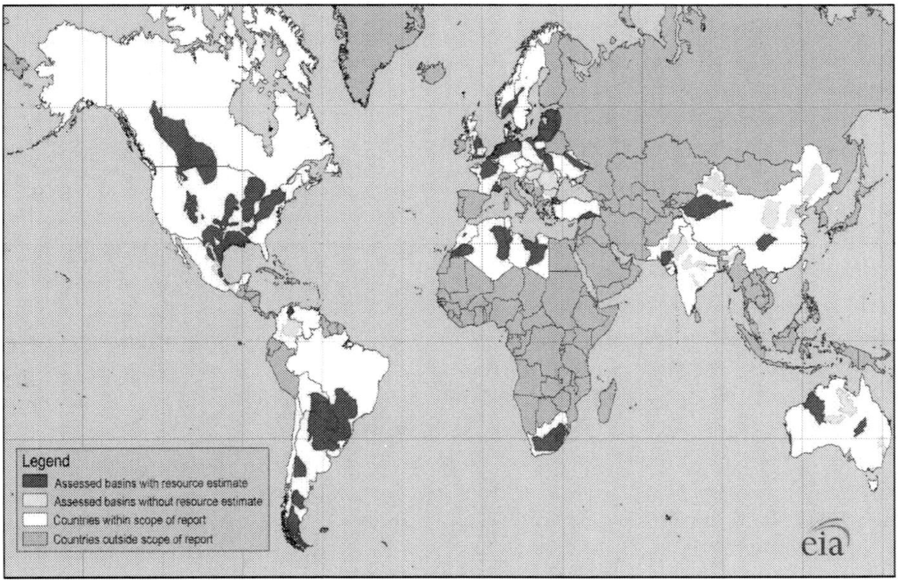

Fig. 8. Sedimentary basins worldwide containing significant shale gas resources.
Source: U.S. Energy Information Administration.

The resource numbers are enormous. Recent figures in some of the oil and gas trade journals suggest that shale gas reserves worldwide could be greater than 6,000 TCF, or more than ten to twenty times the amount of gas estimated for the Marcellus Shale, and possibly much higher. The amount of drilling, core sampling and well testing that has been taking place in most of these countries so far has been minimal, and data are sparse. Thus, most resource estimates are little more than educated guesses. In many countries, the geological thickness and extent of the organic-rich, black shale units are not known, so even educated guesses are not possible. Countries like Poland and Germany, who import most of their natural gas, are interested in developing domestic shale gas resources as a path to energy independence. However, environmental concerns over shale gas and particularly hydraulic fracturing are giving them pause. Environmental politics in the European Union are much stronger than in America, and sensitivities are higher. Countries are looking to the United States for environmental data and best management practices. Many are disappointed with the slow progress of shale gas environmental research in the United States.

4. Environmental concerns with shale gas

Development of shale gas is an industrial activity. Heavy machinery and serious equipment are needed to install the pad, drill down to the appropriate depths, and create and frac the long horizontal boreholes necessary for economic gas recovery. It involves a lot of material, including gravel, water, sand and chemicals for the pad and the hydraulic fracturing operations, along with many trucks to deliver all this to the well site. Installing the well creates noise, mud and dust, and requires a large crew of hard-working people, who live at the site for several days to weeks at a time. The drilling operations run 24/7, and create a neighborhood nuisance with their work lights, constant racket, smelly exhaust and endless activity. Having one of these sites near a home, school or business can be distracting, inconvenient, annoying, and disruptive.

Not all of the environmental impacts of shale gas production are known. A significant period of "baseline data" must be collected so the starting conditions are documented. Many of the parameters needed to determine environmental impacts have not been measured, because there has not been funding or time available to measure them.

Not all of the known environmental impacts are addressed under current regulations. Because the shale gas phenomenon is occurring in places that are not considered traditional oil and gas states, regulations that were largely designed for dealing with small drilling operations are inadequate for the scale of activity associated with shale gas development.

Not all of the current regulations are being properly enforced. State agencies don't have enough personnel to be everywhere all the time to enforce laws across extensive shale gas plays. Despite the huge upsurge in applications for drilling permits, the state oil and gas agencies have been largely unable to add a significant number of additional personnel because of tight budgets and other constraints.

Environmental impacts can be short-term or long term. Short-term impacts are related to well construction, and include things like water withdrawals, flowback fluid disposal, lights and noise from the drilling operations, effects of water impoundments on wildlife, and air pollution. Most of these go away once the well is constructed and all the equipment moves off, but they can be fairly intense during the drilling process. Long-term impacts are related to the well and drill pad occupying the landscape, and include concerns like habitat fragmentation, groundwater contamination from leaks or spills, the potential introduction of invasive species, and the process of ecological succession as the open drill pad slowly fills back in with vegetation. These factors are somewhat more difficult to quantify, and certain concerns, like invasive species, may not show up for some time. The short term impacts tend to be more acute, and the long term more chronic. Assessing both of these types of impacts is important for understanding the overall environmental effects of the gas well.

An additional unknown is cumulative impact, which stems from the planned development of the resource. Environmental effects from individual wells add up as more and more wells are placed within a tract of land, eventually taking conditions across a threshold and causing impacts much greater than the individual wells alone. At the rate the resource is being developed, tracts of land that were only going to contain a few wells have mushroomed into dozens. A problem with cumulative impacts is that it is difficult to tell when a threshold has been crossed until it is too late. Assessments need to be made of the number of wells an ecosystem or a watershed can tolerate per unit area.

The five things that are the most susceptible to the potential environmental impacts of shale gas development are air, water, landscapes, habitat and ecosystems. Quantifying the impacts of shale gas wells on these receptors could help significantly with the improvement of both environmental monitoring and management practices to minimize problems.

Air pollution during well construction and hydraulic fracturing is a concern because of the large numbers of diesel engines needed to run heavy equipment. Some drill rigs are now electric-hydraulic, drawing power from on-site generators running on natural gas supplied by a nearby well instead of diesel. Air can also be contaminated by exhaust from compressor stations needed to boost the gas up to pipeline pressures.

Water issues include potential impacts to water supply, water quality, and damage to small watersheds. These were described in a USGS publication by Soeder and Kappel (2009). The large volumes of water needed for a staged hydraulic fracture have the potential to impact local supplies. Many drillers now build large, central impoundments that they fill during times of high streamflow and low demand. Standards for the quality of water used in a frac have also been relaxed since drillers realized that swelling clays are generally not a problem in thermally-mature gas shales. They no longer require drinking water quality supplies for frac fluid. Recycled flowback and raw water from local streams is often used. Recycling the recovered high salinity flowback fluid into the next frac has lessened water quality concerns considerably. Still, the risk of spills and leaks, the potential for toxic metals and radionuclides to oxidize and leach from drill cuttings, and the movement of stray gas through aquifers remain water quality concerns. Small watersheds risk potential damage as drillers build five-acre drill pads and service roads. A great deal of equipment, supplies and vehicles have to be transported into and out of a drill site. Construction of roads alongside small streams often does not take stream hydrology into account, changing the flow regime and altering the aquatic ecosystem. Other watershed concerns include the potential for chemical spills, seepage of contaminants through shallow groundwater, erosion and sediment issues, and worries that high salinity flowback water could cause major mortality in aquatic ecosystems if released into a stream.

Habitat and ecosystem impacts near shale gas wells are both short-term and long-term. Short term impacts are related to the construction process itself, caused by the effects of lights, noise, activity levels, vehicle movements, and chemical exposure on local flora and fauna. Long term effects include re-occupation of the drillpad area by displaced plants and animals, species succession, potential impacts of structures and facilities on ecosystems, habitat fragmentation, and the possible establishment of invasive species. Many of these impacts are being assessed in a study being carried out in Pennsylvania (Soeder, 2010).

One of the popular concerns expressed about hydraulic fracturing is that the fractures may break upward into overlying aquifers, and contaminate groundwater with formation brines and dangerous chemicals. There are a number of physical reasons that make this highly unlikely; including the length of time the fracturing fluid is under pressure, the volume of fluid injected, the behavior of stress fields near the surface, and flow gradients once the well is in production. It is doubtful that the fluid will climb a mile or more against the force of gravity to contaminate a freshwater aquifer.

Geophysical data support the notion that hydraulic fracture heights remain well below freshwater aquifers. A technique called "microseismic monitoring," originally developed by

DOE and Sandia National Laboratory is used to determine the positions of hydraulic fractures in the ground. The method uses a string of sensitive microphones known as "geophones" that are suspended in a borehole near the frac location. The crackling sound emitted by the breaking rock is detected by the geophones and the arrival times of the sound at the different sensors are carefully measured. These data are then used to triangulate the progression of the frac over time. The vertical geophone string lowered into a well is said to be accurate to within a few cm on the height of the fracture. A graph from Fisher (2010) using microseismic data to compare the height of Marcellus Shale hydraulic fractures with the depth of the deepest aquifer reportedly producing drinking water on a county by county basis is shown in figure 9. In no case do the fracture heights approach within several thousand vertical feet (km) of the aquifers.

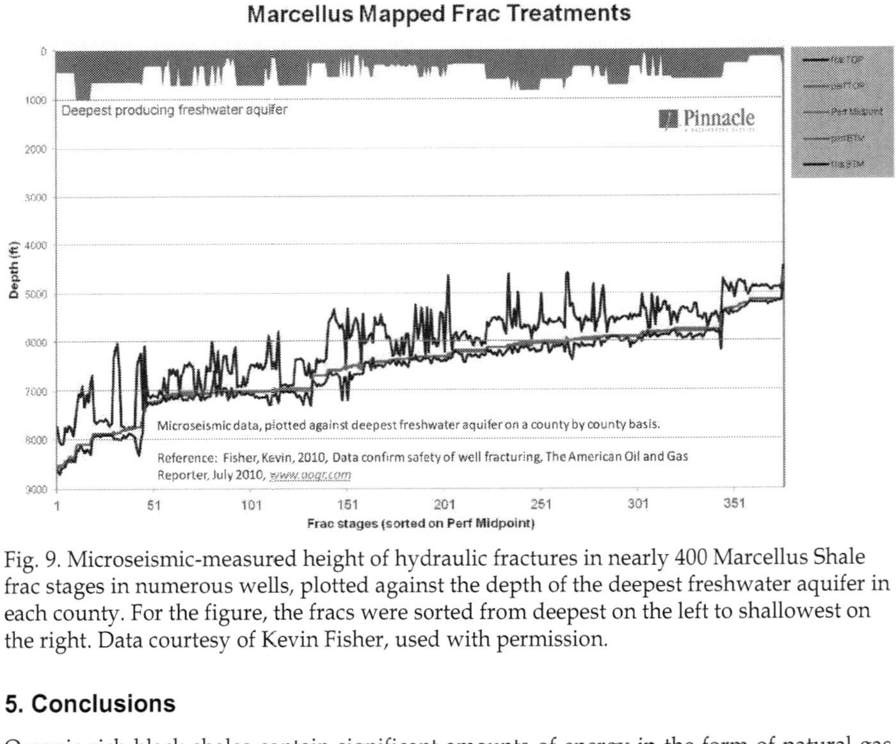

Fig. 9. Microseismic-measured height of hydraulic fractures in nearly 400 Marcellus Shale frac stages in numerous wells, plotted against the depth of the deepest freshwater aquifer in each county. For the figure, the fracs were sorted from deepest on the left to shallowest on the right. Data courtesy of Kevin Fisher, used with permission.

5. Conclusions

Organic-rich black shales contain significant amounts of energy in the form of natural gas, which may be large enough to make the United States energy independent for the first time since the 1950s, and finally bring to an end the so-called "energy crisis" of the 1970s. The size of the gas resource has been known or suspected for many years from government studies like the Eastern Gas Shales Project, but the technology needed to economically recover the gas was not developed until the 1990s. Mitchell Energy persisted with gas production attempts on the Barnett Shale in the Fort Worth Basin of Texas until they finally found a successful combination of horizontal drilling and staged hydraulic fracturing that allowed the recovery of large amounts of shale gas at economic costs. Range Resources was the first to apply the Mitchell-developed technology on the Marcellus Shale in 2007, and started the current play. Shale gas development continues in the U.S. and worldwide.

The shale gas recovery process is not without environmental concerns. Environmental impacts on the Marcellus Shale include potential effects on air, water, ecosystems and habitat, some of which are known and others of which are still being studied. Improved drilling practices, such as frac fluid recycling, are reducing these impacts. There is reason to believe that all environmental impacts and indicators will be identified eventually, and properly regulated.

6. References

Blackwell, D. D., & Richards, M. (2004). *Geothermal Map of North America,* 1 sheet, scale 1:6,500,000. American Assoc. Petroleum Geologists (AAPG), 1444 S. Boulder, Tulsa, OK, 74119, USA

Bolyard, T.H. (1981) A summary and evaluation of the eastern gas shales program cored wells in the Appalachian Basin, Report prepared for U.S. Department of Energy under contract DE-AM21-78MC08216, by Science Applications, Inc., Morgantown, WV, September 1981, 32 p.

Bruner, Kathy R. & Smosna, Richard (2011). A Comparative Study of the Mississippian Barnett Shale, Fort Worth Basin, and Devonian Marcellus Shale, Appalachian Basin, Report *DOE/NETL-2011/1478* prepared for U.S. Department of Energy by URS, Inc., Morgantown, WV, USA, 106 p. Available from http://www.netl.doe.gov/technologies/oil-gas/publications/brochures/DOE-NETL-2011-1478 Marcellus-Barnett.pdf

Coleman, J.L., Milici, R.C., Cook, T.A., Charpentier, R.R., Kirshbaum, M., Klett,T.R., Pollastro, R.M. & Schenk, C.J. (2011). Assessment of undiscovered oil and gas resources of the Devonian Marcellus Shale of the Appalachian Basin Province, 2011, *USGS Fact Sheet 2011–3092,* U.S. Geological Survey, Reston, VA, 2 p., Available from: http://pubs.er.usgs.gov/publication/fs20113092

Durham, L.S. (2010) Marcellus Gave No 'Big Play' Hints, *AAPG Explorer,* Vol. 31, No. 4, April 2010, p. 40-42

Duda, J.R, Salamy, S.P., Aminian, K. & Ameri, Samuel (1991) Pressure analysis of an unstimulated horizontal well with type curves, *Journal of Petroleum Technology,* August 1991, p. 988.

Engelder, T., & Lash, G. 2008, Marcellus Shale play's vast resource potential creating stir in Appalachia: *The American Oil and Gas Reporter,* May 2008, 7 p. Available from: www.aogr.com

Engelder, T., 2009, Marcellus 2008: report card on the breakout year for gas production in the Appalachian Basin: *Fort Worth Basin Oil & Gas Magazine,* August 2009, p. 18-22.

Energy Information Administration (2011) Annual Energy Outlook 2011, *Report DOE/EIA-0383(2011),* U.S. Energy Information Administration, EIA Energy Information Center, Washington, D.C., USA. Available from: http://www.eia.gov/forecasts/aeo/MT_naturalgas.cfm

Fisher, Kevin (2010) Data confirm safety of well fracturing: *The American Oil and Gas Reporter,* Vol. 53, No. 7, July 2010; p. 30-33 (www.aogr.com)

Horton, Andrea Ice (1982) 95 stimulations in 63 wells - DOE reports on comparative analysis of stimulation strategy in eastern gas shales, *Northeast Oil Reporter,* February 1982, p. 63-73

Hosterman, J.W. & Whitlow, S.I. (1980) Munsell color value as related to organic carbon in Devonian shale of the Appalachian basin: *USGS Open-File Report 80-660,* U.S. Geological Survey, Reston, VA, 9 p.

Kulander, B.R., et al. (1977) Fractographic logging for determination of precore and core-induced fractures, Nicholas Combs No. 7239 well, Hazard, KY: *U.S. Department of Energy report, MERC/CR-7713,* Morgantown, WV.

Lash, Gary (2008) Stratigraphy and fracture history of Middle and Upper Devonian succession, western New York - significance to basin evolution and hydrocarbon potential, In: *Pittsburgh Association Petroleum Geologists 2008 Spring Field Trip Guide,* 88 p., Greentree, PA 15242-0352, USA

National Petroleum Council (1980) Unconventional gas sources, Volume III: Devonian shale, Report prepared for the U.S. Secretary of Energy, June 1980, National Petroleum Council, Washington, D.C. 20006, 252 p., accessed online March 16, 2009. Available at http://www.npc.org/

Rabinovich, Abraham (2004), *The Yom Kippur War: The Epic Encounter That Transformed the Middle East,* Schocken Books (Random House), New York, NY, USA, 543 p. (ISBN:0805241760)

Randolph, P.L. (1983) Porosity and permeability of Mesa Verde core from the U.S. DOE Multiwell Experiment, Garfield County, Colorado, paper SPE/DOE 11765, In: *Proceedings of the 1983 SPE/DOE Joint Symposium on Low Permeability Gas Reservoirs,* Denver, CO, March 13-16, 1983.

Schrider, L. A. & Wise, R.L. (1980) Potential new sources of natural gas, *Journal of Petroleum Technology,* April 1980, p. 703-716.

Soeder, D. J. (1986) Laboratory drying procedures and the permeability of tight sandstone core, *SPE Formation Evaluation,* Vol. 1, No. 1, p. 16-22.

Soeder, D. J. (1988) Porosity and permeability of eastern Devonian gas shale, *SPE Formation Evaluation,* Vol. 3, No. 2, p. 116-124, DOI 10.2118/15213-PA.

Soeder, D. J. & Kappel, W.M. (2009) Water resources and natural gas production from the Marcellus Shale, *USGS Fact Sheet 2009-3032,* U.S. Geological Survey, Reston, VA, 6 p. Available from: http://pubs.usgs.gov/fs/2009/3032/

Soeder, D. J. (2010) The Marcellus Shale: resources and reservations, *Eos, Transactions, American Geophysical Union,* Vol. 91, No. 32, 10 August 2010, p. 277-288

Stamm, Nancy (2011). Geologic Names and Paleontologic Databases (GEOLEX), In: *National Geologic Map Database,* accessed 17 October 2011, U.S. Geological Survey, Reston, VA. Available from: http://ngmdb.usgs.gov/Geolex/geolex.html

Wrightstone, G. R. (2011) Bloomin' algae! How paleogeography and algal blooms may have significantly impacted deposition and preservation of the Marcellus Shale, presentation, Geological Society of America, Northeastern/North Central Section Meeting, Pittsburgh, PA, In: *GSA Abstracts with Programs,* Vol. 43, No. 1, p. 51.

Yergin, Daniel (1991) *The Prize: The Epic Quest for Oil, Money, and Power,* Simon & Schuster, New York, NY, USA, 912p. (ISBN: 0671502484)

2

Geochemical Dynamics of the Natural-Gas Hydrate System in the Sea of Marmara, Offshore Turkey

Livio Ruffine et al.*
IFREMER, Centre de Brest, Département des Géosciences Marines, Plouzané, France

1. Introduction

Natural-gas hydrate systems are solid-state light-hydrocarbon accumulations which are encountered in the permafrost and the continental margins. They are stable under high-pressure and low-temperature conditions and represent the major hydrocarbon volume on earth (Kvenvolden, 1988). Gas hydrates consist of a polycrystalline structure where a light hydrocarbon is trapped within a water lattice. The nature of the hydrocarbons is strongly related to their origin which is either microbial (also called biogenic) or thermogenic. Microbial gas-hydrate systems contain hydrocarbons produced by bacteria and archaea. There are primarily methane with a very small amount of ethane and eventually propane (Max, 2003). Others non-hydrocarbon compounds like hydrogen sulphur and carbon dioxide are also present. In the case of microbial gases, the hydrates are formed at or near the gas production area. Owing to the very high-methane content, these hydrates are commonly called methane-hydrate systems.

Beside the microbial gas-hydrate systems, there are hydrate accumulations which are connected to a petroleum system (Max, 2003). Thus, the hydrate-forming gases have a thermogenic origin. They come from deep subsurface reservoirs of hydrocarbons, may contain hydrocarbons from methane to C_{6+}, and migrate to the shallow sediment where the pressure and the temperature are favourable for hydrate nucleation and growth. Heavier hydrocarbons like isobutane, isopentane, neopentane can be trapped into the water lattice.

* Olivia Fandino[1], Joël Etoubleau[1], Sandrine Chéron[1], Jean-Pierre Donval[1], Yoan Germain[1], Emmanuel Ponzevera[1], Vivien Guyader[1], Bernard Dennielou[1], Giuseppe Etiope[2], Luca Gasperini[3], Bortoluzzi Giovanni[3], Pierre Henry[5], Céline Grall[1,5], Çagatay M. Namik[6], Charlou Jean-Luc[1] and Géli Louis[1]

[1]*IFREMER, Centre de Brest, Département de Géosciences Marines, Plouzané, France*
[2]*Istituto Nazionale di Geofisica e Vulcanologia, Sezione Roma, Roma, Italy*
[3]*Université Europeenne Bretagne, UBO IUEM, Plouzane, France*
[4]*ISMAR, CNR, Bologna, Italy*
[5]*Aix-Marseille Univ., CEREGE, France*
[6]*Istanbul Technical University, Istanbul, Turkey*

Since the beginning, investigations on natural-gas hydrates connected to or in relation with a petroleum system have held a prominent position in geosciences as they have been considered as a geohazard for the petroleum industry (Kvenvolden, 1999; Kvenvolden, 2000; Sloan and Koh, 2008). Now natural-gas hydrate systems are seen as a potential energy resource, and investigations are shifting mainly towards the study of the gas-hydrate production. This corresponds to a second period in the young history of natural-gas hydrate systems. However, the production of all hydrate accumulations is not economically viable and the scientific insights gained up to now from studies of gas hydrates as a geohazard is of immense value for the understanding of the behaviour of such complex systems. Therefore, applied research on gas hydrates will still be focused on both fields of interest. Beside, the industrial interests, there is a debate on the destabilisation of natural-gas hydrate systems and their contribution to climate change (Henriet and Mienert, 1998; Krey et al., 2009; Maslin et al., 2010; Yamamoto et al., 2009). In fact, if this huge amount of methane bound into the hydrates is released and reach the atmosphere, it would certainly impact on the climate by raising the temperature as methane is a greenhouse gas. However, there is an ongoing scientific debate about the likelihood for important quantity of methane being released from oceanic hydrates to reach the atmosphere (Reagan and Moridis, 2008).

Whatever the scientific and industrial purposes, the study of natural-gas hydrate systems now involves more and more multidisciplinary approach. Fortunately, thanks to the ever-increasing deep-sea technology development, scientists are able to recover well-preserved samples from different types of geological setting, perform several kinds of analysis and obtain more accurate measurements. More specifically, development of seismic techniques also allows a better visualisation of hidden structures underneath the hydrate accumulations (Collett et al., 2010; Paull and Dillon, 2001; Plaza-Faverola et al., 2010). The resulting data of all those techniques and analyses are gathered, integrated and combined to advance our understanding on natural-gas hydrate systems.

Geochemistry plays an important role in the understanding of a natural-gas hydrate system (Paull and Dillon, 2001). It allows the chemical identification of the different materials involved in the hydrate formation as well as the determination of their origin. Such information enables us to constrain the hydrate formation, accumulation and destabilisation processes, and therefore geochemistry contributes to a large part in the description of the system's dynamics. This chapter is dedicated to the application of a series of geochemical analyses to describe a natural-gas hydrate system located on the Western High, in the Sea of Marmara, offshore Turkey. It has been discovered during the Marnaut scientific expedition in 2007 on the R/V L'Atalante. The collected hydrate samples were porous and had a strong smell of oil. The sediment associated with the hydrates was also stained with oil, which strongly suggests a possible link with a petroleum system, possibly with the Thrace basin. Since this expedition in 2007, two others ones dealing with the study of this hydrate system, Marmesonet in 2009 and MARMARA 2010 in 2010, were undertaken. Here, we gather the data from literature, mainly from Bourry's article (Bourry et al., 2009), along with newly generated ones to provide a description of the geochemical dynamics of the Sea of Marmara natural-gas hydrate system.

2. Geological setting

The Sea of Marmara is a 210 km long and 75 km wide intracontinental sea on a waterway between the Mediterranean Sea and the Black Sea. It is lined in both the North and the South by Turkey and is extended over a surface of 11, 500 km^2. The Sea of Marmara is divided into three pull-apart basins, these are from West to East Tekirdag basin, Central basin and Çinarcik basin with a maximum depth of 1,152 m, 1,265 m and 1,273 m respectively. The basins are separated by NE-rending ridges, the Central High and the Western High, which were probably generated by shear strength and transpression controlled by the overall strain of the North Anatolian Fault (NAF) (Okay et al., 1999). The Sea of Marmara is also crossed by branches of the the North Anatolian Fault, a plate boundary between the Anatolian and the Eurasian plates (Le Pichon et al., 2001). This fault zone is one of the most active and the most dangerous amongst the European faults. It is composed of several segments, which have undergone several major earthquakes during the twentieth century. The last major ones are 1999 Izmit and Duzce earthquakes (Barka et al., 2002; Gasperini et al., 2011; Utkucu et al., 2003).

Stratigraphic analysis of Late Quaternary sediments of the Sea of Marmara indicates that it was a freshwater lake during the last glacial period 12 kyr 14C Before Present (BP) (14.7 ka BP) (Aksu et al., 2002; Cagatay et al., 2000; McHugh et al., 2008; Vidal et al.). After its re-connection with the Mediterranean about 12 kyr 14C BP the sea level in the Sea of Marmara rose in tandem with the global ocean. The mixing of the anoxic lake deep water and oxygenated marine surface water caused the precipitation of authigenic carbonates (Reichel and Halbach, 2007). A sapropelic sediment layer was deposited between 11,500 and 7,000 yr BP under suboxic bottom water conditions. Following marine flooding, two sapropelic sediment layers were deposited during 10,600 - 6,400 and 4,750 - 3,200 a 14C BP under suboxic bottom water conditions. (Çağatay et al., 1999, 2000).

Our area of interest is the Western High, situated between Tekirdag and Central basins. It is a complex geological feature along the northern branch of the NAF, characterized by a mud diapir. A carbonate and a hydrate mounds as well as active cold seeps with gas bubbling to the seafloor are the major features which have been identified at this area (Tryon et al., 2010; Zitter et al., 2008).

3. Experimental section

Three gravity cores, MARM_1, MARM_2 and MARM_3, were used for this study as illustrated in Figure 1. MARM_3, with a length of about 100 cm, collected during the Marnaut scientific expedition, was the only core full of gas hydrates that were sampled and analyzed. MARM_1 and MARM_2, with a length of 248 cm and 418 cm respectively, were collected during the MARMARA 2010 cruise, did not contain any massive hydrates. The location of these cores was chosen from multichannel seismic profiles and previous studies of the site. MARM_1 and MARM_2 were immediately cut into segments of 1 m length and stored at 4 °C prior to being sampling. MARM_1 was used as reference core as it was located away from the gas vent and did not show any visual evidence of neither hydrates nor released gas bubbles. Gas bubbles were also sampled directly from a gas vent on the seafloor close to MARM_2 by the manned submarine Nautile during the Marnaut cruise and has been studied by Bourry et al. (2009).

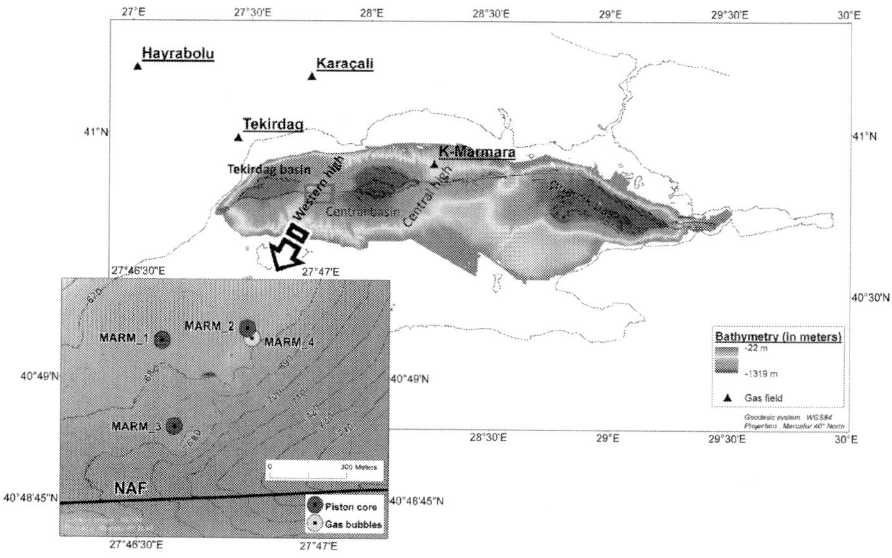

Fig. 1. General map of the Sea of Marmara indicating the locations of major features, the coring sites on the Western High and the natural gas fields.

3.1 Core analyses

On shore, the cores MARM_1 and MARM_2 were scanned using a medical X-ray CT Scanner Philips, model MX 8000 3rd generation. The complete scan for each entire 1 m length section was performed in about 10 minutes and consists of more than 1,000 slides of 2 mm thick took every 1 mm. The core sections were then split in two halves, working half and archive half, for further sampling and analyses. The archive halves were analysed using a X-ray fluorescence (XRF) core scanner from Aavatech to determine variations in its bulk chemical composition (Richter et al., 2006).

3.2 Gas hydrate sampling and analyses

The gas-hydrate specimens had a diameter of about 8 cm. As shown in Figure 2, the hydrates had a well-defined yellow colour and were mixed with a small amount of sediment. The hydrates as well as the sediment were completely stained with oil (Figure 2), with a yellow colour which is a consequence of the presence of the oil. All recovered pieces were stored in liquid nitrogen at 77 K to avoid further decomposition before being analysed in the laboratory.

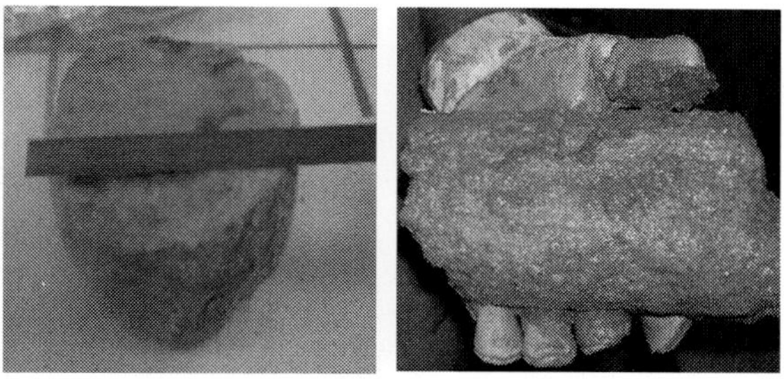

Fig. 2. Pictures of the recovered gas-hydrate specimens. The hydrates were massive and mixed with oily sediment.

For the analyses of the hydrate-bound gases, small pieces of hydrates were allowed to dissociate in a 20 mL glass tube which was previously evacuated. The gases were extracted from the hydrate samples using a gas extraction set-up (Charlou et al., 2004), and the hydrate water was collected and stored in another 20 mL glass tube. The resulting gases were also transferred into 10 mL glass tubes at a pressure ranging between 2 and 3 bars. Gas composition determination was carried out at the Laboratoire de Géochimie et Métallogénie of IFREMER.

For gas compositional analysis, the glass tubes were directly connected to a 6-port sampling valve of a 7890A Agilent gas chromatograph coupled with an Agilent 5975C quadrupole mass spectrometer detector (GC-MS). A custom-made instrument was used. It is equipped with a TCD-FID-mass detector and two capillary columns, Pora PLOT Q and molecular sieve, which can be put in series for the analysis of both hydrocarbons and permanent gases such as nitrogen and oxygen from the air. The analyses were performed with a temperature program running from 313 K to 500 K. Helium was used as a carrier gas. Both the TCD and FID were used for quantitative analyses while the mass spectrometer detector was used for a qualitative purpose. In the latter case, the compounds were identified using the NIST02 mass spectral library and only those giving a confidence in the identification exceeding 90 % have been selected and listed. The response of the TCD and FID were calibrated using a commercially-prepared gas mixture from Air Liquide. Considering the reproducibility of the gas chromatography measurements, the uncertainty in the calibration and the correction to eliminate the admixture of air, we estimated the accuracy in the composition measurement to be within ± 3 % of the molar fraction. The lower detection limit is estimated to 0.05 nmol for the hydrocarbons.

Isotopic measurements were carried out with a gas chromatography-isotope ratio-mass spectrometer (GC-IR-MS) at Isolab Laboratory, the Netherlands. The $\delta^{13}C$ values and the δD are reported as parts per thousand (‰) relative to the Vienna PeeDee Belemnite Standard (VPDB) and the Vienna Standard Mean Ocean Water (VSMOW) respectively. The uncertainty in $\delta^{13}C$ and δD are given as ± 0.1 to 0.3 ‰ and ± 1 to 3 ‰ respectively.

3.3 Pore-fluid sampling and analyses

For each section of MARM_1 and MARM_2, the pore fluids have been sampled for chemical analyses using the Rhyzon® soil moisture samplers. This is an innovative water extraction technology on which the sampler consists of a hydrophilic, porous polymer tube of 2.5 mm in diameter and 50 mm length. The tube is introduced into the sediment from one end, and is permanently connected to a vacuum tube of ~10 mL from the other end where the pore fluid is collected. The samplings were performed with a resolution of about 20 or 30 cm. Alkalinity was measured on ship directly after the sampling. Thus, 1 mL of pore fluid was sampled and analysed by direct titration with ultrapure 0.1 N HCl in an open cell.

The major dissolved elements were analysed by ionic chromatography using a Dionex ICS-2000 instrument equipped with ACS autosampler. For the analyses of sulphate and chloride, an AS-17C column of 250 mm in length and 4 mm in diameter was used with a 4 mm ASRS suppressor. The detection limit was 1.5 ppm and 5 ppm for the sulphate and the chloride, respectively. For all cations, an Ionpac CS12A column of 250 mm length and 4 mm in diameter was used with a CAES suppressor. The detection limit was 0.1 ppm for all species. All peak areas were quantified against equivalently diluted International Association for Physical Sciences of Oceans (IAPSO) standard seawater; the latter was analysed at the beginning of each run. The estimated accuracy on the concentration measurements is within ± 3 % of the mole fraction for all ions.

The strontium isotope ratio measurements were carried out using a Neptune multi-collector inductively coupled plasma mass spectrometer (MC–ICPMS) from Thermo Scientific. A volume of 125 μl of sample was taken and mixed up with 1 mL of nitric acid 5 N in a SAVILLEX beaker placed on a hotplate at 373 K overnight. The residual phase was reacidified with 0.5 mL of nitric acid at 3 N, then loaded onto a column containing a preconditioned crown ether cation exchange resin Sr Spec from EICHROM. The column was previously washed with 2 mL of distilled water (Milli-Q), then 1 mL of nitric acid 3 N. The elution was done with 2.5 mL of nitric acid 3 N and the Strontium collection with 4 mL of nitric acid 0.01 N followed by 1 mL of distilled water. After a new evaporation step, the solid-state sample was dissolved in 0.5 mL of a mixture of nitric acid 2%. The row results were normalized using the NIST 987 standard and we came up with accuracy better than 0.00004.

3.4 Sediment sampling and analyses

The sediment was sampled within 1 hour after splitting the core sections. Slides of 2 cm thickness were taken with a resolution of 10 cm. The collected samples have been analysed using both X-ray fluorescence and X-ray diffraction techniques (XRF, XRD). X-ray fluorescence was used in order to quantify the bulk chemical composition, *i.e* the present major, minor and trace elements in order to complement the XRF core-scanner analysis whereas X-ray diffraction enabled the identification of the various mineral phases as well as the evaluation of their ratio. The preparation of the sediment samples was as follows: the samples were finely crushed in powder and then dried overnight in an oven at 110 ° C. For the identification of the mineral phases by XRD, a part of the prepared powder was directly taken, put in a sample holder where the surface sample was made flat by using a glass slide, then analysed.

The remaining powder was transformed into both fusion beads and compressed powder pellets for XRF analysis. The fusion beads allow for accurate measurements of major and minor element content (*i.e.* mass concentration > 0.1 %) while the compressed powder pellet is used for the analysis of trace elements (mass concentration < 0.1 %). The fabrication of the fusion beads was as follows: the powder was calcinated under air at 1,050 ° C in a platinum cup for 1h 30 (Loss of Ignition, LOI which corresponds to the loss of carbonate and sulphite ions). After removal and cooling, the sample was crushed a second time. A precise amount of 0.50000 g was taken and mixed with exactly 9.000 g of flux of 90 % lithium tetraborate (LiB_4O7) and 10 % of lithium fluoride (LiF). Then, 500 mL of lithium bromide solution (250g/L LiBr) is added by pipetting, and the mixture was dried again. It was then placed in a gold-platinum crucible and put in the oven at 1,050 ° C for 15 min. Finally, the resulting solid solution was removed from the oven and re-homogenized to remove all bubbles. The crucible was put back in the oven for 10 min. at 1,050 °C and subsequently cooled at ambient temperature. After cooling, the fusion beads were ready for analysis. The powder pellets were made by mixing 4 g of sediment sample with 0.4 g of wax (Hoechst type C). The resulting mixture was introduced into a mould and was compacted using a hydraulic press at 10 T/cm^2.

The wavelength dispersive X-ray fluorescence spectrometer used here was a S8 TIGER, BRÜKER-AXS. The X rays are absorbed by the sample, which thereafter emits wavelengths characterizing its composition. The instrument was calibrated using certified reference material (El Maghraoui et al., 1998). The X-ray diffraction analysis was performed with a diffractometer D8 ADVANCE BRÜKER-AXS with Bragg-Brentano geometry, using CuKα radiation with Ni filter at 35 kV and 25 mA and angle from 5° to 70° 2θ.

4. Results and discussion

4.1 Composition and origin of the hydrate-bound gases and the gas bubbles

Bourry *et al.* (2009) carried out isotopic, compositional and structural analyses of both the hydrate sample and the gas bubbles collected on the Western High. This work has provided the basic genetic characteristics of the hydrates and the related gases. The results from this work showed that the hydrate-bound gases and the gas bubbles have a thermogenic origin. With about 90 % methane and the remaining gases being heavier hydrocarbons and carbon dioxide, the molecular composition of the gas bubbles was representative of most of conventional natural gas. Our new measurements on the hydrate sample are in agreement with their values for both the isotopic and molecular compositions (Table 1). The hydrate-bound gases contain 65.2 % methane, and a high amount of propane and isobutane (respectively 18.6 % and 9.2 %). The mass detector of the GC-MS allowed us to detect cyclopentane, cyclohexane, 3-Methylpentane, 2-3-DiMethylbutane, benzene, Methyl-Cyclopentane and Trans 1-4-DiMethyl-cyclohexane in the hydrate-bound gases. Those compounds have not been quantified and are presumably in small amount; therefore they are not taken into account in the hydrate composition in Table 1. As can be calculated from Table 1, the C1/(C2 + C3) ratio is slightly above 3 and 24 for the hydrates and the gas bubbles, respectively. This is also indicative of thermogenic gases. These results are in coherence with the yellow tint of the recovered hydrate specimens, reflecting the petroleum fingerprint. Figure 3 summarises the genetic characterization of both the hydrate-bound gases and the gas bubbles. The CD diagram, based on the carbon and hydrogen isotopic composition of the methane (Schoell, 1983; Whiticar, 1994), indicates a thermogenic gas

	Gas hydrate sample			Gas bubbles (after Bourry et al., 2009)	
Components	% Mole fraction	$\delta^{13}C$ (‰ PDB)	$\delta^{13}D$ (‰ SMOW)	% Mole fraction	$\delta^{13}C$ (‰ PDB)
C1	65.2	-43.7	-222	90.90	-44.4
C2	1.5	-23.5	-106	1.23	-25.7
C3	18.6	-21.6	-146	2.50	-21.1
iC4	9.2	-27.8	-163	0.93	-28
nC4	0.18	-	-	0.15	20.1
neoC5	0.049	-25.2	-199	0.0034	-
iC5	0.037	-26.2	-	0.31	-25.3
nC5	-	-	-	0.010	-18.9
C6+	0.22	-	-	0.0017	-
CO_2	5.1	+27.3	-	3.90	+29.1

Table 1. Molecular and isotopic compositions of the hydrate-bound gases and the gas bubbles

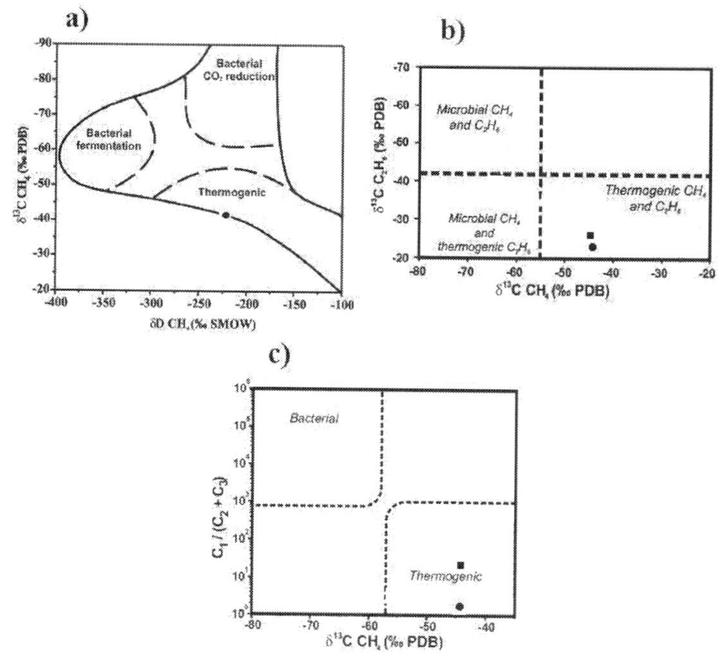

Fig. 3. a) Plot of hydrogen and carbon isotope compositions for the gas-hydrate sample indicating a thermogenic origin of methane (diagram modified after Schoell, 1983), b) Relationship between the stable carbon isotope composition ($\delta^{13}C$) of CH_4 vs. C_2H_6 for the hydrate-bond gases and the gas bubbles. CH_4 and C_2H_6 have a thermogenic origin (diagram modified after Bernard, 1978), c) Relationship between methane carbon isotope composition vs. ethane and propane molecular composition (diagram modified after Bernard, 1978). (•) gas-hydrate sample, (▪) gas bubbles.

which is close to the boundary with the geothermal, hydrothermal and crystalline zone, while both Bernard diagrams (Bernard et al., 1978) plot the hydrate-bound gases and the gas bubbles well inside the thermogenic zone. Thus, the natural gas of the Western High comes from at least one hydrocarbon reservoir.

The unique petroliferous area present nearby is the Thrace basin. It is the major gas producing province of Turkey which extends to the northern part of the Sea of Marmara (Gürgey, 2009; Gürgey et al., 2005; Hoşgörmez and Yalçin, 2005; Hoşgörmez et al., 2005). Therefore it is likely that the natural gas comes from deep-subsurface hydrocarbon reservoirs of this basin and feeds the Natural-Gas Hydrate Occurrence Zone (NGHOZ) of the Western High. Accordingly the gas bubbles and the hydrate-bound gases could share the same source rock.

A large amount of research investigations based on geology, petroleum geology, stratigraphy, geophysics and petroleum geochemistry have been carried out to better understand and describe the Thrace basin petroleum system (Coskun, 1997; Coskun, 2000; Gürgey, 2009; Gürgey et al., 2005; Hoşgörmez and Yalçin, 2005; Hoşgörmez et al., 2005; Şen Şamil et al., 2009). Hoşgömez et al. (2005) have combined isotopic and molecular composition analyses of light alkanes along with basin modelling in order to determine the source rocks of the hydrocarbon fields discovered there. Their results agree and complement those of Gürgey et al. (2005). Overall, they showed that (1) the thermogenic gas in the Thrace basin could come from two different types of source rock: a mature source which consists of a mixed type of organic matter (marine and terrestrial), and another which is an early mature source with marine organic matter, (2) the gas can be classified in three groups: Group I consists of pure thermogenic gases generated by a mixed type of organic matter. Group II is the result of mixing of a fraction of the thermogenic gases from group I with microbial gases. Group III is a mixture of at least two different thermogenic gases, one being the thermogenic gases of Group I and the other one is a marine source rock, (3) there are two potential source rocks for the studied natural gas fields: the Hamitabat formation (Middle Eocene) and the Mezardere formation (Upper Oligocene), (4) the Hamitabat formation is the source rock of K-Marmara natural-gas field, which consists of pure thermogenic gases while the Mezardere formation is the source rock of thermogenic portion of the natural-gas fields containing both thermogenic and microbial gases.

The gas bubbles from the Western High might be genetically linked to one of these source rocks, or be independent of the onshore petroleum systems. Some similarity in molecular composition has been highlighted with the studied gas fields, especially with the K-Marmara natural-gas field (Bourry et al., 2009). The latter consists of pure thermogenic gases (Figure 4b). However, when applying the Chung's diagram (Chung et al., 1988) to both the hydrate-bound gases and the gas bubbles, it appears that they consist of a mixture of thermogenic and microbial gases (Figure 4a) as we came up with a concave down curve instead of a straight line as for pure thermogenic gases. Amongst all natural-gas fields studied by Hoşgörmez and Yalçin (2005), three fields, Tekirdag, Karacali and Hayrabolu, contain a mixture of thermogenic and microbial methane like the gas bubbles (Figure 4b). All three fields also have the same source rock which is the Mezardere formation. Therefore they or their source rock may also be considered as a potential candidate for the source of the hydrate-forming gases and the gas bubbles.

Figure 4b shows that the gas bubbles and the hydrate-bound gases have a source rock which is isotopically heavier (~-15 ‰) than the natural-gas fields (-21 ‰). The difference between the two projected source rocks could be due to the heterogeneities in the precursor kerogen,

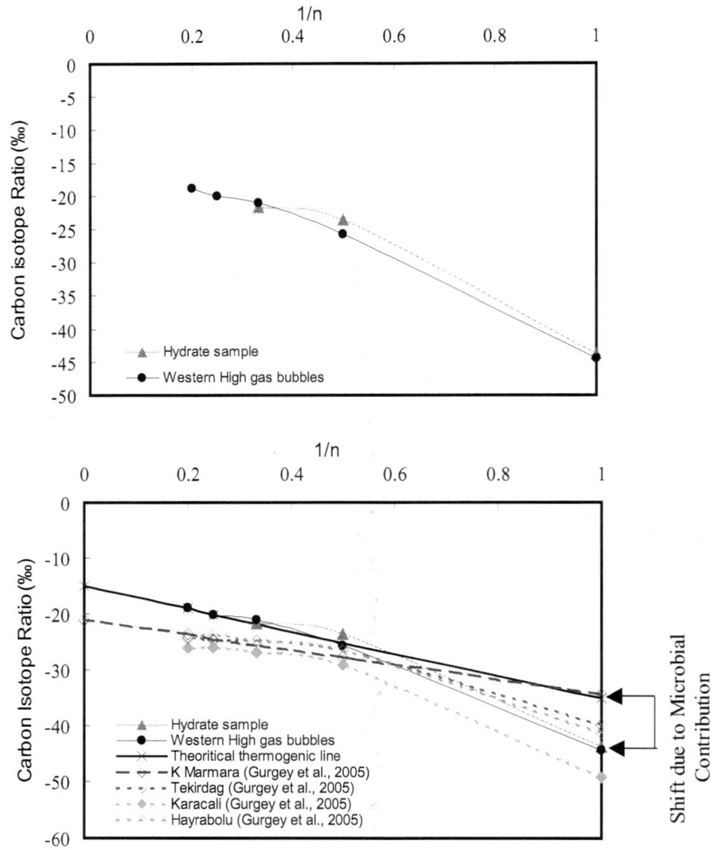

Fig. 4. Gas plot model based on the reciprocal of the carbon chain length against the $\delta^{13}C$ of the respective hydrocarbons (after Chung et al., 1988). a) The non-linearity of the curve is indicative of mixing between thermogenic and microbial gases. b) Comparison between the Western High gases and three natural-gas fields of the Thrace basin.

or the gas generation mechanism (Pohlman et al., 2005). One can also see that the projected pure thermogenic methane of all gases matches with the thermogenic methane from the K-Marmara natural gas. Thus, we believe that a portion of the methane of all the natural-gas seeps presented here comes from the same source rather than the K-Marmara natural gas, namely the Hamitabat formation. However, the origin of the remaining portion is still elusive.

Table 1 shows that the isotopic signature of carbon dioxide for the studied gases differs a lot from that of the previously mentioned natural-gas fields and this may be a key for the interpretation of the gas source. The positive value of $\delta^{13}C$ of carbon dioxide for both the hydrates (+27.3 ‰) and the gas bubbles (+29.1 ‰) contrast with the negative values

(-9.9 ‰, -11.5 ‰, -8.5 ‰) of the respectively Tekirdag, Karacali and Hayrabolu natural-gas fields (Hoşgörmez et al., 2005). Hence, it is unlikely that the gas bubbles may come from directly those natural-gas fields. $\delta^{13}C$-enriched CO_2, with values above +5‰, are typically due to secondary methanogenesis following hydrocarbon biodegradation occurring in petroleum reservoirs or during secondary and tertiary migration (Etiope et al., 2009b). The secondary methanogenesis increases the $\delta^{13}C$ of residual CO_2 (previously produced by anaerobic oxidation of heavy hydrocarbons) which may easily exceed +10‰ (Etiope et al., 2009b; Pallasser, 2000). Petroleum biodegradation process is a phenomenon that was neglected in the past, but today it is considered to affect a large fraction of conventional oil reserves (Dimitrakopoulos and Muehlenbachs, 1987; Head et al., 2003; Jones et al., 2007). Such a biodegradation typically modifies also propane and other n-alkanes isotopic values, leading to large isotopic separations between successive n-alkanes and high C2/C3 and/ or iC4/nC4 ratios (Etiope et al., 2009a; Etiope et al., 2009b; Waseda and Iwano, 2008). Therefore, the isotopic and molecular compositions of the gas bubbles and the hydrate bound-gases show that they meet the requirements of seeps which have undergone a biodegradation at the reservoir level along with a secondary methanogenesis (Etiope et al., 2009b).

The application of the model developed by Berner and Faber (Berner and Faber, 1996) to the different natural-gas seeps leads to the same conclusions (Figure 5).

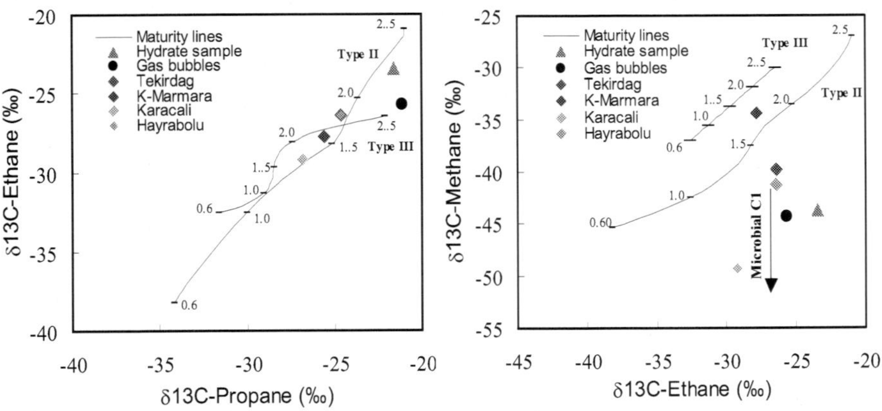

Fig. 5. Genetic characterization of the gases based on the $\delta^{13}C$ of the couples propane- ethane and methane- ethane (modified after Berner and Faber, 1996).

The shift towards heavier $\delta^{13}C$ for the propane indicates that the gases are biodegraded (Figure 5a). The source rock is primarily type II kerogen and the generated gases reflect overmaturity (Ro >1.7). This is consistent with the generation of oil as well as gas in the oil window (Berner and Faber, 1996). Figure 5b presents the level of mixing of microbial methane with the thermogenic one. In the case of both the gas bubbles and the hydrate-forming gases, this shift downwards represents the microbial contribution which is due to the methanogenesis following the biodegradation.

Following the analyses presented above, it is likely that the origin of both the hydrate-bound gases and the gas bubbles studied here is the source rocks supplying the same thermogenic gases that the onshore natural-gas fields. However the hydrocarbons accumulated underneath the Western High underwent a biodegradation, which is not observed for the onshore-fields. It seems that both the Hamitabat and the Mezardere formations could contribute to the supplying of the gases as both formations have reached the expected maturity level for the gas generation. These formations could first supply a reservoir in which biodegradation of oil and methanogenesis occur. The latter process leads to the production of microbial methane at the reservoir level. However the occurrence of these combined processes, biodegradation/ methanogenesis, requires a reservoir at a depth of less than 2000 m and a temperature between 343.15 and 353.15 K (Gürgey et al., 2005).

Presently, it is not possible to link the gas bubbles as well as the hydrate-bound gases to any known offshore reservoirs, and unfortunately there are no available geologic cross sections or published stratigraphic columns for the offshore southern Thrace basin. Petroleum exploration has been made at the southwestern part of the Thrace basin, onshore as well as offshore in Sea of Marmara. After drilling several wells, it turned out that all of them were dry (Sen et al., 2009). A specific study and reconstruction of the Thrace basin sedimentary formations towards the Sea of Marmara should be made, starting from seismic data. It would be essential to verify the position and setting of the formations and the eventual occurrence of reservoirs in the Sea of Marmara.

4.2 Hydrate formation driven by gas/fluid migration and associated mineralization

4.2.1 Gas migration and hydrate formation

The study of the gas bubbles and the hydrate-bound gases revealed that both have primarily a thermogenic origin. By applying the CSM-GEM model (Sloan and Koh, 2008), it has been showed that the gas bubbles are representative of the hydrate-supplying gases (Bourry et al., 2009). The abundances of propane and isobutane along with methane are diagnostic of structure II (Sloan and Koh, 2008), and that has been confirmed by Raman spectroscopy (Bourry et al., 2009). However, the inclusion of heavier hydrocarbons such as the identified cyclo-C_{5+} and Methyl-C_{4+} suggests the possibility of having structure H (Lu et al., 2007; Sassen et al., 1999). Moreover, owing to the presence of a large amount of propane and isobutane, at that depth and temperature conditions, the hydrate accumulation is well inside its thermodynamic stability field. Thus, its destabilisation would occur by a dissolution process in the pore fluid rather than by a dissociation with release of gas bubbles (Sultan et al., 2010). Therefore, it is likely that the gas bubbles sampled nearby are the manifestation on the seafloor of the natural-gas migration from deep-hydrocarbon reservoirs through the fault system of the Western High.

Figure 6 represents the 3D images obtained from the X-ray CT scan analysis of the recovered cores MARM_1 and MARM_2. The uppermost segments of both cores, segments with a length of 48 cm and 18 cm for MARM_1 and MARM_2, respectively, have not been scanned. A network of cracks where gases and fluids can migrate and accumulate characterizes both cores. The first scanned segment of MARM_1 (Figure 6, MARM_1 a) and a')) contains a "light" sediment which was quite difficult to image in 3D. Thus, Figure 6 MARM_1 b) represents the core liner with no possibility of seeing through. The image (a')) represents a

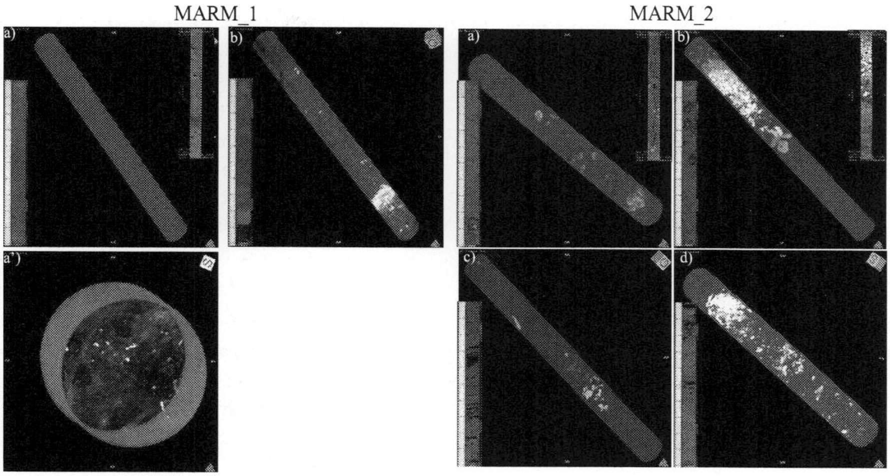

Fig. 6. CT scan images along with core photography of MARM_1 and MARM_2.

front view of the said core-segment which clearly show a lighter sediment and scattered with some concretion-like phases. The second scanned segment of MARM_1 is characterized by a very black sulphidic sediment at its lower part (from 238 cm to the end) which is located right underneath a denser carbonate-rich layer of about 8 cm thick (from 230 cm to the 238 cm) captured in the 3D image (Figure 6, MARM_1 b)). It is most likely that the carbonate-rich level represents the lake/marine transition and the dark sulphidic sediments below are the lacustrine sediments. This is strengthened by the sulphur and the manganese profiles, as well as the iron concentration and the LOI from Table 3. The manganese profile presents two maximum peaks in concentration at 30 cm and 230 cm, respectively. The latter depth also corresponds to the minimum in concentration of iron and a maximum in sulphur in the sediment for MARM_1 (Table 2). Such a behaviour is often related to the incorporation of manganese into the carbonate mineral found at that depth under oxic bottom sea water (Calvert and Pedersen, 1993; Schaller et al., 1997a; Schaller and Wehrli, 1996). The minimum in iron concentration could be explained by the formation of either Fe-colloidal particles as a low level of pyrite was detected (Schaller et al., 1997b). The photography (Figure 6, MARM_1 b)) shows that the dense layer has a lighter grey-colour. Although dense, after splitting the section it turned out that this layer does not present any evidence of visible concretions. As illustrated in both the profiles of Figure 9 and Table 2, the calcium, strontium and barium exhibit a peak at about 230 cm which is probably related to the presence of carbonate mineral phases. Table 3 confirms the presence of calcite at this level. In the case of MARM_2, the positions of the different holes, bivalves and concretions in the photography of each segment are in good agreement with the resulting 3D images. The first section is characterized by the presence of bivalves with a build-up mainly at its lower part, while section 2 and 4 contained a large amount of carbonate concretions which were sampled after splitting the core. MARM_2 was taken near the gas seep and the crack network is a record of intense gas ebullition at this site. Each associated photography also shows a highly disturbed sediment where the concretion deposits are found. Furthermore, overall the XRF analyses showed high concentrations of Si, Al, and Fe which are ubiquitous in clay minerals (Tables 2 and 4). This is strengthened by the XRD analysis where several group of clay, such as kaolinite, smectite, illite and chlorite groups, have been identified in significant amounts (Tables 3 and 5). The results are also consistent with the previous sedimentology work carried out in this area (Bayhan et al., 2001). Usually natural-gas migration through faults in such a clay-rich marine sediment leads to the formation of massive, nodular or vein-filling hydrates (Holland et al., 2008; Paull and Dillon, 2001).

4.2.2 Geochemical processes associated with the hydrate accumulation

Hydrate accumulation in marine sediment is also associated with several biogeochemical processes (Boudreau, 1997; Luff and Wallmann, 2003; Schulz and Zabel, 2000; Wallmann et al., 2006). The analysis of selected elements from both the pore fluids and the sediment often provides useful indication on the occurrence of these processes, and accordingly helps in constraining the origin of the pore fluid which is involved in the formation and accumulation of the natural-gas hydrates. Figure 7 represents the pore-fluid profiles of selected dissolved-species. The analysis of the bottom seawater have not been made, however it is reasonable to assume that the composition is very close to the value found in the upper part of the cores. In Figure 8, the alkalinity and strontium isotope profiles of the pore fluid are shown along with the Mg/Ca and Sr/Ca ratios in the sediment and the pore fluid. The two cores exhibit very different pore-fluid profiles reflecting the heterogeneity of the system. The chloride concentration range from 475 mM to 655 mM for MARM_2 with a large scattering while MARM_1 presents a relatively constant concentration value with depth (~ 610 mM). The scattering for MARM_2 suggests the occurrence of hydrate dissociation and the value of 475 mM is indicative of pore-fluid freshening due to this dissociation. For both cores, the dissolved sulphate decreases with depth. Along with methane, sulphate in pore-fluid is involved in the early diagenesis of marine sediments. Two chemical reactions lead to a depletion of sulphate concentration in anoxic conditions (Berner, 1980; Bhatnagar et al., 2008; Bhatnagar et al., 2011; Borowski et al., 1996; Borowski et al., 1999; Reeburgh, 1976):

- The microbial sulphate reduction (Martens and Val Klump, 1984). This process can be summarized in the following reaction:

$$2\,(CH_2O) + SO_4^{2-} \rightarrow 2\,HCO_3^{-} + H_2S \tag{1}$$

- The anaerobic oxidation of methane (AOM) by a consortium of microorganisms which consumes methane for their energy (Boetius et al., 2000) such as:

$$CH_4 + SO_4^{2-} \rightarrow HCO_3^{-} + HS^{-} + H_2O \tag{2}$$

Equation (2) occurs at the sulphate-methane interface (SMI) in the sediment column which is characterized by high upward flux of methane and downward flux of seawater sulphate. This is the case on the Western High where methane-rich gas bubbles and natural-gas hydrates have been sampled. MARM_2 shows a linear decrease in the sulphate profile which is typical of the occurrence of sulphate reduction coupled with the anaerobic oxidation of methane (Hensen et al., 2003; Joye et al., 2004; Orcutt et al., 2004; Pohlman et al., 2008; Treude et al., 2003). The SMI is located at about 120 cm and corresponds to the depth where the sulphate concentration drops sharply to a value close to zero. Thus, below the SMI corresponds to a sulphate-free zone but rich in methane. The strong steep sulphate gradient from the seafloor to the SMI is a consequence of important upwards hydrocarbon fluxes (Borowski et al., 1996; Hensen et al., 2003). This is in agreement with the gas seep discovered there, as seep is a visible clue of intense upwards methane migration. The AOM process is very often associated with carbonate precipitation because it induces an increase in the total alkalinity (TA), which is directly related to the carbonate and sulphite ion concentrations of the pore fluid (see equation II). Once the pore-fluid saturation is reached, the carbonate formation begins and this is generally accompanied with variations in both pore-fluid and sediment profiles of Ca, Mg, Sr. Accordingly, the study of the latter elements provides useful information on the carbonate diagenetic processes (Karaca et al., 2010; Ussler III et al., 2000). While it is observed a quite constant pore-fluid composition with depth for MARM_1, MARM_2 exhibits a decrease in Ca down to the SMI, then an increase up to about 250 cm depth followed by another decrease. The concentration of Mg constantly increases with depth for MARM_2. Moreover, one can see in Figure 8 that the TA of MARM_1 is constant with depth. However for MARM_2, the TA linearly increases to a shallow inflection at the SMI, then keeps increasing gradually with depth. The above lines of evidence suggest the occurrence of authigenic mineralization accompanied with mainly Ca removal from the pore fluid down to the SMI. This is consistent with carbonate precipitation. Here, the Mg profile exhibits an increase in concentration with depth. Mg is much more concentrated in modern seawater than Ca, thus the concentration-depth profile of the former is less sensitive to carbonate precipitation. Figure 8 shows the Mg/Ca in both the sediment and the pore fluid as a function of depth for both cores, in both the sediment and the pore fluid. Beside the three points higher than 1 in the sediment which is presumably due to the presence of bivalves, these Mg/ Ca curves are nearly the conjugate of each other for both cores, *i.e* the ratio decreases in the pore fluid when it increases in the sediment. This is a consequence of authigenic carbonate precipitation that is typical for marine hydrate and methane-rich regions. Above the SMI, Ca in pore fluids drops sharply by calcite precipitation giving rise to the increase of Mg/Ca ratio in pore fluid and the decrease in the sediment, while below, the precipitation of Mg-rich minerals leads to the reciprocal behaviour.

The more common Mg-rich minerals encountered on a methane-rich sediment are the dolomite, high-Mg calcite and siderite. However, dolomite represents less than 5 % weight of the mineral identified in MARM_2 whereas siderite has not been identified. Therefore, the Mg has probably been incorporated in the calcite. The increase with depth of the TA from the SMI to the lowest part of MARM_2 (Figure 8) could be a manifestation of upwards migration of fluids with high alkalinity, like hydrocarbon-reservoir brines migrating along with the gas bubbles. This would mean that a part of the pore fluid come from depth. The analysis of the radiogenic isotope of strontium can help us in constraining the sources and sinks of the fluid because its isotopes do not fractionate during geochemical processes (Kastner et al., 2008). As illustrated in Figure 8, MARM_1 has a fairly constant strontium isotope value with depth which is close to the modern seawater value (~0.7092). For MARM_2, the fluid becomes less radiogenic with depth by a substantial amount. The value at its lowest part seems to be stabilized at ~0.70877. Therefore, these results strengthen the hypothesis of a deep source contribution of fluid. However, we do not have any other isotopic value available to compare with in order to better constraint the pore-fluid origin. An accurate determination of strontium concentration with depth combined with measurements on other geologic materials like the carbonates would help in gaining insights into the deep fluid source. That has to be considered for the continuation of this study.

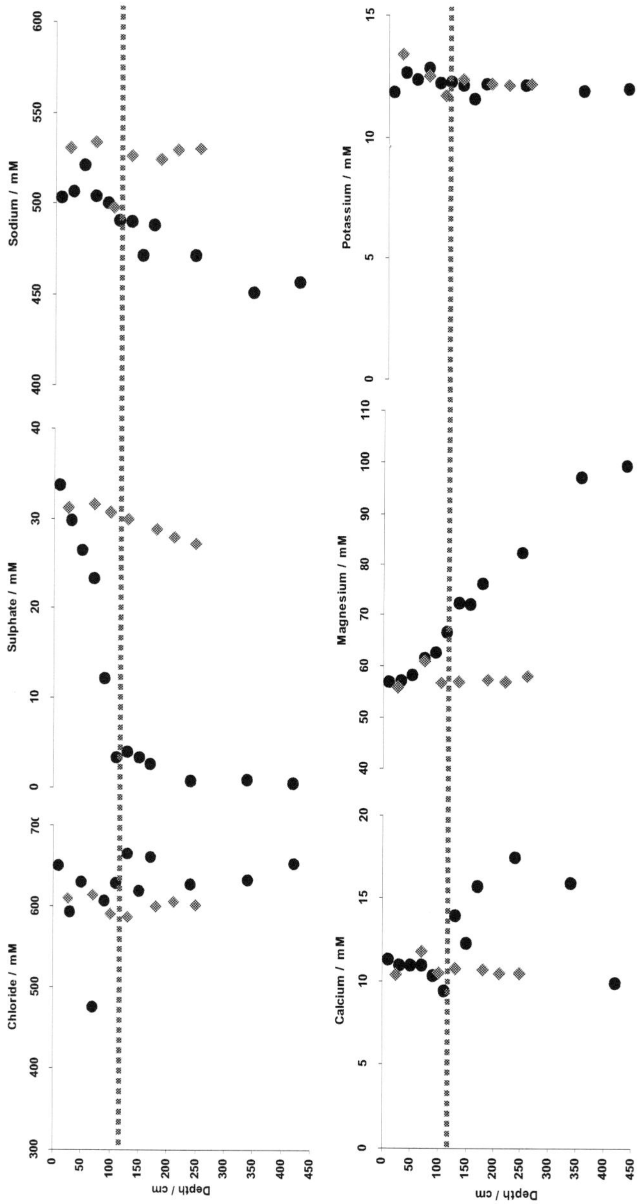

Fig. 7. Pore-fluid concentration profiles of selected dissolved-elements. The red diamonds correspond to MARM_1 and the black dots to MARM_2. The dashed blue line corresponds to the SMI.

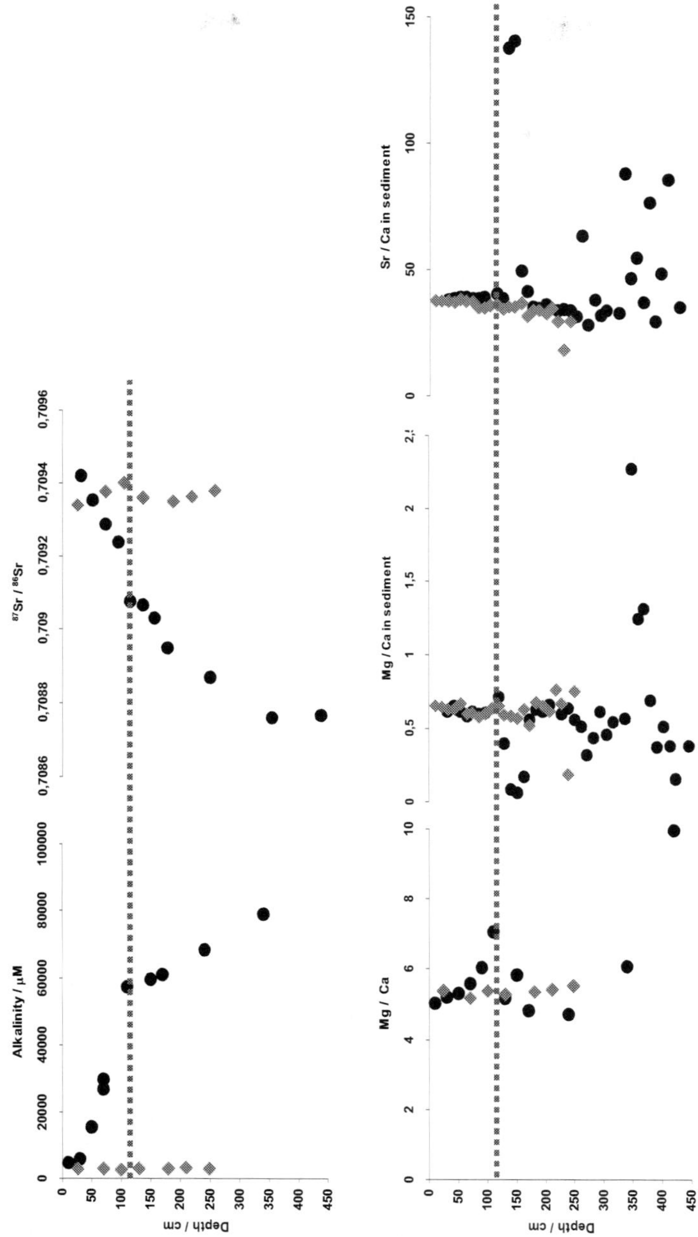

Fig. 8. Alkalinity and strontium isotope profiles in the pore fluid, and Mg/ Ca and Sr/ ca in both the pore fluid and the sediment.

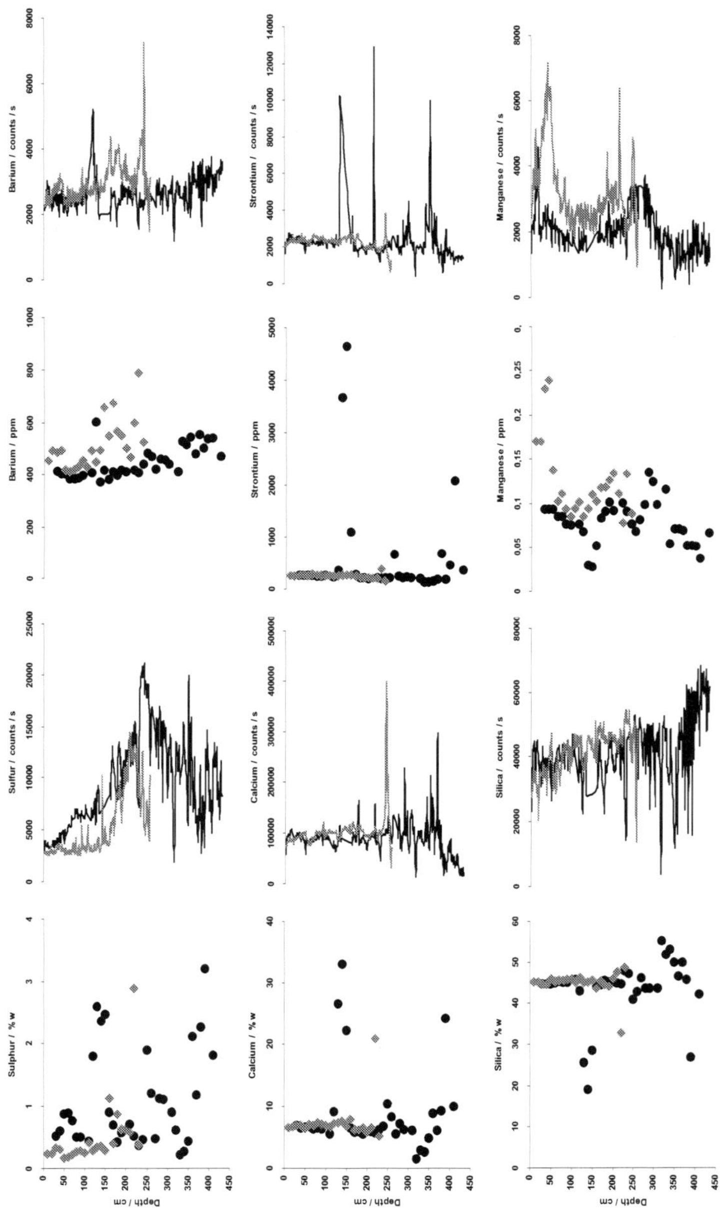

Fig. 9. Sedimentary profiles of selected elements determined with a laboratory XRF instrument (symbols, where the red diamonds are for MARM_1 and black dots for MARM_2) and a XRF core scanner (full line).

Depth / cm	SiO_2 %	Al_2O_3 %	Fe_2O_3 %	MnO %	CaO %	MgO %	K_2O %	Na_2O %	TiO_2 %	P_2O_5 %	SO_4 %	LOI %	Total %
10	45.04	15.71	6.91	0.17	6.56	4.32	2.67	3.20	0.67	0.12	0.24	14.53	100.14
20	45.06	15.68	6.93	0.17	6.65	4.28	2.69	3.09	0.67	0.12	0.23	14.57	100.15
30	44.72	15.60	6.96	0.23	6.95	4.36	2.68	2.99	0.66	0.13	0.32	14.84	100.43
40	44.50	15.53	6.89	0.24	6.94	4.39	2.61	2.98	0.65	0.13	0.30	14.69	99.85
60	46.03	16.18	7.05	0.14	6.35	4.27	2.82	2.90	0.70	0.12	0.16	13.61	100.33
70	45.41	15.83	6.91	0.10	7.16	4.28	2.72	2.72	0.69	0.12	0.17	14.04	100.16
80	45.58	15.91	6.95	0.11	6.92	4.22	2.71	2.67	0.69	0.12	0.21	14.02	100.10
90	45.43	15.38	6.83	0.09	7.33	4.23	2.68	2.81	0.67	0.12	0.28	14.65	100.51
100	45.65	15.18	6.78	0.09	7.08	4.25	2.68	2.84	0.68	0.12	0.29	14.50	100.13
110	46.01	15.55	6.68	0.09	6.94	4.43	2.72	2.76	0.67	0.12	0.25	14.07	100.30
120	45.03	15.40	6.51	0.10	6.70	4.40	2.77	3.19	0.65	0.13	0.41	14.75	100.03
130	46.05	15.47	6.76	0.09	7.29	4.32	2.71	2.66	0.69	0.11	0.28	14.32	100.75
140	45.05	15.10	6.67	0.09	7.45	4.38	2.65	2.56	0.67	0.12	0.35	14.65	99.74
150	45.04	15.01	6.75	0.11	7.61	4.35	2.57	2.51	0.68	0.11	0.36	15.04	100.16
160	45.52	15.28	6.96	0.10	6.85	4.30	2.66	2.62	0.69	0.11	0.29	14.51	99.90
170	43.54	14.48	6.80	0.12	7.96	4.14	2.47	2.57	0.68	0.11	1.13	15.89	99.88
180	45.37	15.31	7.17	0.12	6.32	4.28	2.70	2.67	0.71	0.11	0.39	14.83	99.99
190	44.38	15.08	7.35	0.13	6.24	4.08	2.64	2.73	0.69	0.11	0.86	15.68	99.97
200	44.01	15.30	7.28	0.13	6.60	4.08	2.67	2.50	0.66	0.12	0.64	16.16	100.14
210	45.90	15.33	7.27	0.11	5.83	4.44	2.76	2.56	0.68	0.12	0.62	14.52	100.14
220	47.70	14.93	7.01	0.08	6.51	4.35	2.74	2.50	0.70	0.11	0.57	13.42	100.61
230	32.78	10.16	4.58	0.13	20.85	3.96	1.46	1.73	0.46	0.13	2.88	21.52	100.62
240	48.61	16.09	7.25	0.09	5.20	3.92	3.13	2.57	0.65	0.13	0.38	11.80	99.82

Depth / cm	Cr (PPM)	V (PPM)	Co (PPM)	Ni (PPM)	Cu (PPM)	Zn (PPM)	Ga (PPM)	As (PPM)	Rb (PPM)	Sr (PPM)	Y (PPM)	Zr (PPM)	Nb (PPM)	Mo (PPM)	Pb (PPM)	Ba (PPM)	Ce (PPM)	La (PPM)	Sc (PPM)	Br (PPM)
10	172	152	22	124	40	99	13	7	122	246	20	104	11	<2	35	452	93	38	22	127
20	167	156	22	126	41	98	13	9	122	249	20	105	12	<2	32	491	82	38	19	120
30	165	154	23	132	41	101	14	12	123	261	21	107	12	<2	35	487	81	43	19	121
40	170	149	24	134	41	99	13	10	122	258	21	107	12	<2	37	493	84	45	17	113
60	201	154	21	139	41	103	14	10	126	241	21	112	12	<2	35	416	83	44	21	98
70	168	152	22	129	41	104	14	10	123	268	21	110	12	<2	38	412	69	39	23	100
80	165	152	22	126	41	100	14	10	123	256	21	112	12	<2	33	422	87	44	19	93
90	169	139	21	120	37	94	14	10	119	254	20	111	12	3	27	432	94	38	18	109
100	171	135	21	116	35	95	12	9	121	247	20	115	12	<2	28	453	91	45	14	119
110	169	133	20	115	35	95	14	7	120	247	20	120	14	<2	31	431	86	43	18	107
120	157	135	20	108	37	96	14	8	117	243	19	138	17	<2	34	491	89	31	17	133
130	178	139	21	114	36	96	15	7	118	249	20	109	12	3	27	446	74	38	20	105
140	172	143	20	118	36	96	14	7	119	261	21	109	11	<2	27	492	89	40	21	108
150	179	154	22	119	38	96	13	8	116	267	21	109	11	5	28	656	64	39	20	104
160	182	151	21	125	42	112	15	10	121	252	21	114	11	5	28	546	80	42	18	104
170	180	170	22	125	48	92	14	14	115	251	21	111	12	17	27	674	77	43	21	121
180	190	174	24	131	43	96	14	13	121	213	22	113	12	16	27	567	84	37	18	111
190	188	193	24	138	44	94	13	18	120	212	21	111	12	39	27	550	82	44	18	123
200	184	221	25	149	47	99	14	16	125	214	22	112	12	85	29	503	79	40	22	110
210	204	187	26	174	42	100	14	18	130	202	23	119	13	80	29	466	88	43	19	89
220	203	169	20	153	34	91	13	25	120	192	23	124	12	17	24	598	97	42	17	73
230	156	106	15	136	30	65	7	11	83	377	17	79	9	8	19	785	57	35	15	53
240	210	149	20	125	43	98	15	10	133	154	25	119	12	<2	24	523	106	45	20	64

Table 2. Present major, minor and trace elements detected by X-Ray fluorescence for MARM_1.

Depth / cm	aragonite	quartz	calcite	dolomite	anorthite	albite	halite	pyrite	alpha-soufre	pyroxène	anorthoclase	orthoclase	illite/muscovite	kaolinite	chlorite
10	<5	>20	>20	<5	<5	<5	>5 - <20	<5	<5	none	<5	<5	>5 - <20	>5 - <20	>5 - <20
20	<5	>20	>20	<5	<5	<5	>5 - <20	<5	<5	none	<5	<5	>5 - <20	>5 - <20	>5 - <20
30	<5	>20	>5 - <20	>5 - <20	<5	<5	>5 - <20	<5	<5	none	<5	<5	>5 - <20	<5	<5
40	<5	>20	>5 - <20	>5 - <20	<5	<5	>5 - <20	<5	<5	none	<5	<5	>5 - <20	<5	<5
60	<5	>20	>20	<5	<5	<5	<5	<5	<5	none	<5	<5	>5 - <20	>5 - <20	>5 - <20
70	<5	>20	>20	<5	<5	<5	<5	<5	<5	none	<5	<5	>5 - <20	>5 - <20	>5 - <20
80	<5	>20	>20	<5	<5	<5	<5	<5	<5	none	<5	<5	>5 - <20	>5 - <20	>5 - <20
90	<5	>20	>20	<5	<5	<5	<5	<5	<5	none	<5	<5	>5 - <20	>5 - <20	>5 - <20
100	<5	>20	>20	<5	<5	<5	<5	<5	<5	none	<5	<5	>5 - <20	>5 - <20	>5 - <20
110	<5	>20	>20	<5	<5	<5	<5	<5	<5	none	<5	<5	>5 - <20	>5 - <20	>5 - <20
120	<6	>20	>20	<5	<5	<5	<5	<5	<5	none	<5	<5	>5 - <20	>5 - <20	>5 - <20
130	<5	>20	>20	<5	<5	<5	<5	<5	<5	none	<5	<5	>5 - <20	>5 - <20	>5 - <20
140	<5	>20	>20	<5	<5	<5	<5	<5	<5	none	<5	<5	>5 - <20	>5 - <20	>5 - <20
150	<5	>20	>20	<5	<5	<5	<5	<5	<5	none	<5	<5	>5 - <20	>5 - <20	>5 - <20
160	<5	>20	>20	<5	<5	<5	<5	<5	<5	none	<5	<5	>5 - <20	>5 - <20	>5 - <20
170	<5	>20	>5 - <20	<5	>5 - <20	>5 - <20	<5	<5	<5	none	<5	<5	>5 - <20	>5 - <20	>5 - <20
180	<5	>20	>5 - <20	<5	>5 - <20	>5 - <20	<5	<5	<5	none	<5	<5	>5 - <20	>5 - <20	>5 - <20
190	<5	>20	>5 - <20	<5	>5 - <20	>5 - <20	<5	<5	<5	none	<5	<5	>5 - <20	>5 - <20	>5 - <20
200	<5	>20	>5 - <20	<5	>5 - <20	>5 - <20	<5	<5	<5	none	<5	<5	>5 - <20	>5 - <20	>5 - <20
210	<5	>20	>5 - <20	<5	>5 - <20	>5 - <20	<5	<5	<5	none	<5	<5	>5 - <20	>5 - <20	>5 - <20
220	<5	>20	>5 - <20	<5	>5 - <20	>5 - <20	<5	<5	<5	none	<5	<5	>5 - <20	>5 - <20	>5 - <20
230	<5	>5 - <20	>20	<5	<5	<5	<5	<5	<5	none	<5	<5	<5	<5	<5
240	<5	>20	>5 - <20	<5	>5 - <20	>5 - <20	<5	<5	<5	none	<5	<5	>5 - <20	>5 - <20	>5 - <20

Table 3. Identification and semi-quantitative evaluation of the various mineral phases present in sediment of MARM_1.

Depth / cm	SiO_2 %	Al_2O_3 %	Fe_2O_3 %	MnO %	CaO %	MgO %	K_2O %	Na_2O %	TiO_2 %	P_2O_5 %	SO_4 %	LOI %	Total %
30	44.69	15.91	6.79	0.09	6.95	4.30	2.67	2.91	0.67	0.12	0.52	14.57	100.20
40	44.76	16.05	7.00	0.09	6.57	4.32	2.69	2.86	0.69	0.12	0.61	14.17	99.93
50	44.70	15.82	6.80	0.09	6.76	4.16	2.68	3.08	0.69	0.11	0.87	14.36	100.14
60	44.96	15.96	6.87	0.09	6.76	3.97	2.68	2.82	0.69	0.11	0.89	14.39	100.19
70	45.38	16.17	6.96	0.09	6.42	3.96	2.70	2.84	0.70	0.11	0.77	14.13	100.23
80	45.20	15.87	6.90	0.08	6.50	3.88	2.69	2.77	0.70	0.11	0.51	14.71	99.92
90	45.16	15.89	6.83	0.08	6.44	3.90	2.69	2.78	0.69	0.11	0.51	14.93	100.01
110	45.81	16.20	6.85	0.08	5.61	4.03	2.77	2.84	0.70	0.12	0.44	14.60	100.03
120	42.94	15.00	6.50	0.07	9.23	3.68	2.49	2.66	0.67	0.11	1.80	14.65	99.80
130	25.52	9.02	3.90	0.03	26.65	2.18	1.28	1.54	0.40	0.07	2.60	25.85	99.04
140	19.10	6.84	2.96	0.03	33.06	2.13	0.78	1.03	0.31	0.08	2.36	30.51	99.17
150	28.48	10.05	4.27	0.05	22.31	3.85	1.34	1.47	0.45	0.10	2.47	24.96	99.80
160	44.25	15.51	7.09	0.08	6.59	3.72	2.54	2.49	0.70	0.11	0.91	16.11	100.10
170	44.55	15.43	7.11	0.09	5.87	3.71	2.54	2.65	0.71	0.12	0.71	16.92	100.43
180	45.59	15.72	7.23	0.10	6.01	3.69	2.65	2.52	0.73	0.11	0.43	15.19	99.97
190	45.03	15.67	7.77	0.09	5.53	3.68	2.62	2.54	0.72	0.12	0.58	16.03	100.39
210	44.99	15.46	7.47	0.10	6.04	3.63	2.58	2.53	0.71	0.12	0.71	16.15	100.49
220	44.63	15.61	7.60	0.09	5.70	3.63	2.60	2.54	0.70	0.13	0.52	16.47	100.21
230	47.78	15.15	6.87	0.08	6.27	3.50	2.64	2.39	0.70	0.13	0.37	14.51	100.38
240	47.27	14.88	6.56	0.07	6.87	3.51	2.56	2.31	0.70	0.13	0.46	14.64	99.96
250	40.90	13.65	6.61	0.08	10.48	3.34	2.24	2.20	0.60	0.12	1.89	17.73	99.84
260	42.71	14.10	6.67	0.10	8.39	3.69	2.38	2.56	0.62	0.12	1.22	17.27	99.83
270	46.12	15.69	7.27	0.13	5.58	3.44	2.67	2.48	0.67	0.13	0.48	15.76	100.43
280	43.58	15.10	7.11	0.12	7.28	3.37	2.51	2.41	0.66	0.12	1.13	16.74	100.14
290	43.67	15.30	7.40	0.10	6.31	3.43	2.55	2.51	0.67	0.12	1.12	17.12	100.31
310	43.72	15.36	7.45	0.12	6.13	3.46	2.55	2.53	0.67	0.12	0.90	17.17	100.20
320	55.20	17.08	6.42	0.05	1.50	3.40	2.98	2.32	0.73	0.14	0.62	9.93	100.36
330	51.77	17.10	7.20	0.07	2.93	3.64	3.16	2.28	0.67	0.13	0.21	11.05	100.20
340	53.09	16.94	6.84	0.07	2.66	3.49	3.09	2.41	0.70	0.13	0.28	10.51	100.20
350	49.91	15.60	7.03	0.07	4.87	3.38	2.79	2.28	0.69	0.13	0.44	12.98	100.17
360	46.67	14.31	5.95	0.05	8.89	3.34	2.55	2.12	0.61	0.13	2.11	13.45	100.18
370	50.01	15.04	6.10	0.05	6.17	3.16	2.64	2.32	0.67	0.13	1.18	12.90	100.37
380	45.63	13.97	5.81	0.05	9.36	3.59	2.45	2.15	0.62	0.13	2.27	14.26	100.30
390	26.88	8.50	4.32	0.04	24.25	3.81	1.18	1.48	0.40	0.11	3.20	25.39	99.55
410	42.23	13.31	6.50	0.07	10.08	3.88	2.27	2.27	0.62	0.12	1.81	17.18	100.34

Depth / cm	Cr (PPM)	V (PPM)	Co (PPM)	Ni (PPM)	Cu (PPM)	Zn (PPM)	Ga (PPM)	As (PPM)	Rb (PPM)
30	173	170	21	127	44	101	14	16	124
40	183	171	22	138	45	103	13	18	125
50	179	171	20	132	43	101	14	15	126
60	186	175	20	127	42	104	14	17	128
70	183	174	21	124	42	101	14	17	127
80	183	173	20	122	41	99	13	17	125
90	175	170	20	120	40	99	14	14	125
110	181	187	20	122	43	101	14	16	127
120	184	193	19	120	44	96	14	17	119
130	125	197	13	87	37	65	9	14	73
140	86	168	11	75	30	54	6	9	59
150	125	164	14	94	35	72	9	7	83
160	188	192	22	127	48	96	14	18	118
170	198	201	23	134	53	95	13	15	119
180	201	200	21	130	44	96	15	13	122
190	195	211	25	143	49	97	13	23	123
210	195	243	23	144	47	94	13	13	120
220	192	230	26	148	45	97	14	12	124
230	202	173	20	141	35	94	14	17	123
240	205	164	19	126	32	87	13	14	115
250	184	159	25	172	35	93	15	13	120
260	189	147	20	132	30	79	11	9	109
270	184	198	23	151	42	97	16	10	129
280	186	227	25	141	42	94	14	11	119
290	192	229	25	141	46	94	14	11	122
310	186	255	26	152	49	97	16	12	124
320	203	151	20	136	41	96	16	15	132
330	192	158	18	123	39	100	17	13	143
340	189	159	18	127	38	97	16	17	136
350	197	161	21	141	40	95	15	20	127
360	244	156	18	127	34	85	14	17	113
370	226	153	20	142	39	90	14	16	119
380	182	152	17	122	33	82	13	14	111
390	123	141	15	106	25	63	9	9	76
410	193	168	18	138	31	85	13	15	109

Depth / cm	Sr (PPM)	Y (PPM)	Zr (PPM)	Nb (PPM)	Mo (PPM)	Pb (PPM)	Ba (PPM)	Ce (PPM)	La (PPM)	Sc (PPM)	Br (PPM)
30	266	21	112	12	5	34	414	81	37	22	121
40	253	22	116	13	9	36	404	80	38	22	112
50	264	21	114	13	11	35	403	83	37	20	124
60	264	22	118	13	16	35	385	88	37	16	112
70	248	22	119	13	17	33	385	88	37	18	114
80	251	21	121	13	25	29	388	88	37	22	118
90	253	21	123	14	26	30	399	95	47	21	124
110	226	21	131	15	32	32	407	93	39	18	124
120	355	20	116	13	48	30	603	76	37	22	123
130	3667	11	9	7	61	16	371	72	30	9	84
140	4638	7	9	6	42	14	418	44	19	13	67
150	1108	13	58	8	28	20	381	57	35	17	84
160	272	20	107	11	47	27	410	77	39	20	127
170	206	21	114	12	47	28	397	78	38	18	144
180	209	21	114	12	28	25	419	77	32	18	115
190	201	22	115	12	46	28	412	77	33	19	132
210	205	21	114	12	54	27	419	88	35	18	135
220	195	22	113	12	69	28	407	82	42	20	140
230	212	22	124	13	26	26	440	85	43	16	111
240	215	21	119	12	9	23	483	90	47	19	104
250	663	19	99	11	68	25	470	83	37	20	116
260	236	19	99	11	66	25	420	68	35	15	149
270	212	21	115	12	96	29	460	77	47	21	123
280	232	21	106	11	84	25	457	87	38	21	121
290	212	20	108	11	101	26	441	80	32	16	136
310	201	21	111	11	115	26	410	80	45	16	142
320	132	26	134	14	3	25	526	89	51	17	107
330	137	25	120	12	3	24	515	101	47	20	105
340	145	25	134	13	5	24	543	102	45	16	96
350	182	24	133	13	22	24	480	102	50	20	103
360	683	22	109	11	13	23	555	96	41	17	84
370	183	25	132	12	5	24	503	101	49	20	93
380	456	22	113	11	8	21	537	90	44	19	92
390	2080	12	42	8	34	17	540	54	32	18	92
410	355	20	110	11	44	24	470	83	31	19	113

Table 4. Present major, minor and trace elements detected by X-Ray fluorescence for MARM_1.

Depth / cm	aragonite	quartz	calcite	dolomite	anorthite	albite	halite	pyrite	alpha-soufre	pyroxène	anorthoclase	orthoclase	illite/muscovite	kaolinite	chlorite
30	<5	>20	>20	<5	<5	>5 - <20	>5 - <20	<5	none	<5	none	none	>5 - <20	>5 - <20	>5 - <20
40	<5	>20	>20	<5	<5	>5 - <20	>5 - <20	<5	none	<5	none	none	>5 - <20	>5 - <20	>5 - <20
50	<5	>20	>20	<5	<5	>5 - <20	>5 - <20	<5	none	<5	none	none	>5 - <20	>5 - <20	>5 - <20
60	<5	>20	>20	<5	<5	>5 - <20	>5 - <20	<5	none	<5	none	none	>5 - <20	>5 - <20	>5 - <20
70	<5	>20	>20	<5	<5	>5 - <20	>5 - <20	<5	none	<5	none	none	>5 - <20	>5 - <20	>5 - <20
80	<5	>20	>20	<5	<5	>5 - <20	>5 - <20	<5	none	<5	none	none	>5 - <20	>5 - <20	>5 - <20
90	<5	>20	>20	<5	<5	>5 - <20	>5 - <20	<5	none	<5	none	none	>5 - <20	>5 - <20	>5 - <20
110	<5	>20	>20	<5	<5	>5 - <20	>5 - <20	<5	none	<5	none	none	>5 - <20	>5 - <20	>5 - <20
120	>5 - <20	>20	>20	<5	<5	>5 - <20	>5 - <20	<5	none	<5	none	none	>5 - <20	<5	<5
130	>20	>5 - <20	>5 - <20	<5	<5	<5	>5 - <20	<5	<5	<5	none	none	>5 - <20	<5	<5
140	>20	>5 - <20	>5 - <20	<5	<5	<5	<5	<5	none	<5	none	none	<5	<5	<5
150	>5 - <20	>20	>5 - <20	<5	<5	<5	<5	<5	none	>5 - <20	none	none	>5 - <20	<5	<5
160	<5	>20	>5 - <20	<5	<5	>5 - <20	>5 - <20	<5	none	<5	none	none	>5 - <20	>5 - <20	>5 - <20
170	<5	>20	>5 - <20	<5	<5	>5 - <20	>5 - <20	<5	none	<5	none	none	>5 - <20	>5 - <20	>5 - <20
180	<5	>20	>5 - <20	<5	<5	>5 - <20	>5 - <20	>5 - <20	none	<5	none	none	>5 - <20	>5 - <20	>5 - <20
190	<5	>20	>5 - <20	<5	<5	>5 - <20	>5 - <20	>5 - <20	none	<5	none	none	>5 - <20	>5 - <20	>5 - <20
210	<5	>20	>5 - <20	<5	<5	>5 - <20	>5 - <20	>5 - <20	none	<5	none	none	>5 - <20	>5 - <20	>5 - <20
220	<5	>20	>20	<5	<5	>5 - <20	>5 - <20	>5 - <20	none	<5	none	none	>5 - <20	>5 - <20	>5 - <20
230	<5	>20	>20	<5	<5	>5 - <20	>5 - <20	>5 - <20	none	<5	none	none	>5 - <20	>5 - <20	>5 - <20
240	<5	>20	>20	<5	<5	>5 - <20	>5 - <20	>5 - <20	none	<5	none	none	>5 - <20	>5 - <20	>5 - <20
250	<5	>20	>20	<5	<5	>5 - <20	>5 - <20	>5 - <20	none	<5	none	none	>5 - <20	>5 - <20	>5 - <20
260	<5	>20	>5 - <20	<5	<5	<5	>5 - <20	<5	none	>5 - <20	none	none	>5 - <20	<5	<5
270	<5	>20	>20	<5	<5	>5 - <20	>5 - <20	>5 - <20	none	<5	none	none	>5 - <20	>5 - <20	>5 - <20
280	none	>20	>20	<5	<5	>5 - <20	>5 - <20	>5 - <20	none	<5	none	none	>5 - <20	>5 - <20	>5 - <20
290	<5	>20	>20	<5	<5	>5 - <20	>5 - <20	>5 - <20	none	<5	none	none	>5 - <20	>5 - <20	>5 - <20
310	<5	>20	>20	<5	<5	>5 - <20	>5 - <20	>5 - <20	none	<5	none	none	>5 - <20	>5 - <20	>5 - <20
320	none	>5 - <20	<5	<5	>5 - <20	>20	<5	<5	none	<5	>5 - <20	>5 - <20	>5 - <20	<5	<5
330	none	>20	>20	<5	<5	>5 - <20	>5 - <20	>5 - <20	none	<5	none	none	>5 - <20	>5 - <20	>5 - <20
340	none	>20	>20	<5	<5	>5 - <20	>5 - <20	>5 - <20	none	<5	none	none	>5 - <20	>5 - <20	>5 - <20
350	none	>20	>20	<5	<5	>5 - <20	>5 - <20	>5 - <20	none	<5	none	none	>5 - <20	>5 - <20	>5 - <20
360	>5 - <20	>20	<5	<5	<5	<5	<5	<5	none	>5 - <20	none	none	>5 - <20	<5	<5
370	none	>20	>20	<5	<5	>5 - <20	>5 - <20	>5 - <20	none	<5	none	none	>5 - <20	>5 - <20	>5 - <20
380	<5	>20	>5 - <20	<5	<5	<5	>5 - <20	<5	none	>5 - <20	none	none	>5 - <20	<5	<5
390	>5 - <20	>20	>5 - <20	<5	<5	<5	<5	<5	none	>5 - <20	none	none	>5 - <20	<5	<5
410	none	>20	>5 - <20	<5	<5	<5	>5 - <20	<5	none	>5 - <20	none	none	>5 - <20	<5	<5

Table 5. Identification and semi-quantitative evaluation of the various mineral phases present in sediment of MARM_2.

5. Conclusion

Natural-gas hydrate systems have been studied for decades from both fundamental and industrial interests. The fundamental researches were stimulated by those who want to advance knowledge in the carbon cycling of the ocean, the importance of hydrates for living methanogen-communities or its role in marine slope failure and climate change. Beside, because for a long time it has been considered as a geohazard for the petroleum industry, the applied researches were chiefly orientated to considering natural-gas hydrate systems as a geological structure which needs to be localized, described and taken in consideration into their seafloor-installation optimisation software. Now the game is changing and the major industrial impetuses come from future applications to methane production and/ or carbon-dioxide sequestration. However, whatever the issue to deal with, it is essential to use a versatile approach combining field data, lab experiments and modelling if we want to achieve a high level of understanding of the system. It is only after completing this work that we will be able to provide reliable prediction on the evolution of a natural-gas hydrate system upon either heating by hot reservoir fluid flowing through a pipeline, large-scale CO_2 injection or large-volume methane extraction by natural or anthropogenic perturbation.

In this chapter, we intended to concisely present the geochemical formalism relevant to the study of natural-gas hydrate systems, and its contribution to the overall understanding of such systems. Based on the Western High example in the sea of Marmara, we have proposed a geochemical description of the dynamics of a hydrate accumulation zone by constraining as far as we can the origins of both the hydrate-forming gases and the associated pore-fluid. Therefore, three intimately related materials have been used in this approach: 1) the hydrate samples, especially the gases which are bound into the hydrates and which seeps to the seafloor, 2) the pore fluid which consists of the hydrate-coexisting phase and the source of the hydrate-water, 3) the sediment which is the geological matrix in which the hydrates are formed and interacts.

At the end of our study, the following conclusions have been drawn:

- The natural gas feeding the NGHOZ is of thermogenic origin and has undergone methanogenesis following biodegradation in subsurface. This natural gas certainly shares the same source rock than the predominantly-thermogenic gas fields of the Thrace basin. However, up to now there are no discovered gas fields meeting the methanogenesis following biodegradation requirements in the Thrace basin.
- The gas, especially the methane, is not only involved in the hydrate formation, but also in the associated AOM process. In fact, part of the methane reacts with the dissolved sulphate above the hydrate accumulation zone. This reaction releases carbonate ions which leads to the precipitation of authigenic carbonates under saturation of the pore fluid. The resulting pore-fluid and sediment profiles reflect the spatial heterogeneity of this geochemical process.
- The X-ray CT scan analysis of the core segments in combination with the XRF and the XRD analyses of targeted sediment samples enabled us to evaluate the spatial distribution of bivalves and carbonate concretions. The results show that bivalves are buried in the upper part of the sediment where the gas bubbles have been sampled. A high concentration of carbonate concretions is also found at the same site but distributed below.

Finally, this example enables us to better appreciate the geochemical contribution in the overall understanding of the natural-gas hydrate system. However, it is important to keep in mind that each natural-gas hydrate system is unique, and only a multidisciplinary investigation combined with a multi-analyses approach can lead to a reliable interpretation.

6. Acknowledgment

The authors acknowledge the crews of both MARMARA2010 cruises as well as the PSO for ICP-MS analyses and the hospital "La Cavale Blanche" for the X-ray scan analyses. Sylvain BERMELL, Mickael ROVERE, Angélique ROUBI, Gilbert FLOCH, Pr. Marie-Madeleine BLANC-VALLERON are gratefully acknowledged for their contributions.

7. References

Aksu, A., Hiscott, R., Kaminski, M., Mudie, P., Gillespie, H., Abrajano, T. and Yaar, D., 2002. Last glacial-Holocene paleoceanography of the Black Sea and Marmara Sea: stable isotopic, foraminiferal and coccolith evidence. Marine Geology, 190(1-2): 119-149.

Barka, A., Akyuz, H., Altunel, E., Sunal, G., Cakir, Z., Dikbas, A., Yerli, B., Armijo, R., Meyer, B. and De Chabalier, J., 2002. The surface rupture and slip distribution of the 17 August 1999 Izmit earthquake (M 7.4), North Anatolian fault. Bulletin of the Seismological Society of America, 92(1): 43.

Bayhan, E., Ergin, M., Temel, A. and Keskin, S., 2001. Sedimentology and mineralogy of surficial bottom deposits from the Aegean-Canakkale-Marmara transition (Eastern Mediterranean): effects of marine and terrestrial factors. Marine Geology, 175(1-4): 297-315.

Bernard, B.B., Brooks, J.M. and Sackett, W.M., 1978. Light-hydrocarbons in recent texas continental-shelf and slope sediments. Journal of Geophysical Research-Oceans and Atmospheres, 83(NC8): 4053-4061.

Berner, R., 1980. Early diagenesis: A theoretical approach. Princeton Univ Pr.

Berner, U. and Faber, E., 1996. Empirical carbon isotope/maturity relationships for gases from algal kerogens and terrigenous organic matter, based on dry, open-system pyrolysis. Organic Geochemistry, 24(10-11): 947-955.

Bhatnagar, G., Chapman, W.G., Dickens, G.R., Dugan, B. and Hirasaki, G.J., 2008. Sulfate-methane transition as a proxy for average methane hydrate saturation in marine sediments. Geophysical Research Letters, 35(3).

Bhatnagar, G., Chatterjee, S., Chapman, W.G., Dugan, B., Dickens, G.R. and Hirasaki, G.J., 2011. Analytical theory relating the depth of the sulfate-methane transition to gas hydrate distribution and saturation. Geochemistry Geophysics Geosystems, 12.

Boetius, A., Ravenschlag, K., Schubert, C.J., Rickert, D., Widdel, F., Gieseke, A., Amann, R., Jorgensen, B.B., Witte, U. and Pfannkuche, O., 2000. A marine microbial consortium apparently mediating anaerobic oxidation of methane. Nature, 407(6804): 623-626.

Borowski, W.S., Paull, C.K. and Ussler, W., 1996. Marine pore-water sulfate profiles indicate in situ methane flux from underlying gas hydrate. Geology, 24(7): 655-658.

Borowski, W.S., Paull, C.K. and Ussler, W., 1999. Global and local variations of interstitial sulfate gradients in deep-water, continental margin sediments: Sensitivity to underlying methane and gas hydrates. Marine Geology, 159(1-4): 131-154.

Boudreau, B., 1997. Diagenetic Models and their Implementation, 414 pp. Springer-Verlag, Berlin.

Bourry, C., Chazallon, B., Charlou, J.L., Donval, J.P., Ruffine, L., Henry, P., Geli, L., Cagatay, M.N., Inan, S. and Moreau, M., 2009. Free gas and gas hydrates from the Sea of

Marmara, Turkey Chemical and structural characterization. Chemical Geology, 264(1-4): 197-206.

Cagatay, M.N., Gorur, N., Algan, O., Eastoe, C., Tchapalyga, A., Ongan, D., Kuhn, T. and Kuscu, I., 2000. Late Glacial-Holocene palaeoceanography of the Sea of Marmara: timing of connections with the Mediterranean and the Black Seas. Marine Geology, 167(3-4): 191-206.

Calvert, S. and Pedersen, T., 1993. Geochemistry of recent oxic and anoxic marine sediments: Implications for the geological record. Marine Geology, 113(1-2): 67-88.

Charlou, J., Donval, J., Fouquet, Y., Ondreas, H., Knoery, J., Cochonat, P., Levache, D., Poirier, Y., Jean-Baptiste, P. and Fourre, E., 2004. Physical and chemical characterization of gas hydrates and associated methane plumes in the Congo-Angola Basin. Chemical Geology, 205(3-4): 405-425.

Chung, H., Gormly, J. and Squires, R., 1988. Origin of gaseous hydrocarbons in subsurface environments: theoretical considerations of carbon isotope distribution. Chemical Geology, 71(1-3): 97-104.

Collett, T., Johnson, A., Knapp, C. and Boswell, R., 2010. natural gas hydrates energy resource potential and associated geologic hazards. AAPG Memoir 89.

Coskun, B., 1997. Oil and gas fields—transfer zone relationships, Thrace basin, NW Turkey. Marine and Petroleum Geology, 14(4): 401-416.

Coskun, B., 2000. North Anatolian Fault–Saros Gulf relationships and their relevance to hydrocarbon exploration, northern Aegean Sea, Turkey. Marine and Petroleum Geology, 17(6): 751-772.

Dimitrakopoulos, R. and Muehlenbachs, K., 1987. Biodegradation of petroleum as a source of 13 C-enriched carbon dioxide in the formation of carbonate cement. Chemical Geology, 65(3-4): 283-291.

El Maghraoui, M., Joron, J.L., Etoubleau, J., Cambon, P. and Treuil, M., 1998. Determination of forty major and trace elements in GPMA magmatic rock reference materials using X-ray fluorescence spectrometry (XRF) and instrumental neutron activation analysis (INAA). Geostandards Newsletter, 23(1): 59 - 68.

Etiope, G., Feyzullayev, A. and Baciu, C., 2009a. Terrestrial methane seeps and mud volcanoes: A global perspective of gas origin. Marine and Petroleum Geology, 26(3): 333-344.

Etiope, G., Feyzullayev, A., Milkov, A.V., Waseda, A., Mizobe, K. and Sun, C.H., 2009b. Evidence of subsurface anaerobic biodegradation of hydrocarbons and potential secondary methanogenesis in terrestrial mud volcanoes. Marine and Petroleum Geology, 26(9): 1692-1703.

Gasperini, L., Polonia, A., Bortoluzzi, G., Henry, P., Le Pichon, X., Tryon, M., Cagatay, N. and Geli, L., 2011. How far did the surface rupture of the 1999 Izmit earthquake reach in Sea of Marmara? Tectonics, 30.

Gürgey, K., 2009. Geochemical overview and undiscovered gas resources generated from Hamitabat petroleum system in the Thrace Basin, Turkey. Marine and Petroleum Geology, 26(7): 1240-1254.

Gürgey, K., Philp, R., Clayton, C., Emiro lu, H. and Siyako, M., 2005. Geochemical and isotopic approach to maturity/source/mixing estimations for natural gas and associated condensates in the Thrace Basin, NW Turkey. Applied Geochemistry, 20(11): 2017-2037.

Head, I., Jones, D. and Larter, S., 2003. Biological activity in the deep subsurface and the origin of heavy oil. Nature, 426(6964): 344-352.

Henriet, J.P. and Mienert, J., 1998. Gas hydrates: relevance to world margin stability and climate change. Geological Society Pub House.

Hensen, C., Zabel, M., Pfeifer, K., Schwenk, T., Kasten, S., Riedinger, N., Schulz, H.D. and Boettius, A., 2003. Control of sulfate pore-water profiles by sedimentary events and the significance of anaerobic oxidation of methane for the burial of sulfur in marine sediments. Geochimica et Cosmochimica Acta, 67(14): 2631-2647.

Hoşgörmez, H. and Yalçin, M., 2005. Gas-source rock correlation in Thrace basin, Turkey. Marine and Petroleum Geology, 22(8): 901-916.

Hoşgörmez, H., Yalçin, M., Cramer, B., Gerling, P. and Mann, U., 2005. Molecular and isotopic composition of gas occurrences in the Thrace basin (Turkey): origin of the gases and characteristics of possible source rocks. Chemical Geology, 214(1-2): 179-191.

Jones, D., Head, I., Gray, N., Adams, J., Rowan, A., Aitken, C., Bennett, B., Huang, H., Brown, A. and Bowler, B., 2007. Crude-oil biodegradation via methanogenesis in subsurface petroleum reservoirs. Nature, 451(7175): 176-180.

Joye, S.B., Boetius, A., Orcutt, B.N., Montoya, J.P., Schulz, H.N., Erickson, M.J. and Lugo, S.K., 2004. The anaerobic oxidation of methane and sulfate reduction in sediments from Gulf of Mexico cold seeps. Chemical Geology, 205(3-4): 219-238.

Karaca, D., Hensen, C. and Wallmann, K., 2010. Controls on authigenic carbonate precipitation at cold seeps along the convergent margin off Costa Rica. Geochemistry Geophysics Geosystems, 11: 19.

Kastner, M., Claypool, G. and Robertson, G., 2008. Geochemical constraints on the origin of the pore fluids and gas hydrate distribution at Atwater Valley and Keathley Canyon, northern Gulf of Mexico. Marine and Petroleum Geology, 25(9): 860-872.

Krey, V., Canadell, J.G., Nakicenovic, N., Abe, Y., Andruleit, H., Archer, D., Grubler, A., Hamilton, N.T.M., Johnson, A., Kostov, V., Lamarque, J.-F., Langhorne, N., Nisbet, E.G., O'Neill, B., Riahi, K., Riedel, M., Wang, W. and Yakushev, V., 2009. Gas hydrates: entrance to a methane age or climate threat? Environmental Research Letters, 4(3).

Kvenvolden, K., 1999. Potential effects of gas hydrate on human welfare. National Acad Sciences, pp. 3420-3426.

Kvenvolden, K.A., 1988. Methane hydrate - A major reservoir of carbon in the shallow geosphere. Chemical Geology, 71(1-3): 41-51.

Kvenvolden, K.A., 2000. Gas hydrate and humans. Annals of the New York Academy of Sciences, 912(1): 17-22.

Le Pichon, X., Sengor, A.M.C., Demirbag, E., Rangin, C., Imren, C., Armijo, R., Gorur, N., Cagatay, N., de Lepinay, B.M., Meyer, B., Saatcilar, R. and Tok, B., 2001. The active Main Marmara Fault. Earth and Planetary Science Letters, 192(4): 595-616.

Luff, R. and Wallmann, K., 2003. Fluid flow, methane fluxes, carbonate precipitation and biogeochemical turnover in gas hydrate-bearing sediments at Hydrate Ridge, Cascadia Margin: Numerical modeling and mass balances. Geochimica et Cosmochimica Acta, 67(18): 3403-3421.

Martens, C.S. and Val Klump, J., 1984. Biogeochemical cycling in an organic-rich coastal marine basin 4. An organic carbon budget for sediments dominated by sulfate reduction and methanogenesis. Geochimica et Cosmochimica Acta, 48(10): 1987-2004.

Maslin, M., Owen, M., Betts, R., Day, S., Dunkley Jones, T. and Ridgwell, A., 2010. Gas hydrates: past and future geohazard? Philosophical Transactions of the Royal Society a-Mathematical Physical and Engineering Sciences, 368(1919): 2369-2393.

Max, M., 2003. Natural gas hydrate: in oceanic and permafrost environments. Kluwer Academic Pub.

McHugh, C.M.G., Gurung, D., Giosan, L., Ryan, W.B.F., Mart, Y., Sancar, U., Burckle, L. and Cagatay, M.N., 2008. The last reconnection of the Marmara Sea (Turkey) to the World Ocean: A paleoceanographic and paleoclimatic perspective. Marine Geology, 255(1-2): 64-82.

Okay, A.I., Demirbag, E., Kurt, H., Okay, N. and Kuscu, I., 1999. An active, deep marine strike-slip basin along the North Anatolian fault in Turkey. Tectonics, 18(1): 129-147.

Orcutt, B.N., Boetius, A., Lugo, S.K., MacDonald, I.R., Samarkin, V.A. and Joye, S.B., 2004. Life at the edge of methane ice: microbial cycling of carbon and sulfur in Gulf of Mexico gas hydrates. Chemical Geology, 205(3-4): 239-251.

Pallasser, R., 2000. Recognising biodegradation in gas/oil accumulations through the 13C compositions of gas components. Organic Geochemistry, 31(12): 1363-1373.

Paull, C. and Dillon, W., 2001. Natural gas hydrates: occurrence, distribution, and detection. American Geophysical Union, Washington, DC.

Plaza-Faverola, A., Westbrook, G.K., Ker, S., Exley, R.J.K., Gailler, A., Minshull, T.A. and Broto, K., 2010. Evidence from three-dimensional seismic tomography for a substantial accumulation of gas hydrate in a fluid-escape chimney in the Nyegga pockmark field, offshore Norway. Journal of Geophysical Research-Solid Earth, 115.

Pohlman, J., Canuel, E., Chapman, N., Spence, G., Whiticar, M. and Coffin, R., 2005. The origin of thermogenic gas hydrates on the northern Cascadia Margin as inferred from isotopic (13C/12C and D/H) and molecular composition of hydrate and vent gas. Organic Geochemistry, 36(5): 703-716.

Pohlman, J.W., Ruppel, C., Hutchinson, D.R., Downer, R. and Coffin, R.B., 2008. Assessing sulfate reduction and methane cycling in a high salinity pore water system in the northern Gulf of Mexico. Marine and Petroleum Geology, 25(9): 942-951.

Reagan, M.T. and Moridis, G.J., 2008. Dynamic response of oceanic hydrate deposits to ocean temperature change. Journal of Geophysical Research-Oceans, 113(C12).

Reeburgh, W.S., 1976. Methane consumption in Cariaco Trench waters and sediments. Earth and Planetary Science Letters, 28(3): 337-344.

Reichel, T. and Halbach, P., 2007. An authigenic calcite layer in the sediments of the Sea of Marmara - A geochemical marker horizon with pale oceanographic significance. Deep-Sea Research Part Ii-Topical Studies in Oceanography, 54(11-13): 1201-1215.

Richter, T.O., van der Gaast, S., Koster, B., Vaars, A., Gieles, R., de Stigter, H.C., De Haas, H. and van Weering, T.C.E., 2006. The Avaatech XRF Core Scanner: technical description and applications to NE Atlantic sediments. Geological Society, London, Special Publications, 267(1): 39-50.

Schaller, T., Christoph Moor, H. and Wehrli, B., 1997a. Sedimentary profiles of Fe, Mn, V, Cr, As and Mo as indicators of benthic redox conditions in Baldeggersee. Aquatic Sciences-Research Across Boundaries, 59(4): 345-361.

Schaller, T., Moor, H.C. and Wehrli, B., 1997b. Reconstructing the iron cycle from the horizontal distribution of metals in the sediment of Baldeggersee. Aquatic Sciences-Research Across Boundaries, 59(4): 326-344.

Schaller, T. and Wehrli, B., 1996. Geochemical-focusing of manganese in lake sediments—an indicator of deep-water oxygen conditions. Aquatic Geochemistry, 2(4): 359-378.

Schoell, M., 1983. Genetic characterization of natural gases. Aapg Bulletin, 67(12): 2225-2238.

Schulz, H.D. and Zabel, M., 2000. Marine geochemistry. Springer.

Sen, S., YIllar, S. and Erdal Kerey, 2009. Allochthonous blocks misidentified as the basement: Implication for petroleum exploration in the SW Thrace Basin (Turkey). Journal of Petroleum Science and Engineering, 64(1-4): 55-66.

Şen Şamil , Selin Yillar and Kerey Erdal 2009. Allochthonous blocks misidentified as the basement: Implication for petroleum exploration in the SW Thrace Basin (Turkey). Journal of Petroleum Science and Engineering, 64(1-4): 55-66.

Sloan, E.D. and Koh, C.A., 2008. Clathrate hydrates of natural gases. CRC.

Sultan, N., Marsset, B., Ker, S., Marsset, T., Voisset, M., Vernant, A.M., Bayon, G., Cauquil, E., Adamy, J., Colliat, J.L. and Drapeau, D., 2010. Hydrate dissolution as a potential mechanism for pockmark formation in the Niger delta. Journal of Geophysical Research-Solid Earth, 115.

Treude, T., Boetius, A., Knittel, K., Wallmann, K. and Jorgensen, B.B., 2003. Anaerobic oxidation of methane above gas hydrates at Hydrate Ridge, NE Pacific Ocean. Marine Ecology-Progress Series, 264: 1-14.

Tryon, M.D., Henry, P., Cagatay, M.N., Zitter, T.A.C., Geli, L., Gasperini, L., Burnard, P., Bourlange, S. and Grall, C., 2010. Pore fluid chemistry of the North Anatolian Fault Zone in the Sea of Marmara: A diversity of sources and processes. Geochemistry Geophysics Geosystems, 11.

Ussler III, W., Paull, C. and Fullagar, P., 2000. 2. Pore-Water Strontium Isotopes from the Leg 171B Drilling Transect down the Blake Spur.

Utkucu, M., Nalbant, S.S., McCloskey, J., Steacy, S. and Alptekin, O., 2003. Slip distribution and stress changes associated with the 1999 November 12, Duzce (Turkey) earthquake (M (w)=7.1). Geophysical Journal International, 153(1): 229-241.

Vidal, L., Ménot, G., Joly, C., Bruneton, H., Rostek, F., Çagatay, M., Major, C.O. and Bard, E., Hydrology in the Sea of Marmara during the last 23 ka: implications for timing of Black Sea connections and sapropel deposition.

Wallmann, K., Aloisi, G., Haeckel, M., Obzhirov, A., Pavlova, G. and Tishchenko, P., 2006. Kinetics of organic matter degradation, microbial methane generation, and gas hydrate formation in anoxic marine sediments. Geochimica et Cosmochimica Acta, 70(15): 3905-3927.

Waseda, A. and Iwano, H., 2008. Characterization of natural gases in Japan based on molecular and carbon isotope compositions. Geofluids, 8(4): 286-292.

Whiticar, M.J., 1994. Correlation of natural gases with their sources. In: Magoon, L. B.; Dow, W. G. (Eds), The petroleum System: from Sources to Trap, AAPG Memoir, vol.60: pp 261-283.

Yamamoto, A., Yamanaka, Y. and Taiika, E., 2009. Modeling of methane bubbles released from large sea-floor area: Condition required for methane emission to the atmosphere. Earth and Planetary Science Letters, 284(3-4): 590-598.

Zitter, T., Henry, P., Aloisi, G., Delaygue, G., Çagatay, M., Mercier de Lepinay, B., Al-Samir, M., Fornacciari, F., Tesmer, M. and Pekdeger, A., 2008. Cold seeps along the main Marmara fault in the Sea of Marmara (Turkey). Deep-Sea Research Part I, 55(4): 552-570.

3

Geopolitics of Gas in South America

Ana Lía del Valle Guerrero
National University of the South, Department of Geography and Tourism
Argentina

1. Introduction

When speaking of energy in this first decade of the century, references are still mandatory oil or natural gas, as both accounts for about 60% of the global energy matrix. This energy demand global and regional scale has two outstanding features: on the one hand the growth of demand is increasing and, on the other hand supply is stagnant and in some cases declining. According to a 2007 report by the U.S. Energy Department is expected that the global primary energy demand increase, between 2004 and 2030 - by 57% which implies duplicate the energy consumption in the brief period of 26 years.

To meet this consumer expectations oil production should increase by 42%, natural gas 65% (reaching a figure of 163 trillion cubic feet or equivalent to 29,000 million barrels of oil) and coal by 74%. These increases in theory are almost impossible to materialize into reality because there is a continued contraction in supply, except for occasional new discoveries as in the case of Brazil, where extraction is difficult, and also requires large investments, and time for deployment.

The evolution and perspectives of Latin American regional gas market integration is usually analyzed from the point of view of Economics. This paper instead tries to explain its present state and perspectives by means of a geopolitical approach. The main conclusions (that result from a Master in Policies and Strategy thesis elaborated by the author, entitled "The strategic role of gas as a critical variable of the national and regional energy system"[1]) enable to advance in the analysis of this regional problematic in which the introduction of global actors determined great scale changes. Beginning in 2004 as a binational crisis centered in Argentina and Chile, it became a regional and global one after the active participation of France and Russia.

The development of countries like Argentine and Brazil depend to a large degree on the provision of gas from Bolivia, a country with great amounts of reserves but scarce levels of infrastructure investments. Thus, though long term contracts have already been signed, the actual volume of supply remains uncertain.

Since the Argentina crisis of 2004, with regional effects that still continue, the security of regional energy supply, emerges as a key issue to be considered in the Continental Security

[1] Mg Ana Lía del Valle Guerrero, thesis of Masters in Policies and Strategies "the strategic roll of the gas like critical variable of the national and regional energy system", 2006, National University of the South, Bahía Blanca, Argentina.

Agenda. In spite of the existence of abundant reserves of gas in different countries of the region, like Venezuela, Bolivia and Peru, several facts have caused the gas regional crisis that this paper intends to explain from a geopolitical point of view. The particular situation of the resource gas will be analyzed, in the region whit respect its use and availability, emphasizing the effect of political decisions that causes the regional gas crisis.

The methodology of work based its arguments on non experimental investigation of corelational transactional type, by means of which, it is possible to analyze the problems of power supplying produced in the region - from 2004 to 2011- the period analyzed in the temporary scale.

The critical conditions of gas on regional scale derive from the decisions with respect to the use of a resource, conditioned by its territorial specificity - it is in a country and not in another one. Beside, the use of gas is more subjected to political decisions of the States that own the resource than to economic decisions, since some countries still base their decisions of supply of the resource on unresolved historical and geopolitical conflicts.

The main objective of this paper is to make a valuation and to have a deep knowledge of conflicts around energy-focuses on natural gas- from a South American geopolitical perspective; which emphasizes the interactions society/territory, in the regional scale. It is done from a - holistic and multi-factor - geopolitical perspective that tries to surpass the predominant economics vision.

Knowing that the gas reserves in the region are important and the internal consumption of South American countries is different. It is undeniable the necessity to work out a regional strategy with the objective of satisfying demands in order to avoid futures conflicts. For this it is required to make infrastructure investments and to get financial sources, which need a greater degree legal security that guarantees the participation of international organisms.

In South America, the geopolitics of gas presents a superposition of scenes that contrast projects with realities, and integration processes together with increasing conflicts that generate instability and fragmentation. Consequently, to promote a strategic alliance among States, approaching the issue from a global perspective, is a true priority for South America. This would have to be built with the participation of those actors who demand the resource and those who offer it within the region.

If the South American energetic integration were obtained, the region would be in the position to obtain financing its own future energy growth to future, and supplying a part of the needs to the rest of the world. The geopolitical analysis allows to understand the gas crisis in South America as product of a changing reality and to interpret the complexity and dynamics of existing social, political and economic processes. It also allows understanding the uncertainties and the possibility of new geopolitical situations in a regional space in which the reality is being constructed in daily routine.

This chapter consists of three parts, the first performing a global approach by region, from the economics point of view, in terms of natural gas resource -based on the last report of June 2011- from BP Statiscal Review of World Energy 2011. The second part presents an explanation -from different approaches- of the changes in geopolitics. Finally, the third part -as a synthesis - focuses on the analysis and explanation of the Geopolitics of Gas in South America.

2. The natural gas resource situation in 2011

To get an updated picture of the global energy situation focusing on natural gas, Statiscal BP Review of World Energy 2011 (June 2011), presents some numbers that show the state of resource from different points of view. First, from the point of view of consumption, there was an overall increase of 7.4% (the largest increase since 1984), U.S. presents a historical record of consumption growth of 5.6% in the period 2009 / 2010 and an increase in the same period in Russia (6.3%), China (21.8%) and India (21.5%). However, the greatest increases in consumption at a global level took place in South America led by Peru with 56%, followed by Chile with 51% and Brazil with 33.8% increase in consumption.

2.1 Global approach by region (economic point of view)

From the point of view of production, there was a similar growth in consumption of 7.3% globally. Some highlights are on growth 2009/2010: 30.7% increase in the production of Qatar and the largest at a global level growth is 108.4% in production in Peru. Furthermore, there is also respect to global production in 2010 that the United States (19.3%) moved from the first placed on the production on Russia (18.4%) and the significant decline in production in Canada (5%) for the fourth consecutive year.

In addition, increases of 10% in terms of trade, with a 22.6% increase in shipments of LNG, are highlighted in particular Qataris LNG shipments with a growth of 53.2% and the increase of Russian exports by pipeline. The graphs in Figure 1 shows that, from a historical point of view, taking into account the variations in the period 1980-2010 there is a declining trend in the Middle East, beyond being the largest producing region and possess the highest ratio reserves / production. The reserves of the other regions do not show large variations in its production, giving a total horizon reserve of 58.6 years.

It should be emphasized that the countries that occupy the top 3 worldwide in terms of proved reserves: Russia (23.9%) Iran (15.8%) and Qatar (13.5%) accounting for 53.2% of world reserves of natural gas and if we add the 15 first positions 84.5% of total reserves. In this context, the South American region represents only 4% of world reserves, with a ratio production / reserves of 45.9 years.

At regional level and analyzing especially the situation in the South American region as shown in table 1, we can see a stagnation in consumption in Argentina but you must remember that internal consumption in Argentina is from the 70's the highest of South and Central America and even the second highest in the world which represents 1.4% of world consumption, so it cannot grow more. It is also remarkable how growing consumption of natural gas in the period -2009/2010- in several countries in the region for example, in Peru (56.0%), Chile (51%) and Brazil (33.8%) .

The table 2 shows how are the proved reserves at regional level and notes that in the ratio reserves / production Peru appears whit reserves for 48.8 years followed by Brazil (28.9) and Bolivia (19.5). It should be noted that there is not data available for Venezuela who has the largest reserves in the region but not yet exploited. They are used for reinjected into oil production. Concerning Argentina has proved reserves to 8.6 years, although other studies consider a lower number of only 6 years.

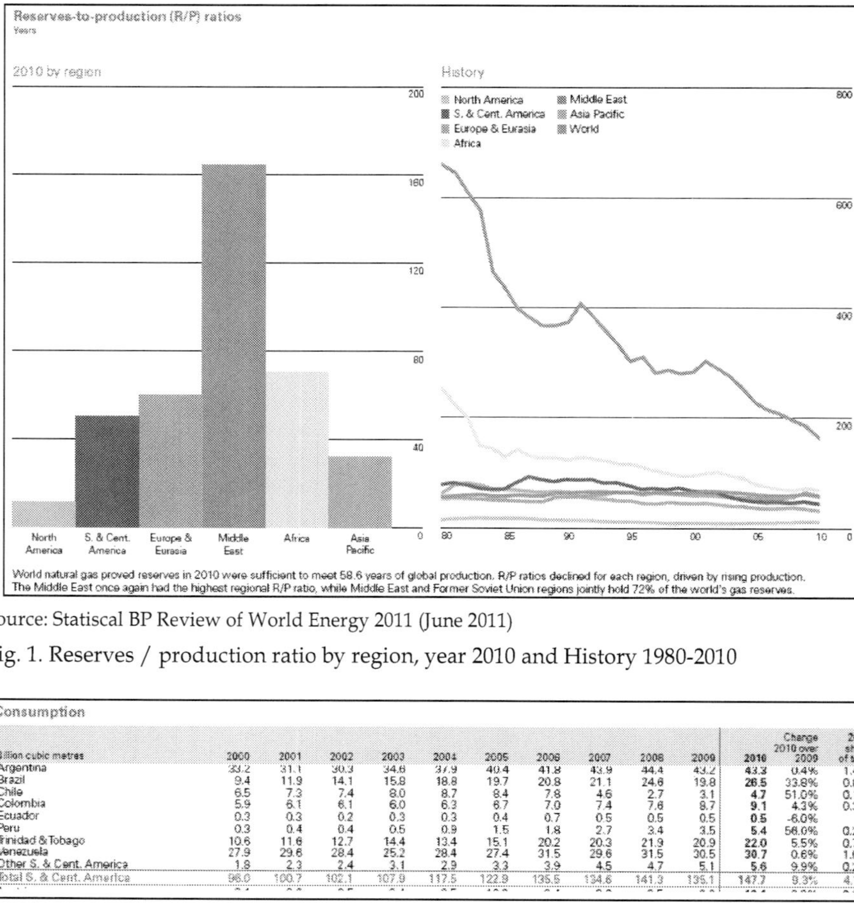

Source: Statiscal BP Review of World Energy 2011 (June 2011)

Fig. 1. Reserves / production ratio by region, year 2010 and History 1980-2010

Consumption

Billion cubic metres	2000	2001	2002	2003	2004	2005	2006	2007	2008	2009	2010	Change 2010 over 2009	2010 share of total
Argentina	33.2	31.1	30.3	34.6	37.9	40.4	41.8	43.9	44.4	43.2	**43.3**	0.4%	1.4%
Brazil	9.4	11.9	14.1	15.8	18.8	19.7	20.8	21.1	24.6	19.8	**26.5**	33.8%	0.8%
Chile	6.5	7.3	7.4	8.0	8.7	8.4	7.8	4.6	2.7	3.1	**4.7**	51.0%	0.1%
Colombia	5.9	6.1	6.1	6.0	6.3	6.7	7.0	7.4	7.6	8.7	**9.1**	4.3%	0.3%
Ecuador	0.3	0.3	0.2	0.3	0.3	0.4	0.7	0.5	0.5	0.5	**0.5**	-6.0%	*
Peru	0.3	0.4	0.4	0.5	0.9	1.5	1.8	2.7	3.4	3.5	**5.4**	56.0%	0.2%
Trinidad & Tobago	10.6	11.6	12.7	14.4	13.4	15.1	20.2	20.3	21.9	20.9	**22.0**	5.5%	0.7%
Venezuela	27.9	29.6	28.4	25.2	28.4	27.4	31.5	29.6	31.5	30.5	**30.7**	0.6%	1.0%
Other S. & Cent. America	1.8	2.3	2.4	3.1	2.9	3.3	3.9	4.5	4.7	5.1	**5.6**	9.9%	0.2%
Total S. & Cent. America	96.0	100.7	102.1	107.9	117.5	122.9	135.5	134.6	141.3	135.1	**147.7**	9.3%	4.7%

Source: Statiscal BP Review of World Energy 2011 (June 2011)

Table 1. Natural gas consumption in the region of Central and South America

Proved reserves

	At end 1990 Trillion cubic metres	At end 2000 Trillion cubic metres	At end 2009 Trillion cubic metres	At end 2010 Trillion cubic feet	At end 2010 Trillion cubic metres	At end 2010 Share of total	At end 2010 R/P ratio
Argentina	0.7	0.8	0.4	**12.2**	**0.3**	0.2%	8.6
Bolivia	0.1	0.7	0.7	**9.9**	**0.3**	0.2%	19.5
Brazil	0.1	0.2	0.4	**14.7**	**0.4**	0.2%	28.9
Colombia	0.1	0.1	0.1	**4.4**	**0.1**	0.1%	11.0
Peru	0.3	0.2	0.4	**12.5**	**0.4**	0.2%	48.8
Trinidad & Tobago	0.3	0.6	0.4	**12.9**	**0.4**	0.2%	8.6
Venezuela	3.4	4.2	5.1	**192.7**	**5.5**	2.9%	*
Other S. & Cent. America	0.2	0.1	0.1	**2.3**	**0.1**	*	22.4
Total S. & Cent. America	5.2	6.9	7.5	261.8	7.4	4.0%	45.9

Source: BP Review of World Energy Statiscal 2011 (June 2011)

Table 2. Proved reserves of natural gas in the region of Central and South America

Concerning to production, stands out the change in the relationship 2009/2010 to Peru with an increase of 108, 4% followed by Brazil with 23% and 16.8% in Bolivia, while Argentina shows the main decrease with a -3.0% despite continuing representing 1.3% of world production.

Production*

Billion cubic metres	2000	2001	2002	2003	2004	2005	2006	2007	2008	2009	2010	Change 2010 over 2009	2010 share of total
Argentina	37.4	37.1	36.1	41.0	44.9	45.6	46.1	44.8	44.1	41.4	**40.1**	-3.0%	1.3%
Bolivia	3.2	4.7	4.9	6.4	9.8	11.9	12.9	13.8	14.3	12.3	**14.4**	16.8%	0.4%
Brazil	7.5	7.7	9.2	10.0	11.0	11.0	11.3	11.2	13.7	11.7	**14.4**	23.5%	0.5%
Colombia	5.9	6.1	6.2	6.1	6.4	6.7	7.0	7.5	9.1	10.5	**11.3**	7.2%	0.4%
Peru	0.3	0.4	0.4	0.5	0.9	1.5	1.8	2.7	3.4	3.5	**7.2**	108.4%	0.2%
Trinidad & Tobago	14.5	15.5	18.0	26.3	27.3	31.0	36.4	39.0	39.3	40.6	**42.4**	4.4%	1.3%
Venezuela	27.9	29.6	28.4	25.2	28.4	27.4	31.5	29.5	30.0	28.7	**28.5**	-0.7%	0.9%
Other S. & Cent. America	3.4	3.5	3.4	3.1	3.1	3.4	4.1	3.9	3.7	3.2	**2.9**	-9.9%	0.1%
Total S. & Cent. America	100.2	104.5	106.7	118.7	131.7	138.6	151.1	152.5	157.6	151.9	161.2	6.2%	5.0%

Source: BP Review of World Energy Statiscal 2011 (June 2011)

Table 3. Natural gas production in the region of Central and South America

Finally, can be performed a comprehensive analysis based on data provided by the report of Statiscal BP Review of World Energy June 2011 to know the current situation by region, considering the following parameters for 2010: reserves tested, production, consumption and ratio reserves / production.

First, in figure 1 and then in table 4,we emphasized that the most favored region considering all parameters is the Middle East with a horizon of the world reserves of 40.5% , a consumption (11.5%) below the production (14.4%) and a ratio reserves / production over 100 years. In this region there are countries that occupy the top positions in terms of reserves, Iran (2 º), Qatar (3 º), Saudi Arabia (4 º), total nearly 63% of reserves in the region.

This distribution of reserves clearly shows two concurrent phenomena, concentration and fragmentation. By one hand there is a spatial concentration of proved reserves in a small

Region	Proved reserves in % 2010	Production in % 2010	Consumption in % 2010	Ratio reserves / production in years
North America	5,3	26	26,9	12,0
South and Central America	4,0	5,0	4,7	45,9
Europe and Eurasia	33,7	32,6	35,8	60,9
Middle East	40,5	14,4	11,5	+100
Africa	7,9	6,5	3,3	70,5
Asia Pacific	8,7	15,4	17,9	32,8

Source: Authors' calculations based on data from BP Statiscal Review of World Energy 2011

Table 4. Comparison by regions, 2010.

number of countries in the Middle East region, while on the other hand, this also implies a spatial fragmentation in the distribution of global resources, since the gas resource has territorial specificity, is in a territory and not in others, it gives power to its possessor.

In above table 4 in different colors can be seen the degree of difficulty faced by each region, in red the most engaged and in yellow the best positioned and green ones that may be future resource providers. This situation also represents global and regional asymmetries between the countries that own the resource and those which must import and therefore competition between producers and consumers. From the point of view of future conflict scenarios outlined in the first place, the North American region with proved reserves of 5.3% of the world (second lowest), with production and consumption within the highest in the world 26% and 26.9% respectively and the lowest ratio reserves / production in all regions of 12 years.

United States and Canada are in this region and occupy 1st and 3rd places respectively -in terms of worldwide production- (remember the mentioned at the beginning of this chapter about declining production in Canada during the last 4 years as an additional problem). Both scenarios show that in spite of being top producers are also important consumers United States in the first place and Canada in the 6th. This situation, combined with this region possess the lower reserves from all regions, makes envision a future of growing dependence on supplies from this region to other regions. Consequently, also generates vulnerability from the point of view of energy security. Particularly, United States would be most affected, considering its status as a global economic power where energy is indispensable to maintain growth.

Other region that can be observed -with future problems- corresponding to Asia Pacific it has 8.7% of global reserves and consumption (17.9%) above the production (15.4%) with a ratio reserves / production of 32.8 years, one of the lowest worldwide. A further problem that complicates this situation is the tendency to growth consumption in the region. This area has two of the most populous countries in the world, such as China and India, whit economies in continuous growth.

From the point of view of reserves, in this region are countries that occupy positions 12th to 15th at the world. They are: Indonesia, Australia, China and Malaysia. China is in 7th place and Indonesia 9th from production and from consumption worldwide China is ranked 4th, Japan 5th and India 8th. Certainly and beyond any discussion mentioning Japan, China and India in the same region, means to mention the leading economic powers and the necessity of rising consumption will surely, as well as problems to ensure a fluid supply.

European region and Eurasia at regional level are in an apparent balance between production (32.6%) and consumption (35.8%), with a ratio reserves / production over 60 years. However, it is remarkable that the whole region depends on the supply of a single large supplier, Russia and to a lesser degree from Norway, which creates vulnerability and dependence of most of European countries on Russian gas.

Africa region shows consumption (3.3%) below its production (6.5%) and a high ratio reserves / production (70.5 years). Economic and political history of the region shows dependency on their colonizers and indiscriminate exploitation of their resources. Therefore, the possibility of placing on the market for its production is limited by previous situations of

States involved because these issues will determine future policy decisions in relation to the use of gas resources, which will go beyond just economic issues.

Finally, analyzing data from the region of Central and South America is one of the regions with lower consumption (4.7%) and production (5%) while the ratio reserves / production is in third place with 45.9 years. It may be noted further that, the main gas reserves in the region belonging to Venezuela remains untapped and that new oil discoveries in Brazil have not yet went into production. Moreover, Bolivia, the second country in reserves in the region, is not exploiting its full potential for lack of investment in infrastructure and Peru began to do as noted in the production growth of 108.4% over the period 2009/2010. All the foregoing indicates that production capacity is still not fully used in relation to its reserves and consumption.

To end, making a comparison among regions at a global level, it can be conclude that in spite of the region of Central and South America presents the lowest share of global reserves with 4% of total generation that has the greatest potential production because their main reserves have not yet been exploited and it presents a low consumption and a ratio reserves / production high. Therefore, this region is positioned as one of the future areas supplier of the resource gas.

Consistent with this point of view, which is not mostly accepted on these days, appeared in the journal Foreign Policy, an opinion of Amy Myers Jaffe[2], director of the Forum of Energy's Baker Institute, Rice University entitled "Goodbye, OPEC" where she argues that OPEC will lose much of its power in the late 2020 because ".. the Americas, not the Middle East, will then the global energy capital," says this change will be technological and political factors (growing regional instability in Africa as it generates drops in production). In the case of South America, include technological changes such as new horizontal drilling techniques for the Shale gas in Argentina with third global reserves, and plus advances for extraction in the pre-salt layer deep, in Brazil. Undoubtedly, this situation has not gone unnoticed by the big powers like China and the United States; they have already started negotiations with Brazil for the future use of their resources and have increased their presence in the region, as well as Iran, France, Spain and Russia among others.

It is argued that, since for any country is crucial to ensure plentiful supplies of energy, any action that could reduce the flow would threaten vital interests of national security, thus considered oil and gas as strategic natural resources at a moment of rising global energy demand against a stagnant offer. In the region, each state need to guarantee the supply of energy to its population, but this makes it necessary to reach agreements to solve both conflicts among countries and internal conflicts as in Argentina (the government has to decided either to supply energy to its population, or to export to Chile) and in Bolivia (government has to fulfill contracts either with Brazil or with Argentina). Therefore, to ensure providers is a priority for States to guarantee its Energy Security. To understand how the countries can obtain this security, geopolitics is an important factor but first it is necessary to know its evolution so that is explain in the next point.

---

[2] Published in el NuevoHerald.com "*Oppenheimer: The new oil center in the world*" Thursday September 29, 2011.Accessed on line http://elnuevoherld.com/2011/09/28/v-print/1034152/oppenheimer-el-nuevo-ce...consultado30/09/2011

3. Geopolitics and their different approaches

First part of work was focused on reserves, production, consumption and trade since an economic point of view. This second part shows the geopolitical point of view as another way to approach reality. In order to understand needs to be done a brief summary of the changes in the field of study of Geopolitics and the causes that motivated them, so you could understand the Geopolitical of Latin-American Gas.

3.1 Geopolitical

A definition that contains the basic features of Geopolitics is the following:

"Geopolitics is the science that studies the mutual relations, influences and actions between the State and Space, to provide knowledge or political solutions" (Marini, 1985,44). Additionally, "The geopolitical space is the geographical area within which act reciprocally geographic and political factors to be studied and resolved" (Marini, 1985, 45)

In summary, Geopolitics focused politically geographical space, through a harmonious relationship between Geography and Politics. Based on the article developed by Hutschenreuter (2008)[3] about the evolution of the concept Geopolitics one can extract some central ideas. "... From the twenties, but particularly since the thirties, geopolitics, became an idea and practice of territorial expansion based on ultra-national, racial and military (Nazism) and was associated with a form of exercise of state and interstate power that dragged the nation to the status of war. Then, the scheme involved a granite bipolar geopolitical division of nations into spaces or spheres of influence ..."(Hutschenreuter, 2008:2)

3.2 Geoeconomics

However, in contrast to years following World War II, when the term (but not its exercise) was banned from the discussions, at the beginning of the new order international Geopolitics began a shy return from the seventies and eighties, with the rise of regional blocks returned from the hand of the geoeconomics. (Hutschenreuter, 2008:3). In the 90 highlights the Geoeconomics, focused more on trade and economic issues than political and territorial. Geoeconomics, establishes relationships between Geography and Economy, through the prominence of use of natural resources from a country by political action of States. Such if the great powers dispute for control of energy sources and those considered critical minerals like oil and gas. The variety, volume and quality of these natural resources are the primary basis for assessing the economic and political power of Nation[4].

Later, geopolitics became a multi-thematic discipline more than a specialized type of geopolitics, a change that virtually meant the dissolution of the discipline. That is, the denaturation of a discipline that always characterized by its multidisciplinary nature,

[3] Albert Hutschenreuter. *The gates of the twentieth century geopolitics.* This introductory article is part of an ongoing investigation. Working paper 04-In Program CAEI Geopolitical Argentine Center of International Studies.
Online www.caei.com.ar/es/programas/geopolitica/04.pdf

[4] Briano, Justo "*Geopolitics and American Geostrategy.*" Military Club Ed. Buenos Aires. 1972. p: 79.

though without ever losing its substance which was originally the incidence and consequence of the geographical factor in the political development of States. (Hutschenreuter, 2008:3)

But geopolitics is not only faced the phenomenon of distorted, but the same disregard, as the most conclusive sign of globalization is manifested in the weakening of the territorial factor in the world of states, for instance, in the disappearance of the national borders, the death of distance, the erosion of physical sovereignty of States, not including a geopolitical territory. Thus gave rise to a new form of state increasingly central and territorial, decentralized and virtual: the disappearance of geographical and territorial state would disappear one of the historical causes of the clashes, which had been the struggle for land or spaces. (Hutschenreuter, 2008:4). However, geopolitics was also affected by globalization characterized by the weakening of territorial factor, and the disappearance of national borders, who transformed it into geopolitics without territory.

3.3 Critical geopolitics

Critical Geopolitics arises then as a new geopolitics, which focuses on political power and ignores the geographical factor. The critical geopolitics is interested in all kinds of rivalries over territory and tries, from discourse analysis, see how enters practices of public and political actors in past and present scenarios. The nature "criticism" of this vision represents a reflection on the ways in which the geopolitical issues are represented and circumstances. Rather than traditional geopolitics research focused on geostrategy in terms of planning and results critical geopolitics is aimed to the examination of the ideas that precede them, their circumstances and their expression at the level of state discourse ... and provides a platform that aims to be valid for all levels of spatial analysis, not only international affairs but in its intrasocial globalization framework (Barton, 1997:17-18 Castro, 2006:191)[5].

At the last stage, with the advent of globalization, the idea of deterritorialization and the development of critical geopolitics is remove one of the fundamental basis of geopolitics, territorial component. Within this context of change appears the Posgeopolitics, marked by the consequences of the implementation of neoliberal economic model that can penetrate into other spaces through a deregulation of markets that facilitates the opening of economies and get the same results but without the use of military power. It is also more a relationship between Economics and Politics, with the Territory.

3.4 New geopolitics of energy

In this context of change of discipline arises another change related to the focus on a particular conflict, but linking global geographical space with individual one who has a specific natural resource with political decisions of resource use, that is known as the New Geopolitics of Energy proposes to go beyond geoeconomics issue, Michael Klare (2003), argues that ". .. in this world after the cold war, wars for resources are not acts of God, or unrelated, on the contrary: they are part of a broader geopolitical and interconnected. Future wars will be for the possession and control of resources needed by modern industrial

[5] Lyndon, A, Hiernaux, D (eds.) 2006-*Treaty of Human Geography*, Chap.8, Castro, P. Geography and Geopolitics, pp. 187-199.Ed.Antrophos, Mexico.

societies to function. Then emerges in this framework, a new geography of conflict, which has mapped a strategic area where political boundaries do not count but the concentrations of resources[6]. "

In a later book Klare (2008:18)[7] argues that "... The global fear energy shortage is accompanied by a New Geopolitics of Energy. Within the international system of power that is being constituted can expect that the struggle for energy exceeds all other considerations, international leaders to go to extremes to secure energy supplies in their countries. the oil (and natural gas) will being primarily a business class that is bought and sold on the international market to become a preeminent strategic resource in the world, whose acquisition, production and distribution will absorb more and more time, effort and attention of governments and military leaders. "

Another component to emphasize in this New Geopolitics of Energy is the resurgence of the State's presence in front of business and private interests in the pursuit and domain of energy resources that would be one of the core dynamics of world affairs in the coming decades. In particular case of South America, appears as a resurgence of nationalism, as is shown by process of resources nationalization in Bolivia, Venezuela and Ecuador.

The proposal to analyze South American Geopolitics of Gas, through a holistic or integrative geopolitical point of view, arises to recover the contributions of these different tendencies in the discipline. Therefore, this paper recognizes the existence of classical geopolitics components, like presence of States and their political decisions in the area; Geoeconomics with its emphasis on presence of natural gas resource. Also analyzes discourses of Presidents, Ministers of Economy, Energy or Defense and political decisions of States as propose Critical Geopoliticis. Finally, from the perspective of the New Geopolitics of Energy recovers the analysis of each component in the global level and put them at the particular context of the South American region.

From geopolitical and strategic point of view, acquire relevance relations between the States involved and the political decisions of use from resource natural gas. The location of the resource also acquires importance, in a certain territory and not in others - there lies the specificity of the territory - and to this, its value as a strategic resource has to be added the high degree of incidence in the economic development of a given country due to its direct use of energy in industrial, residential and commercial activities. Coinciding Isbell argues that "... today, the raw material for energy-particularly oil and potentially natural gas, have emerged as one of the important variables in the global geopolitical context. Although Latin America has relatively few energy resources, compared with Middle East or Russia, but in their own regional context, could achieve self-sufficiency and energy independence. This region represent a direct source of energy supply for the United States and play a very important role in the world energy geopolitical game"(Isbell, 2008:2)[8].

From a geopolitical perspective and according to modern Strategy, a concept that handles international relationships focus on power relations among States, involving relations

---

[6] Klare, M (2003) "*Resource Wars. The future scenario of global conflict,*" pgs: 261/262, Ed Trends. Spain.

[7] Klare, M (2008) *Planet thirsty, dwindling resources. The New Geopolitics of Energy*. Urano. Barcelona España.

[8] I sbell, P (2008) *Energy and Geopolitics in Latin America,* Working Paper No. 12/2008, Real Instituto Elcano. España.

between means and ends[9]. That generates dependency relationships where the power is perceived in terms of degree as States depend on the purpose of other actors. Possesses the resources or technology gives power to those who possess them and generates a dependence relationship of those consumers' countries.

3.5 New geopolitics of energy in South America

This context explains that, new geopolitics of energy in South America, have key players in global scale, such is the case of Iran, France, China, (which own the resource, technology and capital needed for investment in infrastructure) and they appear together with regional actors such as Venezuela, Bolivia and Brazil. There is also the apparent contradiction between go away from the United States to avoid dependency and the need to approach to others global players with whom they also create dependency relationships by transfer of technology to expand their bases of power (as in the case of relations France - Brazil, which aims to develop nuclear submarines to control their new discoveries in the pre-salt).

Also, from a geopolitical perspective, the achievement of energy for region self-sufficiency in the medium and long term, is becoming more evident if you observe the conflicts that arose between the countries of the region from 2004, for example, Argentina, Chile, Bolivia, Argentina, Bolivia, Brazil, among others and the solution was search for the supplies from outside the region by use of LNG tankers, ships and regasification. Situation led the region to have a dependency on supply from abroad, which creates a situation of energy vulnerability, in some cases, dependence on others or both simultaneously; such is the case of Argentina.

In this context, there are two geopolitical views on the relationship between Energy and Security. The first, related to State security and international competition to secure the possession of resources that are considered strategic because that empowered the State. The second is located above the national level and worries about security, development and welfare of society at regional and global levels through processes that promote energy integration. Beyond what is desirable, the reality is the prevalence of first point of view[10].

In the case of South American in particular, process of nationalization of resources generated a territorialization of politics, produced as a result of decisions and political alliances between states- centered not on the market but political power that is exercised on a territory - leading to territorial dynamics of uncertainty and instability in energy at regional level, which threatens regional development and improved quality of life of the people involved in the process.

Usually, when you mention the strategic role of gas as a critical variable in the regional energy system, one thinks about the scarcity of the resource or infrastructure investments and passes unnoticed geopolitical component as determining the critical value of the resource, while decisions of resource use in South America is determined based on existing

[9] Delamer, G. (2005) "*Strategy, for Policy, Company and Security* ". Naval Institute Publications. Buenos Aires. Argentina.

[10] Adapted from Elsa Cardozo NEW SOCIETY 204 (2006) *Regional democratic governance and the role of (dis) integrated energy* pp.: 136 to 149.
Online http://dialnet.unirioja.es/servlet/articulo?codigo=2380895.

political issues rather than on economic grounds. Also, do not forget that the State's role is to act as a guarantor of energy security which would enable safety and welfare of the societies involved.

Into this new global energy order, proposed on basis of possession gas or oil are two main actors of the Cold War but with different roles. Main loser, Russia emerges as a leading energy supplier to Europe and Asia. While the United States, winner of this confrontation appears increasingly dependent on external suppliers of energy and therefore more dependent and vulnerable. In this new geopolitical scenario, add two actors with the fastest growing economies in the world, China and India becoming significant competitors in the search for energy reserves, to ensure global growth. In this context, the presence of China and Russia in the South American region is already notorious, especially in its relations with Venezuela, Bolivia and Brazil.

Also, in South American scenario has been produced the two major changes in the growth of global reserves. First, the great discoveries of hydrocarbons in Brazil, in the pre-salt area, that position not only as a country that achieved its energy self-sufficiency but also as an energy resource surplus country, added to its growing economic and political presence as an emerging country with one of the largest surface extensions and large population, adds a significant improvement in global positioning. Also, the final certification of oil reserves in Venezuela in 2010 (remembering that in many cases is associated with gas field) who listed this country first in the world reserves, above Saudi Arabia.

Principal countries providers of resources try to maximize the benefits and advantages it gives them this position within the new world energy order, which some call "resource nationalism" that could be defined as the management of energy flows according to the vital interests of the state, trying to convert the latent power and political advantage. Then the goals are still available to the respective governments and in some cases are used as instruments of foreign policy (Klare 2008: 34-35). These companies are increasingly involved in all stages of the circuits of production, transportation and marketing of products, either gas or oil.

A resurgence of behaviors "statist" regarding energy resources is a consequence of dependence the States from hydrocarbons in a context of shortage of production that generates threats to energy security and life quality of people in the most developed, while highlights conditions of vulnerability and dependence on the major consuming countries. All these economies-traditional and emerging grow up and compete for the same energy sources.

Further reality to be faced when speaking of the New Geopolitics of Energy is that it will remain dependent on increasingly scarce resources and higher demand since despite the efforts of scientists around the world trying to substitute these resources. Reality is that none energy consumer country can replace these resources on a large scale. Regard the report of the U.S. Department of Energy 2007 as these fuels will continue to meet global energy needs -even in 2030- around the 87% (DoE / EIA, IEO 2007, Table 2, p.85).

Another scenario arises; the countries that have the resources but are not in conditions of defend it against pressure from rich countries but poor in energy resources. This situation is clear in Africa. However, in South America face that possibility also Brazil move forward in its nuclear program and develops a nuclear submarine -with French support- to protect his recent discoveries in the presal from external aggression. Related to this situation and the possibility of signing strategic alliances with States that have technology, weapons or knowledge arises in the region presence of global players such as Russia, France, Iran and China among others. Faced with this change in region United States reactive the 4th Fleet to protect its interests in the region with a power that, any Latin American country can withstand.

Competition for energy is intensified and considers that, owning energy is a factor that rivals with the military power; Russia is a clear example of this situation. The real possibility of transforming the natural gas from a strategic resource to a tactical one (which could be used as a tool of economic domination over other countries) was observed specifically on 1 January 2006, when Russian President Vladimir Putin put in check the European continent. In one of the coldest winters in its history, cut off gas supplies to Ukraine, the main distributor of resources from Russia, but far from its political ideology. Remembering the Cold War tactics, it became apparent that gas resource use represented an element of pressure and power - not military but economic - hence the idea of natural resource use as a strategic weapon.

A growing demand against a standing offer, both at global, regional or local level, certainly generate greater competition among energy-consuming nations, and those owners or producers of resources. Meanwhile, not all countries will be able to acquire these resources, high costs generated by the shortage, with the resulting consequences on the quality of life of the population.

Meanwhile, other countries with resource but which has less support have taken other measures such the proposal of Venezuela and Bolivia to create a "natural gas OPEC" or non-traditional alliances like the one held between China and Russia to reduce the weight of the United States. Experts suggest that every new barrel that is added to global reserves will be more difficult and expensive to extract than the last, will be deeper underground, farther from the coast in hazardous environments or in regions most susceptible the conflict, more hostile. It is probable that this same scenario is repeated in most existing fuels, including coal, natural gas and uranium.

If one analyzes from an energy perspective the reality of Latin America, focusing on gas resources, South American countries are a territorial space characterized by contradictory conditions on one side, have a variety of renewable energy resources and nonrenewable, that would mean the possibility of regional energy self-sufficiency, whereas on the other hand, reality shows that several countries in the region have a situation of dependence on external supply sources located at large distances and in other cases, energy vulnerability becomes more apparent as we move into the twentieth century, product demand growth against a standing offer for lack of investments in infrastructure and exploration tasks.

The most urgent energy problems it faces today the South American region is to supply natural gas fluid, predominant in the regional energetic matrix. That is aggravated by the absence of long-term public policies in this area, or the historical unpredictability that have occurred in some countries, such is the case of Argentina, who behaved like a country with large gas field and leading to overexploitation. Therefore, it becomes evident the need for strategic decisions for the region considering to achieve energy security to ensure the development and welfare of future generations, compared to a situation of vulnerability and dependence on energy in the region.

4. Geopolitics of gas in South America

The changes in South American energy sector between 2004 and 2011 - focused on the use of gas – show situations that move from a bilateral conflict to a regional issue with the subsequent incorporation of global actors outside the region. These processes of change can be glimpsed through the application of geopolitical approach from different sides who allow us to perceive the complex power relations that exist between actors outside the region of global level - such as Russia, Iran and France - with key regional players - Brazil, Venezuela and Bolivia, through a series of non-traditional partnerships that arise as a result of changes in regional and global scenario where contrasting new integration processes globally with growing conflicts causing instability and fragmentation in regional and global level.

4.1 New geopolitics of energy -focused on gas natural - in South America

The geopolitical analysis allows us to understand the South American Gas Crisis as a result of a reality - changing and uncertain, and to interpret the complexity of the existing dynamic in the social, political and economic level. We also understand the power relations that are established for the emergence of new geopolitical situation in regional and global space, where reality is being constructed every day.

South America has abundant renewable energy resources and nonrenewable energy self-sufficiency that would enable the region and avoid dependence on uncertain supplies from abroad. However, the reality shows a situation in which, for historical and geopolitical conflicts existing inadequate policies and decisions, lack of long-term vision in the region have created situations of vulnerability and dependency and the possibility of disappear energy self-sufficiency in the region beyond the existence of reserves in quantity and variety of pipeline constructed and supply agreements signed between the countries

The sum of different factors detailed in this paper allows us to anticipate in the short term, a new crisis in the region in the period 2015-2020, if the necessary measures are not taken to reverse the current situation and creates a lack of energy security in the region. Geopolitical and Strategy, allows a differential analysis of energy security scenarios that goes beyond dominant economic view because they are political and not economic decisions which sometimes define the actions taken by States. In Latin America, particularly in South America, there are four main geopolitical trends concerning the use of natural gas:

1. An increased use of gas resources in relation to the rising price of oil and its minor effects on the environment and the massive installation of combined cycle power plants.
2. Increasing LNG trade growth both globally and regionally, making it possible to emerge from the constraints involving pipelines and that also allows the resource to get anywhere in the world.
3. Nascent proposal, supported by Venezuela and Bolivia, to form a gas group OPEC-style.

4. Potential trend, but ever closer, is the possibility that natural gas becomes a commodity like oil.

4.2 Analysis of the situation by country

Analyzing the case of South America, Chile has been one of the countries that quickly came to use of LNG from the supply conflicts with Argentina in 2004 and in 2010 led to the arise the possibility of exporting gas to Argentina through existing pipelines from the country and exporting country imported despite having no recourse but regasification plants installed in its territory.

In the region, geopolitical and strategic situation that arises is the need for each State to ensure the flow of energy supplies to its population. This requires the signing of agreements to overcome conflicts of interest that arise between players in the region, and internally within each state, for example, in Argentina provide its population or exported to Chile, Bolivia meet contracts with Brazil or Argentina as well as ensuring the domestic supply. Most South American countries are faced with the challenge of implementing policies and strategies that put them safe from possible energy crisis. Against this background, the various governments chose to perform different actions:

ARGENTINA: made a complex pattern of staggered withholding exports of oil and derivatives that led to overexploitation of resources and lack of investments. Added to this Argentina's energy matrix has 90% of depends on oil and gas. However, the two productions fall, almost continuously, since 1998. Probably, Argentina is one of the countries most affected by the energy crisis and move from being an exporter to being a net importer of energy, especially LNG. Currently, subsidies to the energy sector grew 63% in 2010, the National Administration expense subsidies spent amounted to $ 48,032 million, which represents a 47% more than what was spent in 2009. The other bet is on the exploration and exploitation of thigt gas and shale gas reserves discovered in a "megafields" of unconventional gas in Neuquen basin, the argentine President announced in December 2010, that they would expand the gas reserves of 6 to 16 years. Their main problems to be in production are infrastructure investments, the highest price, use only for industries and pollution that it generates.

CHILE: has been one of the countries that quickly came to the use of LNG from the conflict with Argentina supply in 2004 and in 2010 led to the emergence of the possibility of exporting gas to Argentina through existing pipelines from the country and exporting country imported despite having no resource but regasification plants installed in its territory. Also, 53% of the energy matrix is represented by the hydroelectric system is affected by droughts that prevented from operating at full capacity. Response to the crisis has been the development of coal plants by 2015 it is expected that this technology represents almost 30% of the matrix of the SIC, the main electrical system of the country (three times the current share). Another strategy has been the installation of regasification plants that allow to bring resouce from anywhere in the world and is already thinking about export natural gas to Argentina. On November 13 in 2009, Chile[11] was signed the decree that

[11] Tecnoil journal report of January 27, 2011 as reported by the Argentina Association of Budget and Financial Administration (ASAP)

allows the sale of liquefied petroleum gas (LPG) to Argentina and from there to other countries such us Uruguay, Paraguay and Brazil, the sale could be from 2010 or 2011, when Argentina bring its legal framework. Argentina would thus supplier to consumer.

PERU: tries to change his energetic matrix and stop using the petroleum products to exploit the Camisea natural gas, oil field discovered by Shell in 1983. Furthermore, the idea of an energy ring that communicates with Chile and from there to Argentina was dropped and resumed some contacts with Petrobras regarding the installation of refineries in the country. However, this year Peru exported 165 000 metric tons (MT) of natural gas (LNG) that will occur in the plant operating Melchorita Peru LNG consortium, it is projected that this year there will be about 52 shipments LNG around the world. Besides natural gas production has increased from 327 million cubic feet per day to around 700 million, confirming its strongest growth in global production of 108% between 2009/2010.

BRAZIL: it is one of the least affected by rising oil prices since they already achieved self-sufficiency oil and also with the new discoveries of oil will be a surplus from the energetic point of view and have greater potential to prevent energy crises and regional own once you get rid of the volatile gas shipments from Bolivia. It has an energy matrix with 46.6% of energy from renewable resources (the average is 13%). Added to that, the use of sugar cane is the second energy source after oil and this is another resource that leads us away from dependency. Natural gas consumption in 2010 increased 35.5% increase compared to 2009 with a daily average of 52.9 million cubic meters, said the association of gas distributors in the country. One factor for this growth was the heaviest use of thermoelectric plants using natural gas as fuel, since the increase in consumption in this sector was 171%. Among regular, the largest increase in industry which increased its consumption by 20.15% while household consumption grew 7.2% and 6.26% commercial[12].

BOLIVIA: it has the second largest reserve of gas in the region, and its largest supplier. Political instability and legal uncertainty prevented him from consolidating his position for the lack of investment in infrastructure to exploit that resource. Including conflict has domestic supply since the nationalization of hydrocarbons by itself does not generate wealth, foreign investments are needed if the country does not have the financial resources to become profitable and generate revenue from there that can be distributed among the population. Bolivian Fiscal Oilfields (YPFB Corporation) will spend more than $ 43.2 million in exploration work and exploitation of hydrocarbons in the perspective of a 15% increase in oil production, natural gas and associated liquids in the department Cochabamba. The state oil company president, Carlos Villegas[13], announced that from this amount 24 million dollars is earmarked for operating tasks in order to optimize local production and more than $ 18.3 million in exploration for new reserves hydrocarbons.

VENEZUELA: it has the largest reserve of gas in the region but there is the feature that does not export gas to the region due to the lack of pipelines and regasification plants including

[12] On January 27, 2011 was signed this agreement and the Argentine Minister de Vido said that "the goal is to achieve greater exchange. Integration is very important, symmetry, which is going to go to work. It will at some point if a country is about something and the other will need to see how they complement. In this context, it might sell or buy gas, depends on the situation. For example, trade with Brazil we work very well "

[13] According to the publication of the journal Tecnoil of February 8, 2011.

the paradox that imports gas from Colombia. PdVSA reduced 61% their investment goal for 2011 according to the provisions of the Oil Sowing Plan 2010-2015 which was designed in 2009, after changing the deadlines and goals of the original 2005 project. The tax burden from the Venezuelan oil industry on the remains, and is reflected in the forecasts of investment and social contribution. The current plan includes investments of 252 billion dollars, of which PDVSA will provide 197 billion dollars to move toward four strategic objectives: accelerating development of the Orinoco Oil Belt to add 2.8 million barrels a day production for the year 2030, the development of offshore gas to satisfy the domestic market and export to key markets, furthering the country's overall socialist development and territorial balance, and the absolute sovereignty over the oil and gas resources[14].

In summary, concrete actions are only observed in some countries such as Chile and Brazil, with the difference that their interest is to secure its own energy supply, regardless of the recognition of the need for a strategy for achieving Supraregional Energy Security to allow the development of the region. Consequently, the two countries with the largest reserves of gas but no investment in infrastructure, increasing political instability and without legal certainty, such is the case of Bolivia and Venezuela to varying degrees, are creating vulnerability in the region such as Brazil and Argentina, dependent on gas from Bolivia, but with no assurance of continuous supply, despite the long-term contracts already signed. Argentina meanwhile, decreased exports to Chile and increased imports.

5. Conclusion

The survey of the energy potential of South American countries is a necessary step for a sustainable exploitation of resources to enable the use of all energy alternatives available to the region. Vulnerability is the counterpart of energy security to any country aspires and to avoid it is necessary joint actions conducive to regional integration.

The Americas region has enough variety and quantity of renewable and nonrenewable resources in the region without having to rely on sources outside the region. Victor Bronstein[15], Director of the Center for Energy Studies, Politics and Society (CEEPyS) argues that "Latin America as a region and beyond the particular situation of each country energy self-sufficient. Then, the process of integration is important because the threshold of a new global energy crisis, the struggle will be for the price, but by the appeal "Ensuring a reliable supply of energy in America is the responsibility of governments. Therefore, governments should intervene to manage risks and avoid supply disruptions and as in reference to costs. The measures range from subsidies to producers or consumers of energy, to the signing of international conventions. Ensuring Energy Security is a State obligation, but recognition that this cannot be obtained in isolation and affirm the need to firm agreements.

In sum, the geopolitical landscape of energy in the South American region in 2011, focused on the gas resource, shows important changes:

ARGENTINA loses its position as a leading gas exporter in the region, as well as their ability to self-sufficiency due to declining reserves and lack of investment in exploration

[14] According to a report in the journal Tecnoil of January 11, 2011.

[15] Roca, M. (2010) "We are on the verge of a global energy crisis, Victor B." In DEF .Ed TAEDA, No. 64 p: 88.

and infrastructure tasks. His current dependence of LNG is evident when one considers that consumed 8 million cubic meters per day in 2009 with an estimated 270 to 320 million dollars. Can be reached the paradox of imported gas (LNG) from Chile, his old top consumer. Purchases of LNG began in 2008 with 8 ships as a transitional situation for the winter season, but in 2010 reached 21 ships and by 2011 anticipates the arrival of another 46 ships. On the last trip of the president to Qatar signed an agreement to import 20 million m3 per day to replace Venezuelan fuel oil, Infrastructure Minister Julio de Vido said the Qatar LNG is cheaper because it is transported in freight double-sized vessels and is better in environmental terms than Venezuelan fuel oil. Argentina has gone from being an exporter of energy to be a country dependent on imported energy resources (LNG, fuel oil, fuel) in 2010 was a net importer of oil after two decades of self-sufficiency and is vulnerable from point of view of the high costs to be paid to ensure the provision of these resources, oil purchases abroad increased from 76.8% of 1858 million in 2009 to 3,283,000,000 in 2010, according to a report published by the Sectorial Economic Research Center (ICE). Moreover, the alternative of unconventional gas exploration potential still have not been tested for their reserves, or made the necessary investments in exploration and production tasks.

BRAZIL is the biggest winner in the region, thanks to huge discoveries of oil and gas resources in the shelf made during 2007 / 8 and with a plan in place that will allow the rapid acquisition of the resource in 2015 would become a surplus country in energy resources (gas, oil, hydro, alternative energy). However, despite being oil self-sufficiency since 2007, in the short term shows vulnerable by their dependence on the uncertain supply of gas from Bolivia, in this regard, José Sergio Gabrielli, president of Petrobras, said wells in 6 years, new discoveries in the Brazilian continental shelf go into production, which would be achieved, in 2015 - get the gas self-sufficiency and achieving independence from unsafe gas supplies from Bolivia, thus exceeding its vulnerability.

CHILE, can also be a winner, because, before the gas supply crisis in 2004 adopted the necessary measures, like betting to LNG regasification plants and install that let in 2010, be able to feed itself and even export surplus the resource. His condition is a clear example of a country dependent but not vulnerable, given its regasification plant and sourcing safe and varied, but away from countries that do not generate vulnerability. In addition, he resumed investment in hydropower and coal and also took preventive measures energy rationing. VENEZUELA, BOLIVIA AND ECUADOR, are energy-exporting countries and nationalizing their reserves to increase its negotiating leverage with large multinationals in order to earn more income derived from oil and gas, but generating conflicts over legal certainty for investors, as a result, they do not make the necessary investments in infrastructure and this affects their growth potential.

BOLIVIA is another of the losers in the region, being a central actor for the regional energy supply started to have serious difficulties in complying with treaties agreed with Brazil and Argentina. The nationalization of the hydrocarbons generated that the companies concerned in their interests not to perform the necessary investments in infrastructure to ensure higher production that would allow her time to meet the agreed volumes of supply contracts. This is a special case of vulnerability and dependence because it depends on the one hand, of investments in infrastructure made by other third countries and the sale of its gas to countries like Argentina and Brazil. Despite having signed long term supply contracts with these countries, cannot be enforced by the lack of investment in infrastructure, thus being in a vicious cycle that cannot leave despite having abundant reserves of gas resource.

VENEZUELA is also far from being the lead actor that was in 2008 with his proposal of the Southern Gas Pipeline, the fall in oil prices reduced its income. Also, climate issues as a severe drought seriously affected its hydroelectric resource dependent and uncovered its shortcomings and lack of investment in infrastructure to avoid a dependence on climate, the more even with large oil resources. However, the fact he mainly oil dependence creates extreme vulnerability of the country's GDP from the exportation resources. Oil Minister Rafael Ramirez said that Venezuela has a volume of country's oil reserves 297,000 million barrels, which would be the first country with the largest proved oil reserves, ahead of Arabia, with 266,000 million barrels, and countries like Iran or Kuwait, according to the Organization of Petroleum Exporting Countries (OPEC).However, it has problems of lack of investment in infrastructure as well Bolivia. Despite having the largest gas reserves in America these are not exploited and maybe it should think in the short and medium term export through LNG ships since the idea of South Pipeline had many obstacles to become possible. So far, only Venezuela provides fuel oil to Argentina and despite being the country's largest gas reserves in America, NO PROVIDE GAS TO THE REGION.

The entry of global players in the region, including Russia, France, China, Iran and Spain (with investments in Venezuela and Argentina through Repsol-YPF), due to the presence of abundant natural resources demonstrates the need for consumers to ensure the provision of them to not be vulnerable and also need for resource holders to ensure the ability to effectively protect, through deterrent military equipment. It seeks to obtain the transfer of technology that provides France and supported by Russia and China. This led to an increase in military spending in the region.

The application of geopolitical analysis allowed understand the South American gas crisis as a result of a changing reality and interpret the complexity and dynamics of social, political (resurgence of nationalism) and existing business and also understand the uncertainties and the possibility of emergence of new geopolitical situation in a regional space in which reality is being constructed daily. Knowledge of political, economic and geopolitical has shown, therefore, the important to draw scenarios about possible developments in energy markets and security of a regular supply companies.

Monitoring the energy situation in the country and the region began in earlier papers by the author allow to suggest that the conflict has expanded from a binational problem, focusing on conflicts of gas supply between Argentina and Chile, to a regional conflict who joined Bolivia, Venezuela and Brazil with the intention of the South American Gas Pipeline now discarded, which currently includes global players outside the South American region, as in case of Russia, France, Iran, and a look around the U.S. on the region. This allows to affirm the existence today of a situation of energy dependency and vulnerability in the region that must be reversed because it has sufficient resources to achieve its self-sufficiency energetic and also be exporting.

Concluding, while the reality is that only existing pipeline in operation - with the advantages that this implies - are in Bolivia, the reality is that contracts are not met and pipelines cannot be filled unless there are investments in infrastructure to ensure supply. There is the contradiction, the region is less vulnerable if choose to import gas from Qatar and other countries, beyond the distance and dependence that it generates, than if import the resources from countries with highest gas reserves in the region, Venezuela and Bolivia. As a result, promote a strategic alliance between States, since a global perspective is not only advisable, is a real priority for South America. This partnership should be built with the participation of State actors who claim the resource and those that offer it within the region. Achieving the South American energy integration, the region would be in the enviable position of energy self-sustaining future growth, and supply some of the needs of the rest of the world. At this point another challenge, not least the integration, is exporting hydrocarbons, but at the same time to develop, with the consequent improvement of quality of life of people from a specific resource territories, the energy.

6. References

Briano, J. (1972) "*Geopolitics and American Geostrategy.*" Military Club Ed. Buenos Aires.

Delamer, G. (2005) "*Strategy. For policy, the Company and Security* ". Naval Institute Publications. Buenos Aires. Argentina.

Escribano Francés, G. (2006), "*Energy Security: concept, scenarios and implications for Spain and the EU*", Real Instituto Elcano, Working Paper 33/2006.

Guerrero, A. (2006a) *The geopolitics of South American gas. Policies, Territories and Resources.* Posted on abstracts and CD, 13 pages. 8th International Meeting in Humboldt, Colon, Entre Rios, Argentina, from 25 to 29 September 2006.

Guerrero, A. (2006b) *Where does the energy integration in the Southern Cone? Contributions from a geopolitical perspective.* Published on CD, 10 pages. International Symposium on Latin America and the Caribbean, 18,19 and 20 October 2006. Faculty of Economic Sciences of the UBA.

Guerrero, A. (2006c) Masters Thesis Policy and Strategies *The strategic role of gas as a critical variable in the national and regional energy system*, UNS, 2006.

Guerrero, A. (2008) *South American energy security. Public policies and strategies* on CD ISBN 978-950-1149-6. In VIII International Symposium and the First Congress, CEINLADI, University of Buenos Aires. Faculty of Economics, 15, 16 and October 17, 2008.

Guerrero, A. (2009a) *Contribution Geopolitical analysis of regional energy scenarios. New trends, new scenarios* in 12th Meeting of Latin American Geographers 3 to April 7, 2009, Montevideo Uruguay. ISBN 978 - 9974-8194-0-5 published in the network, www.egal2009.

Guerrero, A. (2009b) 24th World Gas Conference 2009, The global energy challenge: reviewing the Strategies for Natural Gas. Panellist at Expert Forum: Geopolitics and Future Impact of Globalisation on Sustaintable Gas Supply and Trade. Paper:

Understanding Latin American Gas Crisis. A Geopolitical Approach. 5 to 9 October 2009. Published on CD by IGU Committee.

Guerrero, A. (2010a) "*International relations, power relations. New geopolitical scenarios and new actors in the process of energy integration regionally and globally.* " IV Inter-Oceanic Congress of Latin American Studies. X Seminar IV Argentine-Chilean Cono Sur Seminar Social Sciences, Humanities and International Relations. Univ.Nac. de Cuyo, Mendoza, Argentina, 10,11 and 12 March 2010. Publication of Proceedings on CD - ISBN 978-987-9441-40-4.

Guerrero, A. (2010b) *The critical geopolitical approach as a tool for analysis of new nuclear energy scenarios. Regional and global actors in conflict.* II and IX SYMPOSIUM INTERNATIONAL CONGRESS OF LATIN AMERICA AND THE CARIBBEAN 2010.UBA.Argentina October.

Guerrero, A. (2011) III Latin American Congress of Energy Economics; ELAEE2011, Paper: *Energy vulnerability of the countries of Latin America. Case study: The focus of critical geopolitics as a tool for analysis of new energy scenarios focusing on gas integration.*18 and April 19, 2011, Buenos Aires, Argentina. 20 pages.

Hutschenreuter, A (2008). *The gates of the twentieth century geopolitics.* This introductory article is part of an ongoing investigation. Working paper 04-In Program CAEI geopolitical Argentine Center of International Studies. Online www.caei.com.ar/es/programas/geopolitica/04.pdf

Isbell, P (2008) *Energy and Geopolitics in Latin America,* Working Paper No. 12/2008, Real Instituto Elcano. España.

Klare, M (2003) "*Resource Wars. The future scenario of global conflict.*" Ed Trends. Spain.

Klare, M (2008)"*Planet thirsty, dwindling resources. The New Geopolitics of Energy* Urano.Barcelona.España.

Lyndon, A, Hiernaux, D (eds.) 2006-*Treaty of Human Geography,* Chap.8, Castro, P. Geography and Geopolitics, pp. 187-199.Ed.Antrophos, Mexico.

Palenzuelos, E. (Dir.) (2008) "*Oil and gas in the global geo-strategy.*" Editorial Akal, Economía. Madrid. España.

Statiscal BP Review of World Energy 2011 (June 2011)

DoE / EIA, International Energy Outlook 2007 (Washington DC)International Energy Agency (IEA), World Energy Outlok 2007 (IEA, Paris, 2007).

"*Oppenheimer: The new oil center in the world*" Published in the NuevoHerald.com Thursday September 29, 2011. Online http://elnuevoherld.com/2011/09/28/v-print/1034152/oppenheimer-el-nuevo-ce...consultado30/09/2011

Cardozo, E (2006) NEW SOCIETY 204 *Regional democratic governance and the role of (dis) integrated energy* pp.: 136 to 149. Online http://dialnet.unirioja.es/servlet/articulo?codigo=2380895.

Roca, M. (2010) *We are on the verge of a global energy crisis,* Victor B. In DEF Ed TAEDA No. 64 p: 88.

Qatar gas instead of fuel oil in Venezuela. (2011. 22 January). Clarín, Section The country p: 8.

For the first time in 20 years the country was a net importer of fuels (2011, 23 January). Clarín, Country section, p: 20.

Venezuela face a difficult economic year with lower incomes (2011, January 23). Clarín, World section, p: 25.

The coming year will be spent twice to import gas by ship (August 21, 2011). Clarín, Country section, p: 23.

4

Geological and Geochemical Setting of Natural Hydrocarbon Emissions in Italy

Giovanni Martinelli[1], Stefano Cremonini[2] and Eleonora Samonati[2]
[1]*ARPA Emilia Romagna, Reggio Emilia Department, Reggio Emilia,*
[2]*University of Bologna, Dipartimento di Scienze della Terra e Geologico-Ambientali, Bologna, Italy*

1. Introduction

Hydrocarbons are hosted in underground geological formations and they can slowly migrate under the action of the lithostatic load and tectonic activity. Spontaneous hydrocarbon emissions can be detected on the earth surface and have historically drawn man's attention, and have even been the subject of health or religious cults. The natural hydrocarbon emissions were already well known in the ancient world, in particular during the classical age. Aristotle recalled the kitcgchen of the persian Kings fed by natural fossil fuel seeps (Montesauro Veronesi 1585). Lucretius described flammable gaseous emissions in his *DeRerum Naturae.* Pliny (23 A.D.-79 A.D.) wrote in his *Historia Naturalis* said that in the Syrian city of Commagene a pond existed expelling a burning loam called "malta" (Bianconi, 1840). In about the whole of Italy the various levels of interest paid to hydrocarbons by local inhabitants and in economic and industrial structure are recognizable in the historical and archeological documentation. During the Renaissance the mud volcano cluster of Sassuolo (meaning *"Boulder of the Oil"*) near Modena, the largest in Italy, was particularly famous due to the possible medical properties of the their brackish waters (Scicli 1972) . That area was thus the first to be explored and exploited in the second half of the 19th century. The first database compilers acquired field information from the knowledge of salutiferous and religious cults widely known in the ancient specialized literature (Bacci 1571). Local detailed maps and lists of hydrocarbon seepages were compiled in the period 1850-1950 with the purpose of addressing the drilling strategies. The modern studies in Italy begun with the description of many hydrocarbon emissions recorded by Camerana and Galdi (1911; Biasutti 1907) in Emilia Region . Successively the recognition was extended to the whole of Italy (e.g. Camerana et al., 1926). This study phase lasted up to the end of the 1940s (Zuber 1938; Idem 1940) and the last traces of this approach to the research can be found up to 1969 (Martinis 1969; Reeves 1953). The advent of modern geophysical prospection methods (e.g. Accademia Nazionale dei Lincei and Ente Nazionale Idrocarburi 1948) and the growth of new study trends such as the isotopes geochemistry in the two last decades of the past century (Mattavelli et al. 1983; Lindquist 1999; Casero 2004, Bertello et al. 2010) lowered the importance of those former empirical methodologies. Thus no traces of interest can be found in the last modern handbooks dedicated to the petroleum sciences (ENI 2009). Furthermore, the growing of the anthropogenic impact on the landscape (eg. roads and towns building) erased a great deal of natural evidence of hydrocarbon occurrences. The scientific literature has only recently has renewed its attention to the databases of gas or oil natural emissions and a possible loss of knowledge of sites related to hydrocarbon occurrence. In recent years Martinelli and Judd (2004), Etiope et al. (2009, and references therein) etc. recovered information on the occurrence and chemical composition of gaseous hydrocarbons bubbling in mud volcanoes. Furthermore current scientific literature has devoted attention to some spontaneous gaseous non mud-volcanic emissions as well but a large part of the small or low flow rate methane and oil emissions has been not listed. Since the recovery of the old geographical information and modern geochemical data set can help to achieve a better understanding of the geological phenomena related to deep fluid accumulation and migration (Minissale et al. 2000; Capozzi and Picotti 2010), faulting linked to the crustal stress field , natural greenhouse emissions (Etiope et al. 2009), etc., the present paper is devoted to a first recovery attempt of the available historical as well as recent information on the natural hydrocarbon emissions and to its comparison with the updated findings on Italy's geological features. A map of hydrocarbon gas seepages has been made and commented together with available analytical data on natural hydrocarbon emissions (see below). Hydrocarbon seepages drove the hydrocarbon exploration strategies and allowed for the discovery of important gas and oil rock sources. Most of the hydrocarbon accumulations are found in the foreland and in foothill areas whereas they are less frequent in mountain chain areas because of tectonic activity or of high heat flow areas. By moving in a subduction zone from the back arc tensive area, through the main thrust area, to the foredeep-foreland area, compression became dominant and newly formed sedimentary sequences were subjected to strong subsidence and compaction. In this kind of ambient the abundant organic matter and its chemical alteration produced hydrocarbons that tend to be squeezed towards the surface, mostly along fault systems. The main gas accumulations are located along a strip parallel to the Apenninic chain (Fig. 1). In particular, in the foredeep main biogenic gas accumulations occur due to high subsidence, sinsedimentary tectonics and turbidite sedimentation. In the Apennine chain gas of thermogenic origin is prevalent due to intense tectonic activity (Mattavelli and Novelli, 1988). Most of the Adriatic and Sicilian oils are high density while the northern Apennine oils are lighter, probably because of a more effective thermal differentiation. Heavy oils originated from Mesozoic rocks while those in the chain have a more diverse origin (Pieri and Mattavelli, 1986). A comparison with the available upper crustal sections reveals main escape conduits along faulted rock volumes. Surface hydrocarbon occurrences are represented by gas and oil seeps and mud volcanoes. At time gas seeps are accompanied by cold or warm water springs due to gas interactions with less deep groundwater circulation paths. Mud volcanoes are well-known gaseous seepages bubbling in a liquid consisting of clay minerals and brackish water. They are chiefly related to areas of tectonic compression characterized by thick sedimentary sequences. Their occurrence is limited to the continental Appeninic chain and Sicily. Some hydrocarbons seeps, sinkholes and mud volcanoes were reported offshore within a few kilometers of the coast and their origin has been recognized to be similar to continental hydrocarbon emissions (Curzi et al. 1998; Camerlenghi and Pini 2009; Fusi et al 2006;Praeg et al. 2009; Holland et al.2003).

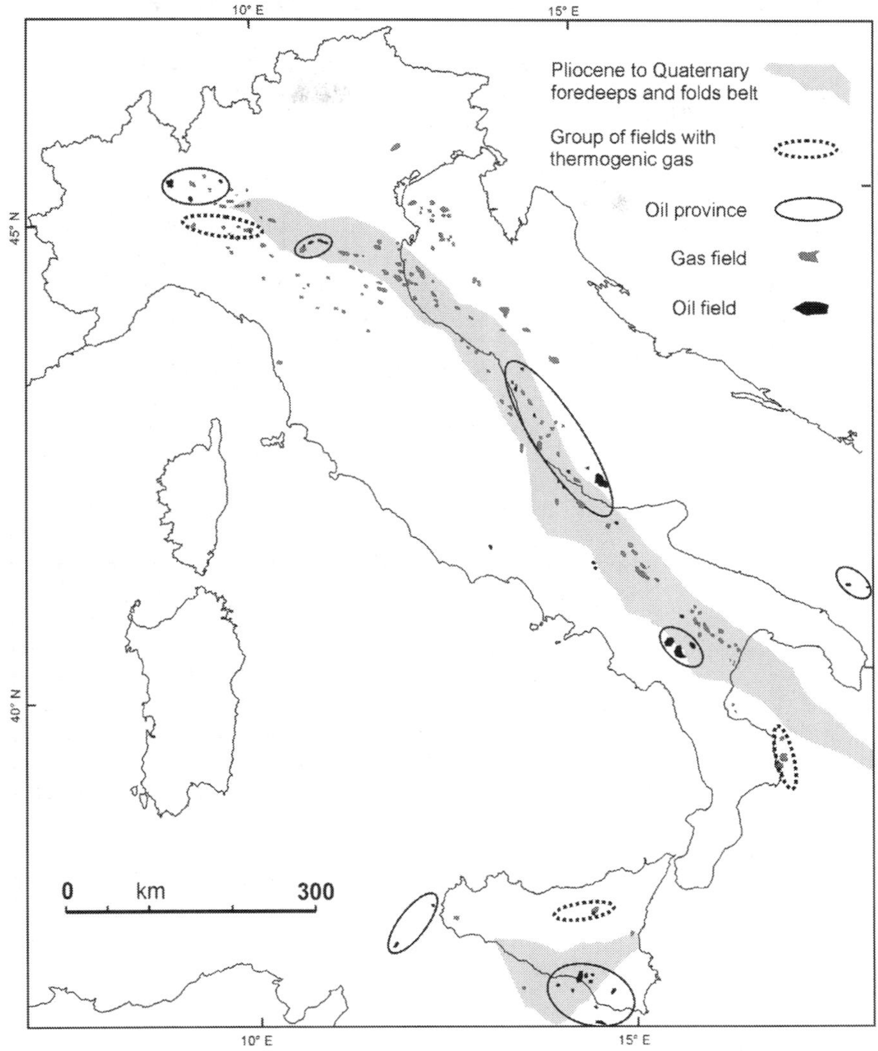

Fig. 1. Distribution of main hydrocarbon reservoirs in Italy
(after Casero 2004, Sheet 1 and Bertello et al. 2010; redrawn).

2. Geological framework

Two main tectonic events are responsible for Italy's present geological setting: 1) extensional tectonics from the Jurassic to the Early Cretaceous; 2) compressional tectonics from the Cretaceous to the Quaternary. The extensional tectonics occurred during the separation of Africa and Europe determining the origin of a new ocean (Tethis). Therefore, the Italian peninsula and Sicily, as a part of an African microplate (Adria or Apulia indenter), was affected by subsidence and fragmentation as an isostatic response to crustal thinning of the passive continental margin. Triassic to early Cretaceous carbonate sediments were mainly deposited during the extensional tectonics that preceded and accompanied the oceanic opening. Euxinic conditions were present before and during this passive margin regime in the Middle-Late Triassic (pre-rifting stage). The Triassic source rocks are formed by thick black limestones and shales and were deposited in lagoons or mainly in narrow discontinuous troughs, originated by rifting and/or transcurrent movements (Catalano and D'Argenio 1982). Compressional tectonics due to the convergence of African and European plates began in the Cretaceous and caused the Alpine orogeny. The Alps and then the Apennines were thus formed and the Italian peninsula took shape. The origin of these chains was complex and occurred during various tectonic phases characterized by different vergences. The Alpine structures, north of the Insubric line, were piled up onto the European continental margin according to a north-vergence and the southern Alps and Apennines were pushed onto the African margin. Southern Alps and the Apennines were formed more recently than the Alps and their origin was mainly due to the Neogene tectonic events (Castellarin et al. 1992; Vai and Martini 2001). In particular, the earliest compressive phases, which occurred during the Aptian-Albian and Cenomanian-Turonian, were accompanied by anoxic events. Nevertheless, Cretaceous organic rich sediments were characterized by a more widespread distribution but distinctly thinner sequences in comparison with Triassic and Jurassic anoxic facies. The deposition of the terrigenous sediments mainly during compressional tectonics was a consequence of the generation of the new mountain belts. These deposits are chiefly formed by thick Tertiary turbidites, deposited in elongated basins parallel to the Apennines chain (Mutti and Ricci Lucchi 1972). Anoxic facies were not recognized in Tertiary sequences but the preservation of organic matter deposited in the external part of the turbidites was favored by a rapid burial in the more active subsiding areas (Mattavelli and Novelli, 1988). Thus, the origin and distribution of gas fields in Italy was linked to the Neogene tectonic and sedimentary events related to the Southern Alps and the Apennines surrection. During the Neogene three main tectono-sedementary domains characterized the general framework of Italy: Southern Alps; Apennine chain with its foredeep, the related foreland. Most of the Italian gas fields were discovered in the Neogene turbiditic sequences of the Plio Pleistocene. Condensate gas fields have been also found in the deep Mesozoic carbonate rocks of Northern Italy and some gas accumulations of Central and Southern Italy were found in Late Cretaceous limestones.

3. Liquid hydrocarbons source rocks

At least five important source rocks have been recognized which are distributed in age from Mesozoic to Pleistocene. Three of them were deposited during Mesozoic crustal extension and are mainly oil-prone. The deposition of organic-rich sediments in restricted basins began during the Middle-Late Triassic and Early Jurassic extensional phases pre-dating the break-up of Pangea. Discontinuous anoxic basins developed in the southern Alps, southern Apennines and Sicily (Pieri and Mattavelli, 1986). Hydrocarbon occurrences associated with these sources are usually found in complex carbonate structures along the Apennines thrust-and-fold belt and in the foreland. Two other important source rocks were generated in the foredeep terrigenous units which formed during the Alpine and Apennine Cenozoic orogenesis (Casero 2004; Bertello et al., 2010). The older source rocks are thermogenic gas-prone and are found in the highly tectonized Oligo-Miocene foredeep wedges: gas

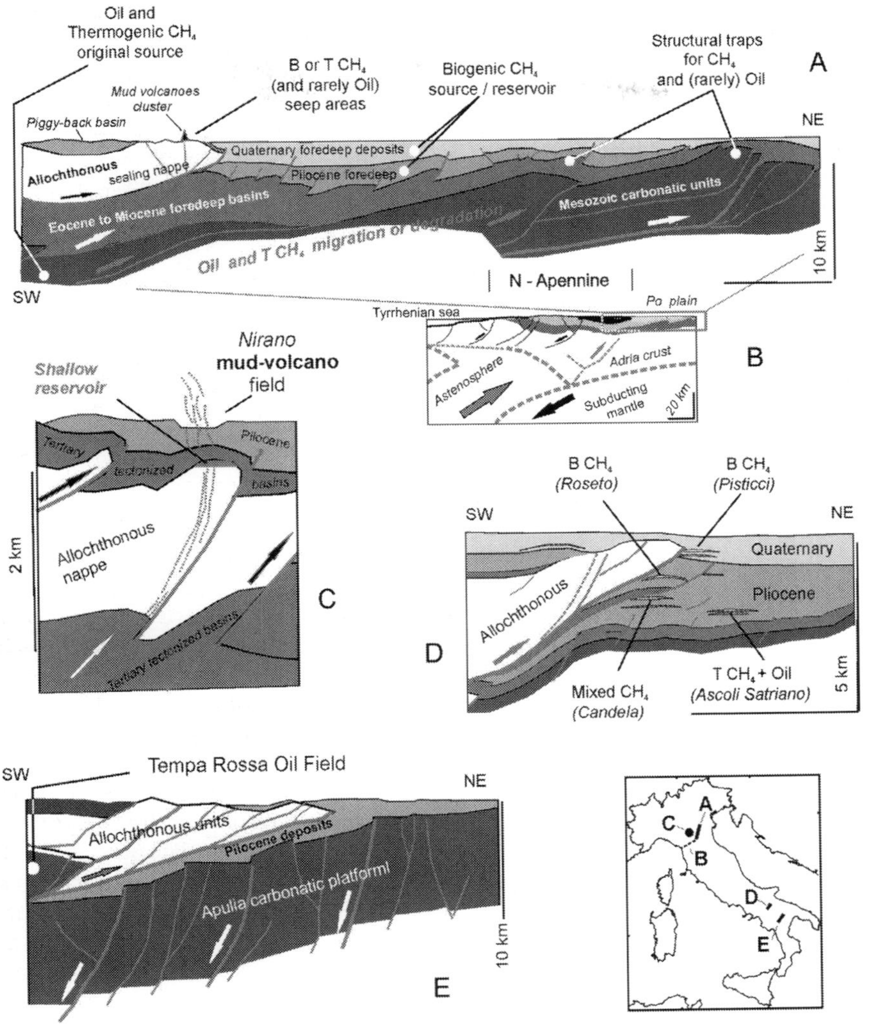

Fig. 2. Schematic transects of gas setting across the Apennine Chain (related location are shown in the vignette lying in the lower right corner). A) Ideal, simplified stratigraphic section (after Boccaletti and Martelli 2004, redrawn) showing an average outline of the relationships existing among lithology, tectonics, oil and methane types in the northern Apennine and related foredeep. B) Outline of the structural setting of the northern Appennine Chain (after Picotti and Pazzaglia, fig. 1, redrawn). C) An explaining model for a mud volcano activity in the northern Apennine (*Nirano* case) (after Bonini 2007, fig 5 and 9, redrawn). D) Hydrocarbon trends geological setting in the *Candela-Roseto* gas field (after Casero 2004, Plate 4, section 4a bis, redrawn). E) Geological setting of the *Tempa Rossa* oil field (after Bertello at al. 2010, fig. 6C, redrawn).

occurrences associated with the gas source are mainly concentrated along the northern Apennines margin, in Calabria and Sicily. The younger source rocks are biogenic gas-prone and are located in the outer and recent Plio-Pleistocene foredeeps of Po plain and northern Adriatic Sea. (eg. Casero 2004; Ministero Sviluppo Economico and Assomineraria 2008). About 95% of Italian oils were generated from source rocks related to the first described group (Mattavelli et al 1993). Anyway, no significant reservoired hydrocarbons can be correlated to these deposits due to migration processes linked to the subsequent tectonism . In the Tertiary era the organic content of the flysch shales also generated a minor amount of oil in the northern and southern Apennines. Maturation of the above-mentioned Mesozoic and Tertiary source rocks was induced by regional tectonic factors characterizing the different structural settings. The Late Neogene tectonism had a major role in the fold and thrust belt, both for the burial and maturation of source sediments under thick thrust sheets and for the development of hydrocarbon traps (Mattavelli and Novelli, 1990). However oil was generated during Jurassic-Palaeogene times from Late Triassic sources and could possibly have been preserved by early migration in traps at the top of the carbonate sequence (Casero et al., 1991). Heavy oils are prevalent in the foreland and foredeep domains, whereas light oils prevail in the thrust belt. Thermogenic gas was also generated during oil maturation (Mattavelli et al.1993). (Fig. 2)

4. Gas source rocks

Most of the Italian natural gases have been generated through bacterial fermentation and/or low temperature thermochemical reactions in immature Plio-Pleistocene sediments of the Apennine foredeep (Mattavelli and Novelli 1988). Bacterial gas is characterized by almost pure and isotopically light methane (Mattavelli et al., 1983). Its generation and accumulation is essentially favoured by high sedimentation rates, the deposition of alternating sands and shales, and synsedimentary tectonics, with the early genesis of structural traps (Pieri and Mattavelli 1986). The distribution of discovered original gas reserves is shown in Fig. 1. Thermogenic gas is confined to the thrust belt structural domain, whereas bacterial gas is distributed in the Pliocene-Pleistocene reservoirs of the external thrust belt and of the foredeep. The rapid burial and turbiditic sedimentation associated with very early compressional tectonics represented the ideal conditions for the formation and accumulation of biogenic gases. The peculiarity of the Apennine foredeep is the high percentage of biogenic gas which is great part of the total amount of hydrocarbons discovered over the past half a century in Italy (Mattavelli and Novelli, 1988). A lesser amount of gases was produced by thermal degradation of organic matter at great depths (in general >5000 m) either in the foredeep or in the thrust belts, where a considerable increase in temperature, caused by the emplacement of the thrust sheets, fostered the generation of thermogenic gases. Nevertheless, the tectonic movements, active during the entire Neogene, represent a limiting factor for the preservation of such generated gases. The small quantity (10%) and the uniform make-up of the gases (99% biogenic gases) discovered in the foreland are strictly related to the peculiar characteristics of this tectonic regime. In the immature Tertiary sediments the reduced thickness of the terrigenous deposits generated only a limited amount of bacterial and /or diagenetic gases. On the other hand in Mesozoic sediments, mainly formed by thick carbonate sequences, the possible present thermogenic gases were lost by diffusion through poorly efficient cap rocks (Mattavelli and Novelli, 1988; Buttinelli et al. 2011). Geochemical and geological evidences indicates that migration and accumulation of gaseous hydrocarbons took place mostly during the Plio-Pleistocene. In particular , migration is still active in the gas fields of the northern Apennine foredeep (Dal Piaz 1959), owing to the presence of thin impervious layers. In this area, in fact, a kind of steady state equilibrium has been reached between losses through diffusion and the continuous supply of newly generated natural gases.

5. Types of gaseous hydrocarbons

Biogenic, mixed, and thermogenic gases were found in Italy. Biogenic gases are usually found in Plio-Pleistocene sediments and are characterized by almost pure and isotopically light methane. The more negative isotopic values accompanied by the absence of heavier homologues were observed in Pleistocene shallow reservoirs. The chemical and isotopic composition of these gases is considered as evidence for their *in situ* formation through bacterial or diagenetic processes (Schoell 1980, 1983; Mattavelli, et al. 1983). Mixed gases were discovered in reservoirs from Middle Pliocene to Cretaceous and are characterized by a wide range of mixing proportions between biogenic and thermogenic gases. Thermogenic gases migrated from deeper layers and mixed in different amounts with shallower biogenic gases. Thermogenic gases are generally found in the pre-Pliocene reservoirs they are characterized by $^{13}C/^{12}C$ values ranging from -31 to -51‰. Italian thermogenic gases reflect all stages of maturation of organic matter. Condensate and dry thermogenic gases are enriched in heavy carbon ($^{13}C/^{12}C$ -31 to -36‰) and Deuterium which indicate a generation from highly mature source rocks (Mattavelli and Novelli, 1988).

6. Hydrocarbon accumulations

Biogenic gas pools were found in shallow marine sands and foredeep turbiditic multi layer sands and sandstones involved in thrust folds and their source is in the interbedded clays. Thermogenic gas pools are in turbiditic sandstones involved in thrust folds in foothill areas. The gas generated at great depth in the flysch, migrated laterally-updip along the inner flank of the folds. Liquid hydrocarbons are reservoired in carbonatic series in foothills and foreland domains. In the foothills belts ther traps are thrust folds. In the foreland the oils are stored in carbonates involved in paleostructures of different nature (Casero 2004; Bertello et al. 2010;). Some biogenic gases originated at a very shallow depth (i.e. less than 100 m) mask the exact localization of deep reservoirs and justify the need of geophysical and geochemical prospection to better constrain deep gas accumulations. Sometimes a mixing between very shallow biogenic gas and deep originated methane occurs generating mixing phenomena (Cremonini et al. 2008). In the subaerial environment the same biogenic gases generate and/or use shallow/surficial systems of fractures and faults as escape paths (Bonori et al. 2000; Castellarin et al. 2006; Cremonini et al. 2010; Cremonini 2010a) and some authors suggested also the possibility of identifying subaerial pockmarks (Curzi et al. 1987; Marabini et al. 1987; Cremonini 2010b). Usual pockmark morphologies are known to exist on the

Adriatic Sea bottom along the Meso Adriatic Depression (Curzi and Veggiani 1985; Curzi et al. 1987; Praeg et al. 2009; Geletti et al. 2008;Mazzotti et al.,1987; Trincardi et al. 2011c), on the Sardinian continental shelf (Dalla Valle and Gamberi 2011) and at the bottom of Lake Garda in northern Italy (Violante and Michetti 2010). Unfortunately, for all of those features no data concerning the seeping gas are available. Other well known morphologies linked to sure shallow gas seepage on the Northern Adriatic sea floor are generating small carbonatic mounds and layers (Conti et al. 2002; Panieri 2006), but also no analytical data are available for these.

7. Comments on database and maps

Historical scientific literature (Camerana et al. 1926; Zuber 1940; Martinis 1969, Fig. 8 and previous references therein; Martinelli 2007) have also reported the location of natural hydrocarbon occurrences not always considered in the recent scientific literature (Martinelli and Judd 2004). In any case, all the available sampling points or historically recognized points have been mapped and shown in figures 3a-d. The location data related the previous figures are recorded in Tables 1 to 4 . Due to the fact that the geographical location of the majority of the considered points cannot be gleaned from any other original edited source, the related coordinates were graphically extrapolated and as a consequence each georeferenced location must be understood as the barycentre point of a circular area in a possible location having a radius of up to 5 km in length. The related municipality quoted in the Tables 1 to 4 is the biggest one existing near the location point. Hence, they will be merely indicative even if significant on the scale of the present study. Seepages occurring in Italy are represented by : i) dry gas emissions; ii) gas bubbling in mud-volcanic waters; iii) gas bubbling in ground waters; iv) oil spills; v) asphalts and bitumen; vi) solid waxes. Figure 3 collects the related locations subdivided into four subsets, i.e. gas (Fig. 3a), Oil (Fig. 3b), Solid (Fig. 3c) and Mud-Volcanoes (Fig. 3d). The whole seepage set vs. the structural map of Italy is provided in Fig. 4. The available updated natural gas analyses are collected in Table 5 and their location and kind are shown in Fig. 5. When more than one site reported in the scientific literature was found within the same municipality then the most significant and representative of them or the main centre itself was selected as being representative. The analytical data were kept by the reference sources quoted in Table 5 (Borgia et al. 1988; Minissale et al. 2000; Duchi et al. 2005; Etiope 2007; Heinicke et al. 2010). In some cases, the original analytical strings have been completed by means of data published by the authors referenced in the reported list. Analytical data usually refer to dry gases and to the gases bubbling in mud volcanoes. Some low-depth wells (<200m) drilled close to natural gaseous emissions have been considered as well as some wells characterized by the certain representativeness of local seepages. All the sampling points have been georeferenced. Analytical data have been plotted and are shown in Figure 6. The graphs obtained indicate that only a minority of considered gases has biogenic origin, while all the others are thermogenic or mixed thermo-biogenic. Due to fact that analytical data obtained from the gases sampled in deep industrial wells highlight the same proportions of biogenic and thermogenic gases, we can conclude that the surface seepages are representative of a deep hydrocarbon setting and, in principle, could be still exploited as indicators of deep-seated reservoirs.

No.	Region	Province	Municipality	Place name	φ WGS 84	λ WGS 84	Reference
1	Piemonte	Alessandria	Casale Monferrato		45 08 14	08 27 03	Camerana et al., 1926
2	Piemonte	Alessandria	**Gabiano		*45 09 38	*08 11 44	Martinis 1969, Fig. 8
3	Piemonte	Alessandria	**Casale Monferrato		*45 09 52	*08 23 55	Martinis 1969, Fig. 8
4	Lombardia	Pavia	Casteggio	Casteggio	45 00 43	09 07 24	Camerana et al., 1926
5	Lombardia	Pavia	Salice Terme	Salice Terme	44 54 51	09 01 38	Camerana et al., 1926
6	Lombardia	Pavia	Rivanazzano Terme	Rile dell'olio	44 55 37	09 00 51	Camerana et al., 1926
7	Lombardia	Pavia	**Rocca Susella		*44 55 25	*09 06 49	Martinis 1969, Fig. 8
8	Trentino-A.A.	Trento	**Brentonico	Castione	*45 48 46	*10 57 07	Martinis 1969, Fig. 8
9	Trentino-A.A.	Trento	**Ala	S. Margherita	*45 44 48	*11 04 16	Martinis 1969, Fig. 8
10	Veneto	Belluno	**Feltre		*45 58 29	*11 49 15	Martinis 1969, Fig. 8
11	Friuli V. G.	Udine	**Ovaro		*46 28 34	*12 52 53	Martinis 1969, Fig. 8
12	Emilia Romagna	Bologna	Casalfiumanese	Cà Bordona	44 17 55	11 37 16	Martinelli 2007, tab 6.1
13	Emilia Romagna	Bologna	Castel del Rio	Molinaccio	44 12 56	11 30 14	Martinelli 2007, tab 6.1
14	Emilia Romagna	Bologna	Castiglione dei Pepoli	Creda	44 08 35	11 09 51	Martinelli 2007, tab 6.1
15	Emilia Romagna	Bologna	Gaggio Montano	**5** sites: Rovine Tommasi, Saldine, Cà Masera, Molinazzo,Cà di Riccio	44 11 53	10 56 01	Martinelli 2007, tab 6.1
16	Emilia Romagna	Bologna	Grizzana Morandi	Cà Bellavista	44 15 28	11 09 08	Martinelli 2007, tab 6.1
17	Emilia Romagna	Bologna	Lizzano in Belvedere	Grecchia	44 09 42	10 53 38	Martinelli 2007, tab 6.1
18	Emilia Romagna	Bologna	Monterenzio	Casa Domenicali	44 19 31	11 24 16	Martinelli 2007, tab 6.1
19	Emilia Romagna	Bologna	Porretta Terme	**2** sites: Cà Salgastri, Sasso Cardo.	44 09 15	10 58 32	Martinelli 2007, tab 6.1
20	Emilia Romagna	Bologna	S. Benedetto Val di Sambro	Castel dell'Alpi	44 12 56	11 14 04	Martinelli 2007, tab 6.1
21	Emilia Romagna	Bologna	Sasso Marconi	Reno river bed	44 23 44	11 14 53	Martinelli 2007, tab 6.1
22	Emilia Romagna	Bologna	Savigno	Monte Falò	44 23 27	11 04 29	Martinelli 2007, tab 6.1
23	Emilia Romagna	Ferrara	Cento	Corpo Reno - Casa "Il Gas"	44 45 22	11 18 08	Martinelli 2007, tab 6.1
24	Emilia Romagna	Ferrara	Comacchio	Valli del Mezzano	44 40 37	12 01 44	Cremonini et al. 2008
25	Emilia Romagna	Forlì-Cesena	Bagno di Romagna	**3** sites: Terme, Cimitero, S. Martino di Larciano	43 50 02	11 57 33	Martinelli 2007, tab 6.1
26	Emilia Romagna	Forlì-Cesena	Bertinoro	Rio Salso	44 08 56	12 07 58	Martinelli 2007, tab 6.1
27	Emilia Romagna	Forlì-Cesena	Castrocaro Terme	Bollirone	44 10 20	11 56 51	Martinelli 2007, tab 6.1
28	Emilia Romagna	Forlì-Cesena	Galeata	**2** sites: Casa Tolice, Rio Suasia	43 59 48	11 54 43	Martinelli 2007, tab 6.1
29	Emilia Romagna	Forlì Cesena	Portico e S. Benedetto	Portico di Romagna	44 01 31	11 46 57	Martinis 1969, Fig. 8

No.	Region	Province	Municipality	Place name	φ WGS 84	λ WGS 84	Reference
30	Emilia Romagna	Forlì-Cesena	Rocca San Cascaino	**3** sites: Case Budria, Casalecchio, Fosso di Rinaldo,	44 03 30	11 50 30	Martinelli 2007, tab 6.1
31	Emilia Romagna	Forlì-Cesena	Sogliano al Rubicone	Torrente Torchio	44 00 18	12 18 01	Martinelli 2007, tab 6.1
32	Emilia Romagna	Forlì-Cesena	Tredozio	Busca	44 04 41	11 44 35	Martinelli 2007, tab 6.1
33	Emilia Romagna	Modena	Castelvetro di Modena	S. Luigi	44 30 12	10 56 35	Martinelli 2007, tab 6.1
34	Emilia Romagna	Modena	Fanano	Chiesa di Trignano	44 12 39	10 50 29	Martinelli 2007, tab 6.1
35	Emilia Romagna	Modena	Lama Mocogno	**3** sites: Barigazzo, Lagadelle, Case di Sotto,	44 18 25	10 43 47	Martinelli 2007, tab 6.1
36	Emilia Romagna	Modena	Maranello	**2** sites: Torre Maina, La Govana	44 31 31	10 51 59	Martinelli 2007, tab 6.1
37	Emilia Romagna	Modena	Marano sul Panaro	Prediera	44 27 21	10 57 58	Martinelli 2007, tab 6.1
38	Emilia Romagna	Modena	Montefiorino	**4** sites: Macognano, Farneta, Il Fuoco, Cà Medole	44 21 31	10 37 24	Martinelli 2007, tab 6.1
39	Emilia Romagna	Modena	Montese	Cà Boschi	44 16 06	10 56 27	Martinelli 2007, tab 6.1
40	Emilia Romagna	Modena	Palagano	Casa Bottega	44 19 14	10 38 52	Martinelli 2007, tab 6.1
41	Emilia Romagna	Modena	San Possidonio	Fondo Bordina	44 53 30	10 59 45	Martinelli 2007, tab 6.1
42	Emilia Romagna	Modena	Medolla		44 50 55	11 04 14	Gasperi, Pellegrini 1981
43	Emilia Romagna	Modena	Sassuolo	**3** sites: Gozzano, Salsa di sotto, Salvarola	44 32 30	10 46 54	Martinelli 2007, tab 6.1
44	Emilia Romagna	Modena	Serramazzoni	**2** sites: Pozzi dell'olio. Campodolio	44 25 33	10 47 15	Martinelli 2007, tab 6.1
45	Emilia Romagna	Modena	Sestola	**5** sites: Troncoscaglia, Bandita, Cà Boldrini, Fontanine, Trignano	44 13 50	10 46 14	Martinelli 2007, tab 6.1
46	Emilia Romagna	Parma	Bardi	**3** sites: Ormei, Volpi, Tosca	44 37 55	09 43 53	Martinelli 2007, tab 6.1
47	Emilia Romagna	Parma	Berceto	**7** sites: Scorza, Costa d'Asino, Molinari, Borgallo, Castellonchio, Macchie di Monte Marino, Lagodignano	44 30 38	09 59 22	Martinelli 2007, tab 6.1
48	Emilia Romagna	Parma	Berceto	Berceto	44 30 39	09 59 24	Martinis 1969, Fig. 8
49	Emilia Romagna	Parma	Berceto	Castellonchio	44 32 59	10 00 25	Martinis 1969, Fig. 8
50	Emilia Romagna	Parma	Collecchio	Cà Ginestra	44 45 06	10 12 54	Martinelli 2007, tab 6.1
51	Emilia Romagna	Parma	Corniglio	**3** sites: Miano, Prella, Rividulano	44 28 33	10 05 18	Martinelli 2007, tab 6.1
52	Emilia Romagna	Parma	Corniglio	Grammatica	44 26 21	10 05 15	Martinis 1969, Fig. 8
53	Emilia Romagna	Parma	Fornovo Taro	**4** sites: Riccò, Ozzano, Case Folli, Vallezza	44 41 29	10 05 49	Martinelli 2007, tab 6.1
54	Emilia Romagna	Parma	Fornovo Taro	Vallezza	44 39 36	10 09 14	Martinis 1969, Fig. 8

No.	Region	Province	Municipality	Place name	φ WGS 84	λ WGS 84	Reference
55	Emilia Romagna	Parma	Medesano	**4** sites: Miano, Casa Goletta, Casa Brozzi, Sant'Andrea Bagni	44 45 23	10 08 27	Martinelli 2007, tab 6.1
56	Emilia Romagna	Parma	Neviano degli Arduini	**2** sites: Case Cavandola, Villa	44 34 56	10 18 57	Martinelli 2007, tab 6.1
57	Emilia Romagna	Parma	Salsomaggiore	Centopozzi	44 48 58	09 58 43	Martinelli 2007, tab 6.1
58	Emilia Romagna	Parma	Valmozzola	Valmozzola	44 34 08	09 53 01	Martinis 1969, Fig. 8
59	Emilia Romagna	Piacenza	Agazzano	**2** sites: Casa Boriona, Cà Ragaiona	44 56 48	09 31 12	Martinelli 2007, tab 6.1
60	Emilia Romagna	Piacenza	Bobbio	**4** sites: Piancasale, Case Canneto, Ponte San Martino, S. Salvatore	44 46 02	09 23 12	Martinelli 2007, tab 6.1
61	Emilia Romagna	Piacenza	Castell'Arquato	Villa S. Lorenzo	44 51 04	09 52 02	Martinelli 2007, tab 6.1
62	Emilia Romagna	Piacenza	Farini d'Olmo	**3** sites: Troncamorso, Case Tornara, Case Chiappetti	44 42 49	09 34 11	Martinelli 2007, tab 6.1
63	Emilia Romagna	Piacenza	Gazzola	Casa Mirabello	44 57 32	09 32 50	Martinelli 2007, tab 6.1
64	Emilia Romagna	Piacenza	Lugagnano Val d'Arda	Velleia: Velleia	44 47 07	09 43 18	Martinelli 2007, tab 6.1
65	Emilia Romagna	Parma	Palanzano	Palanzano	44 26 08	10 11 32	Martinis 1969, Fig. 8
66	Emilia Romagna	Piacenza	Podenzano	Cà dei Gatti	44 57 24	09 41 08	Martinelli 2007, tab 6.1
67	Emilia Romagna	Parma	Travesetolo	Torre di Rivazzano	44 37 11	10 20 44	Martinis 1969, Fig. 8
68	Emilia Romagna	Piacenza	Travo	Campo dei Re (o Statto)	44 51 36	09 32 36	Martinelli 2007, tab 6.1
69	Emilia Romagna	Piacenza	Vigolzone	Carmiano	44 54 49	09 40 08	Martinelli 2007, tab 6.1
70	Emilia Romagna	Ravenna	Brisighella	**3** sites: Cà Domenico, Cà Poriva, Monticello	44 13 29	11 46 33	Martinelli 2007, tab 6.1
71	Emilia Romagna	Ravenna	Riolo Terme	Rio vecchio	44 16 31	11 43 19	Martinelli 2007, tab 6.1
72	Emilia Romagna	Reggio Emilia	Correggio	Correggio	44 46 16	10 46 50	Martinelli 2007, tab 6.1
73	Emilia Romagna	Reggio Emilia	Reggio Emilia	S. Bartolomeo	44 41 52	10 37 51	Martinelli 2007, tab 6.1
74	Emilia Romagna	Reggio Emilia	Toano	Quara	44 22 35	10 33 34	Martinelli 2007, tab 6.1
75	Emilia Romagna	Reggio Emilia	Vezzano sul Crostolo	**2** sites: La Vecchia, Casola Canossa	44 36 03	10 32 46	Martinelli 2007, tab 6.1
76	Emilia Romagna	Reggio Emilia	Viano	Fattoria del Lupo	44 32 37	10 37 09	Martinelli 2007, tab 6.1
77	Emilia Romagna	Reggio Emilia	Villa Minozzo	**2** sites: Casa Salata, Cà dell'Onestà	44 21 54	10 28 03	Martinelli 2007, tab 6.1
78	Emilia Romagna	Rimini	Viserba	seafloor	44 05 18	12 32 01	Martinelli 2007, tab 6.1
79	Toscana	Firenze	Firenzuola	Pietramala	44 07 14	11 22 49	Martinis 1969, Fig. 8
80	Toscana	Firenze	Montespertoli		43 38 36	11 04 28	Camerana et al., 1926, p 276
81	Toscana	Lucca	Viareggio	**Torre del Lago	*43 49 23	*10 20 36	Martinis 1969, Fig. 8
82	Toscana	Arezzo	**Pieve S. Stefano		*43 42 59	*12 03 59	Martinis 1969, Fig. 8
83	Toscana	Pisa	Pisa	Fornaci (e varie località)	43 40 49	10 20 28	Camerana et al., 1926, p 274

No.	Region	Province	Municipality	Place name	φ WGS 84	λ WGS 84	Reference
84	Toscana	Firenze	**S. Casciano Val di Pesa		*43 36 26	*11 09 21	Martinis 1969, Fig. 8
85	Toscana	Siena	**Poggibonsi		*43 30 00	*11 10 10	Martinis 1969, Fig. 8
86	Toscana	Livorno	**Collesalvetti		*43 30 41	*10 28 34	Martinis 1969, Fig. 8
87	Toscana	Pisa	Volterra		43 25 03	10 55 35	Martinis 1969, Fig. 8
88	Toscana	Siena	Siena	Montarioso	43 21 01	11 18 37	Camerana et al., 1926, p 276
89	Toscana	Siena	**Casole d'Elsa		*43 19 02	*11 02 51	Martinis 1969, Fig. 8
90	Toscana	Siena	**Monteroni d'Arbia		*43 15 38	*11 24 06	Martinis 1969, Fig. 8
91	Toscana	Livorno	**Suvereto		*43 04 55	*10 44 57	Martinis 1969, Fig. 8
92	Toscana	Grosseto	**Roccastrada		*43 02 56	*11 10 29	Martinis 1969, Fig. 8
93	Toscana	Grosseto	**Santa Fiora		*42 51 29	*11 34 39	Martinis 1969, Fig. 8
94	Toscana	Grosseto	Grosseto	Fondo Casone	42 49 17	11 05 58	Camerana et al., 1926, p 275
95	Toscana	Grosseto	Grosseto	Fondo Tripoli	42 42 59	11 05 14	Camerana et al., 1926, p 275
96	Toscana	Grosseto	**Manciano		*42 33 35	*11 28 40	Martinis 1969, Fig. 8
97	Emilia Romagna	Rimini	**Montefiore Conca		*43 54 06	*12 34 21	Martinis 1969, Fig. 8
98	Emilia Romagna	Rimini	**Novafeltria		*43 53 28	*12 17 35	Martinis 1969, Fig. 8
99	Marche	Ancona	**Agugliano		*43 32 50	*13 24 52	Martinis 1969, Fig. 8
100	Marche	Macerata	**Montecosaro		*43 19 26	*13 37 18	Martinis 1969, Fig. 8
101	Marche	Macerata	**Petriolo		*43 10 27	*13 22 25	Martinis 1969, Fig. 8
102	Marche	Fermo	**Lapedona		*43 06 05	*13 44 17	Martinis 1969, Fig. 8
103	Marche	Macerata	**Sernano		*43 01 50	*13 18 38	Martinis 1969, Fig. 8
104	Marche	Fermo	**Montefortino		*42 54 45	*13 17 38	Martinis 1969, Fig. 8
105	Marche	Ascoli Piceno	**Force		*42 55 57	*13 31 04	Martinis 1969, Fig. 8
106	Abruzzo	Teramo	**Civitella del Tronto	Valle Tronto a N Civitella	*42 47 45	*13 31 29	Martinis 1969, Fig. 8
107	Marche	Ascoli Piceno	Offida		42 56 13	13 42 01	Zuber 1940
108	Marche	Ascoli Piceno	near Maltignano	Villa Passo	42 49 57	13 41 13	Camerana et al., 1926
109	Umbria	Perugia	**Gubbio		*43 26 44	*12 33 16	Martinis 1969, Fig. 8
110	Umbria	Perugia	Perugia		43 08 50	12 23 59	Martinis 1969, Fig. 8
111	Umbria	Terni	**Montecastrilli		*42 39 49	*12 24 38	Martinis 1969, Fig. 8
112	Lazio	Rieti	**Montopoli Sabina		*42 13 14	*12 40 12	Martinis 1969, Fig. 8
113	Lazio	Roma	**Ostia		*41 44 57	*12 23 19	Martinis 1969, Fig. 8
114	Lazio	Frosinone	**Pontecorvo		*41 27 60	*13 35 28	Martinis 1969, Fig. 8
115	Abruzzo	Pescara	**Collecorvino		*42 28 13	*14 02 05	Martinis 1969, Fig. 8

No.	Region	Province	Municipality	Place name	φ WGS 84	λ WGS 84	Reference
116	Abruzzo	L'Aquila	**Ofena		*42 18 38	*13 46 57	Martinis 1969, Fig. 8
117	Abruzzo	Pescara	**Bussi sul Tirino		*42 14 34	*13 48 16	Martinis 1969, Fig. 8
118	Abruzzo	L'Aquila	Avezzano	Pozzone di Paterno -Fucino	42 03 06	13 28 51	Ciotoli et al. 1998
119	Abruzzo	L'Aquila	**Trasacco		*41 58 33	*13 34 25	Martinis 1969, Fig. 8
120	Abruzzo	L'Aquila	**Pescocostanzo		*41 53 01	*14 04 03	Martinis 1969, Fig. 8
121	Molise	Campobasso	**Mafalda		*41 58 05	*14 43 01	Martinis 1969, Fig. 8
122	Molise	Isernia	**Rocca Sicura		*41 41 15	*14 13 17	Martinis 1969, Fig. 8
123	Molise	Isernia	**Macchiagodena		*41 35 24	*14 22 56	Martinis 1969, Fig. 8
124	Molise	Campobasso	**S. Paolo Matese		*41 26 39	*14 30 42	Martinis 1969, Fig. 8
125	Molise	Campobasso	**Sepino		*41 22 23	*14 37 06	Martinis 1969, Fig. 8
126	Campania	Benevento	**Morcone		*41 19 51	*14 44 34	Martinis 1969, Fig. 8
127	Campania	Avellino	**Andretta		*40 57 26	*15 21 09	Martinis 1969, Fig. 8
128	Campania	Avellino	**Bagnoli Irpino		*40 49 04	*15 00 40	Martinis 1969, Fig. 8
129	Basilicata	Potenza	Tramutola	Cavolo and Agri rivers junction	40 20 53	15 47 05	Camerana et al., 1926
130	Basilicata	Matera	Nova Siri	Fontana di Sant'Alessio	40 08 57	16 32 29	Camerana et al., 1926
131	Basilicata	Potenza	Rapolla	Colle S. Lucia	40 58 37	15 40 19	Camerana et al., 1926
132	Basilicata	Matera	**Pisticci		*40 25 51	*16 31 31	Martinis 1969, Fig. 8
133	Basilicata	Matera	**Scanzano Ionico		*40 16 12	*16 40 27	Martinis 1969, Fig. 8
134	Basilicata	Matera	**Colobraro		*40 10 37	*16 22 21	Martinis 1969, Fig. 8
135	Calabria	Cosenza	**Montegiordano		*40 01 50	*16 30 43	Martinis 1969, Fig. 8
136	Calabria	Cosenza	**Corigliano Calabro		*39 41 04	*16 25 31	Martinis 1969, Fig. 8
137	Calabria	Cosenza	S. Vincenzo la Costa		39 21 53	16 09 03	Martinis 1969, Fig. 8
138	Calabria	Crotone	**Verzino		*39 20 16	*16 50 13	Martinis 1969, Fig. 8
139	Calabria	Crotone	**Cutro		*38 59 17	*16 59 07	Martinis 1969, Fig. 8
140	Calabria	Crotone	**Isola Capo Rizzuto		*38 59 39	*17 06 04	Martinis 1969, Fig. 8
141	Calabria	Reggio Calabria	**Rosarno		*38 30 28	*16 00 36	Martinis 1969, Fig. 8
142	Calabria	Reggio Calabria	**Staiti		*37 58 56	*16 02 52	Martinis 1969, Fig. 8
143	Sicilia	Messina	Mistretta	Castel di Lucio, Rocca Pizzutella	37 55 46	14 21 46	Camerana et al., 1926
144	Sicilia	Enna	Cerami	Monania	37 48 42	14 30 31	Camerana et al., 1926
145	Sicilia	Enna	Troina	verso Bronte, Pianezze	37 47 10	14 36 10	Camerana et al., 1926
146	Sicilia	Palermo	Caltavuturo	Pagliuzza	37 49 19	13 53 28	Camerana et al., 1926
147	Sicilia	Agrigento	Bivona	Censo, Casa il Censo	37 37 11	13 26 22	Camerana et al., 1926

No.	Region	Province	Municipality	Place name	φ WGS 84	λ WGS 84	Reference
148	Sicilia	Palermo	Bagni di Sclafani		37 49 18	13 51 17	Camerana et al., 1926
149	Sicilia	Enna	Piazza Armerina	Piazza Armerina	37 23 05	14 21 52	Camerana et al., 1926
150	Sicilia	Catania	**Bronte		*37 45 21	*14 53 50	Martinis 1969, Fig. 8
151	Sicilia	Enna	**Cerami		*37 46 18	*14 28 05	Martinis 1969, Fig. 8
152	Sicilia	Palermo	**Polizzi Generosa		*37 50 05	*13 59 04	Martinis 1969, Fig. 8
153	Sicilia	Trapani	**Santa Ninfa		*37 44 43	*12 51 05	Martinis 1969, Fig. 8
154	Sicilia	Agrigento	**Sant'Elisabetta		*37 27 21	*13 33 13	Martinis 1969, Fig. 8
155	Sicilia	Enna	**Valguarnera Caropepe		*37 28 13	*14 24 20	Martinis 1969, Fig. 8
156	Sicilia	Catania	**Raddusa		*37 29 56	*14 31 07	Martinis 1969, Fig. 8
157	Sicilia	Catania	**Paternò		*37 35 52	*14 53 35	Martinis 1969, Fig. 8
158	Sicilia	Catania	**Catania		*37 24 48	*15 00 42	Martinis 1969, Fig. 8
159	Sicilia	Siracusa	**Lentini		*37 25 35	*14 52 16	Martinis 1969, Fig. 8
160	Sicilia	Catania	**Palagonia		*37 17 36	*14 42 55	Martinis 1969, Fig. 8
161	Sicilia	Catania	**Vizzini		*37 09 60	*14 47 21	Martinis 1969, Fig. 8
162	Sicilia	Siracusa	**Noto		*36 55 32	*15 01 18	Martinis 1969, Fig. 8
163	Sicilia	Caltanissetta	**Caltanissetta		*37 23 04	*14 04 09	Martinis 1969, Fig. 8
164	Sicilia	Agrigento	**Racalmuto		*37 19 24	*13 47 30	Martinis 1969, Fig. 8
165	Sicilia	Agrigento	**Canicattì		*37 20 07	*13 54 42	Martinis 1969, Fig. 8
166	Sicilia	Agrigento	**Ravanusa		*37 16 52	*13 59 57	Martinis 1969, Fig. 8
167	Sicilia	Agrigento	**Campobello di Licata		*37 15 31	*13 51 53	Martinis 1969, Fig. 8
168	Sicilia	Agrigento	**Palma di Montechiaro		*37 12 49	*13 44 08	Martinis 1969, Fig. 8
169	Sardegna	Carbonia-Iglesias	**Iglesias		*39 20 36	*08 34 10	Martinis 1969, Fig. 8

** = the main municipality nearest to the point location (not necessarily coinciding with the real seep location Municipality).
* = approximate value obtained by means of original map georeferencing.

Table 1. List of natural GAS seepages in Italy.

order No.	Region	Province	Municipality	Place name	φ WGS 84	λ WGS 84	Note	Reference
1	Piemonte	Alessandria	**Cuccaro Monferrato		*45 00 28	*08 27 07		Martinis 1969, fig 8
2	Piemonte	Cuneo	**S. Vittoria d'Alba		*44 43 27	*07 55 23		Martinis 1969, fig 8
3	Lombardia	Pavia	**Zavattarello		*44 53 16	*09 16 47		Martinis 1969, fig 8
4	Trentino-A.A.	Trento	Taio	Mollaro	46 19 06	11 11 02		Martinis 1969, fig 8
5	Veneto	Belluno	**Forno di Zoldo		*46 21 44	*12 13 25		Martinis 1969, fig 8
6	Emilia Romagna	Piacenza	**Coli		*44 44 37	*09 25 25		Martinis 1969, fig 8
7	Emilia Romagna	Piacenza	**Morfasso		*44 43 21	*09 40 04		Martinis 1969, fig 8
8	Emilia Romagna	Piacenza	** Val d'Arda		*44 50 07	*09 48 37		Martinis 1969, fig 8
9	Emilia Romagna	Parma	**Bore		*44 43 17	*09 49 39		Martinis 1969, fig 8
10	Emilia Romagna	Parma	**Bardi		*44 37 41	*09 46 12		Martinis 1969, fig 8
11	Emilia Romagna	Reggio Emilia	**Castelnuovo Monti		*44 28 01	*10 28 44		Martinis 1969, fig 8
12	Emilia Romagna	Modena	**Maranello		*44 30 31	*10 50 08		Martinis 1969, fig 8
13	Emilia Romagna	Bologna	**Pianoro		*44 22 46	*11 18 54		Martinis 1969, fig 8
14	Emilia Romagna	Bologna	**S. Lazzaro di Savena	(S. Ruffillo)	*44 27 38	*11 26 06		Martinis 1969, fig 8
15	Emilia Romagna	Forlì-Cesena	**Forlì		*44 15 18	*12 06 25	(Not reliable)	Martinis 1969, fig 8
16	Toscana	Arezzo	**Pieve S. Stefano		*43 40 18	*12 11 15		Martinis 1969, fig 8
17	Toscana	Siena	**Pienza		*43 06 58	*11 38 14		Martinis 1969, fig 8
18	Marche	Macerata	**Sarnano		*43 00 10	*13 11 36		Martinis 1969, fig 8
19	Marche	Macerata	**Porto S. Elpidio	Fontespina	*43 17 33	*13 49 39	1.7km Offshore	Martinis 1969, fig 8

order No.	Region	Province	Municipality	Place name	φ WGS 84	λ WGS 84	Note	Reference
20	Lazio	Frosinone	**Monte S. Giovanni Campano		*41 37 28	*13 33 53		Martinis 1969, fig 8
21	Lazio	Frosinone	**Ceccano		*41 33 32	*13 18 45		Martinis 1969, fig 8
22	Lazio	Frosinone	**Amaseno		*41 27 22	*13 22 21		Martinis 1969, fig 8
23	Lazio	Frosinone	**Arce		*41 31 59	*13 35 06		Martinis 1969, fig 8
24	Abruzzo	Pescara	**Torre dei Passeri		*42 15 13	*13 54 02		Martinis 1969, fig 8
25	Abruzzo	Pescara	**Caramanico Terme		*42 10 52	*14 01 49		Martinis 1969, fig 8
26	Abruzzo	L'Aquila	**Sulmona		*42 01 07	*13 55 06		Martinis 1969, fig 8
27	Abruzzo	Chieti	**Scerni		*42 05 20	*14 32 32		Martinis 1969, fig 8
28	Campania	Avellino	**Guardia Lombardi		*41 00 30	*15 13 59		Martinis 1969, fig 8
29	Basilicata	Potenza	**Baragiano		*40 40 48	*15 37 28		Martinis 1969, fig 8
30	Basilicata	Potenza	**Marsico Nuovo		*40 26 45	*15 45 56		Martinis 1969, fig 8
31	Basilicata	Potenza	**S. Chirico Raparo		*40 12 00	*16 05 41		Martinis 1969, fig 8
32	Basilicata	Potenza	**Sant'Arcangelo		*40 13 58	*16 17 41		Martinis 1969, fig 8
33	Basilicata	Cosenza	**Alessandria del Carretto		*40 00 01	*16 23 38		Martinis 1969, fig 8
34	Calabria	Cosenza	**Campana		*39 24 26	*16 51 40		Martinis 1969, fig 8
35	Calabria	Crotone	**Belvedere di Spinello		*39 14 59	*16 53 19		Martinis 1969, fig 8
36	Sicilia	Enna	**Cerami		*37 50 56	*14 25 28		Martinis 1969, fig 8
37	Sicilia	Catania	**Bronte		*37 48 04	*14 47 40		Martinis 1969, fig 8
38	Sicilia	Palermo	**Petralia Sottana		*37 44 28	*14 08 17		Martinis 1969, fig 8
39	Sicilia	Caltanissetta	**Villalba		*37 39 53	*13 48 37		Martinis 1969, fig 8
40	Sicilia	Caltanissetta	**Serradifalco		*37 25 35	*13 51 29		Martinis 1969, fig 8
41	Sicilia	Catania	**Paternò		*37 29 12	*14 51 07		Martinis 1969, fig 8
42	Sicilia	Catania	**Vizzini		*37 08 03	*14 52 07		Martinis 1969, fig 8
43	Sicilia	Ragusa	**Ragusa		*37 00 43	*14 42 09		Martinis 1969, fig 8
44	Sicilia	Ragusa	**Modica		*36 55 10	*14 50 45		Martinis 1969, fig 8
45	Sicilia	Ragusa	**Ispica		*36 46 06	*14 57 10		Martinis 1969, fig 8

** = the main municipality nearest to the point location (not necessarily coinciding with the real seep location Municipality).
* = approximate value obtained by means of original map georeferencing.

Table 2. List of natural oil occurrence in Italy.

No.	Region	Province	Municipality	Place name	φ WGS 84	λ WGS 84	Reference
1	Piemonte	Cuneo	La Morra		44 38 23	07 56 01	Martinis 1969, fig 8
2	Piemonte	Cuneo	Bene Vagienna		44 32 44	07 49 59	Martinis 1969, fig 8
3	Trentino-A.A.	Trento	**Molina di Ledro		*45 50 44	*10 43 33	Martinis 1969, fig 8
4	Lombardia	Brescia	**Bedizzole		*45 32 03	*10 27 05	Martinis 1969, fig 8
5	Lombardia	Brescia	**Brescia		*45 33 04	*10 13 39	Martinis 1969, fig 8
6	Trentino-A.A.	Trento	**Pozza di Fassa		*46 26 02	*11 44 55	Martinis 1969, fig 8
7	Trentino-A.A.	Rovereto	**Pamarolo		*45 56 27	*11 00 58	Martinis 1969, fig 8
8	Trentino-A.A.	Trento	Taio	Mollaro	46 17 39	11 04 18	Martinis 1969, fig 8
9	Trentino-A.A.	Trento	**Pinzolo		*46 08 50	*10 50 34	Martinis 1969, fig 8
10	Veneto	Belluno	**Lozzo di Cadore		*46 29 47	*12 21 49	Martinis 1969, fig 8
11	Veneto	Belluno	**Valle Agordina		*46 18 10	*12 07 28	Martinis 1969, fig 8
12	Trentino-A.A.	Trento	**Fiera di Primiero		*46 13 51	*11 51 44	Martinis 1969, fig 8
13	Veneto	Treviso	**Crespano del Grappa		*45 53 19	*11 48 13	Martinis 1969, fig 8
14	Friuli V. G.	Udine	**Moggio Udinese		*46 27 26	*13 07 03	Martinis 1969, fig 8
15	Friuli V. G.	Udine	**Socchieve		*46 26 27	*12 45 50	Martinis 1969, fig 8
16	Friuli V. G.	Pordenone	**Clauzetto		*46 18 50	*12 55 14	Martinis 1969, fig 8
17	Friuli V. G.	Udine	**Reana del Roiale		*46 06 52	*13 17 15	Martinis 1969, fig 8
18	Friuli V. G.	Udine	Resiutta		46 23 33	13 13 07	Martinis 1969, fig 8
19	Friuli V. G.	Udine	**Tarcento		*46 12 53	*13 12 46	Martinis 1969, fig 8
20	Emilia-Romagna	Bologna	Savigno	M. Falò	44 23 27	11 04 29	Bombicci 1881
21	Toscana	Siena	**Colle Val d'Elsa		*43 22 52	*11 03 48	Martinis 1969, fig 8
22	Marche	Pesaro-Urbino	**Auditore		*43 48 45	*12 33 19	Martinis 1969, fig 8
23	Marche	Pesaro-Urbino	**Fermignano		*43 39 55	*12 40 41	Martinis 1969, fig 8
24	Marche	Ancona	**Genga		*43 26 03	*12 57 21	Martinis 1969, fig 8
25	Marche	Ancona	**Fabriano		*43 17 60	*12 52 40	Martinis 1969, fig 8
26	Marche	Macerata	**Pioraco		*43 12 33	*12 59 27	Martinis 1969, fig 8
27	Abruzzo	L'Aquila	**Avezzano		*42 00 13	*13 24 57	Martinis 1969, fig 8
28	Lazio	Roma	**Vallepietra		*41 56 07	*13 11 52	Martinis 1969, fig 8
29	Lazio	Frosinone	**Guarcino		*41 51 35	*13 19 37	Martinis 1969, fig 8
30	Abruzzo	L'Aquila	**Civitella Roveto		*41 56 14	*13 27 31	Martinis 1969, fig 8
31	Lazio	L'Aquila	**Villa Valle Roveto		*41 49 35	*13 34 37	Martinis 1969, fig 8

No.	Region	Province	Municipality	Place name	φ WGS 84	λ WGS 84	Reference
32	Lazio	Frosinone	**Veroli		*41 43 20	*13 31 20	Martinis 1969, fig 8
33	Lazio	Frosinone	**Sora		*41 44 05	*13 39 26	Martinis 1969, fig 8
34	Lazio	Frosinone	**Rocca d'Arce		*41 35 16	*13 38 19	Martinis 1969, fig 8
35	Lazio	Frosinone	**Pastena		*41 29 15	*13 29 25	Martinis 1969, fig 8
36	Lazio	Frosinone	**Castrocielo		*41 30 55	*13 40 50	Martinis 1969, fig 8
37	Lazio	Frosinone	**Pontecorvo		*41 24 15	*13 39 25	Martinis 1969, fig 8
38	Lazio	Rieti	**Amatrice		*42 38 01	*13 14 15	Martinis 1969, fig 8
39	Abruzzo	L'Aquila	**Scoppito		*42 32 01	*13 21 35	Martinis 1969, fig 8
40	Abruzzo	L'Aquila	**S. Stefano di Sessanio		*42 18 09	*13 39 09	Martinis 1969, fig 8
41	Abruzzo	L'Aquila	**Avezzano		*42 10 33	*13 23 17	Martinis 1969, fig 8
42	Abruzzo	L'Aquila	**Celano		*42 07 30	*13 28 47	Martinis 1969, fig 8
43	Abruzzo	L'Aquila	**Celano		*42 04 06	*13 34 14	Martinis 1969, fig 8
44	Abruzzo	L'Aquila	**Raiano		*42 06 03	*13 47 13	Martinis 1969, fig 8
45	Abruzzo	L'Aquila	**Pratola Peligna		*42 05 28	*13 52 52	Martinis 1969, fig 8
46	Abruzzo	Pescara	**Caramanico Terme		*42 06 49	*14 02 29	Martinis 1969, fig 8
47	Abruzzo	Chieti	**Fara San Martino		*42 04 32	*14 06 05	Martinis 1969, fig 8
48	Marche	Ascoli Piceno	**Acquasanta Terme		*42 43 48	*13 26 19	Martinis 1969, fig 8
49	Abruzzo	Teramo	**Isola del Gran Sasso		*42 31 13	*13 32 46	Martinis 1969, fig 8
50	Abruzzo	L'Aquila	**Collelongo		*41 52 27	*13 33 41	Martinis 1969, fig 8
51	Molise	Isernia	**S.Angelo del Pesco		*41 52 13	*14 15 15	Martinis 1969, fig 8
52	Molise	Isernia	**Isernia		*41 42 21	*14 18 58	Martinis 1969, fig 8
53	Molise	Campobasso	**Boiano		*41 33 03	*14 28 34	Martinis 1969, fig 8
54	Campania	Avellino	**Savignano Irpino		*41 12 49	*15 13 54	Martinis 1969, fig 8
55	Campania	Avellino	**Cardito		*41 08 41	*15 00 59	Martinis 1969, fig 8
56	Puglia	Foggia	**Anzano di Puglia		*41 07 10	*15 17 53	Martinis 1969, fig 8
57	Campania	Avellino	**Caposele		*40 48 31	*15 16 38	Martinis 1969, fig 8
58	Campania	Salerno	**Acerno		*40 44 58	*15 03 12	Martinis 1969, fig 8
59	Campania	Salerno	**Colliano		*40 42 48	*15 16 58	Martinis 1969, fig 8
60	Campania	Salerno	**Castelcivita		*40 27 55	*15 14 27	Martinis 1969, fig 8
61	Campania	Salerno	**Bellosguardo		*40 24 38	*15 19 08	Martinis 1969, fig 8

No.	Region	Province	Municipality	Place name	φ WGS 84	λ WGS 84	Reference
62	Campania	Salerno	**Sacco		*40 21 20	*15 26 38	Martinis 1969, fig 8
63	Campania	Salerno	**Trentinara		*40 22 19	*15 06 18	Martinis 1969, fig 8
64	Campania	Salerno	**Stio		*40 18 14	*15 14 36	Martinis 1969, fig 8
65	Campania	Salerno	**Vallo della Lucania		*40 15 25	*15 21 57	Martinis 1969, fig 8
66	Campania	Salerno	**Castiglione dei Genovesi		*40 43 38	*14 51 44	Martinis 1969, fig 8
67	Campania	Salerno	**Sanza		*40 11 44	*15 37 21	Martinis 1969, fig 8
68	Basilicata	Potenza	**Maratea		*40 03 11	*15 39 09	Martinis 1969, fig 8
69	Basilicata	Potenza	**Castelsaraceno		*40 10 34	*15 57 11	Martinis 1969, fig 8
70	Basilicata	Matera	**Rotondella		*40 08 23	*16 28 56	Martinis 1969, fig 8
71	Sicilia	Messina	**Tripi		*38 01 24	*15 04 40	Martinis 1969, fig 8
72	Sicilia	Messina	**Montalbano Elicona		*37 59 16	*14 57 30	Martinis 1969, fig 8
73	Sicilia	Catania	**Maletto		*37 49 40	*14 53 35	Martinis 1969, fig 8
74	Sicilia	Palermo	**Petralia Sottana		*37 47 36	*14 06 14	Martinis 1969, fig 8
75	Sicilia	Palermo	**Sclafani Bagni		*37 52 32	*13 50 50	Martinis 1969, fig 8
76	Sicilia	Palermo	**Montemaggiore Belsito		*37 51 45	*13 43 48	Martinis 1969, fig 8
77	Sicilia	Palermo	**Vicari		*37 50 34	*13 28 26	Martinis 1969, fig 8
78	Sicilia	Palermo	**Corleone		*37 51 25	*13 20 10	Martinis 1969, fig 8
79	Sicilia	Palermo	**Camporeale		*37 54 14	*13 01 12	Martinis 1969, fig 8
80	Sicilia	Trapani	**Alcamo		*37 55 49	*12 56 20	Martinis 1969, fig 8
81	Sicilia	Palermo	**Palazzo Adriano		*37 42 33	*13 23 14	Martinis 1969, fig 8
82	Sicilia	Agrigento	**Bivona		*37 35 24	*13 33 21	Martinis 1969, fig 8
83	Sicilia	Catania	**Palagonia		*37 19 34	*14 38 16	Martinis 1969, fig 8
84	Sicilia	Catania	**Licodia Eubea		*37 11 32	*14 42 20	Martinis 1969, fig 8
85	Sicilia	Siracusa	**Sortino		*37 09 47	*15 03 09	Martinis 1969, fig 8
86	Sicilia	Ragusa	**Giarratana		*37 00 32	*14 48 24	Martinis 1969, fig 8
87	Sicilia	Ragusa	**Ragusa		*36 54 56	*14 46 25	Martinis 1969, fig 8
88	Sicilia	Ragusa	**Scicli		*36 48 54	*14 42 12	Martinis 1969, fig 8
89	Sicilia	Ragusa	**Modica		*36 48 47	*14 48 24	Martinis 1969, fig 8
90	Sicilia	Siracusa	**Pachino		*36 42 28	*15 04 05	Martinis 1969, fig 8
91	Lombardia	Varese	Besano	Besano	45 53 23	08 53 24	Martinis 1969, fig 8

Table 3. List of solid hydrocarbon occurrence in Italy.

No	Region	Province	Municipality	Place name	φ WGS 84	λ WGS 84	Note	Reference
1	Emilia Romagna	Bologna	Casalfiumanese	Casa Bubano	44 15 05	11 28 35		Martinelli, Judd 2004, Table 1
2	Emilia Romagna	Bologna	Casalfiumanese	Casa Campagnola	44 20 32	11 35 20		Martinelli, Judd 2004, Table 1
3	Emilia Romagna	Bologna	Casalfiumanese	Case Nuove di Rifiano	44 19 23	11 34 20		Martinelli, Judd 2004, Table 1
4	Emilia Romagna	Bologna	Castel S.Pietro Terme	San Martino in Pedriolo	44 21 13	11 34 22		Martinelli, Judd 2004, Table 1
5	Emilia Romagna	Bologna	Imola	Bergullo	44 18 32	11 44 14		Martinelli, Judd 2004, Table 1
6	Emilia Romagna	Bologna	Imola	Campo di Fondo	44 21 12	11 42 50		Martinelli, Judd 2004, Table 1
7	Emilia Romagna	Bologna	Monterenzio	Mercatale (Dragone Ardito Desio)	44 23 01	11 26 18		Cantelli 1994
8	Emilia Romagna	Bologna	Monterenzio	San Clemente (or Dragone, or Sassuno)	44 20 09	11 27 18		Martinelli, Judd 2004, Table 1
9	Emilia Romagna	Bologna	Ozzano Emilia	Montebugnolo	44 26 38	11 28 25		Martinelli, Judd 2004, Table 1
10	Emilia Romagna	Modena	Fiorano Modenese	Salsa (di Monte Ave) di Fiorano	44 31 46	10 48 33		Camerana, Galdi 1911
11	Emilia Romagna	Modena	Fiorano Modenese	Nirano	44 30 48	10 49 25		Martinelli, Judd 2004, Table 1
12	Emilia Romagna	Modena	Maranello	Puianello	44 28 36	10 52 00		Martinelli, Judd 2004, Table 1
13	Emilia Romagna	Modena	Marano sul Panaro	Ospitaletto	44 26 11	10 52 54		Martinelli, Judd 2004, Table 1
14	Emilia Romagna	Modena	Polinago	Canalina	44 24 49	10 43 42		Martinelli, Judd 2004, Table 1
15	Emilia Romagna	Modena	Sassuolo	La Rovina di Montegibbio (S. del Rio dei Bagni 1)	44 30 45	10 47 19		Camerana, Galdi 1911
16	Emilia Romagna	Modena	Sassuolo	La Rovina di Montegibbio (S. del Rio dei Bagni 2)	44 30 41	10 47 48		Camerana, Galdi 1911
17	Emilia Romagna	Modena	Sassuolo	La Rovina di Montegibbio (S. "dei Cinghiali")	44 30 58	10 47 59		//

No	Region	Province	Municipality	Place name	φ WGS 84	λ WGS 84	Note	Reference
18	Emilia Romagna	Modena	Sassuolo	La Rovina di Montegibbio (archaeological excavation)	44 30 47	10 47 08		Borgatti et al. 2010
19	Emilia Romagna	Modena	Sassuolo	Montegibbio (S. di sotto)	44 30 55	10 46 39		Martinelli, Judd 2004, Table 1
20	Emilia Romagna	Modena	Sassuolo	Montegibbio (Salsa storica o grande)	44 31 07	10 46 45		Camerana et al., 1926
21	Emilia Romagna	Modena	Sassuolo	Montegibbio (S. di sopra)	44 31 15	10 46 43		Camerana et al., 1926
22	Emilia Romagna	Modena	Serra Mazzoni	Centora-Montardone	44 28 07	10 47 42		Martinelli, Judd 2004, Table 1
23	Emilia Romagna	Parma	Lesignano Bagni	Rivalta	44 37 45	10 19 34		Martinelli, Judd 2004, Table 1
24	Emilia Romagna	Parma	Traversetolo	Torre	44 37 13	10 20 19		Martinelli, Judd 2004, Table 1
25	Emilia Romagna	Reggio Emilia	Viano	Casola - Querciola	44 45 44	10 31 38		Martinelli, Judd 2004, Table 1
26	Emilia Romagna	Reggio Emilia	Viano	Regnano	44 33 27	10 34 34		Martinelli, Judd 2004, Table 1
27	Marche	Ancona	Ancona	Serra dè Conti	43 32 33	13 02 12		Martinelli, Judd 2004, Table 1
28	Marche	Ancona	Ancona	Aspio di Ancona	43 32 00	13 30 04		Martinelli, Judd 2004, Table 1
29	Marche	Ancona	Maiolati Spontini	Moie	43 30 10	13 07 48		Martinelli, Judd 2004, Table 1
30	Marche	Ancona	Maiolati Spontini	Contrada Calapigna	43 28 34	13 07 13		Martinelli, Judd 2004, Table 1
31	Marche	Ancona	Monte Roberto	Monte Roberto	43 28 50	13 08 18		Martinelli, Judd 2004, Table 1
32	Marche	Ancona	Osimo	Santo Stefano	43 30 30	13 27 40		Martinelli, Judd 2004, Table 1
33	Marche	Ancona	San Paolo di Jesi	Battinebbia	43 27 27	13 10 03		Martinelli, Judd 2004, Table 1
34	Marche	Ancona	San Paolo di Jesi	Bagno	43 27 14	13 10 26		Martinelli, Judd 2004, Table 1

No	Region	Province	Municipality	Place name	φ WGS 84	λ WGS 84	Note	Reference
35	Marche	Ascoli Piceno	Fermo	Capodarco	43 11 19	13 45 41		Martinelli, Judd 2004, Table 1
36	Marche	Ascoli Piceno	Monte Rinaldo	Contrada Crocchia	43 01 39	13 34 47		Martinelli, Judd 2004, Table 1
37	Marche	Ascoli Piceno	Offida	Offida	42 56 06	13 41 26		Martinelli, Judd 2004, Table 1
38	Marche	Ascoli Piceno	Rotella	Madonna di Montemisio	42 57 14	13 33 38		Martinelli, Judd 2004, Table 1
39	Marche	Ascoli Piceno	Rotella	Contrada Osteria	42 56 47	13 32 33		Martinelli, Judd 2004, Table 1
40	Marche	Ascoli Piceno	Senigallia	Vallone	43 08 18	13 43 22		Martinelli, Judd 2004, Table 1
41	Marche	Macerata	Macerata	Mogliano	43 11 07	13 28 45		Martinelli, Judd 2004, Table 1
42	Marche	Pesaro-Urbino	Isola del Piano	Isola del Piano	43 44 11	12 46 57		Martinelli, Judd 2004, Table 1
43	Marche	Pesaro-Urbino	Petriano	Petriano	43 46 47	12 44 02		Martinelli, Judd 2004, Table 1
44	Marche	Pesaro-Urbino	Saltara	Saltara	43 45 12	12 53 50		Martinelli, Judd 2004, Table 1
45	Abruzzo	Chieti	Frisa	Frisa	42 15 42	14 22 03		Martinelli, Judd 2004, Table 1
46	Abruzzo	Chieti	Poggiofiorito	Poggiofiorito	42 15 19	14 19 24		Martinelli, Judd 2004, Table 1
47	Abruzzo	Pescara	Penne	Picciano	42 28 26	13 59 27		Martinelli, Judd 2004, Table 1
48	Abruzzo	Teramo	Bisenti	Chiovano	42 32 07	13 40 16		Martinelli, Judd 2004, Table 1
49	Abruzzo	Teramo	Cellino Attanasio	Astelina	42 35 08	13 51 34		Martinelli, Judd 2004, Table 1
50	Abruzzo	Teramo	Cellino Attanasio	Pian Palazzo	42 35 08	13 51 34		Martinelli, Judd 2004, Table 1
51	Abruzzo	Teramo	Pineto	Pineto	42 36 29	14 04 02		Martinelli, Judd 2004, Table 1

No	Region	Province	Municipality	Place name	φ WGS 84	λ WGS 84	Note	Reference
52	Abruzzo	Teramo	Torano Nuovo	Frola	42 39 31	13 42 14		Martinelli, Judd 2004, Table 1
53	Campania	Benevento	Castelfranco in Misciano	Malvizza	41 17 49	15 05 06		Martinelli, Judd 2004, Table 1
54	Basilicata	Potenza	Cancellara	Contrada Bòfete	40 43 51	15 55 23		Martinelli, Judd 2004, Table 1
55	Calabria	Cosenza	San Vincenzo la Costa	San Sisti	39 21 50	16 09 04		Martinelli, Judd 2004, Table 1
56	Calabria	Reggio Calabria	Palizzi	Rocchette	37 55 09	15 59 11		Martinelli, Judd 2004, Table 1
57	Sicilia	Agrigento	Aragona	Zorba	37 23 32	13 37 26		Martinelli, Judd 2004, Table 1
58	Sicilia	Agrigento	Cammarata	Cammarata	37 37 57	13 38 13		Martinelli, Judd 2004, Table 1
59	Sicilia	Agrigento	Casteltermini	Casteltermini	37 32 24	13 38 42		Martinelli, Judd 2004, Table 1
60	Sicilia	Agrigento	Cattolica Eraclea	Bissana	37 26 20	13 23 42		Martinelli, Judd 2004, Table 1
61	Sicilia	Caltanissetta	Caltanissetta	Xirbi	37 29 25	14 03 24		Martinelli, Judd 2004, Table 1
62	Sicilia	Catania	Paternò	Simeto	37 33 57	14 54 06		Martinelli, Judd 2004, Table 1
63	Sicilia	Catania	Paternò	Stadio	37 33 50	14 54 11		Martinelli, Judd 2004, Table 1
64	Sicilia	Catania	Paternò	Vallone Salato	37 33 47	14 54 15		Martinelli, Judd 2004, Table 1
65	Sicilia	Enna	Aidone	Aidone	37 24 52	14 26 47		Martinelli, Judd 2004, Table 1
66	Sicilia	Enna	Valguarnera Caropepe	Valguarnera Caropepe	37 29 42	14 23 20		Martinelli, Judd 2004, Table 1
67	Sicilia	Enna	Villarosa	Villarosa	37 35 08	14 10 24		Martinelli, Judd 2004, Table 1
68	Sicilia	Palermo	Lercara Friddi	Lercara Friddi	37 44 51	13 36 12		Martinelli, Judd 2004, Table 1

No	Region	Province	Municipality	Place name	φ WGS 84	λ WGS 84	Note	Reference
69	Sicilia	Palermo	Palazzo Adriano	Palazzo Adriano	37 40 52	13 22 44		Martinelli, Judd 2004, Table 1
70	Puglia	Ionian Sea	Apulian plate	offshore	39 31 05	18 33 02	inferred location	Fusi et al. 2006
71	Puglia	Ionian Sea	Apulian plate	offshore	39 29 07	18 27 46	inferred location	Fusi et al. 2006
72	Puglia	Ionian Sea	Apulian plate	offshore	39 22 05	18 18 05	inferred location	Fusi et al. 2006
73	Calabria	Ionian Sea	Calabrian Outer Arc	offshore	38 49 25	17 23 20		Fusi et al. 2006
74	Calabria	Ionian Sea	Calabrian Outer Arc	offshore: *Pythagoras*	37 48 20	17 16 20		Praeg et al. 2009
75	Calabria	Ionian Sea	Calabrian Outer Arc	offshore: *Madonna dello Ionio*	38 12 00	16 56 00		Praeg et al. 2009
76	Sicilia	Canale di Sicilia	Malta Plateau	offshore	36 36 25	14 37 37	inferred location	Holland et al. 2003
77	Marche	Adriatic Sea		offshore	43 56 08	13 41 54		Camerlenghi, Pini 2009, Fig.4
78	Marche	Adriatic Sea		offshore (*Bonaccia Field*)	43 28 51	14 22 06		Curzi et al. 1998, Fig.1
79	Lombardia	Mantova	Poggio Rusco	Corte Vulcanello	44 56 57	11 09 17	Toponym	Castellarin et al 2006; Cremonini 2010
80	Emilia Romagna	Modena	Finale Emilia	Bollitora (Reno Finalese)	44 50 23	11 22 53	Toponym	Cremonini 2010; Cremonini et al. 2010
81	Emilia Romagna	Modena	Sassuolo	Sarzola	44 30 47	10 48 29	Toponym	//
82	Piemonte	Torino	Verrua Savoia	Verrua Savoia	45 09 42	08 07 16	Miocene fossil m.v.	Clari et al. 2004

Table 4. List of mud volcanoes known in Italy.

No.	Table reference	Region	Province	Municipality	Place name	References
//	//	//	//	//	Atmosphere	
1	//	Emilia Romagna	Bologna	Castel di Casio	Gaggiola	Borgia et al. 1988, Tab. 1
2	G15	Emilia Romagna	Bologna	Gaggio Montano	Gaggio Montano	Duchi et al. 2005 , Tab 5
3	G 15	Emilia Romagna	Bologna	Gaggio Montano	Molinazzo	Borgia et al. 1988, Tab. 1
4	G 16	Emilia Romagna	Bologna	Grizzana M.	Ca Bellavista	Borgia et al. 1988, Tab. 1
5	V 5	Emilia Romagna	Bologna	Imola	Bergullo	Etiope et al. 2007, Tab.
6	V 8	Emilia Romagna	Bologna	Monterenzio	Drag. Sassuno = S. Clemente	Etiope et al. 2007, Tab.
7	G 19	Emilia Romagna	Bologna	Porretta Terme	Cà Salgastri	Borgia et al. 1988, Tab. 1
8	G 19	Emilia Romagna	Bologna	Porretta	Porretta	Borgia et al. 1988, Tab. 1; *Minissale et al. 2000, Tab. 2, 3
9	G 20	Emilia Romagna	Bologna	S. Benedetto V.Sambro	Castel dell'Alpi	Duchi et al. 2005 , Tab 5; *Borgia et al. 1988, Tab. 1
10	G 23	Emilia Romagna	Ferrara	Cento	Corporeno	Etiope et al. 2007, Tab.
11	G 24	Emilia Romagna	Ferrara	Comacchio	Valli Mezzano	Cremonini et al. 2008, Tab. 1
12	G 25	Emilia Romagna	Forlì-Cesena	Bagno di Romagna	Terme di S. Agnese	Duchi et al. 2005 , Tab 5
13	G 27	Emilia Romagna	Forlì-Cesena	Castrocaro	Bolga well	Capozzi and Picotti 2010, Tab.3
14	G 32	Emilia Romagna	Forlì-Cesena	Tredozio	Monte Busca	Etiope et al. 2007 , Tab.
15	G 34	Emilia Romagna	Modena	Fanano	Trignano	Borgia et al. 1988, Tab.1; *Minissale et al. 2000, Tab. 2, 3
16	V 11	Emilia Romagna	Modena	Fiorano Modenese	Nirano	Etiope et al. 2007 , Tab.
17	G 35	Emilia Romagna	Modena	Lama Mocogno	Barigazzo	Borgia et al. 1988, Tab. 1
18	V 13	Emilia Romagna	Modena	Marano Panaro	Ospitaletto	Etiope et al. 2007, Tab.
19	V 12	Emilia Romagna	Modena	Maranello	Puianello	Duchi et al. 2005 , Tab 5

20	G 36	Emilia Romagna	Modena	Maranello	Govana	Duchi et al. 2005 , Tab 5
21	G 39	Emilia Romagna	Modena	Montese	Montese 19	Borgia et al. 1988, Tab.1
22	V 19	Emilia Romagna	Modena	Sassuolo	Montegibbio	Duchi et al. 2005 , Tab 5
23	//	Emilia Romagna	Modena	Serramazzoni	Selva	Borgia et al. 1988, Tab.1
24	G 45	Emilia Romagna	Modena	Sestola	Ca Boldrini Roncoscaglia	Borgia et al. 1988, Tab.1
25	G 53	Emilia Romagna	Parma	Fornovo Taro	Vallezza	Borgia et al. 1988, Tab.1
26	V 23	Emilia Romagna	Parma	Lesignano Bagni	Rivalta	Etiope et al. 2007, Tab.
27	G 51	Emilia Romagna	Parma	Corniglio	Miano	Heinicke et al . 2010; *Duchi et al. 2005 , Tab 5
28	G 57	Emilia Romagna	Parma	Salsomaggiore	Salsomaggiore	Duchi et al. 2005, Tab 5; *Borgia et al. 1988, T. 1
29	V 24	Emilia Romagna	Parma	Traversetolo	Torre	Etiope et al. 2007, Tab.
30	//	Emilia Romagna	Piacenza	Gropparello	Montechino	Etiope et al. 2007, Tab.
31	V 25	Emilia Romagna	Reggio Emilia	Viano	Casola-Querciola	Duchi et al. 2005, Tab 5
32	V 26	Emilia Romagna	Reggio Emilia	Viano	Regnano	Etiope et al. 2007, Tab.
33	//	Emilia Romagna	Rimini	S. Agata Feltria	Caioletto	Duchi et al. 2005, Tab 5
34	G 79	Toscana	Firenze	Firenzuola	Pietramala	Minissale et al. 2000, Tab. 2, 3
35	//	Toscana	Pistoia	Larciano	Larciano	Duchi et al. 2005 , Tab 5
36	V 51	Molise	Teramo	Pineto	Pineto	Etiope et al. 2007, Tab.
37	V 53	Campania	Benevento	Castelfranco Misciano	Malvizza	Etiope et al. 2007, Tab.
38	G 129	Basilicata	Potenza	Tramutola	Tramutola	Etiope et al. 2007, Tab.
39	V 57	Sicilia	Agrigento	Aragona	Maccalube	Etiope et al. 2007, Tab.
40	G 147	Sicilia	Agrigento	Bivona	Censo	Etiope et al. 2007, Tab. 1
41	V 60	Sicilia	Agrigento	Cattolica Eraclea	Bissana	Etiope et al. 2002, Tab. 2
42	V 63	Sicilia	Catania	Paternò	Salinelle di S. Biagio	Etiope et al. 2002, Tab. 2

* = data source.

Table 5A. Analytical data concerning natural gaseous hydrocarbon manifestations in Italy.

No.	Place name	Lat ° ' ""	Long ° ' ""	CH4 %	CO2 %	N2 %	He%	Ar %	δ13C ‰ PDB	δD ‰ SMOW	Origin	Depth (m)
//	Atmosphere	//	//	0,0002	0,03	78,1	0,0005	0,9	//	//	//	//
1	Gaggiola	44 10 47	10 59 46	95,74	0,96	0,81	n.a.	n.a.	-36,7	-141,8	T	w+s, 160
2	Gaggio Montano	44 11 53	10 56 01	99,35	0,29	0,2	0,002	0,002	n.a.	n.a.	n.a.	s
3	Molinazzo	44 12 34	11 01 23	98,11	0,73	0,45	n.a.	n.a.	-32,7	-129,6	T	w+s, 530
4	Ca Bellavista	44 15 28	11 09 08	98,06	1,6	0,13	n.a.	n.a.	-29,1	-142,8	T	w+s, 170
5	Bergullo	44 18 32	11 44 14	98,61	0,27	0,89	< 0,001	0,02	-69,43	-180,2	B	mv
6	Drag. Sassuno S. Clem.	44 20 09	11 27 18	88,85	2,9	4,15	< 0,001	0,01	-58,4	-219	M	mv
7	Cà Salgastri	44 09 15	10 58 32	98,41	0,99	0,54	n.a.	n.a.	-32	-131	T	w+s, 130
8	Porretta	44 09 04	11 58 01	99,62	0,35	*0,35	*0,0021	*0,0039	-31,3	-138,1	T	w, < 100
9	Castel dell'Alpi	44 12 56	11 14 04	94,54	4,57	0,78	< 0.001	0,009	*-37,4	n.a.	n.a.	s
10	Corporeno	44 45 22	11 18 08	66,52	5,1	26,53	< 0.001	0,51	-65,98	-174,1	B	s
11	Valli Mezzano	44 40 37	12 01 44	66,81	16,03	16,73	0,0027	0,41	-76,14	-223	B	s
12	Terme di S. Agnese	43 50 02	11 57 33	95,85	0,32	3,7	0,006	0,03	n.a.	n.a.	n.a.	s
13	Bolga well	44 10 20	11 56 51	91,68	8,32	n.a.	n.a.	n.a.	-75,5	-171	B	w
14	Monte Busca	44 04 41	11 44 35	58,44	0,45	37,96	0,163	0,06	-35,81	-160,9	T	s
15	Trignano	44 12 39	10 50 29	98,58	0,81	0,24	*0,003	*0,0127	-31,4	-141,9	T	w+s,<100
16	Nirano	44 30 48	10 49 25	98,26	0,58	0,97	0.02	0,01	-45,65	-185,5	T	mv
17	Barigazzo	44 18 25	10 43 47	96,49	1,83	0,21	n.a.	n.a.	-31,8	-140,6	T	w+s,<100
18	Ospitaletto	44 26 11	10 52 54	96,62	2,16	1,07	0,0026	1,01	-45,6	-183,3	T	mv
19	Puianello	44 28 36	10 52 00	95,91	2,35	0,67	< 0,001	0,008	n.a.	n.a.	n.a.	mv
20	Govana	44 31 31	10 51 59	89,26	0,03	10,14	< 0.001	0,11	n.a.	n.a.	n.a.	s
21	Montese 19	44 16 06	10 56 27	97,44	0,1	0,13	n.a.	n.a.	-33,2	-140,3	T	w+s, 190
22	Montegibbio	44 30 55	10 46 39	98,34	0,11	1,4	n.a.	0,006	n.a.	n.a.	n.a.	mv
23	Selva	44 23 46	10 47 38	96,41	0,07	0,32	n.a.	n.a.	-40,2	-147,3	T	s
24	Ca Boldrini Roncoscagli	44 13 50	10 46 14	93,41	1,51	0,01	n.a.	n.a.	-38,8	-140,6	T	w+s,<100
25	Vallezza	44 41 29	10 05 49	88,52	0,17	0,5	n.a.	n.a.	-40,6	-150,7	T	w+s,1500
26	Rivalta	44 37 45	10 19 34	98,32	1,24	0,42	0,0034	0,01	-41,38	-180,6	T	mv
27	Miano	44 29 37	10 06 05	98,62	0,44	0,91	0,0019	*0,051	-39,38	-168,4	T	w+s,1040
28	Salsomaggiore	44 48 58	09 58 43	98,13	0,14	0,61	0,02	n.a.	*-48,1	*-184,2	M	s
29	Torre	44 37 12	10 20 17	96,79	2,73	0,4	0,0013	0,01	-39,1	n.a.	n.a.	mv
30	Montechino	44 48 17	09 41 21	95,3	0,05	0,42	0,0017	0,01	-33,98	-132,6	T	w+s,<100

No.	Place name	Lat ° ' ""	Long ° ' ""	CH4 %	CO2 %	N2 %	He%	Ar %	δ13C ‰ PDB	δD ‰ SMOW	Origin	Depth (m)
31	Casola-Querciola	44 45 44	10 31 38	92,16	0,71	5,79	0,003	0,066	n.a.	n.a.	n.a.	mv
32	Regnano	44 33 25	10 34 34	96,78	2,12	0,92	0,0016	0,01	-45,72	-152,6	M	mv
33	Caioletto	43 49 11	12 10 57	97,83	0,13	1,54	n.a.	0,018	n.a.	n.a.	n.a.	s
34	Pietramala	44 07 14	11 22 49	93,49	0,87	0,22	0,002	0,0021	-36,6	n.a.	n.a.	s
35	Larciano	43 50 00	10 53 24	97,6	0,98	1,3	n.a.	0,08	n.a.	n.a.	n.a.	s
36	Pineto	42 36 52	14 03 41	94,13	0,36	5,4	0,0016	0,11	-73,11	-188,2	B	mv
37	Malvizza	41 17 49	15 05 06	95,64	1,66	1,94	0,025	0,03	-59,09	-163,8	M	mv
38	Tramutola	40 18 56	15 47 23	82,61	2,17	15,12	0,0026	0,01	-42,12	-193,8	T	s
39	Maccalube	37 23 32	13 37 26	91,2	0,73	6,46	0,0071	n.a.	-48,07	-189,6	M	mv
40	Censo	37 37 11	13 26 22	86	2,2	9,66	0,0367	n.a.	-35,1	-146	T	s
41	Bissana	37 26 20	13 23 42	96,2	2,9	0,83	0,0501	n.a.	n.a.	n.a.	n.a.	mv
42	Salinelle di S. Biagio	37 33 50	14 54 11	35,1	64,6	0,78	0,0151	n.a.	n.a.	n.a.	n.a.	mv

Table reference: G = data from Table 1 (gas), V = data from Table 4 (mud volcano). ° ' " = sexag degrees. % = by volume. Origin : B = biogenic, M= mixed; T= thermogenic. Depth (m): s = gas seep, mv = mud volcano, w = well. n.a. = not available.

Table 5B. Analytical data concerning natural gaseous hydrocarbon manifestations in Italy.

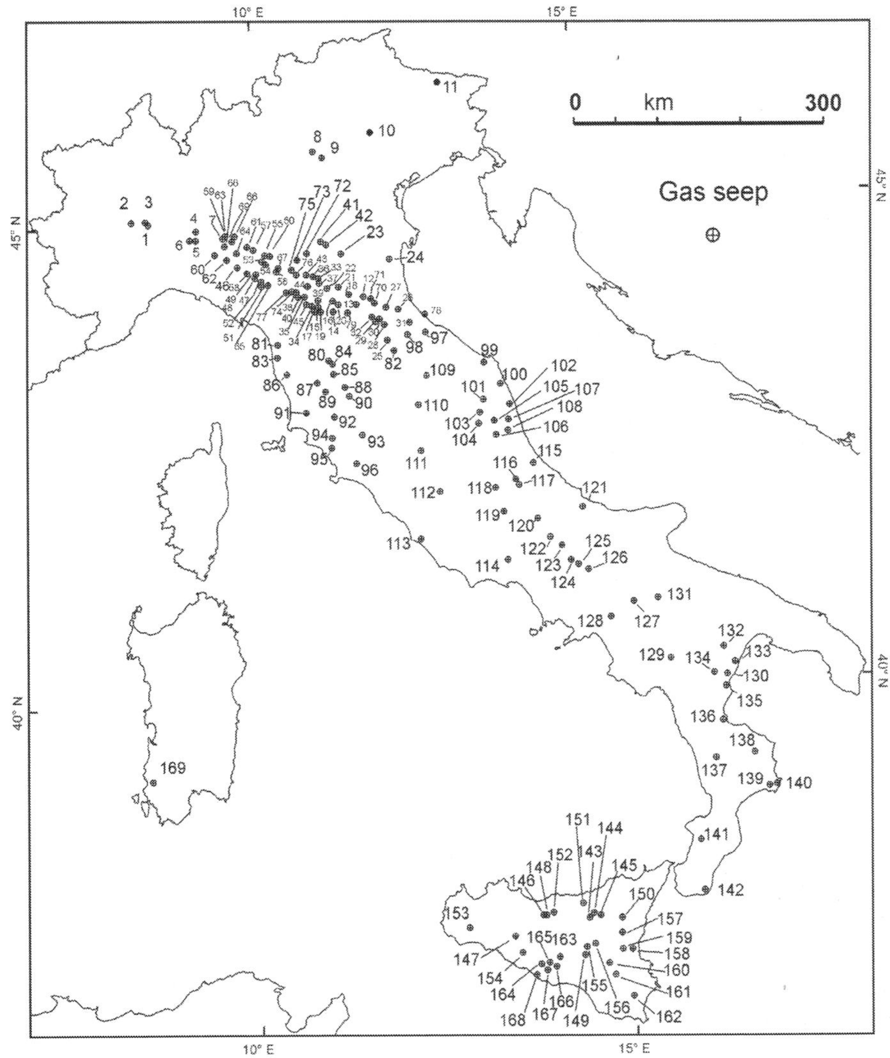

Fig. 3A. Map of natural gas seepages in Italy. Coordinates and related references are given in Table 1.

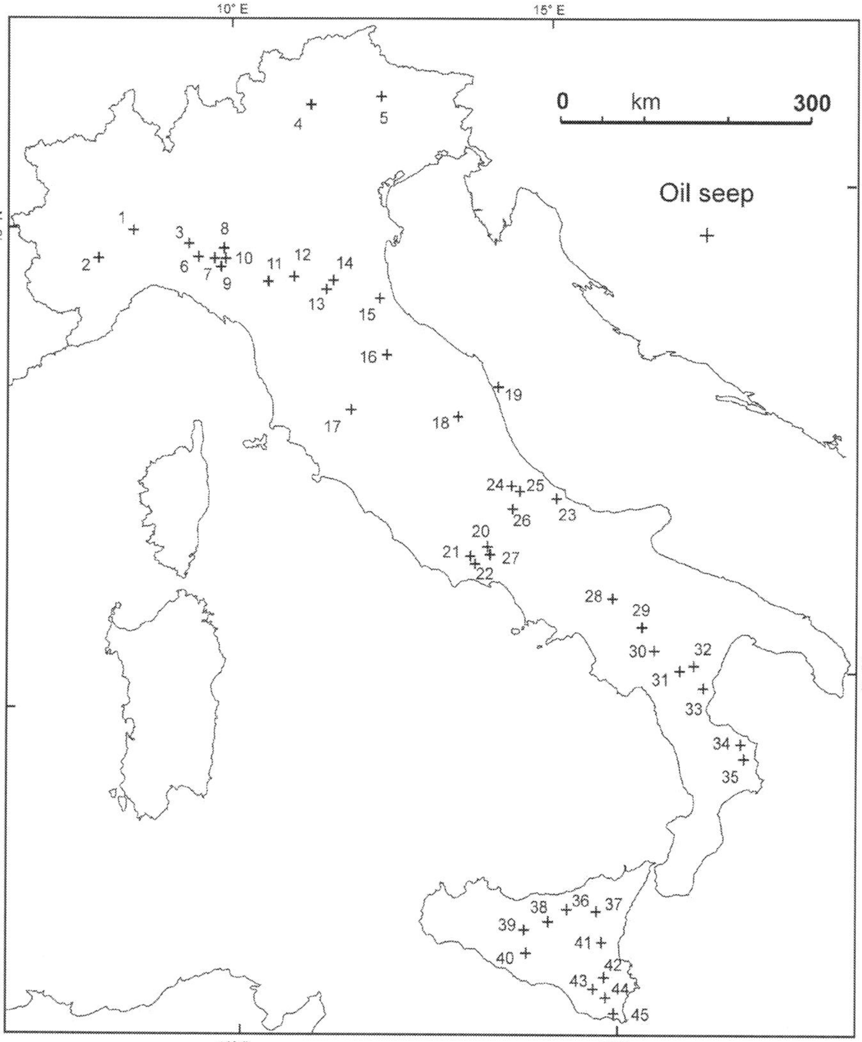

Fig. 3B. Map of natural oil occurrence in Italy. Coordinates and related references are given in Table 2.

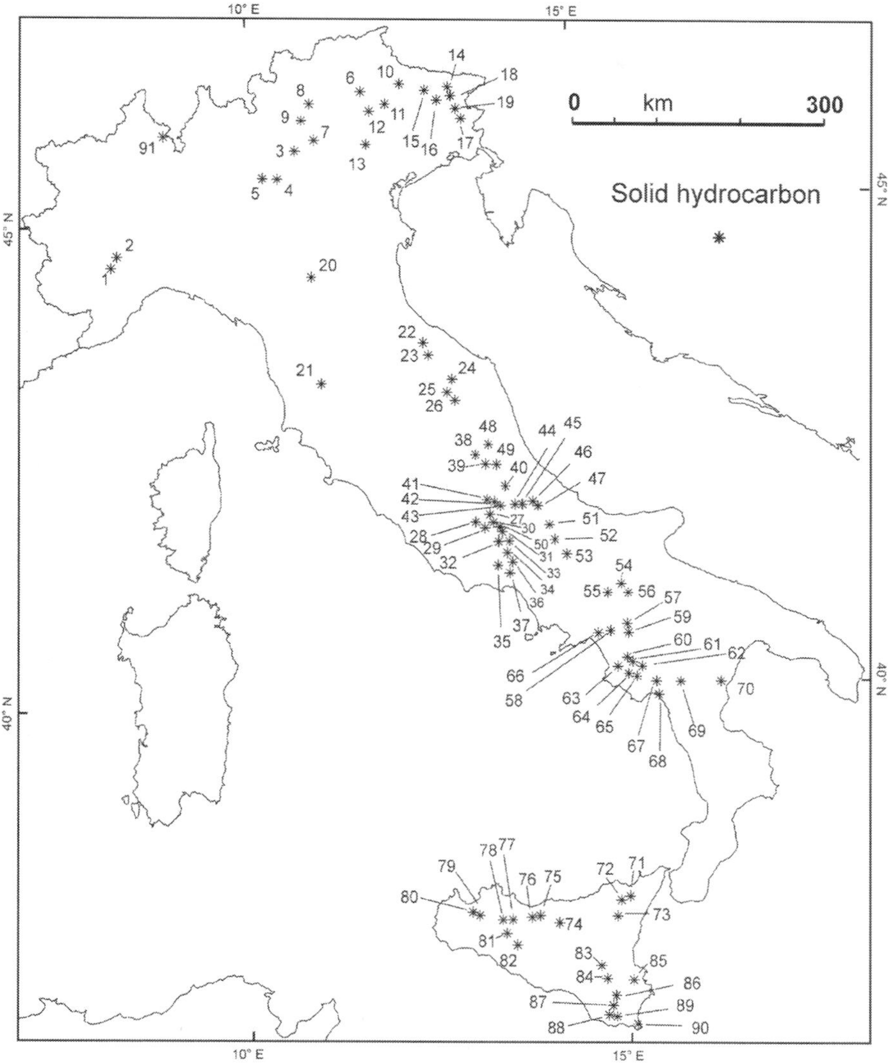

Fig. 3C. Map of natural solid hydrocarbon occurrence in Italy. Coordinates and related references are given in Table 3.

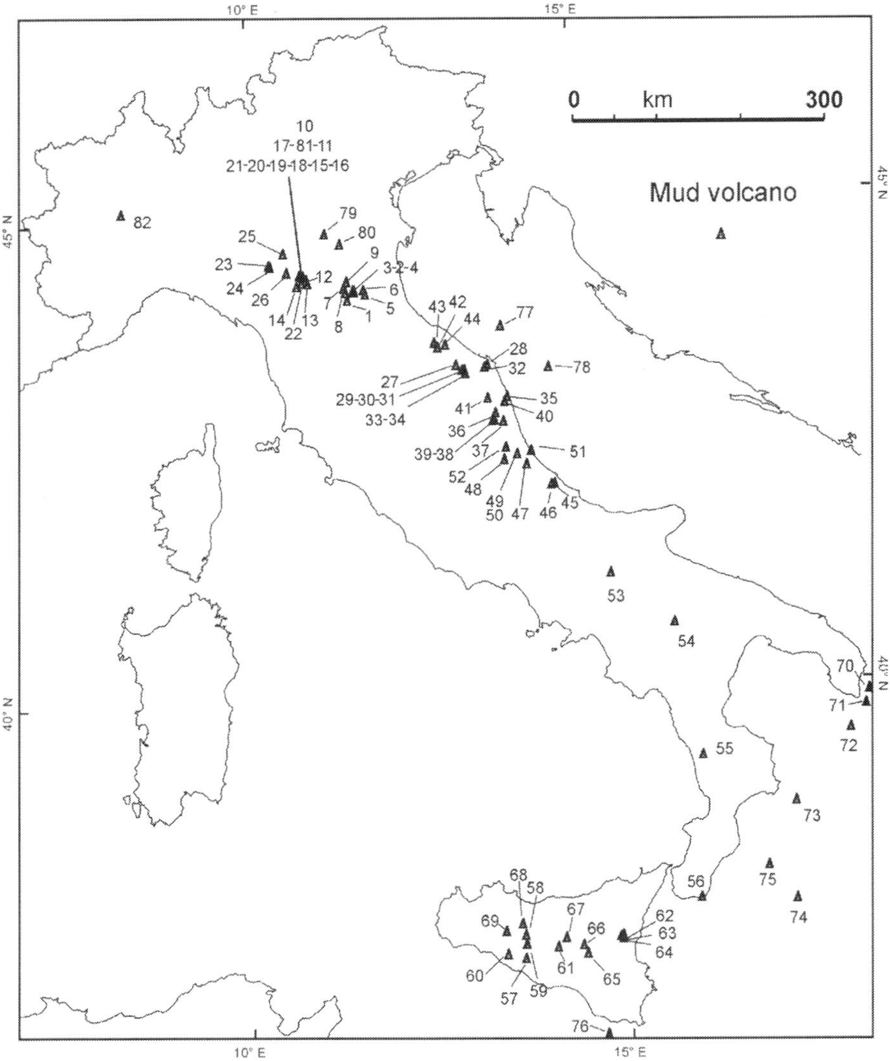

Fig. 3D. Map of mud volcanoes in Italy. Coordinates and related references are given in Table 4.

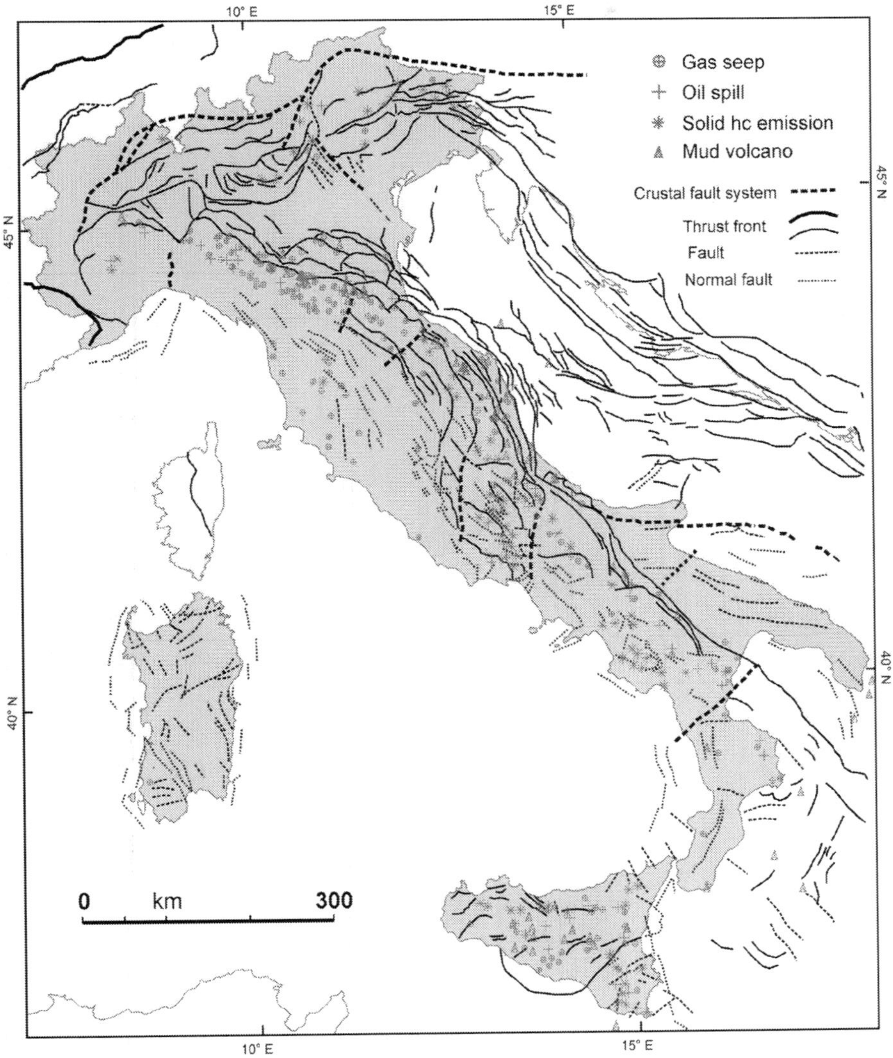

Fig. 4. Relationships existing between Hydrocarbon occurrences and main structural setting of Italy. The structural frame was simplified and redrawn after (CNR 1990; Fantoni and Franciosi 2010, fig. 5).

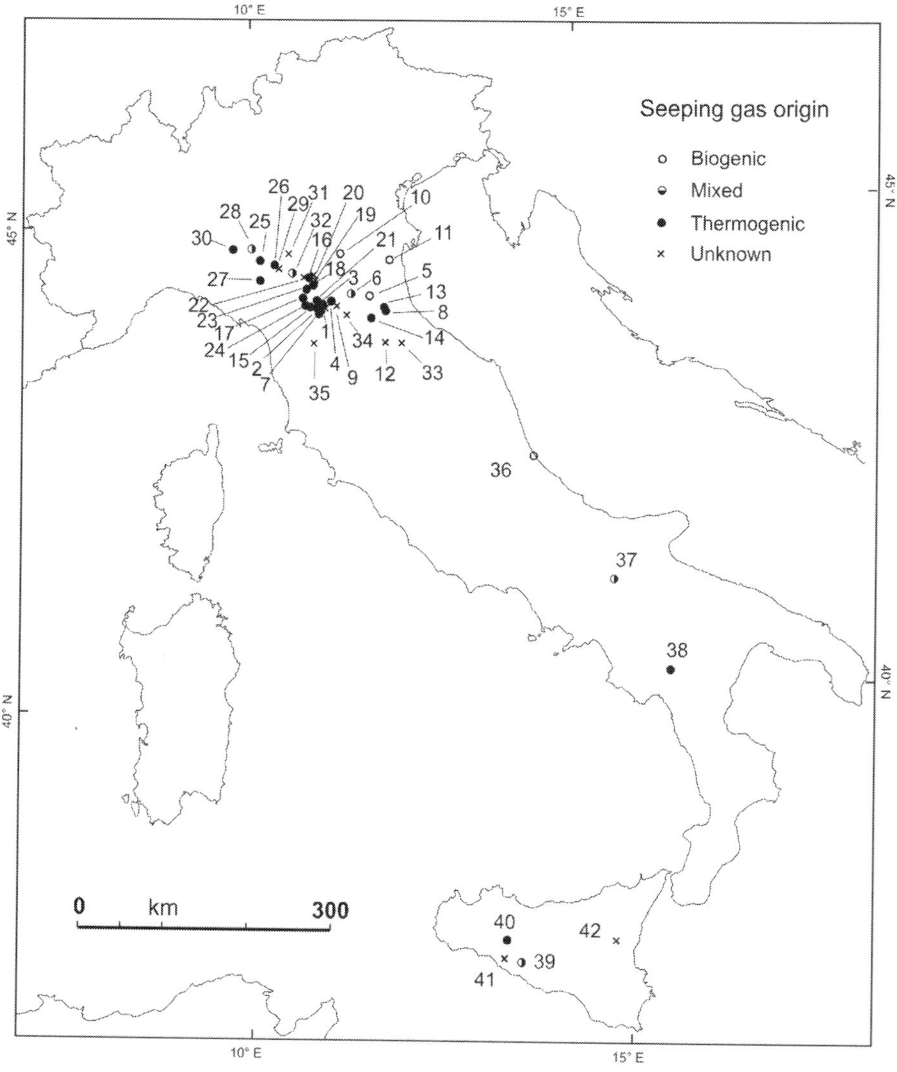

Fig. 5. Location map of gas seepage points in Italy for which analytical data are available in Table 5.

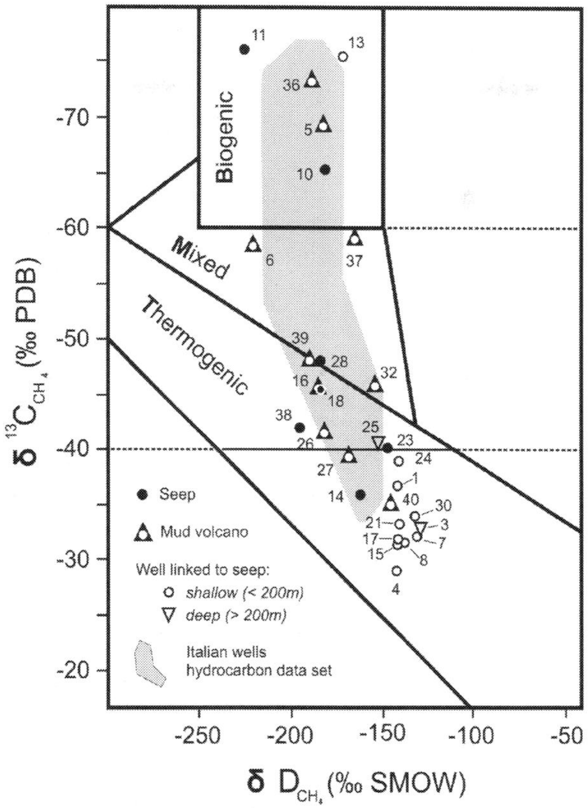

Fig. 6. Simplified Schoell's diagram evidencing biogenic, mixed and thermogenic characters of natural gaseous hydrocarbon seepages in Italy. The cluster of Italian hydrocarbon wells analytical data (gray area) is also shown (after Mattavelli and Novelli 1988).

8. Tectonic stress field in Italy and seepages

A stress field map (Montone et al. 2004) can be used (Fig. 7) for a better understanding of active tectonic processes, to understand the behavior of faults recognized by other methods (CNR 1990) and to infer the origin of surficial manifestations of hydrocarbons (Figs. 3 and 8). The map shows that an extensional regime affects most of the Apenninic belt. Conversely, a compressional (or transpressional) regime characterizes the eastern Alps, the eastern side of the northern Apennines, and the South Tyrrhenian to Northern Sicilian zone. An abrupt change in stress directions marks the transition between northern and southern Apennines, suggesting that the two arcs are characterized by a different tectonic setting and recent evolution. Present stress field probably have not changed significantly in last 10 kyr; thus hydrocarbon seepages are due to rock fracturing, overpressure phenomena and tectonic pumping processes constantly generated in the upper crustal layers. In particular, the intense tectonization of the orogen (Montone et al.,2004; Picotti and Pazzaglia 2008) can

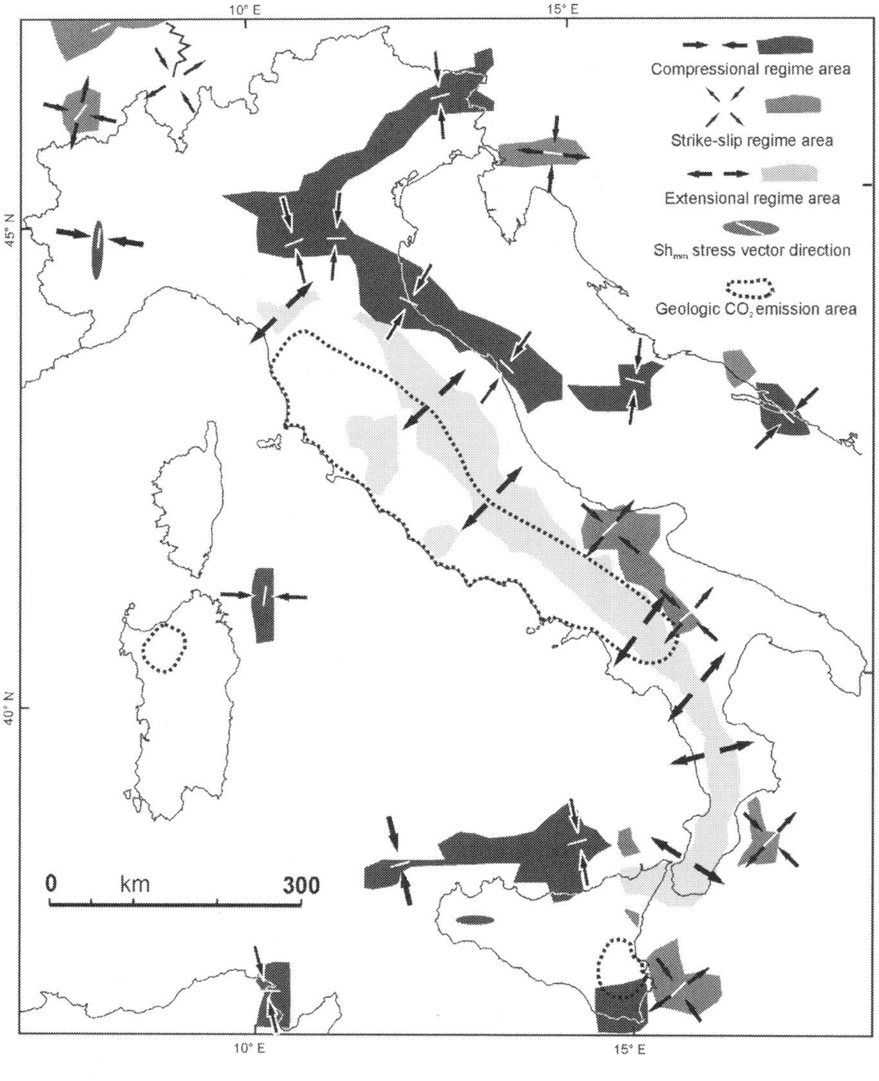

Fig. 7. Stress field data in Italy (after Montone et al.2004, redrawn) and crustal carbon dioxide degassing areas (Frezzotti et al. 2009).

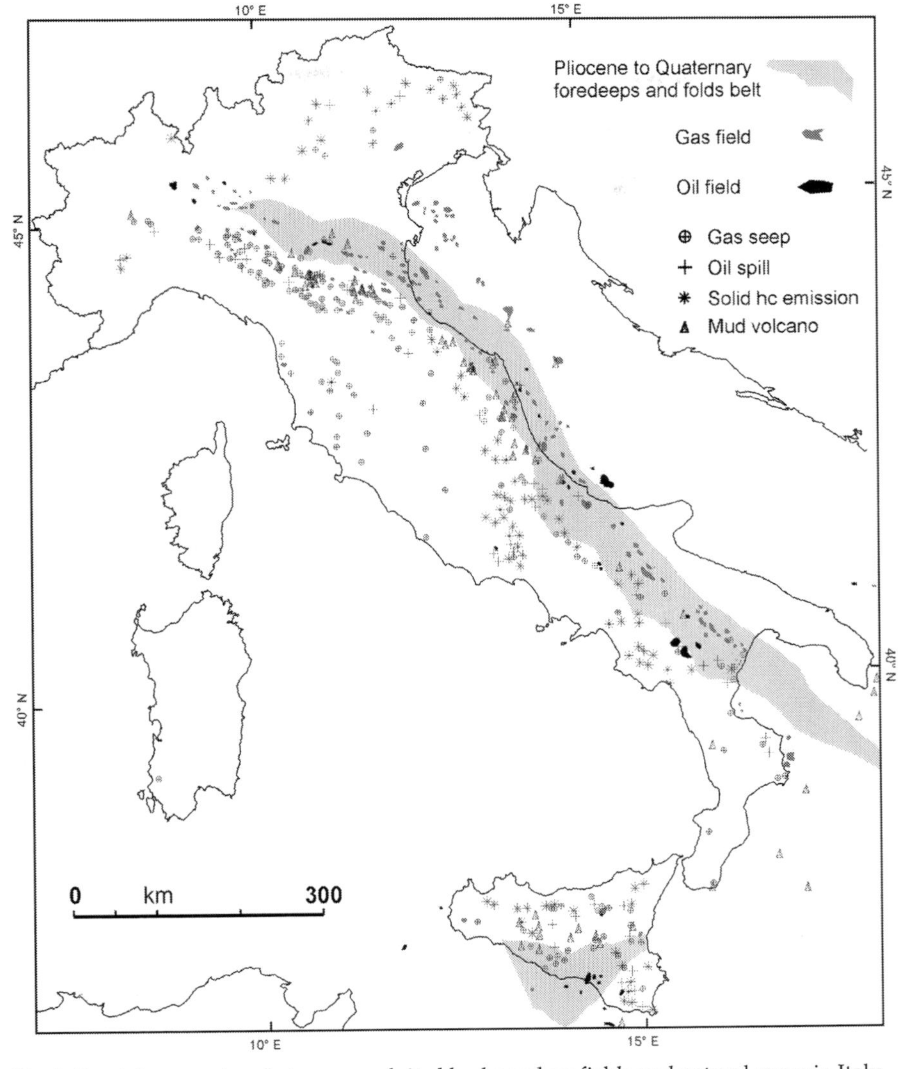

Fig. 8. Spatial comparison between exploited hydrocarbon fields and natural seeps in Italy (from figures 1 and 3).

be responsible for the high number of hydrocarbon seepages of whole Apennines (Martinelli and Judd 2004; Capozzi and Picotti 2010). The distensive behavior of central Italy is probably responsible for the high heat flow values recorded in the Tuscany and Latium areas unfavourable to hydrocarbon accumulations. The western part of the Italian peninsula is characterized by intense CO_2 degassing activity (Minissale et al 2004). Carbon dioxide is chiefly originated by mantle degassing, and by crustal thermometamorphic reactions (e.g.

Chiodini et al. 2004; Frezzotti et al. 2010). Conversely carbon dioxide degassing is lacking in the outer (eastern) flexural domain of the Apennine chain due to higher crustal thickness and to the relatively low geothermal gradient (Buttinelli et al. 2011 and therein references).

9. Conclusions

The mapping of most important gas emissions (Fig.8) shows that the hydrocarbon domain is chiefly located in the core of the raised Apennine belt immediately behind the chain front at the boundary of and its related Plio-Quaternary foredeep, whereas CO_2 emissions are located in the Apennine backdeep area. The geographic distribution of important gas accumulations in Italy does not show a highly significant correlation between surface seepages and the exploited reservoirs and it could also suggest the existence of other still unknown deep reservoirs (Pieri 2001) or their small remnants difficult to be checked up. The majority of the hydrocarbon wells is characterized by biogenic gases, while thermogenic methane is predominant in surface seeping, confirming the sealed condition of most of the biogenic reservoirs and that they still have not experienced the complete evolution of organic matter towards the thermogenic terms induced by pressure and temperature.

10. References

Accademia Nazionale dei Lincei and Ente Nazionale Idrocarburi (Eds.) (1948). I giacimenti gassiferi dell'Europa Occidentale. 3 Voll.

Bacci A. (1571). De thermis libri septem. Romae, p. 425.

Bertello F., Fantoni R., Franciosi R., Gatti V., Ghielmi M., Pugliese A. (2010). From thrust-and-fold belt to foreland: hydrocarbon occurrences in Italy. Vining B. A., Pickering S. C. (eds.), Petroleum Geology: from mature basins to new frontiers - Proceedingsof the 7th Petroleum Geology Conference, The Geological Society, London, 113-126. DOI: 10.1144/0070113.

Bianconi G. (1840). Storia naturale dei terreni ardenti, dei vulcani fangosi, delle sorgenti infiammabili, dei pozzi idropirici e di altri fenomeni geologici operati dal gas idrogene e dell'origine di esso gas. Annali delle Scienze Naturali, Bologna, Voll. 2-4, p. 216.

Biasutti R. (1907). Materiali per lo studio delle salse. I - Le salse dell'Appennino settentrionale. Supplemento alle Memorie Geografiche, 2, pp. 254.

Boccaletti M. and Martelli L. (Eds.) (2004). Note illustrative della Carta sismotettonica della Regione Emilia-Romagna, scala 1:250.000. Firenze 2004, 60 pp.

Bombicci L. (1881). Regione del Torrente Samoggia. In: L'Appennino Bolognese, Bologna, 466-467.

Bonini M. (2007). Interrelations of mud volcanism, fluid venting and thrust-anticline folding: examples from the external northern Apennines (Emilia-Romagna, Italy). Journal of Geophysical Research, 112, B08413, DOI: 10.1029/2006JB004859, 2007.

Bonori O., Ciabatti M., Cremonini S., Di Giovambattista R., Martinelli G., Maurizzi S., Quadri G., Rabbi E., Righi P.V., Tinti S., Zantedeschi E. (2000). Geochemical and geophysical monitoring in tectonically active areas of the Po Valley (N. Italy). Case histories linked to gas emission structures. Geogr. Fis. Dinam. Quat., 23: 3-20.

Borgatti L., Cervi F., Corsini A., Guandalini F., Ronchetti F., Pellegrini M. (2010). Ipotesi sugli eventi distruttivi rilevati nell'abitato romano di Montegibbio. Guandalini F. and Labate D. (eds.), L'insediamento di Montegibbio. Quaderni di Archeologia dell'Emilia Romagna, 26, 95-109.

Borgia G. C., Elmi C., Ricchiuto T. (1988). Correlation by genetic properties of the shallow gas seepages in the Emilian Apennine (Northern Italy). Organic Geochemistry, 13, 319-324.

Buttinelli M., Procesi M., Cantucci B., Quattrocchi F., Boschi E. (2011). The geo-database of caprock quality and deep saline aquifer distribution for geological storage of CO_2 in Italy. Energy, 36, 2968-2983. DOI: 10.1016/j.energy.2011.02.041.

Camerana E. and Galdi B. (1911). I giacimenti petroliferi dell'Emilia. Memorie descrittive della Carta Geologica d'Italia, 14, p. 335, 16 sheets.

Camerana E., Cotese E., Crema C., De Stefani C., Fossa-Mancini E., Lotti B., Novarese V., Porro C., Sacco F., Stella A. (1926) Studi geologici per la ricerca del petrolio in Italia. Memorie Descrittive della Carta Geologica d'Italia, 20, 1-288.

Camerlenghi A., Pini A. (2009). Mud volcanoes, olistostromes and Argille scagliose in the Mediterranean region. Sedimentology, 56, 319-365. DOI: 10.1111/j.1365-3091.2008.01016.x.

Cantelli C. (1994). Una nuova "salsa" nei dintorni di Bologna. Natura e Montagna, 49, 52-53.

Capozzi R. and Picotti V. (2010). Spontaneous fluid emissions in the northern Apennines: geochemistry, structures and implications for the petroleum system. Goffey G.P., Craig J., Needham T.,Scott R. (Eds.), *Hydrocarbon in contractional belts*, Geological Society, London, Special Publications, 348,115-135.

Casero P. (2004). Structural setting of petroleum exploration plays in Italy. In: Crescenti V., D.'Offizi S., Merlino S., Sacchi L., (Eds.), *Geology of Italy*, Special volume of the Italian Geological Society for the IGC 32 Florence-2004, 189-199.

Casero P., Roure F., Vially R. (1991). Tectonic framework and petroleum potential of the southern Apennines. Spencer A. M. (ed.), Generation, accumulation and production of Europe'sHydrocarbons,Spec. Publ. Eur. Assoc.Petrol. Geoscientists, 1, Oxford University Press, 381-387.

Castellarin A., Cantelli L., Fesce A. M., Mercier J.L., Picotti V., Pini G. A., Prosser G., Selli L. (1992). Alpine compressional tectonics in the Southern Alps. Relationships with the N-Apennines. Annales Tectonicae, 6, 62-94.

Castellarin A., Rabbi E., Cremonini S., Martelli L., Piattoni F. (2006). New insights into the underground hydrology of the eastern Po Plain (northern Italy), in Boll. Geof. Teor. Appl., 47, pp. 271-298.

Catalano and D'Argenio (1982). Infraliassic strike-slip tectonics in Sicily and Southern Apennines. Rendiconti Società Geologica Italiana, 5, 5-10.

Chiodini G., Cardellini C., Amato A., Boschi E., Caliro S., Frondini F. (2004). Carbon dioxide Earth degassing and seismogenesis in central and southern Italy. Geophysical Research Letters, 31, L07615.

Ciotoli G., Guerra M., Lombardi S., Vittori E. (1998). Soil gas survey for tracing seismogenic faults: a case study in the Fucino basin, Central Italy. Journal of Geophysical Research, 103, B 10, 23781-23794.

Clari P., Cavagna S., Martire L., Unziker J. (2004). A Miocene mud volcano and its plumbing system: a chaotic complex revisited (Monferrato, NW Italy). Journal of Sedimentary Research, 74, 662-676.

CNR (Consiglio Nazionale delle Ricerche) (1990). Structural model of Italy and gravity map(1:500.000). Prog. Final. Geodinamica, Pubblic. n.114, 6 sheets.

Conti A., Stefanon A., Zuppi G. M. (2002). Gas seeps and rock formation in the Northern Adriatic Sea. Continental Shelf Research, 22, 2333-2344.

Cremonini S. (2010a). A preliminary overview of sinkholes in the Emilia-Romagna Region (Italy). Atti del 2° Workshop Internazionale sui fenomeni di sinkhole *"I sinkholes. Gli sprofondamenti catastrofici nell'ambiente naturale ed in quello antropizzato"*, Roma , 3-4 Dicembre 2009, ISPRA. Roma 2010, 257-281.

Cremonini S. (2010b). Can subaerial pockmarks exist? Atti del 2° Workshop Internazionale sui fenomeni di sinkhole *"I sinkholes. Gli sprofondamenti catastrofici nell'ambiente naturale ed in quello antropizzato"*, Roma , 3-4 Dicembre 2009, ISPRA. Roma 2010, 29-34.

Cremonini S., Etiope G., Italiano F., Martinelli G. (2008). Evidence of possibile enhanced peat burning by deep-originated methane in the Po River delta plain (Italy). Journal of Geology, 116, 401-413.

Cremonini S., Martelli L., Zanutta A. (2010). An Initial approach to the analysis of alluvial plain sinkhole-clusters at Finale Emilia and Reno Finalese (Modena - Italy). Atti del 2° Workshop Internazionale sui fenomeni di sinkhole *"I sinkholes. Gli sprofondamenti catastrofici nell'ambiente naturale ed in quello antropizzato"*, Roma , 3-4 Dicembre 2009, ISPRA. Roma 2010, 453-468.

Curzi P. V., Lenaz R., Marabini S., Vai G. B. (1987). Mesostrutture deformative tipo pockmark in un paleosuolo olocenico di Romagna. Rend. Soc. Geol. It., 10, 38-40.

Curzi P.V. and Veggiani A. (1985). I pockmarks nel mare Adriatico Centrale. L'Ateneo Parmense Acta Nat., 21, 79-90.

Curzi P.V., D'Onofrio S., Gabbianelli G., Panieri G., Ceffa L., Gasparini G., Ricchiuto T. (1998). A possible mechanism of rising gas at a Bonaccia Field pockmark (Central Adriatic Sea). Curzi P.V. and Judd A.G. (Eds.), Abstracts and Guide book of the 5th International Conference on gas in marine sediments, Bologna September 9-12, 174-177.

Dal Piaz G. (1959). Il bacino quaternario polesano-ferrarese e i suoi giacimenti gassiferi. Accademia Nazionale dei Lincei and Ente Nazionale Idrocarburi (Eds.), *I giacimenti gassiferi dell'Europa Occidentale*, I, 433-471.

Dalla Valle G., Gamberi F. (2011). Pockmarks and seafloor instability in the Olbia continental slope (northeastern Sardinian margin, Tyrrhenian Sea). Marine Geophysical Research, 32, 193-205. DOI: 10.1007/s11001-011-9133-1.

Duchi V., Venturelli G., Boccasavia I., Bonicolini F., Ferrari C., Poli D. (2005). Studio geochimico dei fluidi dell'Appennino Tosco-Emiliano-Romagnolo. Boll. Soc.Geol.It., 124, 475-491.

ENI (Ente Nazionale Idrocarburi) (Ed.) (2009). Enciclopedia degli Idrocarburi. Istituto dell'Enciclopedia Italiana, Roma, 5 voll., p.4340.

Etiope G., Caracausi A.,Favara R., Italiano F., Baciu C. (2002). Methane emission from the mud volcanoes of Sicily (Italy). Geophysical Research Letters, 29, DOI: 10.1029/2001GL014340.

Etiope G., Feyzullayev A.,Baciu C. (2009). Terrestrial methane seeps and mud volcanoes: A global perspective of gas origin. Marine and Petroleum Geology, 26, 333-344.

Etiope G., Martinelli G., Caracausi A., Italiano F. (2007). Methane seeps and mud volcanoes in Italy: gas origin, fractionation and emission to the atmosphere. Geophysical Research Letters, 34, L14303.

Fantoni R. and Franciosi R. (2010). Tectono-sedimentary setting of the Po Plain and Adriatic foreland. Rendiconti Fisici Accademia dei Lincei, 21 (suppl. 1), S197-S209. DOI: 10.1007/s12210-010102-4.

Finetti I. R., (ed.) (2008). CROP PROJECT: deep seismic exploration of the central Mediterranean and Italy. Elsevier Atlas in Geosciences, 1, pp. 794.

Frezzotti M.L., Peccerillo A., Panza G. (2009). Carbonate metasomatism and CO_2 lithosphere-astenosphere degassing beneath the Western Mediterranean: an integrated model arising from petrological and geophysical data. Chemical Geology, 262, 108-120. DOI: 10.1016/j.chemgeo.2009.02.015.

Fusi N., Savini A., Corselli C. (2006). Evidence of mud diapirism and coral colonies in the Ionian Sea (Central Mediterranean) from high resolution chirp sonar survey. Annals of Geophysics, 49, 751-765.

Gasperi G. and Pellegrini M. (1981). Note di geologia del comprensorio di pianura della bassa pianura modenese. Povegliano Veronese (VR), 98 p.

Geletti R., Del Ben A., Busetti M., Ramella R., Volpi V. (2008). Gas seeps linked to salt structures in the Central Adriatic Sea. Basin Research, 20, 473-487, DOI: 10.1111/j.1365-2117.2008.00373.x.

Heinicke J., Italiano F., Koch U., Martinelli G., Telesca L. (2010). Anomalous fluid emission of deep borehole in a seismically active area of Northern Apennines (Italy). Applied Geochemistry, 25, 555-571.

Holland C. W., Etiope G., Milkov A. V., Michelozzi E., Favali P. (2003). Mud volcanoes discovered offshore Sicily. Marine Geology, 199, 1-6.

Lindquist S.J. (1999). Petroleum systems of the Po basin province of Northern Italy and the Northern Adriatic Sea: Porto Garibaldi (biogenic), Deride/Riva di Solto (thermal) and Marnoso Arenacea (thermal). USGS Open-file Report 99-50-M, pp. 35.

Marabini S., Lenaz R., Vai G.B. (1987). Pleistocene superiore e Olocene del margine pedeappenninico romagnolo. Rend. Soc. Geol. It., 10, 33-37.

Martinelli G. (2007). Attività degassanti nella Regione Emilia-Romagna. In : SSR-RER (Servizio Sanitario Regionale-Regione Emilia-Romagna) (Ed.), *Il Radon ambientale in Emilia-Romagna.* Contributi alla Prevenzione nei luoghi di vita e di lavoro, 51, 57-62.

Martinelli G., Judd A. (2004). Mud volcanoes of Italy. Geological Journal, 39, 49-61.

Martinelli G., Panahi B. (Eds.) (2005). Mud volcanoes, geodynamics and seismicity. NATO Science Series IV. Earth and Environmental Sciences, 51, p.288.

Martinis B. (1969). Manifestazioni petrolifere. In: E.N.I. (Ed.), Enciclopedia del petrolio e del gas naturale, vol 6, 1251-1265.

Mattavelli L. and Novelli L. (1988). Geochemistry and habitat of natural gases in Italy. Organic Geochemistry, 13, 1-13.

Mattavelli L. and Novelli L. (1990). Geochemistry and habitat of the oils in Italy. The American Association of Petroleum Geologists Bullettin, 74, 1623-1639.

Mattavelli L., Pieri M., Groppi G. (1993). Petroleum exploration in Italy: a review. Marine and Petroleum Geology, 10, 410-425.

Mattavelli L., Ricchiuto T., Grignani D., Shoell M. (1983). Geochemistry and habitat of natural gases in Po Basin, Northern Italy. The American Association of Petroleum Geologists Bullettin, 67, 2239-2254.

Mazzotti L. , Segantini S., Tramontana M., Wezel F. C. (1987). Characteristic of pockmarks on the Jabuka Trough floor (Central Adriatic Sea). Bollettino di Oceanografia Teorica ed Applicata, 5, 237-250.

Minissale A. (2004). Origin, transport and discharge of CO_2 in Central Italy. Earth-Science Reviews, 66, 89-141. DOI: 10.1016/j.earscirev.2003.09.001.

Minissale A., Magro G., Martinelli G., Vaselli O., Tassi G.F. (2000). Fluid geochemical transect in the Northern Apennines (central-northern Italy); fluid genesis and migration and tectonic implications. Tectonophysics, 319, 199-222.

Ministero dello Sviluppo Economico and Assomineraria (2008). Italy: structural setting of petroleum exploration plays - Petroleum systems. 1 Sheet.

Montesauro Veronesi D. (1585). Aristotelis Stagiritae Opera omnia, : Quibus nunc primùm accesserunt Politicorum seu De republica, liber nonus & decimus; item Theologiae seu Philosophiae mysticae libri quatuordecim.

Montone P., Mariucci M.T., Pondrelli S., Amato A. (2004). An improved stress map for Italy and surrounding regions (central Mediterranean). Journal Geophysical Research, 109, B10410, DOI: 10.1029/2003JB002703,2004.

Mutti E. and Ricci Lucchi F. (1972). Le torbiditi dell'Appennino settentrionale: introduzione all'analisi di facies. Memorie Società Geologica Italiana, 11, 161-199.

Panieri G. (2006). Foraminiferal response to an active methane seep environment: a case study from the Adriatic Sea. Marine Micropaleontology, 61, 116-130.

Picotti V, and Pazzaglia F. J. (2008). A new active tectonic model for the construction of the Northern Apennines mountain front near Bologna (Italy), Journal of Geophysical Research, 113, BO8412, doi:10.1029/2007JB005307, 2008.

Pieri M. (2001). Italian petroleum geology. In: Vai G.B. and Martini P.I. (Eds.), Anatomy o fan froge. The Apennines and Adjacent Mediterranean basins. Kluwer Academic, Dordrecht, 533-549.

Pieri M. and Mattavelli L. (1986). Geologic Framework of Italian Petroleum Resources. AAPG Bulletin, 70, 103-130.

Pieri M. and Mattavelli L. (1986). Geologic framework of italian petroleum resources. The American Association of Petroleum Geologists Bullettin, 70, 103-130.

Praeg D., Ceramicola S., Barbieri R., Unnithan V., Wardell N. (2009). Tectonically-driven mud volcanism since the late Pliocene on the Calabrian accretionary prism, central Mediterranean Sea. Marine and Petroleum Geology, 26, 1849-1865.

Reeves F. (1953). Italian oil and gas resources. American Association Petroleum Geologists Bullettin, 37, 601-653.

Schoell M. (1980). The hydrogen and carbon isotopic composition of methane from natural gases of various origins. Geochimica et Cosmochimica Acta, 44, 649-661.

Schoell, M. (1983). Genetic characterization of natural gases. American Association Petroleum Geology Bulletin, 67, 2225-2238.

Scicli A. (1972). L'attività estrattiva e le risorse minerarie della Regione Emilia-Romagna, Modena, 626 p.

Trincardi F., Argnani A., Correggiari A. (Eds.) (2011c). Foglio NK 33-5 (Pescara) con Note Illustrative. ISPRA (Istituto Superiore per la Protezione e la Ricerca Ambientale)-Servizio Geologico d'Italia, Carta geologica dei mari Italiani alla scala 1: 250.000, Firenze, 2 Sheets, pp.167.

Vai G. B. and Martini I. P. (Eds.) (2001). Anatomy of an Orogen. The Apennines and adjacent Mediterranean basins. Kluwer Academic, Dordrecht, p. 633.

Violante, C. and Michetti, A.M. (2010). Campagna Perla-G2010. Rapporto interno IAMC_CNR , 15p. (available at: http://eprints.bice.rm.cnr.it/1473/1/Campagna_Perla%2DG_2010.pdf).

Zuber S. (1938). Italy. Dunstan A. E., Nash A. V., Brooks B. T., Tizard H. (eds.), *The Science of petroleum*. Oxford University Press, London, vol.1,189-192.

Zuber S. (1940). Manifestazioni e prospettive petrolifere dell'Italia alla luce degli accertamenti dell'ultimo trentennio (Pubblicato col cortese concorso della presidenza dell'Azienda Generale Italiana Petroli AGIP) . Italgraf, Roma, 135 p., [21] c. di tav.

Part 2

Natural Gas Production

5

The Expansion of Unconventional Production of Natural Gas (Tight Gas, Gas Shale and Coal Bed Methane)

Alejandro Alonso Suárez
National Energy Commission,
Spain

1. Introduction

Finding new oil reserves is becoming more and more difficult (al-Husseini, 2007). A recent study (Aleklett, 2007) alerted that global oil production has very probably passed its maximum, so the World will have reached the Peak of the Oil Age (also known as Hubbert Peak). The rate of discoveries of new oil reserves is less than the present rate of consumption, and already 5 out of every 6 oil producing countries have declining production. World natural gas proved reserves in 2010 were sufficient to meet 59 years of global production while proved oil reserves were sufficient to meet 46 years of global production (BP, 2011).

Natural gas production was expected to begin declining at the United States at the beginning of the XXI century, and in fact, conventional gas production peaked 15 years ago.

However, in the last decade, US internal gas production has been maintained and even increased in 2008. More surprisingly, US domestic gas resources have increase a 40% since 2006. Most of the growth in natural gas production in the United States and in other developed countries like Australia is coming from unconventional sources of natural gas that were considered until recently as non recoverable resources.

New and advanced exploration, well drilling and completion technologies are allowing increasingly better access to non conventional gas resources at competitive prices, so unconventional gas is becoming more "conventional". As most of discoveries are taking place in the US, they have a tremendous impact on global gas markets.

As a prove of the interest of the industry on unconventional gas, in December 2009, ExxonMobil announced it would buy XTO Energy, the bigger independent gas producer in US, one of the companies operating in the big Barnett Shale basin (Texas), for 31,000 million dollars.

Unconventional gas is defined by the International Energy Agency as gas that is more technologically difficult or more expensive to produce that conventional gas. Unconventional gas resources can be divided into coal bed methane, tight gas and shale gas. Other huge unconventional resources are gas hydrates (methane molecules trapped in crystalline water compounds), but they are not expected to provide a major contribution to gas production in the next 20 years.

A generally accepted industry definition is reservoirs that do not produce economic volumes of natural gas without assistance from massive stimulation treatments or special recovery processes and technologies, such as multiple fracturing.

This paper presents a review of the state of art and recent developments of unconventional gas production (tight gas, shale gas and coal bed methane): technology, pioneers projects and bigger unconventional fields, actual production, costs trends and potential reserves.

2. Types of non conventional natural gas

Fossil fuels - including crude oil and natural gas – were originated from organic matter over hundreds of millions of years. The organic matter, buried under layers of mud in anaerobic conditions (without oxygen), gradually decomposed and was converted into petroleum (natural gas or crude oil) by the combined action of heat and high pressure,

Gas generated in the process migrates towards the surface (the atmosphere) unless it encounters various types of geological traps in the form of porous rock formations tightly sealed by overlying layers of impermeable rock.

Natural gas yields can be broadly classified into conventional and non conventional reservoirs:

- In a **conventional reservoir**, natural gas is trapped in a limited boundary within porous rocks by impermeable rocks above (seal) that prevents it from escaping to the surface. Usually have as well a down-dip water contact. This led to the formation of conventional gas deposits, resembling gas-soaked sponges. The natural gas can be "associated" (mixed with oil) or "non-associated" found in reservoirs that do not contain oil. The majority of gas production is non-associated.
  Non-associated gas is recovered from the formation by an expansion process. Wells drilled into the gas reservoir allow the highly compressed gas to expand through the wells in a controlled manner, to be captured, treated and transported at the surface. This expansion process generally leads to high recovery factors from conventional, good-quality gas reservoirs. If, for example, the average pressure in a gas reservoir is reduced from an initial 300 bar to 60 bar over the lifetime of the field, then approximately 80% of the gas initially in place will be recovered.
- **Unconventional reservoirs** are more continuous, consisting of a stacking of sedimentary layers with low permeability that are charged with gas.
  They usually require advanced technology such as horizontal wells or artificial stimulation in order to be economically productive; recovery factors are much lower — typically of the order of 15 - 20% of the gas in place.

The unconventional gas sources include shale gas, gas locked in insulated rock pores (tight gas), coal bed methane and gas hydrates. Important characteristics of gas reservoirs are porosity - the volume of the pore spaces, and permeability - how connected the pore spaces are.

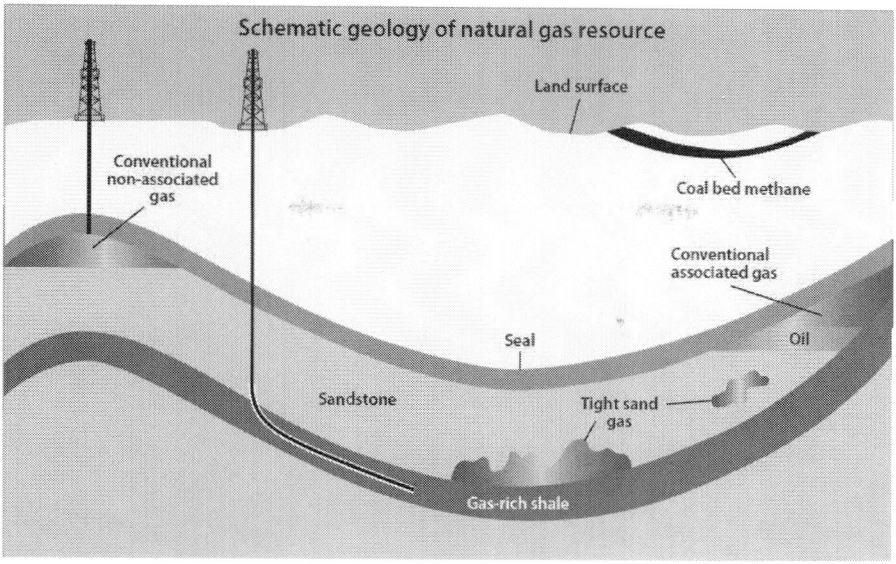

Fig. 1. Types of gas yields. Source US Energy Information Administration, 2011.

2.1 Tight gas

Tight Gas is natural gas found in reservoirs with low porosity and low permeability. Tight reservoirs are usually a sandstone formation that is unusually non-porous, although carbonate rocks can also be tight-gas producers.

Tight gas is trapped in geological formations similar to conventional gas, so the distinction between conventional and tight gas is not so clear in some basins.

The standard definition for a tight-gas reservoir is a rock with matrix porosity of 10% or less and permeability of 0.1 millidarcy or less.

Low permeability is primarily due to the fine-grained nature of the sediments, compaction, or infilling of pore spaces by carbonate or silicate cements.

Gas production from a tight-gas well will be low compared with gas production from conventional reservoirs (MIT, 2010). A lot of wells have to be drilled to get most of the oil or gas out of the ground in unconventional reservoirs. Tight gas requires multiple fracturing in order for any significant amount of gas to be available.

Exploration for tight gas sand reservoirs differs from conventional reservoirs in that tight gas sands are continuous, consisting of a stacking of sedimentary layers that are charged with oil or gas much in the same way that an aquifer is charged with water. Conventional reservoirs have much more limited boundaries, and well as a down-dip water contact, which is absent from the continuous reservoirs. Also, the great preponderance of continuous reservoirs are charged with gas, rather than crude oil.

While some continuous reservoirs may be found at a relative shallow depth, many are located at greater depths of 3000 m.

The key to producing this vast resource is locating areas and drilling wells where natural fractures abound "a sweet-spot". The distribution, orientation, and density of these fractures is key to proper planning and well scheduling in tight gas reservoirs. Identifying the best locations for tight-gas well completions requires a host of reservoir evaluation technologies, including seismic interpretation, logging and well testing.

Unless natural fractures are present, almost all tight reservoirs must be fracture-stimulated to obtain economic production rates.

2.1.1 Tight gas sands resources

Although tight gas reservoirs are scattered across the planet, most are concentrated in North America, Russia and China.

Estimates put the total volume of these unconventional resources at 310,000 bcm. Until the past decade, tight gas was considered non-economic to produce, but improvements in technology have clearly made a big difference in the amount of unconventional gas that can be recovered economically.

Only a 10% recovery rates for this type of reservoir guarantees reserves of 30,000 bcm, a not-significant amount compared with the 187,000 bcm of proved conventional gas reserves for the total World (Kuuskraa & Stevens, 2009). However, unlocking them raises a number of challenges, both technical and financial.

Historically, tight gas has been the most significant source of unconventional gas production in the United States, and is likely to remain so for some time. Tracking tight gas production in the World can be difficult because it can exist in a continuum with conventional gas.

2.2 Shale gas

Shale Gas is natural gas that is produced from reservoirs predominantly composed of shale (a fine-grained sedimentary rock which is easily breakable into thin, parallel layers), rather than from more conventional sandstone or limestone reservoirs.

Shale has low matrix permeability, so gas production in commercial quantities requires fractures to provide permeability. While a gas well in a Qatari reservoir might produce 4 million m^3/day, a shale gas well (without stimulation) might produce only 0.15 million m^3/day.

Shale gas has been produced for years from shale with natural fractures; the shale gas boom in recent years has been due to modern technology in hydraulic fracturing to create extensive artificial fractures around well bores. Horizontal drilling is often used with shale gas wells, with lateral lengths up to 3000 meters within the shale, to create maximum borehole surface area in contact with the shale.

Unlike conventional gas production, shale gas potential is not confined to limited traps or structures, and may exist across large geographic areas. Total natural gas resource potential for gas shale at US has been estimated to be from 10 to 25 Tm^3 of recoverable gas resources (EIA, 2011), compared with the actual volume of 7.7 Tm^3 of proved gas reserves at US (BP, 2011).

At present, most commercial shale-gas development is in the US and Canada. To date, tight gas, as well as other unconventional resources, remains underexploited outside North America.

2.3 Coal bed methane

Coal Bed Methane (CBM) refers to gas that contains a high percentage of methane and comes from coal deposits underground. CBM is natural gas generated and stored in coal seams. Coal seams have a dual porosity system comprising micropores, which exist in the coal matrix, and a system of natural fractures called cleats, which are the macropores. In a CBM gas reservoir, water completely permeates coal beds, and its pressure causes the methane to be adsorbed onto the grain surfaces of the coal.

The ability for coal reservoirs to store coal-bed methane depends on pressure of the reservoir, composition of coal, coal rank, and structure of the micropores, molecular properties of adsorbed gas and temperature of the reservoir.

To produce CBM, the water must be drawn off first, lowering the pressure so that the methane will desorbs from the coal and then flow to the well bore. As the amount of water in the coal decreases, gas production increases.

Coal beds tend to have low permeability, so that fluids do not flow easily through them unless the reservoirs are stimulated, for example, with hydraulic fracturing.

2.3.1 Coal bed methane resources

Coal bed methane is probably widely spread around the world. Large amounts are known to exist, notably in Australia, Canada, China, Germany, India, Indonesia, Poland, Russia and South Africa (Kuuskraa & Stevens, 2009).

Coal Bed Methane resources are to be found at a deep relatively shallow (usually 200 - 1000 meters depth) but unfeasible to be mined. It is important to notice that not all coal is suitable for methane production. Long-term brown coal fields are poor with methane. Anthracite is marked by high gas concentration but it is impossible to extract it because of the high density and quite low field openness. Methane is usually found in sub-bituminous and bituminous coal, located in the middle between brown coal and anthracite.

The production of gas from CBM reservoirs is unconventional compared to production of sandstone or carbonate reservoirs. Since no entirely reliable technology yet exists to assess how much gas a particular coal bed can yield, the methane gas extraction process is often one of trial and error. CBM exploratory wells are drilled in an attempt to find an economically viable quantity of trapped methane. If the trial is successful, other wells are drilled to produce the methane by bringing it up to the surface, where it is processed and transported through pipelines to the market.

2.3.2 Environmental impact

A high production of coal-bed methane carries with it some technological and environmental difficulties and costs. As water pressure provides the necessary pressure to keep the gas confined within the coal, large volumes of co-produced water are necessary to reduce pressure before gas can be brought to the surface. The amount of water produced varies widely among basins with CBM production, obtaining volumes of water of 4 to 64 m^3 per day and well (US Geological Survey, 2006).

Management of co-produced water is a concern because of the large volumes of water involved and because of the composition of the water. This water, which is commonly saline but in some areas can be potable, must be disposed of in an environmentally acceptable manner.

Common water management strategies include discharge into surface drainages, stock ponds, evaporation ponds and infiltration ponds. In some cases, water is re-injected into subsurface rock formations. In cold regions, it is possible to freeze the water in the winter, collect the salts that separate out, and dispose of or utilize them independently of the water, which can be discharged.

2.4 Gas hydrates

Gas hydrates are gas deposits trapped in ice crystals in permafrost and on the ocean floor. The gas resource contained in hydrates is estimated to be larger than all other sources of natural gas combined, but such gas is not commercially producible with today's technologies.

Globally, the total amount of methane sequestered in these deposits probably exceeds 2,000 Tm^3. Most of this methane is trapped in highly disseminated and/or low saturation methane hydrates that are unlikely to ever be a commercially viable gas source.

3. Main technology improvements to exploit unconventional gas resources

The principal drivers of unconventional gas production have been technological advances in drilling and completion. Many of the technical challenges are common to CBM, tight gas, and shale gas.

3.1 Hydraulic fracturing

During the 1980's, the introduction of massive artificial fractures, known as hydraulic fracturing, was proving successful in the Barnett shale. With the evolution of this hydraulic fracturing technology, the application of a simple blend of water and sand treatments and high pump rates demonstrated the potential for wide scale exploitation at commercial rates. This had the effect in the Barnett shale of doubling initial well rates, with commensurate increases in cumulative recovery.

Hydraulic fracturing (also known as fracking) is a technique used to create fractures that extend from the well bore into rock or coal formations. These fractures allow the oil or gas to travel more easily from the rock pores, where the oil or gas is trapped, to the production well.

Typically, in order to create fractures a mixture of water, proppants (sand or ceramic beads) and some specialty high viscosity fluid additives is pumped down the well at high pressures for short periods of time (hours).

Eventually, the formation will not be able to absorb the fluid as quickly as it is being injected. At this point, the pressure created (up to 500 bar) exceeds the rock strength and causes the formation to crack or fracture.

The sand carried by the high viscosity additives is pumped into the fractures to keep them from closing when the pumping pressure is released. The high viscosity fluid becomes a lower viscosity fluid after a short period of time. Both the injected water and the now low viscosity fluids travel back through the man-made fracture to the well and up to the surface.

Ideally, hydraulic fracture treatment design is aimed at creating long, well-contained fractures for maximum productivity. Fracture migration to the bounding layers is akin to failure of the stimulation job. Problems posed by unconfined fracture growth may include massive fluid loss.

The fluids currently used for fracturing a well are water (90%), mixed with sand (9%) and other additives (1%) like potassium chloride or other friction reducing additives. Sometimes, fracturing is initiated with the pumping of an acid treatment (water with some hydrochloric acid), in order to dissolve part of the rock material, so that the rock pores open and fluid flows more quickly into the well.

During the 2000's, the technology evolved to allow single-trip multi-stage hydraulic fracturing systems and zone isolation, which is driving the cost-effective exploitation of difficult reservoirs.

The location of fracturing along the length of the borehole can be controlled by inserting tough inflatable plugs, also known as bridge plugs, below and above the region to be fractured. This allows a borehole to be progressively fractured along the length of the bore, without leaking fracture fluid out through previously fractured regions. The plugs are inserted into the bore deflated, then expanded to seal off the borehole into a small working region. Piping through the upper plug admits fracturing fluid and proppant into the working region.

3.2 Horizontal /directional drilling

Horizontally drilled wells were first drilled in Texas in the 1930's. The technology has been continuously improved and developed; and by the 1980's, horizontal drilling has become a standard industry practice.

In late 1990's, the application of horizontal drilling enabled more aggressive development as multiple transverse fractures could be placed along a horizontal lateral well-bore.

The technology of horizontal well completions was first adapted for shale gas development to provide increased wellbore exposure to the reservoir area while allowing for a reduced number of surface locations in urban areas. Barnett horizontal wells have laterals ranging from 1000 to more than 3000 meters. A horizontal well may cost three times more than a vertical well, but horizontal drilling provides much more exposure to a formation than does a vertical well.

Thus, horizontal wells can reduce the number of wells needed to developing a gas field. In addition, one can significantly reduce the overall number of well pads, access roads, pipeline routes, and production facilities required, thus minimizing impacts to public and overall environmental footprint.

At the late 1990s, only 40 drilling rigs in the US (6% of total active rigs) were capable of onshore horizontal drilling; that number grew to 519 rigs (28% of active rigs) by 2008.

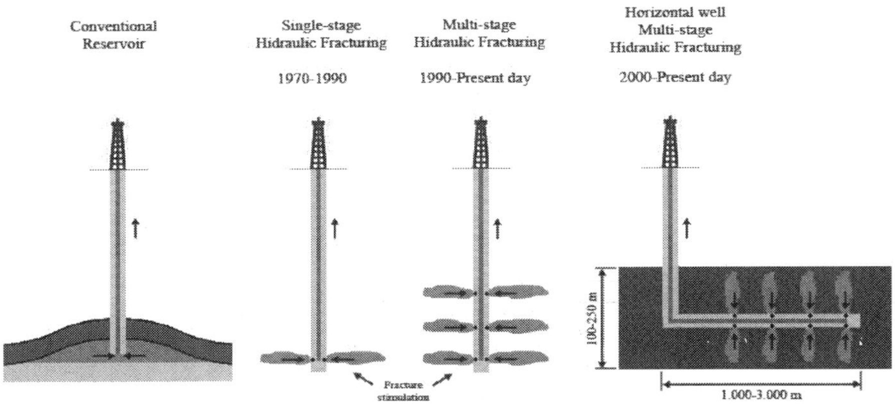

Fig. 2. Main technology improvements to exploit unconventional gas resources.

3.3 Reservoir evaluation, characterization, modeling and sweet- spot detection

A collection of improved exploration and production technologies has aided the development of unconventional gas. Apart from fracking and horizontal drilling, mayor advances have occurred in data acquisition, data processing and integration of seismic data with other geological data.

The tools used by geologist include 3-D seismic, gravity and magnetic surveys, combined with advances in computer power. Ultimately, successful reservoir characterization will result in the ability to accurately develop 3-D representations of these reservoirs. Unconventional gas reservoirs are more difficult to model than conventional gas reservoirs because the flow behavior is transient for much longer periods of time before stabilization. Current logging tools provide measurement of permeability, effective porosity, organic content, gas content, gas saturation, water saturation, clay mineralogy, pressure, temperature, bulk density and other properties.

Understanding of the geo-mechanical and geo-chemical characteristics of the resource rock will add to the ability to locate and characterize natural and artificially created fractures. Developing detection methods for "sweet spot" (areas with natural fracture) is very important to identify where production performance will be most attractive.

At the same time micro seismic technology is used to track and confirm fracture length, width and orientation while carrying out a fracture treatment. This allowed more efficient control of the process, better placement of wells and more economical exploitation.

In parallel, advances in drilling, casing and completion of wells are available through the conventional oil and gas industry. Continuous progress, experience and competition are bringing down the cost of applying these techniques, lowering the economic threshold and making the technology available to more producers.

3.4 Environmental concerns

An environmental concern raised by hydraulic fracturing is the possibility of introducing contaminants into aquifers (Arthur et al, 2008). Hydraulic fracturing does induce new fractures into shale, and can propagate fractures up to 500 - 1000 meters along the bedding plane of a shale formation. The potential for propagating fractures to an overlying aquifer may depend on the depth separating the two.

Frac jobs are always designing to limit the fractures to the height of the gas-producing shale zones, because any fracture propagated to an overlying aquifer could allow water to flow down into the gas-producing portion of the shale, and could significantly hamper gas production.

A frac treatment may pump as much as 10,000 m^3 of fluid for each well. A large proportion of these fluids (60 to 80%) are recovered afterwards by pumping them out of the well. However, fracturing fluids spilled on the ground surface could infiltrate downwards to shallow groundwater. This could pose a risk to superficial aquifers composed of permeable sand.

Unconventional gas shale's resources are typically deeper than coal bed methane formations and are often geologically isolated from drinking water aquifers. The environmental concern may be higher in the shallow coal bed methane formations because these shallow coal beds can also contain high quality water which could meet drinking water standards.

A properly designed and cased well will prevent drilling fluids, hydraulic fracturing fluids, or natural gas from leaking into the permeable aquifer and contaminating groundwater.

Although fracturing operations related to oil and gas production are exempted from the US Safe Drinking Water Act, the item remains highly controversial.

4. The expansion of non conventional gas in North America

4.1 Shale gas basins

4.1.1 The expansion of US Gas Shale production

The extraction of natural gas from shale formations has been occurring in parts of the United States since earliest years of gas development. However, many of these early wells were never able to produce a marketable quantity of natural gas.

Shale gas is actually the fastest growing energy sector in US (US Department of Energy, 2009), driven by the technology improvements and the presence of big shale basins.

The development of shale gas plays has become a "game changer" for the U.S. natural gas market.

In 1996, shale gas wells in the United States produced 8 bcm; by 2006, production raised to 30 bcm, 5.9% of US gas production.

By 2005 there were 14,990 shale gas wells in the US. A record 4,185 shale gas wells were completed in 2007.

In 2008, US Shale gas production was 55 bcm, making up more than 10% of total US production of dry gas

The proliferation of activity into new shale plays has increased shale gas production in the United States to 138 bcm in 2010, or 23% of U.S. dry gas production.

Shale gas reserves have increased to about 1 700 bcm by year-end 2009, when they comprised about 21 percent of overall U.S. natural gas reserves, now at the highest level since 1971. This number does still not counts estimated gas resources in new areas.

4.1.2 Main US gas Shale Basins

Most of the natural gas containing Devonian shale in the US is located around the Appalachian Basin. Devonian shales are formed from the mud of shallow seas that existed about 350 million years ago, during the Devonian period of the Paleozoic era.

The two big Shale Basins are the Barnett Shale in Texas and the Marcellus Shale, but there is also strong growth in Fayetteville and Haynesville in particular, with modest growth in several others. In Canada, main gas shale areas are Horn River and Montney, both in British Columbia.

The seven most promising shale gas plays in North America - referred to by the industry as the 'Magnificent Seven'- are likely to double output by 2020 (US Energy Information Administration, 2011).

Main gas shale formations in US and Canada
Barnett (Texas)
Fayetteville (Arkansas)
Haynesville (Louisiana)
Woodford (Oklahoma)
Marcellus (WV, PA, NY)
Montney (Canada)
Horn River (Canada)

Table 1. Main gas shale formations in US and Canada (the magnificent seven).

The big seven gas shales in North America (USA and Canada) have around 20 Tm^3 of recoverable resources of natural gas, which represent about a 12% of the total estimated gas resources of these fields (Kuuskraa & Stevens, 2009).

4.1.3 The Barnett Shale

The relatively recent focus on shale gas development goes back 20 years, and is highlighted by the success achieved in the Barnett shale in Texas.

The Barnett Shale is a large natural gas reserve encompassing more than 12,000 km^2 and covering at least 17 counties in Fort Worth Basin, Texas. This layer of organically rich sediment ranges from 120 to 240 meters thick and lies about two kilometers below the surface. Experts believe that the Barnett Shale may be one of the largest onshore natural gas fields in the United States, containing around 1,000 bcm of recoverable resources of natural gas.

Mitchell Energy drilled the first gas well in the Barnett Shale in 1981. Large scale hydraulic fracturing was first used in the Barnett in 1986; likewise, the first Barnett horizontal well was drilled in 1992. Actually, there are more than 10,000 natural gas wells, and the Barnett is now the largest producing gas field in the US, and currently produces more than 6% of US natural gas production.

4.1.4 The Marcellus Shale

The Marcellus Shale is a formation covering 250,000 km^2 beneath much of Ohio, West Virginia, Pennsylvania and New York (an area equivalent to half Spain). These states contain also some of the most densely populated areas in the United States. Marcellus has the advantage of being close to the north-east US gas market

The first Marcellus gas production from the well began in 2005. Early estimates indicate that the Marcellus Shale might contain 50 Tm^3 of natural gas resources.

Using some of the same horizontal drilling and hydraulic fracturing methods that had previously been applied in the Barnett Shale, perhaps up to 10% of that gas (5,000 bcm) might be recoverable. That volume of natural gas would be enough to supply the entire United States for about ten years.

Since 2006, when the big potential of the Marcellus was first suspected, many landowners are being approached with offers to lease their land, and a lot of companies are actively drilling or leasing Marcellus Shale properties, so in several years, Marcellus could be producing as much gas as Barnett shale.

A similar process is starting to occur at Fayetteville and Haynesville basins, so we can expect a strong growth of US shale gas production at least for the next 3 to 5 years.

However, the production decline rate of a shale well is much higher than those of a conventional gas wells, which means than continuous drilling is needed to maintain plateau production.

4.2 Tight gas basins

Tight gas is today the largest source of unconventional gas at US, with an annual production around 160 - 180 bcm, that represents about a 30% of total US production of dry gas, with more than 100,000 tight producing wells (US Energy Information Administration, 2011). In Canada, tight gas represents about 15% of gas production.

Tight gas wells started being produced not by major oil companies, for the most part, but by small, independent operators who focus on recovering production from old fields.

First large scale production of tight sands was developed in the 1970's in San Juan Basin. The San Juan Basin, located in north-western New Mexico and south-western Colorado, is a predominately tight basin with 13,000 tight producing wells.

Rapid progress in fracturing techniques led to the launch of tight gas production in the United States, from 35 bcm in 1995 to more than 150 bcm since 2007.

Most of the US's tight-gas proved reserves are in the Rocky Mountain region. Total recoverable resources at US are estimated around 9 Tm^3.

4.3 Coal Bed Methane basins

4.3.1 The expansion of US Coal Bed Methane production

To date, CBM industry has been developed only in United States, Australia, Canada and Colombia.

In United States, coal beds were primarily developed through programs led by the Department of Energy. The Crude Oil Windfall Profits Act (1980) provided tax incentives to develop unconventional fuels including CBM. The definition of deregulated natural gas addressed in this Act included occluded gas - naturally occurring natural gas released from entrapment from the fractures, pores, and bedding planes of coal seams. Also specified were gas produced from deep (greater than 3,000 m), high cost natural gas reservoirs, natural gas dissolved in an over-pressured brine and natural gas produced from Devonian shale. Coal Bed Methane production rapidly increased from 5 bcm in 1990 to 27.1 bcm in 1995 and in 2008 it reached 56 bcm, which accounts for nearly 10% of U.S. gas production. The Wasatch Plateau, Utah, and the Powder River Basin, Wyoming and Montana, are two of the newest, most productive areas of coal-bed methane activity in the United States.

In Australia, CBM production started in 1998 replacing ageing conventional gas fields, due to Government incentives for gas-fired power. In 2008, CBM accounted for around 7% of Australian gas production.

4.3.2 Main US Coal Bed Methane basins

The two big CBM basins are the Powder River Basin in Wyoming and Montana, and San Juan Basin in Colorado and New Mexico.

According to the Potential Gas Committee estimated data, CBM basins in US have around 4.5 Tm^3 of recoverable resources of natural gas.

a. The Powder River Basin CBM

The Powder River Basin is located in north-eastern Wyoming and south-eastern Montana. It is an area of approximately 55,000 km^2 that is underlain by many coal seams. The basin extends over 400 km from Douglas, Wyoming, in the south to Forsyth, Montana, in the north. Extraction of methane gas from the coal seams that underlie the Powder River Basin began in Wyoming in the late 1980s and in Montana in the late 1990s (US Geological Survey, 2000). With advances in technology, development and production of CBM has been increasing substantially since the mid 1990s. The basin is characteristic because of the extraordinary thicknesses of individual seams, between 15 and 67 m; most of this resource is at a depth of 760 m or less.

b. San Juan Basin CBM.

The San Juan basin is approximately 23,300 km^2 in north western New Mexico and south-western Colorado. It is the leading producer of coal bed gas in the world.

Conventional gas exploration began in the early 1900s. The first well was drilled in 1901, but the first commercially successful well was drilled in 1921. Thousands of wells have been drilled in the San Juan basin since then.

CBM development started in earnest in the late 1980s, in the Fruitland Formation of the Northern San Juan Basin. The basin has experienced highly successful CBM production because of the favorable characteristics of the coal seam related to thickness, permeability, gas content, depth, and coal rank in a large area.

5. Unconventional gas production at the United States and EIA projections

Tight gas was the first unconventional production gas developed, and is today the largest source of unconventional gas at U.S. However, shale gas production in the United States grew from 20 to 100 bcm over the 2006-2010 period, and is actually the fastest growing energy sector in U.S.

In 2010, natural gas production in United States was around 590 bcm, including 186 bcm of tight gas (31% of US production), 93 bcm of shale gas (16%) and 51 bcm of coal bed methane (9%).

The IEA Annual Energy Outlook (2011) estimates that US natural gas production will grow from 590 bcm in 2009 to 740 bcm in 2035 (IEA reference case). The increase in natural gas production from 2009 to 2035 results primarily from continued exploration and development of shale gas resources, and shale gas is the largest contributor to production growth.

Shale gas production could reach 350 bcm in 2035 (47% of total U.S. production, nearly triple its 16% share in 2009). Only two years ago, the reference scenario of IAE considered a shale gas production of "only" 130 bcm in 2030.

The projections for tight sands and coal bed methane deposits remains more stable, around 170 and 50 bcm per year, respectively, during all the period [Figure 3].

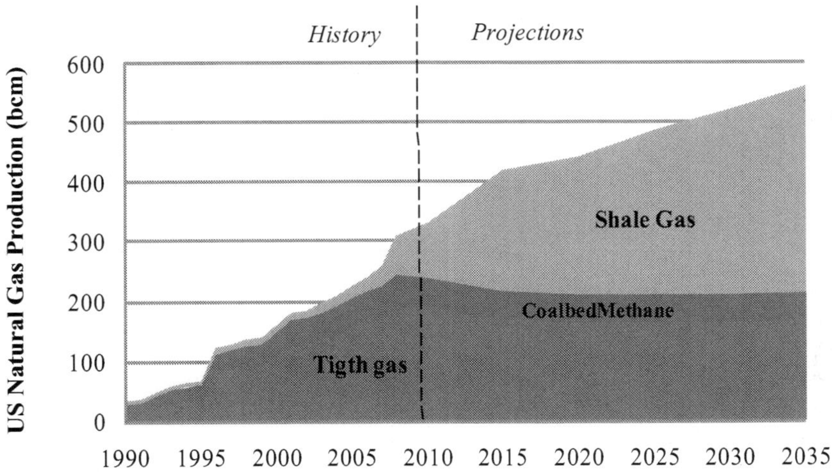

Fig. 3. US Unconventional Gas Production and Projections to 2035. Source: US Energy Information Administration, 2011.

The Energy Information Administration estimates that the US has more than 47 Tm^3 of technically recoverable natural gas, including 6 Tm3 of proved reserves (the discovered and economically recoverable fraction of the original gas in place). Technically recoverable unconventional gas (shale gas, tight sands, and coal bed methane) accounts for 60% of the onshore recoverable resource. At the US production rates for 2010 the current recoverable resource estimate could provide enough natural gas to supply the US for the next 75 years.

The proved reserves of shale gas in North America and the existing LNG infrastructure create the potential of LNG exports to Europe or Asia, which can help Europe to diversify its natural gas supply sources.

However, the high decline rates of unconventional gas will also require continuous investment in new wells to maintain production. The drilling activity is very sensitive to the price of gas. Since June 2008, the drop in natural gas prices have weaken the incentive to drill, but unconventional gas seems to maintain competitive even at low gas prices.

6. Estimations of recoverable resources and production for unconventional gas

Although understanding of the scale of global unconventional gas resources is improving, the complex issues related to unconventional gas resources mean that future production projections are subject to a large degree of uncertainty, particularly in regions where little or no such production has been undertaken to date.

Unconventional gas resources, although believed to be widespread, have not as yet been quantified on a national basis for most countries, apart from the United States and a few other countries.

Due to his early state of play, studies about unconventional gas do not have information about "proved reserves" of unconventional gas. Instead, they generally use estimates about the natural gas resource base, and technically /economically recoverable resources:

- **The natural gas resource base** - The broadest classification of natural gas estimates is generally termed the natural gas resource base. The total natural gas resource base includes the entire volume of natural gas contained and trapped in the earth, before any is extracted and produced.
- **Economically recoverable resources** - Economically recoverable resources are those natural gas resources for which there are economic incentives for production; that is, the cost of extracting those resources is low enough to allow natural gas companies to generate an adequate financial return given current market conditions. However, it is important to note that economically unrecoverable resources may, at some time in the future, become recoverable, as soon as the technology to produce them becomes less expensive, or the characteristics of the natural gas market are such that companies can ensure a fair return on their investment by extracting this gas.
- **Reserves** - Those resources discovered, technically and economically recoverable. In general, reserves can be broken down into two main categories - proved reserves, and other reserves.

- **Proved Reserves** - Proved reserves are those reserves that geological and engineering data indicate with reasonable certainty to be recoverable today with current technology and under current economic conditions.

6.1 World projections of the International Energy Agency (World Energy Outlook 2011).

In the Gas Scenario of the IEA World Energy Outlook 2011, the total gas production grows from an estimated 3.3 Tm^3 in 2010 to 5.1 Tm^3 by 2035, an increase of more than 50%, and more than double gas production in 2000. The average annual growth in gas production is 2% from 2008 to 2020, and then moderates to around 1.6% for the remainder of the Outlook period.

Natural gas production increasingly comes from unconventional sources. Unconventional gas production meets more than 40% of the increase in demand over the period: unconventional gas output worldwide will rise from 367 bcm in 2007 and is projected to reach 1200 bcm in 2035. As a result, the share of unconventional gas in global gas production is expected to rise from 12% in 2008 to nearly 25% in 2035.

Most of the increase in unconventional gas production in the EIA Gas Scenario comes from shale gas and CBM. In particular, the share of shale gas in global gas production reaches 11% in 2035, while CBM reaches 7% and tight gas 6%.

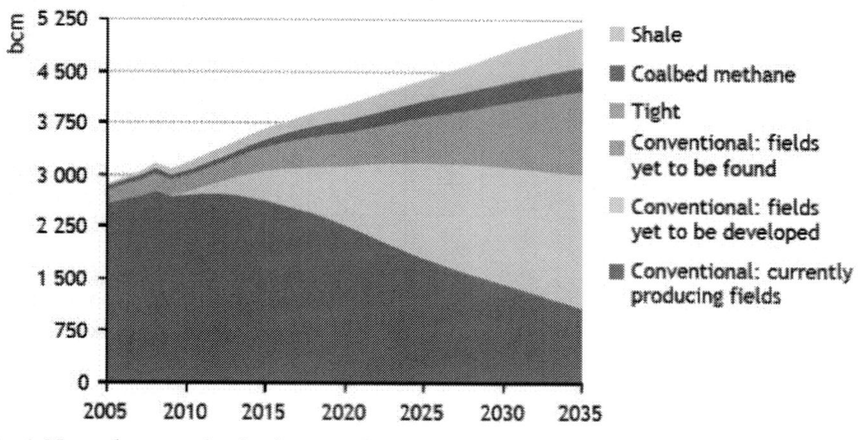

Fig. 4. Natural gas production by type. Source: IEA Energy Outlook 2011.

The natural gas resource base is vast ant widely dispersed geographically, with conventional recoverable resources equivalent to over 120 of current global consumption, while total recoverable resources could exceed 250 years, according to IEA.

Unconventional gas resources, comprising shale gas, tight gas and coal bed methane are estimated to be as large as conventional resources. IEA analysis suggests than plentiful volumes can be produced at costs similar to those in North America (between 3 \$/mmBtu and 7 \$/mmBtu).

Unconventional gas production is currently concentrated in the United States and Canada. By the year 2035, unconventional gas also reaches a significant scale in China (CBM and shale), Russia (tight gas), India (shale) and Australia (CBM).

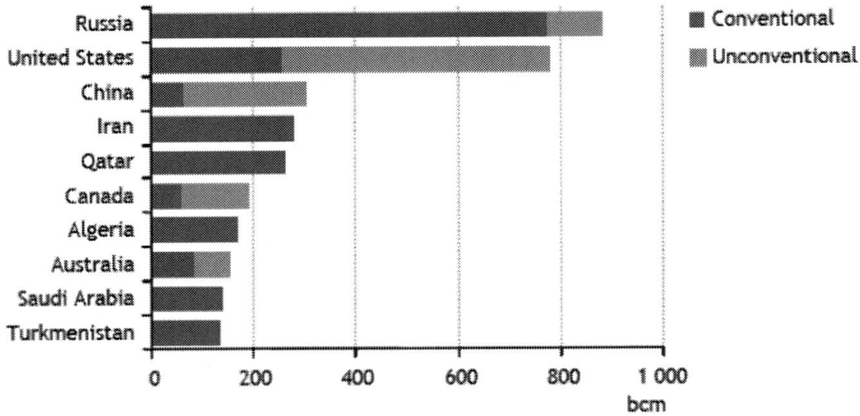

Fig. 5. Largest gas producers by type in 2035. Source: IEA Energy Outlook 2011.

6.2 Assessment of world shale gas resources by US Energy Information Administration (2011)

To gain a better understanding of the potential of international shale gas resources, EIA commissioned an external consultant, Advanced Resources International, Inc. (ARI), to develop an initial set of shale gas resource assessments, excluding tight gas and CBM (EIA, 2011).

Although the shale gas resource estimates will likely change over time as additional information becomes available, the report shows that the international shale gas resource base is vast. The initial estimate of technically recoverable shale gas resources in the 32 countries examined is 163 Tm^3. Adding the U.S. estimate of the shale gas technically recoverable resources of 24 Tm^3 results in a total shale resource base estimate of 187 Tm^3 for the 33 countries assessed, as shown in Table 2.

To put this shale gas resource estimate in some perspective, world proven reserves of natural gas as of January 1, 2010 are about 188 Tm^3 and world technically recoverable gas resources are roughly 453 Tm^3, largely excluding shale gas. Thus, adding the identified shale gas resources increases world technically recoverable gas resources by over 40% to 640 Tm^3.

Russia and Central Asia, Middle East, South East Asia, and Central Africa were not addressed by this report. This was primarily because there was either significant quantities of conventional natural gas reserves noted to be in place (i.e., Russia and the Middle East), or because of a general lack of information to carry out even an initial assessment.

Technically Recoverable Shale Gas Resources (Tm^3)		
North America	United States	24,4
	Canada	11,0
	Mexico	19,3
	Total North America	**54,7**
South America	Argentina	21,9
	Brazil	6,4
	Chile	1,8
	Paraguay	1,8
	Bolivia	1,4
	Uruguay	0,6
	Colombia	0,5
	Venezuela	0,3
	Total South America	**34,7**
Asia	China	36,1
	Australia	11,2
	India	1,8
	Pakistan	1,4
	Total Asia	**50,5**
Africa	South Africa	13,7
	Libya	8,2
	Algeria	6,5
	Tunisia	0,5
	Morocco	0,3
	Western Sahara	0,2
	Mauritania	0,0
	Total Africa	**29,5**
Europe	Poland	5,3
	France	5,1
	Norway	2,4
	Ukraine	1,2
	Sweden	1,2
	Denmark	0,7
	U.K.	0,6
	Netherlands	0,5
	Turkey	0,4
	Germany	0,2
	Lithuania	0,1
	Total Europe	**17,6**
Total	**Total 33 countries**	**187,0**

Table 2. Technically Recoverable Shale Gas Resources. Source: AIE. World Shale Gas Resources: An Initial Assessment, 2011. Note: Only 33 countries are covered by the report. In particular, Russia is not included in the assessment.

6.3 Unconventional gas exploration outside US

The recent US success in developing shale gas has prompted other countries to consider its own unconventional gas potential.

Most of the mayor oil and gas companies are actually making investment in US unconventional gas assets, in order to have access to these gas resources, but also to acquire technical exploration and production experience, and transfer this experience to other regions. The extent to which the boom in unconventional gas production can spread to other countries remains highly uncertain.

In China, India, Australia and Southeast Asia, unconventional gas might begin to make inroads into the supply mix in 5 to 10 years' time. This is probably a conservative projection, and there is potential to higher increases.

6.3.1 China

Shale gas and CMB looks promising in China, although appraisal is still at a very early stage. China currently is not producing shale gas, but it has very similar geological conditions to the US so could boast similar shale gas development potential. In fact, Chinese shale gas resource could be up to 26 - 36 Tm^3 of recoverable gas. Assuming the government supports the industry, China's unconventional resources could begin to have an impact from the middle of the next decade.

6.3.2 Europe

Shale gas exploration in Europe is in its infancy. As a consequence, little is known about Europe's ultimate potential.

There are some potentially major regional shale gas plays in Europe plus a number of others with local potential, but they don't seem equivalent to the American scale. Most promising are the Baltic Depression (mainly in Poland, also in Lithuania), Lower Saxon basin (Northwest Germany) and several areas in UK and Netherland, Sweden and Austria. In Ukraine, the shale gas potential exists in the Lublin Basin and the Dnieper-Donets Basin.

European governments, particularly in Eastern Europe (Poland, Ukraine), see an opportunity to reduce dependence on Russian gas imports.

Poland is seen as the most promising country for shale gas development in Europe, thanks to favorable geological and regulatory environments, and it leads the way to shale gas in Europe. Poland has already issued 221 licenses for exploration of hydrocarbons, of which 63 have been issued for shale gas. Exploration work covers 11% of Poland's area. Lane Energy Poland Sp. was the first company to start drilling in June 2010. Reliable information on the resource base will be probably available in four or five years.

According to the Polish Kosciuszko Institute, Polish shale gas exploration and production costs are 50% higher than in the US; however, gas prices could still be lower than those under long term oil-indexed contracts with Russia.

Some good news may also come from other countries, like Spain, were preliminary studies carried out in northern Spain estimates the presence of about 185 bcm of shale gas in the Gran Enara field in Alava.

By contrast, in **France** environmentalists have been campaigning against the fracking after a few of shale drilling licenses were awarded in the south of France and around Paris. In July 2011, France has approved to outlaw the fracking techniques, because the environmental concerns.

Russia and North African countries could also have important unconventional gas resources, but given the fact that Russia has the World large reserves of conventional natural gas, there may not be sufficient incentive to identify or exploit unconventional gas resources in the near term.

7. Economic impact on gas markets

Natural gas markets are smaller and less mature than oil markets. At present, trade is centered in three distinct regional gas markets – North America, Europe and Asia). Each region has a different market structure resulting from the degree of market maturity, the sources of supply, the dependence on imports and other geographical and political factors. Importantly, these regional markets set natural gas prices in different ways.

- In general, the U.S. has gas-on-gas competition. Robust spot and derivatives markets have developed in, with prices set by the forces of supply and demand.
- The European market relies more heavily on long-term contracts with price terms based on a mix of competing fuels, e.g., fuel oil.
- Asia uses crude oil as a benchmark for natural gas prices and favors long-term contracts. Japan and Korea almost totally dependent on imported LNG. This dependence places a high premium on security of supply and has kept prices in Asia high relative to other regions.

This regionalized and varied structure of natural gas markets stands in contrast to the global oil market. The physical characteristics of oil – a very high energy density at normal conditions of temperature and pressure – make it readily transportable over long distances, at moderate cost. This has allowed the development over time of a global oil market, where multiple supply sources serve markets at transparent spot prices.

In comparison, the physical characteristics of natural gas constrain transportation options. Unlike oil, transportation costs – whether for pipeline gas or liquefied natural gas (LNG) – constitute a significant fraction of the total delivered cost of natural gas (MIT, 2010).

7.1 US market decoupling from oil market

The rapid development of unconventional gas resources in the Unites States and Canada, particularly in the last three years, has transformed the gas market in North America.

Marginal cost of unconventional sources have fallen steeply, at a wellhead cost between 3 and 5 $/mmBtu. An additional benefit is that gas is found in areas with much of the necessary pipeline infrastructure already in place. Many of these areas are also proximal to big population centers thus potentially facilitating transportation to consumers.

This supplement to supply, combined with weak gas demand following the economic crisis, has led to a drop in US gas prices from 13.68 $/mmBtu in July 2008 to prices under 5 $/mmBtu in 2009.

The energy parity between oil and gas prices seems to be already broken at US markets since January 2009: Oil prices have hovered around 100 $/barrel for much of 2010 and 2011 while the U.S. Henry Hub (HH) price has been consistently below 5 $/mmBtu (Figure 6).

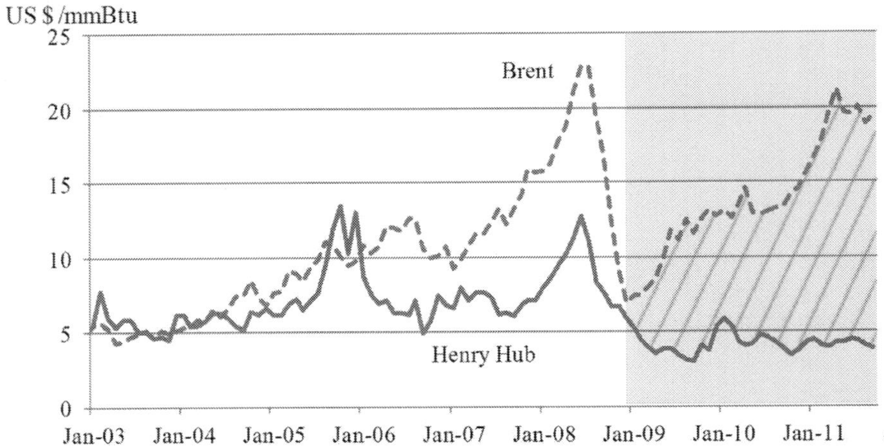

Fig. 6. Evolution of oil (Brent) and U.S. gas prices (Henry Hub). The energy parity between oil and gas prices seems to be already broken since 2009.

The development of unconventional gas resources means that North America gas production may be sustainable at current levels for decades.

In 2008, the trend of increasing import volumes stalled as net imports only met about 13% of overall US natural gas consumption in 2008, which is the lowest percentage since 1997.

A first consequence is that North American gas market has fewer needs to compete for available global supplies, as relatively low prices are expected to discourage imports of LNG, and can also delay the construction of gas pipelines from Alaska.

In contrast, oil prices maintain stronger than gas prices in the economic downturn scenario. While technological advances in finding and producing oil have made it possible to bring oil to the surface from more remote reservoirs at ever increasing depths, the total finding and lifting costs of new oil wells have increased sharply in recent years. Also, new oil discoveries can be insufficient to replace the depletion of existing ones.

When the world economy rebounds, oil global demand will put pressure on oil prices, and oil prices will rise faster than gas prices on the next decade (Stevens, 2010).

7.2 Decoupling US gas market from European gas markets

The European market still has not been affected by the rise of U.S. unconventional gas.

As Figure 7 shows, Europe currently offers prices in 2010 that are more than double those in the U.S. The UK's Natural Balancing Point price (NBP) has been in the 8 to 10 $/mmBtu

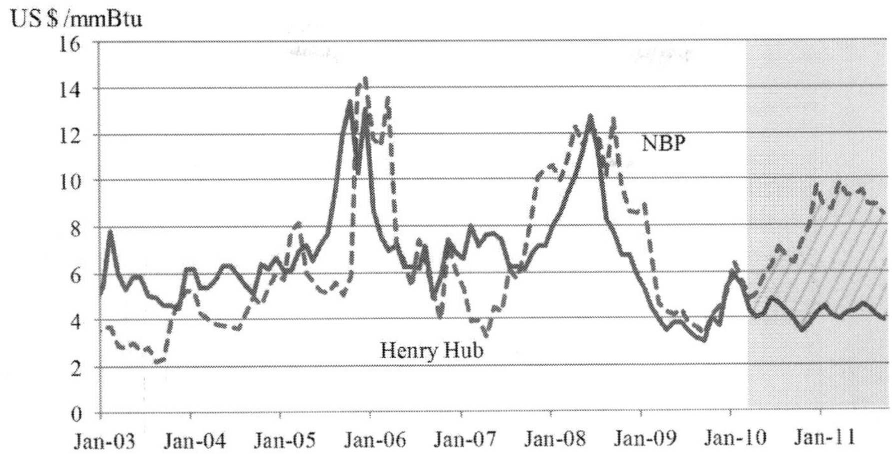

Fig. 7. Evolution of natural gas prices between U.K. market (NBP) and U.S. market (Henry Hub). Decoupling of price is observed since January 2010.

range for the first nine months of this year, while Henry Hub (HH) price has been consistently below 5 $/mmBtu.

Also long term contract (oil-linked) in continental Europe have been at an average premium of $1.60/mmBtu over NBP during 2011.

The global economic and financial crisis, which began in late 2008, has significantly depressed European gas demand. In addition to that, substantial new LNG supply is coming on stream during 2009-10. Even with a depressed demand, LNG imports to Europe increased more than 10% in 2009, particularly to the UK market.

The substantial short term supply gas surplus in Europe is increasing the pressure for change in the price-setting mechanism of European oil-linked long term gas contracts (Stern, 2009). Short term prices at market hubs are at around half of oil-linked levels in 2009.

The premium price paid for LNG in Europe and Asia over gas prices North America has prompted some companies to plan to turn US under-utilized LNG receiving terminals into export plants by building liquefaction trains. This would be attractive if the actual gap between US and Europe /Asian gas prices, that compensate the transport cost, continue in the next years.

While it seems unlikely to avoid oil - price indexation for Europe and Asian long term gas contracts, they will likely start to reflect the gas bearish conditions with decreasing indexation slopes. Buyers across Europe push for a move to hub-based pricing, which will increase price volatility and could push the price level down towards the UK's NBP price. The change could be more rapid if some unconventional fields start to develop in Europe.

8. Conclusions

There are abundant supplies of natural gas in the world, and many of these supplies can be developed and produced at relatively low cost. Consequently, natural gas is set to play a key role in meeting the XXI century world's energy demand and his production is likely to continue to expand.

According to IEA, the total gas production will grow a 50% by 2035. Unconventional gas accounts for 40% of the increase in global supply, with new non-US producers emerging.

In total, unconventional gas resources could add between 60 to 250% to world gas reserves.

Main potential is in US, Canada, China, Russia and Australia.

In North America, shale gas development over the past decade has substantially increased assessments of resources producible at modest cost. The pace of shale technology development has been very rapid over the past few years.

The emergence of shale gas as a potentially major energy resource can have important strategic implications for geopolitics and the energy industry:

- Shale gas resources in the United States will keep natural gas prices relatively low for an extended period of time. Longer periods of lower gas prices will likely result in additional demand for gas from the power generation sector.
- Global conventional natural gas resources are concentrated geographically, with 70% in three countries: Qatar, Iran and Russia. Unconventional supplies could provide a major opportunity for diversification and improved security of supply in some parts of the world.

In favor	Barriers
• Large potential gas resources worldwide. • Technology is already competitive in USA. Major oil and gas companies will contribute to spread the technology around the world. • China, India and Australia have large unconventional resources and can be next countries to develop them in a large scale • It is possible to locate new reservoirs in OCDE countries (close to consumption) • In Europe exploration is underway in some countries (Poland, Ukraine).	• Limited knowledge of location and volume of gas reservoirs. • Incertitude about extraction costs outside USA, and rate of decline of wells. • Lack of interest of some exporting countries (mainly Russia). • Environmental concerns (example: France) • Low gas demand and gas prices will difficult the financing of new projects

Fig. 8. Main factors affecting the development of unconventional gas resources.

- Shale gas exploitation can benefit some European countries, but it will not significantly reduce dependency on gas imports from Russia, at least in short term. However, unconventional gas may create pressure on gas exporters to move away from oil-price indexation.

However, there are also a number of key uncertainties:

- Limited knowledge of location and volume of unconventional gas reservoirs.
- Environmental concerns, mainly in Europe.
- Incertitude about extraction costs outside US, and rate of decline of wells. Because of the shale gas revolution, there are now huge uncertainties for investors at all stages of the gas value chain.

9. References

Al-Husseini, *"Long-Term Oil Supply Outlook: Constraints on Increasing Production Capacity"*, presented at Oil and Money Conference, London, October 2007, Available: http://www.boell-meo.org/download_en/saudi_peak_oil.pdf

Aleklett, K, *"Peak Oil and the Evolving Strategies of Oil Importing and Exporting Countries"*. OECD International Transport Forum. Discussion Paper N° 2007-17, Dec, 2007

Arthur J.D.; Bohm B and Layne M., *"Hydraulic Fracturing Considerations for Natural Wells of the Marcellus Shale"*, presented at Ground Water Protection Council Annual Forum, Cincinnati, Ohio. September 2008.

BP Statistical Review of World Energy. June 2011. Available: http//www.bp.com/statisticalreview

International Energy Agency (IEA). World Energy Outlook 2011.

International Energy Agency (IAE). Golden Age of Gas 2011 (special report)

Kuuskraa V.A. and Stevens S.H., *"World Gas Shales and Unconventional Gas: A Status Report"*, presented at United Nations Climate Change Conference, Copenhagen, December 2009.

Massachusetts Institute of Technology, *Study on the Future of Natural Gas*, 2010

Stern J., *"Continental European Long-Term Gas Contracts: is a transition away from oil product-linked pricing inevitable and imminent?"*, Oxford Institute for Energy Studies, September,2009.

Stevens, Paul; *"The Shale Gas Revolution. Hype and Reality"*. Chatham House Report, 2010

U.S. Department of Energy, *Modern Shale Gas Development in the United States: A Primer.*, April 2009. Available: http// www.netl.doe.gov/

U.S. Geological Survey. November 2000. *Water Produced with Coal-Bed Methane*

U.S. Geological Survey. November 2006. *Coalbed Methane Extraction and Soil Suitability Concerns in the Powder River Basin, Montana and Wyoming*

U.S. Energy Information Administration (EIA). Annual Energy Outlook 2011 with projections to 2035. April 2011. Available: www.eia.doe.gov

U.S. Energy Information Administration. *World Shale Gas Resources: An Initial Assessment of 14 Regions Outside the United States*. April 2011

6

Fluid-Solid Coupling Numerical Simulation on Natural Gas Production from Hydrate Reservoirs by Depressurization

YuanFang Cheng and LingDong Li
China University of Petroleum (East China), China

1. Introduction

Natural gas hydrate is considered to be the most prospective energy in the 21st century for its various advantages, such as extensive distributing, vast amount of quantity, high energy density, and environmental-friendly.[1] Since the end of 1980s in last century, numerous countries have been playing highly attention on the exploration of natural gas hydrate and made a long-term plan from the point of energy stratagem.

The basic characteristics and particular storage style of natural gas hydrate determine its special developing methods which are different from the other traditional oil or gas energy sources. Currently, the exploiting method for natural gas hydrate is to transform natural gas hydrate into free gas initially and then stimulate the gas using some traditional ways. The presenting conventional approaches of developing natural gas hydrate, including thermal activation, pressure releasing, chemical inhibitor, replacement and slurry mining etc. are just some concept models[2] and the research relating to the exploitation of natural gas hydrate is still at the early stage, mainly for the economical and technological limits. Therefore, the large-scale exploitation of this energy isn't widely achieved in the world currently.

Unlike conventional energy sources, the petrophysical properties, some mechanical parameters, and pore pressure etc. would change dynamically due to solid phase hydrate decomposition during natural gas hydrate production. These phenomena is regarded as a very important and extraordinary characteristic for this energy and named as "gas hydrate decomposition effect" in this research.

Four major aspects which are different from traditional exploitation approaches are shown as follows:

1. When gas hydrates dissociate, the mechanical properties will change, which includes reduction in modulus, decrease in formation mechanical strength and loss of the cementation provided by the gas hydrates.
2. The formation of natural gas hydrate can occupy certain pore space and the space for solid phase hydrate would gradually reduce with the hydrate decomposing, while possible flowing spaces and permeability in formation would increase.

3. Water and a large volume of gas can be released because of hydrates decomposition, thereby resulting in pore pressure increase in limited formation pore spaces and the change of pore pressure can impact the formation effective stress distribution;
4. Additionally, absorbing heat during hydrates decomposition would cause the change of formation temperature.

Similar to conventional oil or gas energies, the exploitation of natural gas hydrates by depressurization is a dynamic coupling process between the rock deformation and flowing fluid in porous media. There are two different facets: To begin with, hydrates decomposition can generate phase transition while gas hydrate decomposition effect can lead to the dynamic and complex change of mechanical strength, physical properties and the pore pressure. Furthermore, absorbing heat during hydrates decomposition can give rise to formation temperature change and on the other hand, the formation temperature change will limit the decomposing rate of hydrates. Consequently, the exploitation process of natural gas hydrates is a complicated solid-fluid coupling process affected by multiple factors.

Nowadays, numerical simulation on the productivity of natural gas hydrates is increasingly becoming a spirited issue for scholars. Various numerous stimulation presented generally are only considering two-phase flow between gas and water and the formation permeability change by hydrate decomposition at the isothermal condition, while the change of mechanical properties and the formation temperature due to hydrates decomposition, the overall realization on the fluid-solid coupling seepage features of hydrates reservoir with the particular phase transition changing are ignored. What's more, the present models place a value on productivity analysis rather than the stability of the decomposing zones nearby wells with weakly cemented, low strength and high porosity and permeability.

The objectives of this study are finally to construct a theoretical analysis system for the whole exploiting process of natural gas hydrates production by depressurization with fluid-solid coupling numerical simulation, which are based on physicochemical theory. Moreover, on the basis of the dynamic change law of stresses status and physical properties for hydrates formation and considering some features of dissociation zones nearby wellbore with weakly cemented, low strength, high porosity and permeability etc, this research does relating research on fluid-solid coupling numerical simulation for natural gas hydrate production, which provides theoretical basis and necessary technological support for the industrial production of natural gas hydrate.

2. Establishment of fluid-solid coupling mathematical models

During the development of conventional oil and gas reservoirs, the production can cause the pore pressure decrease, redistribution of skeleton stress for solid phase, thereby resulting in rock skeleton deformation and the change of reservoir physical properties including porosity, permeability etc. Meantime, these changes can influence the seepage process of oil and gas. This phenomenon is called solid-fluid coupling effect.

The exploitation of natural gas hydrates by depressurization is a multiphase, dynamic phase, multi-component, non-isothermal and physical-chemistry seepage process. Besides traditional two-phase (gas and water) seepage mechanism, the process also includes morphology change of hydrates and the change of physical and mechanical properties, pore pressure and temperature of gas hydrates formation, which is caused by hydrates dissociation. Additionally, the mingled interaction between seepage field and solid deformation field is considerable. Consequently, the exploitation of natural gas hydrates by depressurization is a complex process influenced by numerous factors.

2.1 Research methods of fluid-solid coupling

Oil-bearing formation is a geological system which consists of cemented solid rock skeleton, pore space, and inner fluids (gas and water) in multi-phase porous media. Cemented rock skeleton and pore space make up of rock formation that is a basic unit for appraising the level of mechanical performances and deformation. A distinguished feature of solid-fluid coupling in seepage process, namely solid field and fluid filed mingle and include together while hard to differentiate clearly. Thus, fluid phase and solid phase should be seem as overlapping continuous media and the reaction can happen between different phases[3]. This feature contributes the establish of control equations for fluid-solid coupling model need to aim at special physical phenomena .Meanwhile, the effects of fluid-solid coupling are reflected by control equations, namely some items in control equations of fluid flow can describe solid deformation and vice verse.

According to the multiple definition of field, geological system is classified into interacting multiple solid deformation fields and multiphase fluid seepage fields in math. A solid deformation media field and fluid seepage field are connected by fluid system pressure and multiphase seepage fluid is connected by capillary pressure. The physical quantities of describing solid and fluid parameters are reflected by the representative elementary volume (REV) in continuous medium mathematic model.

Thus, porous media can be regarded as a large number of particles with enough porous and skeleton. Therefore, porous particles can define their materials parameters, such as fluid density, solid density and strength etc. Meanwhile, particles can bear stress and fluid pressure, namely particles can be defined as state variables. When the particle is enough small comparing the whole seepage filed, material parameters and variables are the function of points in space and are changing constantly with the continuous change of their positions. Consequently, porous media can virtually be replaced by a kind of hypothetical continuous media.

The analysis of coupling effect between fluid seepage and rock deformation requires to systematically utilize diverse theories , for instance, rock mechanics, permeation fluid mechanics etc, along with the research of coupling rules for seepage field and deformation and establishing relating control equations. Two meanings are included in this study. For one thing, under the effect of reservoir deformation field, this paper present a research about how to obtain its multiphase seepage rules (oil, gas and water); For another, the research about the stress, strain and strength problems etc. for formation rock under the circumstance of seepage field is discussed.

The solid-fluid coupling numerical simulation models for the exploitation of natural gas hydrate by depressurization mainly include four parts: temperature field, seepage field, deformation field and the relations of seepage and deformation coupling model.

As previously mentioned the exploitation of natural gas hydrates by depressurization is a complex solid-fluid coupling process including phase transition and is influenced by diverse elements. Some basic assumptions about solid-fluid coupling numerical simulation models in this study are as follows:

1. Formation rock abide generalized Hook law and Drucker-Prager yield criterion, and deformation of rock matrix is generally small displacement deformation;
2. Seepage system is made up of two phases(water and gas) at saturation without concerning the chemical effect between formation rock and fluid;
3. The flow of each phase in fluid satisfied generalized Darcy's law comparing rock skeleton particles;
4. Relative permeability of two phases (gas and water) and capillary pressure are the single function of water saturation;
5. Temperature only impact the decomposition of hydrates but not rock skeleton and fluid properties;
6. Formation rock skeleton density and solid hydrate density are constant during the process of deformation;
7. The relative velocity between solid hydrate and formation is zero and the relative velocity between gas and fluid is zero either.

2.2 Seepage field equations of fluid-solid model

2.2.1 Kinetic equation of gas hydrates decomposition

Kinetic equation of gas hydrates decomposition is proposed by Kim-Bishnoi[4], as follows:

$$m_g = K_{rd} M_g A_{dec} (f_e - f_g) \tag{1}$$

or

$$m_g = K_{rd} M_g A_{dec} (\varphi_e P_e - \varphi_g P_g) \tag{1a}$$

The intrinsic dissociation rate constant K_{rd} is a function of the system temperature.

$$K_{rd} = K_d^0 \cdot \exp(\frac{\Delta E}{RT}) \tag{2}$$

The specific surface area of the hydrate decomposition, A_{dec}, is given by

$$A_{dec} = \phi S_H A_{HS} \tag{3}$$

The rate of water generated is calculated according to the rate of the gas generated from the hydrate decomposition (m_g) and hydrate number (N_H), then the rate of the hydrate decomposition (m_H) is computed.

$$m_W = N_H M_W \, m_g N_H M_W / M_g \tag{4}$$

$$m_H = -M_H m_g / M_g \tag{5}$$

Where, N_H is the hydrate number, the value of which is 6.0 for methane hydrate.

2.2.2 Energy conservation equations of natural gas hydrate

With regard to heat conduction, convection and supplement of heat from outside and without consideration of kinetic energy and thermal radiation, energy conservation equations would be calculated by the following equation:

$$\begin{aligned}&\frac{\partial}{\partial t}\left[(1-\varphi)\rho_r H_r+\varphi S_h\rho_h H_H+\varphi S_g\rho_g H_g+\varphi S_w\rho_w H_w\right]=\\&=\nabla\cdot(K_c\nabla T)-\nabla\cdot\left(\rho_g\vec{u}_g H_g+\rho_w\vec{u}_w H_w\right)+Q_{in}\end{aligned} \tag{6}$$

The left side of Eq. (6) means the increment of internal energy and the first item at the right side of Eq. (6) reflects energy accessing the unit porous media system by heat conduction, the second item at the right side of Eq. (6) illustrates energy taken away from gas and water through hydrate compositing, the third item at the right side of the equation shows energy supplement to hydrate formation from outside.

Without considering throttling effect of influence, energy conservation equation of hydrate can be signified with temperature and illustrated by the following equation after equation groupies derived:

$$\begin{aligned}&\left[(1-\varphi)\rho_r C_{pr}+\varphi S_H\rho_H C_{pH}+\varphi S_g\rho_g C_{pg}+\varphi S_w\rho_w C_{pw}\right]\frac{\partial T}{\partial t}\\&+\nabla\cdot\left[(\rho_g\vec{u}_g C_{pg}+\rho_w\vec{u}_w C_{pw})T\right]-\nabla\cdot(K_c\nabla T)=-m_H\Delta H_D+Q_{in}\end{aligned} \tag{7}$$

Heat of hydrate decomposition causes formation temperature change and limitation for decomposition rate of hydrate. Therefore, the calculation of temperature field is significant to the research of hydrate reservoir simulation.

2.2.3 Fluid-solid coupling seepage equations of gas hydrate reservoirs

2.2.3.1 Rock matrix continuity equation of gas hydrate reservoirs

Rock matrix continuity equation is as follows:

$$\frac{\partial(\hat{\rho}_s)}{\partial t}+\nabla\cdot(\hat{\rho}_s\vec{v}_s)=0 \tag{8}$$

If rock particles density is ρ_s, then:

$$\hat{\rho}_s=\rho_s(1-\varphi) \tag{9}$$

Assume the density of rock matrix in hydrate reservoir remains unchanged during the deformation, namely ρ_s is constant, then:

$$\frac{\partial(1-\varphi)}{\partial t}+\nabla\cdot((1-\varphi)\vec{v}_s)=0 \tag{10}$$

Eq. (10) reflects the relationship between the porosity changes and matrix displacement velocity.

2.2.3.2 Fluid-solid coupling seepage motion equations of gas hydrate reservoirs

Similar with the normal reservoirs, during production from hydrate reservoirs by depressurization, not only fluid particle seepage take place in the porous media, but the reservoirs will undergo deformations and the rock particles will move for the rigid displacement and deformation displacement.

In the deformed porous media, the real velocity of fluid motion in the fluid-solid coupling seepage is[5]:

$$\vec{v}_{ra} = \vec{v}_a - \vec{v}_s \tag{11}$$

The seepage takes place in the porous media, so the real velocity of fluid relative to the rock skeleton is:

$$\vec{v}_{ra} = Q/(A_p S_a) \tag{12}$$

The Darcy velocity of fluid seepage is:

$$\vec{u}_a = Q/A = -\frac{K_a}{\mu_a}(\nabla P_a - \rho_a g) \tag{13}$$

There is a relationship as follows:

$$\varphi = V_p/V = (A_p L)/AL = A_p/A \tag{14}$$

So, the relationship between Darcy velocity $\vec{u}_a$ and the real velocity relative to the rock $\vec{v}_{ra}$ can be deduced as follows:

$$\vec{u}_a = \varphi S_a \vec{v}_{ra} \tag{15}$$

Then the relationship among the real velocity, Darcy velocity and solid rock skeleton velocity is:

$$\vec{v}_a = \vec{v}_{ra} + \vec{v}_s = \frac{1}{\varphi S_a}\vec{u}_a + \vec{v}_s \tag{16}$$

Eq. (16) is the fluid-solid coupling seepage motion equation.
While the velocity of rock skeleton particle is:

$$\vec{v}_s = \partial u/\partial t \tag{17}$$

Then: $v_{sx} = \partial u_x/\partial t \quad v_{sy} = \partial u_y/\partial t \quad v_{sz} = \partial u_z/\partial t$

2.2.3.3 Fluid-solid coupling seepage continuity equations of gas hydrate reservoirs

Because the object of this study is the law of fluid-solid coupling seepage, we must use the real velocity and the real mass.

To one of the phases in the seepage system α , the mass of per unit volume is: $m_a = \phi\rho_a S_a$;

While the mass flow rate is: $\phi\rho_a S_a \vec{v}_a$;

Then we take constituent i for example, and establish continuity equation.

Assume X_{ig} as the mass fraction of constituent i in the gas phase, X_{iw} as the mass fraction of constituent i in the liquid phase.

According to the principle of mass conservation, the continuity equation of constituent i can be obtained as follows:

$$-\nabla\cdot\left[X_{ig}\varphi\rho_g S_g\vec{v}_g + X_{iw}\varphi\rho_w S_w\vec{v}_w\right] + q_i = \frac{\partial}{\partial t}\left[X_{ig}\varphi\rho_g S_g + X_{iw}\varphi\rho_w S_w\right] \tag{18}$$

The above equation can be written in abbreviation as follows:

$$-\nabla\cdot\left(\varphi\sum_{a=g,w} X_{ia}\rho_a S_a\vec{v}_a\right) + q_i = \frac{\partial}{\partial t}\left(\varphi\sum_{a=g,w} X_{ia}\rho_a S_a\right) \tag{19}$$

For the hydrate reservoirs, there are three phases in the reservoir pore space: liquid phase, gas phase and solid phase. Considering that there will be gas and water products when the solid hydrates decompose, we need to add one output in the continuity equation to reflect the decomposition products of hydrate.

Ignoring the dissolution of gas in the water, we can get the fluid-solid coupling seepage continuity equations of gas phase, fluid phase and solid hydrate in the hydrate reservoirs.

Gas phase:
$$-\nabla\cdot\left(\varphi\rho_g S_g\vec{v}_g\right) + m_g + q_g = \frac{\partial(\varphi\rho_g S_g)}{\partial t} \tag{20}$$

Fluid phase:
$$-\nabla\cdot\left(\varphi\rho_w S_w\vec{v}_w\right) + m_w + q_w = \frac{\partial(\varphi\rho_w S_w)}{\partial t} \tag{21}$$

Solid hydrate:
$$-\nabla\cdot\left(\varphi\rho_H S_H\vec{v}_s\right) - m_H = \frac{\partial(\varphi\rho_H S_H)}{\partial t} \tag{22}$$

Where, Φ is the absolute porosity of reservoir; ρ_g, ρ_w, ρ_H are the densities of gas, water, hydrate, respectively; S_g, S_w, S_H are the saturations of gas, water, hydrate; $\vec{v}_g$, $\vec{v}_w$, $\vec{v}_s$ are the real velocities of gas, water, rock skeleton, respectively; q_g, q_w are the source or sink terms of gas and water, while set source as positive, sink as negative; m_g is the local mass rate of gas generation per unit volume of porous media; m_w is the mass rate of water generated from the hydrate decomposition, and m_H is the mass rate of methane hydrate decomposition.

2.2.3.4 Fluid-solid coupling seepage equations of gas hydrate reservoirs

The seepage of gas phase and water phase follow the Darcy law:

Gas phase: $$\varphi S_g \vec{v}_{rg} = -\frac{K_{rg}}{\mu_g}[K]\left(\nabla P_g + \rho_g \vec{g}\right) \tag{23}$$

Water phase: $$\varphi S_w \vec{v}_{rw} = -\frac{K_{rw}}{\mu_w}[K](\nabla P_w + \rho_w \vec{g}) \tag{24}$$

Where, $\vec{g}$ is the acceleration of gravity, the value is $\begin{bmatrix}0 & 0 & -9.8\end{bmatrix}^T$; $[K]=\begin{bmatrix} K_{xx} & K_{xy} & K_{xz} \\ K_{yx} & K_{yy} & K_{yz} \\ K_{zx} & K_{zy} & K_{zz} \end{bmatrix}$ is the permeability matrix of reservoirs. For the 3D problems, $[K]$ has nine coefficients. Because K_{ij} is equal to K_{ji}, there are six independent coefficients. To plane problems, $[K]$ has three independent coefficients.

The relationships among the real velocity of gas/water phase ($\vec{v}_g$, $\vec{v}_w$), the seepage velocity of gas/water phase relative to the rock skeleton ($\vec{v}_{rg}$, $\vec{v}_{rw}$) and the motion velocity of rock skeleton ($\vec{v}_s$) are as follows:

$$\vec{v}_{rg} = \vec{v}_g - \vec{v}_s \,;\, \vec{v}_{rw} = \vec{v}_w - \vec{v}_s \tag{25}$$

Substituting Eq. (25) into the continuity equations of gas/water phase Eq.(20) and Eq. (21), we get:

$$\frac{\partial\left(\varphi\rho_g S_g\right)}{\partial t} + \nabla\cdot\left(\varphi\rho_g S_g \vec{v}_{rg}\right) + \nabla\cdot\left(\varphi\rho_g S_g \vec{v}_s\right) = m_g + q_g \tag{26}$$

$$\frac{\partial(\varphi\rho_w S_w)}{\partial t} + \nabla\cdot(\varphi\rho_w S_w \vec{v}_{rw}) + \nabla\cdot(\varphi\rho_w S_w \vec{v}_s) = m_w + q_w \tag{27}$$

Combining Eq. (23) , Eq. (26), Eq. (24) and Eq.(27) can deduce the seepage equations of gas/water phase under the influence of reservoir skeleton deformation:

Gas phase: $$\frac{\partial\left(\varphi\rho_g S_g\right)}{\partial t} - \nabla\cdot\left(\frac{K_{rg}\rho_g}{\mu_g}[K]\left(\nabla P_g + \rho_g \vec{g}\right)\right) + \nabla\cdot\left(\varphi\rho_g S_g \vec{v}_s\right) = m_g + q_g \tag{28}$$

Water phase: $$\frac{\partial(\varphi\rho_W S_W)}{\partial t} - \nabla\cdot\left(\frac{K_{rW}\rho_W}{\mu_W}[K](\nabla P_W + \rho_W \vec{g})\right) + \nabla\cdot(\varphi\rho W S_W \vec{v}_s) = m_W + q_W \tag{29}$$

Compared with normal gas/water seepage equations, the fluid-solid coupling seepage equations, (28) and (29), have two extra parameters, m_l and $\nabla\cdot(\varphi\rho_l S_l \vec{v}_s)$ (l=g,w).The first parameter m_l reflects that the production from hydrate reservoirs by depressurization is a physical and chemical seepage process with phases changed; The latter one, $\nabla\cdot(\varphi\rho_l S_l \bar{v}_s)$, reflects the influence of reservoir skeleton deformation on the seepage field, meanwhile, that the porosity and permeability changed with reservoirs stress state reflects the coupling effect of seepage field and deformation field.

To simplify the questions, we assume that the deformation of reservoir skeleton is a steady deformation with small displacement and ignore the inertia force of rock skeleton. Then the individual derivative is similar to the partial derivative. So:

$$\frac{d(\ )}{dx_i}=\frac{\partial(\ )}{\partial x_i},\ \frac{d(\ )}{dt}=\frac{\partial(\ )}{\partial t},\ \nabla=\nabla',\ v_i=\frac{\partial u_i}{\partial t},\ a_i=\frac{\partial v_i}{\partial t}\approx 0,\ \varepsilon_{ij}=\frac{1}{2}\left(u_{i,j}+u_{j,i}\right) \tag{30}$$

Using Eq. (30), the continuity equation of reservoir rock matrix, namely Eq. (10), can be simplified as follows:

$$\frac{\partial(1-\varphi)}{\partial t}+(1-\varphi)\nabla\cdot\bar{v}_s=0 \tag{31}$$

According to the solid small deformation theory, we get:

$$\nabla\cdot\bar{v}_s=\nabla\cdot\left(\frac{\partial\bar{u}}{\partial t}\right)=\frac{\partial}{\partial t}(\nabla\cdot\bar{u})=\frac{\partial}{\partial t}\left(\frac{\partial u_x}{\partial x}+\frac{\partial u_y}{\partial y}+\frac{\partial u_z}{\partial z}\right)=\frac{\partial}{\partial t}\left(\varepsilon_{xx}+\varepsilon_{yy}+\varepsilon_{zz}\right)=\frac{\partial\varepsilon_v}{\partial t} \tag{32}$$

Where, $\bar{u}\left(u_x,u_y,u_z\right)$ is the displacement of rock skeleton, and ε_v is the volume strain of rock skeleton.

Substituting Eq. (32) into Eq. (31), the continuity equation of rock skeleton can be got as follows:

$$\frac{\partial\varphi}{\partial t}+\varphi\frac{\partial\varepsilon_v}{\partial t}=\frac{\partial\varepsilon_v}{\partial t} \tag{33}$$

The third term of the left of Eq. (28) and Eq. (29) can be transformed as follows:

$$\nabla\cdot\left(\varphi\rho_g S_g\bar{v}_s\right)=\nabla\cdot\left(\varphi\rho_g S_g\right)\bar{v}_s+\left(\varphi\rho_g S_g\right)\nabla\cdot\bar{v}_s \tag{34}$$

$$\nabla\cdot\left(\varphi\rho_w S_w\bar{v}_s\right)=\nabla\cdot\left(\varphi\rho_w S_w\right)\bar{v}_s+\left(\varphi\rho_w S_w\right)\nabla\cdot\bar{v}_s \tag{35}$$

The first terms of the right side of Eqs. (34) and (35) are fluid convection flow along with matrix displacement. According to the small displacement theory, these can be ignored. The gas/water two-phase seepage equations (28) and (29) can be simplified further as follows:

Gas phase: $$\frac{\partial\left(\varphi\rho_g S_g\right)}{\partial t}-\nabla\cdot\left(\frac{K_{rg}\rho_g}{\mu_g}[K]\left(\nabla P_g+\rho_g\bar{g}\right)\right)+\left(\varphi\rho_g S_g\right)\nabla\cdot\bar{v}_s=m_g+q_g \tag{36}$$

Water phase: $$\frac{\partial(\varphi\rho_w S_w)}{\partial t} - \nabla\cdot\left(\frac{K_{rw}\rho_w}{\mu_w}[K](\nabla P_w + \rho_w \vec{g})\right) + (\varphi\rho_w S_w)\nabla\cdot\vec{v}_s = m_w + q_w \quad (37)$$

From Eqs. (32) and (33), the follow relationship can be deduced:

$$\frac{\partial(\varphi\rho_g S_g)}{\partial t} = \rho_g S_g \frac{\partial\varphi}{\partial t} + \varphi\frac{(\rho_g S_g)}{\partial t} = \rho_g S_g\left(\frac{\partial\varepsilon_v}{\partial t} - \varphi\frac{\partial\varepsilon_v}{\partial t}\right) + \varphi\frac{(\rho_g S_g)}{\partial t} \quad (38)$$

$$(\varphi\rho_g S_g)\nabla\cdot\vec{v}_s = \varphi\rho_g S_g\frac{\partial\varepsilon_v}{\partial t} \quad (39)$$

Using Eqs. (33), (38) and (39) to simplify Eq. (36), we can get:

Gas phase: $$\rho_g S_g\frac{\partial\varepsilon_v}{\partial t} + \varphi\frac{\partial(\rho_g S_g)}{\partial t} - \nabla\cdot\left(\frac{K_{rg}\rho_g}{\mu_g}[K](\nabla P_g + \rho_g\vec{g})\right) = m_g + q_g \quad (40)$$

Water phase: $$\rho_w S_w\frac{\partial\varepsilon_v}{\partial t} + \varphi\frac{\partial(\rho_w S_w)}{\partial t} - \nabla\cdot\left(\frac{K_{rw}\rho_w}{\mu_w}[K](\nabla P_w + \rho_w\vec{g})\right) = m_w + q_w \quad (41)$$

Plus Eq. (40) and Eq.(41), then it can be got:

$$\begin{aligned} &-\nabla\cdot\left(\frac{K_{rg}\rho_g}{\mu_g}[K](\nabla P_g + \rho_g\vec{g})\right) - \nabla\cdot\left(\frac{K_{rw}\rho_w}{\mu_w}[K](\nabla P_w + \rho_w\vec{g})\right) = \\ &= q_g + q_w + m_g + m_w - (\rho_g S_g + \rho_w S_w)\frac{\partial\varepsilon_v}{\partial t} - \varphi\frac{\partial(\rho_g S_g + \rho_w S_w)}{\partial t} \end{aligned} \quad (42)$$

The relationship between gas phase pressure P_g and water phase pressure P_w is as follows:

$$P_w = P_g - P_c + (\rho_g - \rho_w)g_i h_i \quad (43)$$

Where, P_c is capillary pressure, which is the function of water saturation.

The derivation of Eq.(43) is:

$$\nabla P_w = \nabla P_g - \nabla P_c + (\rho_g - \rho_w)\vec{g} = \nabla P_g - P_c^{'}\nabla S_w + (\rho_g - \rho_w)\vec{g} \quad (44)$$

Where, $\nabla P_c = P_c^{'}\nabla S_w$, and $P_c^{'} = \frac{\partial P_c}{\partial S_w} \le 0$, which can be determined by capillary pressure curve.

Substitute Eq.(44) into Eq.(42) and simplify it, we get a pressure equation with gas phase pressure as its solution variables.

$$-\nabla\cdot\left(\left(\frac{K_{rg}\rho_g}{\mu_g}[K]+\frac{K_{rw}\rho_w}{\mu_w}[K]\right)\nabla P_g\right)=$$

$$=q_g+q_w+m_g+m_w-\left(\rho_g S_g+\rho_w S_w\right)\frac{\partial\varepsilon_v}{\partial t}-\varphi\frac{\partial\left(\rho_g S_g+\rho_w S_w\right)}{\partial t}$$

$$+\nabla\cdot\left(\left(\frac{K_{rg}\rho_g}{\mu_g}[K]+\frac{K_{rw}\rho_w}{\mu_w}[K]\right)\rho_g\vec{g}\right)-\nabla\cdot\left(\frac{K_{rw}\rho_w}{\mu_w}[K]\nabla P_c\right) \tag{45}$$

And:

$$\varphi\frac{\partial\left(\rho_w S_w\right)}{\partial t}=\varphi\rho_w\frac{\partial S_w}{\partial t}+\varphi S_w\frac{\partial\rho_w}{\partial t} \tag{46}$$

Substitute Eq.(44) into Eq.(41) and change the water phase pressure into gas phase pressure, then use Eq.(46) to simplify them and get the following equation.

$$\left(\rho_w\frac{\partial\varepsilon_v}{\partial t}+\varphi\frac{\partial\rho_w}{\partial t}\right)S_w+\varphi\rho_w\frac{\partial S_w}{\partial t}+\nabla\cdot\left(\frac{K_{rw}\rho_w}{\mu_w}[K]P_c^{'}\nabla S_w\right)=$$
$$=q_w+m_w+\nabla\cdot\left(\frac{K_{rw}\rho_w}{\mu_w}[K]\left(\nabla P_g+\rho_g\vec{g}\right)\right) \tag{47}$$

Eqs. (45) and (47) are the general form of fluid-solid coupling seepage equations of gas hydrate reservoir with gas phase pressure and water saturation as their basic solution variables.

2.2.4 Initial and boundary conditions of seepage field equations

Initial and boundary conditions should be complemented to solve seepage field equations of fluid-solid coupling model. Also, for flow equations, the continuity condition and boundary condition must be satisfied in continuous domain.

1. Boundary flow continuity

$$-\vec{n}\frac{\rho_g K_{rg}}{\mu_g}[K]\left(\nabla P_g+\rho_g\vec{g}\right)-\bar{q}_g=0 \tag{48}$$

$$-\vec{n}\frac{\rho_w K_{rw}}{\mu_w}[K](\nabla P_w+\rho_w\vec{g})-\bar{q}_w=0 \tag{49}$$

Where, $\vec{n}$ is bounding surface unit vector; $\bar{q}_g$ and $\bar{q}_w$ are gas and water phase flow per unit area on bounding surface.

2. Constant pressure boundary condition

The reservoir boundary pressure or bottom-hole pressure are known as following[6]:

$$P_G=f_P(x,y,z,t) \tag{50}$$

Where, $f_P(x,y,z,t)$ is the given pressure function on boundary G at moment t.

3. Constant flow boundary condition

This means that the pressure derivative of the reservoir external boundary or the production in internal boundary is known, as following:

$$\frac{\partial P}{\partial n}|G = f_q(x,y,z,t) \tag{51}$$

Where, n is the direction of normal line; $f_q(x,y,z,t)$ is the known function on specify boundary.

4. Initial condition

The initial conditions of seepage field are mainly the initial pressure and saturation of gas hydrate reservoirs, as following[6]:

$$P(x,y,z)|_{t=0} = P_l(x,y,z) \tag{52}$$

$$S|_{t=0} = S_l(x,y,z) \tag{53}$$

Where, $P_l(x, y, z)$ is the known pressure function, and $S_l(x, y, z)$ is the known saturation function.

The water-gas two-phase fluid-solid coupling seepage equations of gas hydrate reservoirs, namely Eqs. (47), can't be solved separately. Several simultaneous equations should be cited, including the kinetic equation of gas hydrates decomposition (1), the continuity equation solid phase hydrate (22), temperature field equations (17) and fluid-solid coupling deformation field equation in the next section. What's more, relevant state equation, assistant equations and the initial and boundary conditions, etc still need to be supplemented.

2.3 Deformation field equations of fluid-solid coupling model

The unknown parameters or unknown item in the seepage equations are affected by rock skeleton deformation, so the fluid-solid coupling seepage equations have highly nonlinear characteristics, and need to simultaneity the deformation field equations to solve the question. For porous media deformation field, the basic component equations mainly include equilibrium equation, geometric equation and constitutive equation, and the corresponding definite conditions. In this portion, the porous media deformation field mathematical model is constructed.

2.3.1 Equilibrium equations of reservoir skeleton

On the basis of the conventional force equilibrium principle of cell cube, we can get the deformation field balance equation with tensor form as follows:

$$\sigma_{ij,j}^{T} + f_i = 0 \tag{54}$$

As the total stress supported by formation are composed of pore pressure and skeleton stress which is known as skeleton effective stress. The deformation and strength properties of reservoirs are controlled by the skeleton effective stress instead of the total stress.

We appoint that the tensile stress is positive and the compressive stress is negative. The Terzaghi effective stress formula is combined with:

$$\sigma_{ij}^{T} = \sigma_{ij} - \alpha \overline{P} \delta_{ij} \tag{55}$$

Where, $\delta_{ij} = [1\ 1\ 1\ 0\ 0\ 0]^T$.

As to water-gas two-phase flow, the calculation of equivalent pore pressure is as follows:

$$\overline{P} = P_g S_g + P_w S_w \tag{56}$$

Zinkiewicz pointed that, for saturated fluid, the Biot coefficient can be calculated by:

$$\alpha = 1 - K / K_s \tag{57}$$

Where, K is rock volume modulus, and K_s is rock skeleton volume modulus.

Substituting the modified effective stress equation (55) into Eq. (54), we can obtain the fluid-solid coupling deformation equilibrium equations based on the effective stress, which are the basic equations to solve the porous media deformation field.

$$\sigma_{ij,j} + f_i - \left(\alpha \delta_{ij} \overline{P}\right)_{,j} = 0 \tag{58}$$

Where, f_i is the gravity term, in which f_x is equal to f_y, and the value of them is zero, f_z is calculated by the formula $\left[(1-\varphi)\rho_s + \varphi S_o \rho_o + \varphi S_w \rho_w\right] g$.

2.3.2 Geometric equations of reservoir skeleton

Based on the solid small deformation theory, the relationship between strain component and displacement component can be described by following geometric equations with tensor form:

$$\varepsilon_{ij} = \frac{1}{2}\left(u_{i,j} + u_{j,i}\right) \tag{59}$$

2.3.3 Constitutive equations of reservoir skeleton

2.3.3.1 Linear elastic constitutive equations

For linear elastic media, stress and strain have a linear corresponding relation. The linear elastic constitutive relation can be expressed by generalized Hook's law with matrix form as follows:

$$\sigma_{ij} = D_e \varepsilon_{ij} \tag{60}$$

In the above equation, D_e is the elastic constitutive matrix, which has 36 elements in 3 D coordinate. Elastic mechanics has proved it is symmetric matrices with only 21 independent constant. For the isotropic orthogonal media, there are only two independent elastic material constants, with elastic modulus E and Poisson's ratio ν signified.

Then the elastic constitutive matrix of rock skeleton can be expressed as:

$$[D_e]=\frac{E(1-\nu)}{(1+\nu)(1-2\nu)}\begin{vmatrix} 1 & & & & & \\ \frac{\nu}{1-\nu} & 1 & & & & \\ \frac{\nu}{1-\nu} & \frac{\nu}{1-\nu} & 1 & & & \\ 0 & 0 & 0 & \frac{1-2\nu}{2(1-\nu)} & & \\ 0 & 0 & 0 & 0 & \frac{1-2\nu}{2(1-\nu)} & \\ 0 & 0 & 0 & 0 & 0 & \frac{1-2\nu}{2(1-\nu)} \end{vmatrix} \tag{61}$$

2.3.3.2 Elastoplastic constructive equations

For elastic-plastic deformation, the strain is not only related to the current stress, but also to the loading history, loaded/unloaded condition, loading paths and microscopic structure of rock deformation, and so on.

In this study, the elastoplastic constructive relation is adopted to describe the relationship between strain and stress, which is expressed by the incremental form as follows:

$$d\sigma_{ij}=D_{ijkl}d\varepsilon_{kl} \tag{62}$$

Eq. (62) can be represented by the matrix form as follows:

$$\{d\sigma\}=\left[D_{ep}\right]\{d\varepsilon\} \tag{63}$$

The elastoplastic matrix can be calculated by the following equation.

$$\left[D_{ep}\right]=[D_e]-\left[D_p\right] \tag{64}$$

The elastic matrix can be got by Eq. (61), and the plastic matrix can be deduced from the elastic-plastic associative flow rule, as follows:

$$\left[D_p\right]=\frac{[D_e]\left\{\frac{\partial F}{\partial \sigma}\right\}\left\{\frac{\partial F}{\partial \sigma}\right\}^T[D_e]}{A+\left\{\frac{\partial F}{\partial \sigma}\right\}^T[D_e]\left\{\frac{\partial F}{\partial \sigma}\right\}} \tag{65}$$

Where, A is the hardening index, and F is the yield function.

Rock is friction type material, the plastic yield of which need to consider both the shear yield and the volume strain yield. Therefore, the yield function can be expressed using the stress invariant as follows:

$$F(I_1, J_2, J_3) = 0 \tag{66}$$

Currently the Mohr-Coulomb criterion and the Drucker-Prager criterion are used for rock material in the engineering territory.

1. Modified Mohr-Coulomb criterion

The Mohr-Coulomb criterion can be deduced using the first stress tensor invariant I_1, the second stress deviator J_2 and Lode stress angle θ_σ, as follows:

$$F = \frac{1}{3} I_1 \sin\varphi + \left(\cos\theta_\sigma - \frac{1}{\sqrt{3}} \sin\theta_\sigma \sin\varphi \right) \sqrt{J_2} - C\cos\varphi = 0 \tag{67}$$

Where:

$$I_1 = \sigma_x + \sigma_y + \sigma_z \tag{68}$$

$$J_2 = \frac{1}{6}\left[\left(\sigma_x - \sigma_y\right)^2 + \left(\sigma_y - \sigma_z\right)^2 + \left(\sigma_z - \sigma_x\right)^2 + 6\left(\tau_{xy}{}^2 + \tau_{yz}{}^2 + \tau_{zx}{}^2\right) \right] \tag{69}$$

$$\theta_\sigma = arctg\left[\frac{1}{\sqrt{3}} \left(2\frac{\sigma_2 - \sigma_3}{\sigma_1 - \sigma_3} - 1 \right) \right] \tag{70}$$

Where, C is the cohesion of rock; φ is internal friction angle; $\sigma_1, \sigma_2, \sigma_3$ are three principle stresses.

2. Drucker-Prager yield criterion

The expression of the Drucker-Prager yield criterion is as follows:

$$F = \alpha I_1 - \sqrt{J_2} + k = 0 \tag{71}$$

Where, a and K are the yield function parameters, the expressions of which are as follows:

$$\alpha = \frac{\sin\varphi}{\sqrt{9+3\sin^2\varphi}}; \ k = \frac{3c\cos\varphi}{\sqrt{9+3\sin^2\varphi}} \tag{72}$$

Comparing the two yield criterions above, the Mohr-Coulomb criterion neglects the effects of the middle principle stress σ_2, and needs to determine the magnitude of each principle stress when it is used, which is not convenient. So we adopt the Drucker-Prager yield criterion in this research.

2.3.4 Initial and boundary conditions of deformation field equations

Assume that the space region occupied by rock skeleton is Ω^d (the value of d is two or three), and the boundary of the region is $\partial\Omega$, in which the displacement boundary is Γ^u

and the stress boundary is Γ^{σ} , what's more, $\partial\Omega = \Gamma^{u} \cup \Gamma^{\sigma}$ and $\Gamma^{u} \cap \Gamma^{\sigma} = 0$. So we get the following boundary conditions:

The known stress boundary: $T_i = \sigma_{ij} n_j \quad \forall \overline{X} \in \Gamma^{\sigma}$ (73)

The known displacement boundary: $u_i = \overline{u}_i \quad \forall \overline{X} \in \Gamma^{u}$ (74)

If the effective stress is adopted, so the expressions are:

Stress boundary: $T_i = \left(\sigma'_{ij} - \alpha \overline{P} \delta_{ij}\right) n_j \quad \forall \overline{X} \in \Gamma^{\sigma}$ (75)

Displacement boundary: $u_i = \overline{u}_i \quad \forall \overline{X} \in \Gamma^{u}$ (76)

2.4 Basic property parameters calculation of gas hydrate reservoirs

2.4.1 Permeability and capillary pressure

The effective permeability of gas hydrate reservoir is the function of hydrate saturation, and it meets the model proposed by Masuda (1997)[7,8], as follows:

$$K = K_0 (1 - S_H)^N \tag{77}$$

Where, K_0 is the absolute permeability of the formation, of which the hydrate saturation is zero; S_H is the hydrate saturation; N is the decline exponent of permeability, which is related to the types of hydrates combined in porous space, and the value is 2~15.

The relative permeability and capillary pressure curve can be obtained from the modified models proposed by Van Genuchten (1980)[9] and Parker (1987)[10].

The relative permeability of the water phase:

$$K_{rw} = K_{rwo} \overline{S}_w^{1/2} \left[1 - (1 - \overline{S}_w^{1/m})^m \right]^2 \tag{78}$$

The relative permeability of the gas phase:

$$K_{rg} = K_{rgo} \overline{S}_g^{1/2} \left[1 - (1 - \overline{S}_{wH}^{1/m})^m \right]^{2m} \tag{79}$$

The capillary pressure between liquid phase and gas phase:

$$P_e = P_{co} \left[(\overline{S}_w)^{-1/m} - 1 \right]^{1-m} \tag{80}$$

Where:

$$\overline{S}_w = \frac{S_w - S_{wr}}{1 - S_{wr} - S_{gr}} \tag{81}$$

$$\bar{S}_{wH} = \frac{S_w + S_H - S_{wr}}{1 - S_{wr} - S_{gr}} \tag{82}$$

$$\bar{S}_g = \frac{1 - S_w - S_H - S_{gr}}{1 - S_{wr} - S_{gr}} \tag{83}$$

Where, $P_{co} = 1KPa$, $m = 0.45$, $S_{wr} = 0.3$, $S_{gr} = 0.05$, $K_{rwo} = 0.5$, $K_{rgo} = 1.0$.

When the saturation of hydrates is 0.4, the relative permeability curve of gas and water phase and the capillary pressure curve are shown as Figs. 2-1 and 2-2.

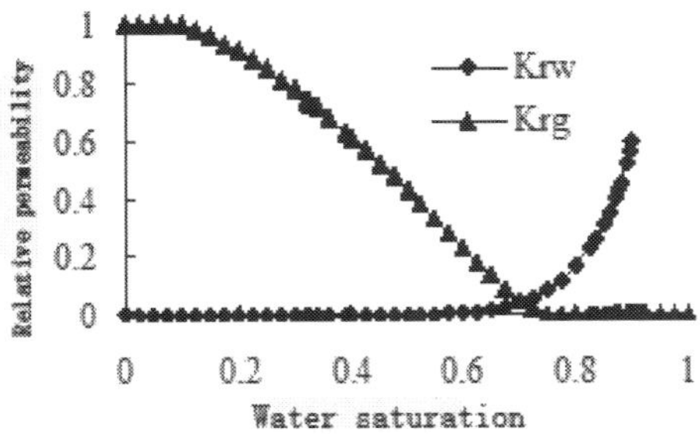

Fig. 2-1. Curve of gas/water relative permeability

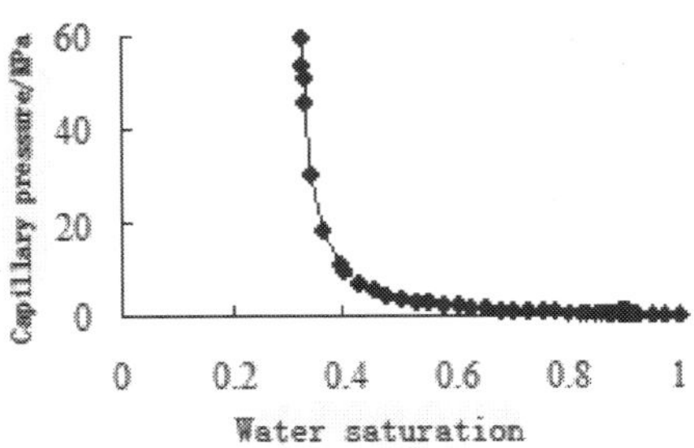

Fig. 2-2. Capillary pressure curve

2.4.2 Thermal physical parameters

2.4.2.1 Effective coefficient of heat conduction

The coefficient of heat conduction for gas hydrate reservoirs is related to the component. Assume that the heat conductions coefficient of rock skeleton, hydrates, water and gas are constant, so the effective coefficient of heat conduction K_c can be calculated by the following equation[11-13].

$$K_c = (1-\varphi)K_r + \varphi(S_H K_H + S_g K_g + S_w K_w) \tag{84}$$

2.4.2.2 Specific heat capacity

The specific heat capacities of rock skeleton and hydrates are seemed as constants, while the specific heat capacities of water and gas change with temperature. When the pressure is 15MPa, and the temperature range from 273.15K~373.15K, we can get the following relationship (American National Standard)[14]:

$$C_{pg} = 90907.40529 - 912.13019T + 3.59612T^2 - 0.00638.T^3 + 4.30251 \times 10^{-6} T^4 \quad (85)$$

$$C_{pw} = 4733.58633 - 3.99122T - 0.00677T^2 \quad (86)$$

$$C_{ph} = 1600 \quad (87)$$

$$C_{ps} = 835 \quad (88)$$

Where, C_{pg} and C_{pw} are the specific heat capacities of methane and water, J/Kg/K, which are related to temperature; C_{ph} and C_{ps} are the specific heat capacities of hydrates and rock skeleton, J/Kg/K.

2.4.2.3 Decomposition heat of methane hydrate

The decomposition heat of methane hydrate can be calculated by Masuda (1999)[15] model as follows:

$$Q_H = -\frac{m_H(c+d \cdot T)}{M_H} \quad (89)$$

Where, Q_H is the decomposition of hydrates; M_H is the molar mass of hydrates; c and d is the experiment coefficients, the values are proposed 56599J/mol and -16.744J/mol/K for methane hydrates[16].

2.4.3 Fugacity coefficient and density of methane

2.4.3.1 Density of methane

The density of methane is calculated by Peng-Robinson state equation as follows:

$$P = \frac{RT}{V-b} - \frac{\alpha(T)}{V(V+b)+b(V-b)} \quad (90)$$

PR state equation is a mostly used cubic equation of state, which can calculate the gas volume and gas compression factor. PR equation can be expressed with compressibility factor Z, as follows:

$$Z^3 + (B-1)Z^2 + (A-2B-3B^2)Z + (B^3 + B^2 - AB) = 0 \quad (91)$$

Where:

$$A = \frac{\alpha(T)P}{R^2T^2}$$

$$B = \frac{bP}{RT}$$

$$\alpha(\mathrm{T}) = \frac{0.457235R^2T_c^2\alpha(T_r,\omega)}{P_c}$$

$$b = \frac{0.07779607RT_c}{P_c}$$

$$\alpha(T_r,\omega) = \left[1+k(1-T_r^{0.5})\right]^2$$

$$k = 0.37464 + 1.54226\omega - 0.26992\omega^2$$

Where, a(T) and b are the parameters of PR equation; T_c is the critical temperature; P_c is the critical pressure, and ω is the acentric factor.

Eq. (91) has three real roots, in which the maximum one is the gas compressibility factor, the minimum one is the liquid compressibility factor.

After obtaining the methane compressibility factor under the conditions of temperature T_i and pressure P_g, the density of methane ρ_g can be calculated by the following equation:

$$\rho_g = \frac{P_g M_g}{ZRT_i} \tag{92}$$

2.4.3.2 Fugacity coefficient of methane

The fugacity coefficient of methane is calculated by PR state equation, as follows:

$$\varphi = \frac{f}{P} = \exp\left[(Z-1)-\ln(Z-B)-\frac{A}{2\sqrt{2}B}\ln\left(\frac{Z+(1+\sqrt{2})B}{Z+(1-\sqrt{2})B}\right)\right] \tag{93}$$

2.4.4 Phase equilibrium equation

The phase equilibrium of methane hydrates, water and methane is calculated by Makogon (1997) model[17-19], and the expression is:

$$\log_{10}(P_e) = A(T-T_o) + B(T-T_o)^2 + C \tag{94}$$

Where, P_e is the equilibrium pressure, Pa; the value of T_0 is 27.15K; A, B, C are experiment coefficients, and the value is 0.0342K^{-1}, 0.0005K^{-2}, 6.4804, respectively.

What's more, the viscosities of water and gas are assumed as constant, which don't change with temperature and pressure.

2.5 Comprehensive dynamic models of property parameters for gas hydrate reservoirs

To resolve fluid-solid coupling numerical model for gas hydrate reservoirs not only needs equations of seepage fluid and deformation field but also needs relevant supplementary equations. The letter mainly include dynamic models of physical and mechanical parameters such as permeability, porosity and elastic modulus ,which reflect the real time change of the property at the process of exploitation for gas hydrate reservoirs.

Factors affecting physical property parameters of conventional reservoirs include volumetric strain of rock, effective stress, temperature, and so on. Usually one or more factors and property parameters including permeability,porosity and elastic modulus are used to build up a certain relation, that is to build up the dynamic models of physical property parameters.

For gas hydrate reservoirs, hydrate decomposition is the most direct and notable factor to affect physical properties. So in order to build up comprehensive dynamic models of physical properties for gas hydrate reservoirs, two factors must be considered: firstly, it must take example from dynamic models of conventional reservoirs to reflect the relationship between physical property parameters including permeability, porosity and elastic modulus and physical quantity including volumetric strain, effective stress and temperature; secondly, it need to highlight the influence of gas hydrate decomposition effect on related physical parameters.

2.5.1 Comprehensive dynamic model of permeability

Experiments of permeability stress sensitivity for loose sandstone done by Yizhong Zhao[20] indicate that where far away from borehole, the permeability is mainly controlled by principal stress and the fit accuracy of permeability and effective stress is high. The model built by them is indicated as Eq. (95); since sensitivity of nearly well reservoirs is mainly controlled by stress difference, relativity between permeability and volumetric strain is better. So for nearly well reservoir we could adopt the model of permeability and volumetric strain, as the following equation (96) shows.

$$K_{\sigma} / K = c \cdot \sigma^2 + d \cdot \sigma + e \tag{95}$$

Where, K_σ is permeability of some stress state, 10^{-3} μm^2; K is the initial permeability, 10^{-3} μm^2; σ is the effective stress, MPa; c, d and e are regression coefficients of experiment.

$$\frac{K_{\sigma}}{K} = \frac{1}{1+\varepsilon_V} \cdot \left(1+\frac{\varepsilon_V}{\varphi_e}\right)^{f\left(\frac{\varepsilon_V}{\varphi_e}\right)^2 + j\frac{\varepsilon_V}{\varphi_e} + l} \tag{96}$$

Where, ε_V is the volumetric strain; φ_e is the initial effective porosity; f, j and l are the regression coefficients of experiment.

Combined with the permeability computation model Eq. (77) for gas hydrate reservoirs and considering the difference of permeability effect mechanism for each part of the reservoir, this study builds up two comprehensive dynamic models of permeability for gas hydrate reservoirs. The formula (97) is fit in reservoirs which are far away borehole; however, the formula (98) is fit in the relative analysis of nearly well reservoirs.

$$K_{\sigma} = K_0(c \cdot \sigma^2 + d \cdot \sigma + e)(1 - S_H)^N \tag{97}$$

$$K_{\sigma} = K_0 \frac{(1-S_H)^N}{1+\varepsilon_V} \cdot \left(1+\frac{\varepsilon_V}{\varphi_e}\right)^{f\left(\frac{\varepsilon_V}{\varphi_e}\right)^2 + j\frac{\varepsilon_V}{\varphi_e} + l} \tag{98}$$

The model built up by this study roundly reflects the influence of gas hydrate decomposition effect and change of reservoirs stress state on permeability.

2.5.2 Comprehensive dynamic model of porosity

JianJun Liu, Chinese academy of sciences Seepage institute, and JishunQin, China University of Petroleum, who conduct a lot of experimental research on the relationship between the porosity of low permeable porous medium and the effective stress, proposed that the index relationship could be used to describe the changing regularity between them.

$$\varphi_{\sigma} / \varphi_e = m \cdot \exp(n \cdot \sigma) \tag{99}$$

Where, Φ_{σ} is the porosity at some stress state;Φe is the initial effective porosity; m and n are experiment coefficients; σ is the effective stress, MPa.

Palmer suggests that the relationship between porosity of unconsolidated sandstone and the effective stress can be described by index relationship, whose form is similar with the formula (99). When Yizhong Zhao studied the reservoir of unconsolidated sandstone, they optimize the index relationship which is similar with (99) as dynamic models of porosity.

Supposing the distribution of porosity is uniform in gas hydrate reservoirs and ignoring the influence of temperature, so the following relationship can be obtained.

$$\varphi_e = \varphi(1 - S_H) \tag{100}$$

Where, Φ is the absolute porosity when the saturation of hydrates is zero; S_H is the saturation of hydrates.

In conclusion, the comprehensive dynamic model of porosity for hydrate reservoirs built for this study is:

$$\varphi_{\sigma} = \varphi(1 - S_H) \times m \cdot \exp(n \cdot \sigma) \tag{101}$$

This model above reflects the effect of hydrate decomposition effect and the influence stress state change on the porosity.

2.5.3 Comprehensive dynamic model of elastic modulus

The research of Guangquan Li, Jianjun Liu etc shows that the relation between elastic modulus and effective stress accords with power function form, as follows:

$$E_{\sigma} / E = a \cdot \sigma^b \tag{102}$$

Where, E_σ is elastic modulus at some stress state, GPa; E is the initial elastic modulus, GPa; a and b are the fitting coefficients of experiment; σ is the effective stress, MPa.

By making sensitivity experiments of elastic modulus for unconsolidated sandstone, Yizhong Zhao points that the regression relation of quadratic polynomial between elastic modulus and effective stress is more accurately, the expression of which is as follows:

$$E_\sigma / E = A \cdot \sigma^2 + B \cdot \sigma + C \tag{103}$$

Where, A, B, C is the fitting coefficients by experiment.

In recent years,through experimental research,the petroleum research institute of Commonwealth Scientific and Industrial Research Organization (CSIRO) and Tohidi, from hydrate research center of Heriot-Watt University point that the decomposition of hydrates will result in the increase of porosity and the decline in the elastic modulus correspondingly, based on which they built the relation between elastic modulus change and the variation of porosity, as follows:

$$\Delta E = -\xi \cdot \Delta\varphi \tag{104}$$

Where, $\xi$is the fitting coefficient by experiment, the value of which is -9.5; ΔE is the decrease of elastic modulus, GPa; $\Delta\Phi$ is the increment of porosity.

Based on the comprehensive dynamic model of property for gas hydrate reservoirs which we built for this research, the mathematical relation between the hydrate decomposition effect and the increase of porosity resulting from the change of stress condition is as follows:

$$\Delta\varphi = \varphi(1 - S_H) \times m \cdot \exp(n \cdot \sigma) - \varphi(1 - S_{Hi}) \tag{105}$$

Where, Φ is the absolute porosity when the hydrate saturation is zero; S_{Hi} is the initial saturation of hydrates; S_H is the current hydrates saturation; σ is the effective stress, MPa; m and n is the experimental matched coefficients.

Combining Eq. (104) into Eq. (105), the change of elastic modulus is as follows:

$$\Delta E = -\xi \cdot \left[\varphi(1-S_H) \times m \cdot \exp(n \cdot \sigma) - \varphi(1-S_{Hi})\right] \tag{106}$$

At the same time, there is the following relationship:

$$E = E_0 + \Delta E \tag{107}$$

Where, E_0 is the elastic modulus before hydrates decomposition, GPa.

Allying Eqs. (106) and (107) with Eq. (103), the comprehensive dynamic model of elastic modulus is built for this study, as follows:

$$E_\sigma = \left\{E_0 - \xi \cdot \left[\varphi(1-S_H) \times m \cdot \exp(n \cdot \sigma) - \varphi(1-S_{Hi})\right]\right\} \cdot \left(A \cdot \sigma^2 + B \cdot \sigma + C\right) \tag{108}$$

This model reflects the influence of hydrate decomposition and stress state change on the elastic modulus.

Because the change of Poisson ratio is little, the Poisson ratio is seen as constant during the fluid-solid coupling numerical simulation.

2.5.4 Comprehensive dynamic model of cohesion

During the exploitation of gas hydrate reservoirs, hydrates decomposition can cause the cementation of the reservoirs loosened and the remarkable change of the cohesive force must be taken into consideration.

In 2005, Clennell, from the petroleum research institute of Commonwealth Scientific and Industrial Research Organisation's (CSIRO), and Tohidi, from Heriot-Watt University hydrate research center, proposed that as the decomposition of hydrates, the reservoirs cementation will become weaker, and the cohesion will decrease continuously. So they built the model of cohesion decrease for reservoirs, which was widely adopted by scholars. The mathematical expression is[21-23]:

$$C = C_0\left[1-1.2\cdot(\Delta\varphi)\right] \tag{109}$$

Where, C is the cohesive force of the reservoirs after hydrates decomposition, MPa; C_0 is the cohesive force before hydrates decomposition, MPa; $\Delta\Phi$ is the increment of porosity as a result of hydrates decomposition.

Eq. (109) is also taken as the dynamic model of cohesive force for the production from hydrate reservoirs by depressurization.

3. Fluid-solid coupling simulation studies on productivity of gas production from hydrate reservoirs by depressurization

Gas production from hydrate reservoirs by depressurization is a complex fluid-solid coupling process with hydrates decomposition, which is affected by the hydrates decomposition effect, fluid-solid coupling and borehole effect. In this process, the reservoir physical parameters shows the complex characteristics, reservoir porosity characteristics change and fluid-solid coupling have the dynamic effect for gas production from hydrate reservoirs by depressurization.

Based on the gas and water fluid-solid coupling mathematical model and the physical and mechanical models build above, this section will solve these models. Taking a gas hydrate reservoir of Gulf of Mexico as an example, we simulate the productivity of gas production from hydrate reservoirs by depressurization. At the same time, the main influence factors to productivity are also researched.

3.1 Fluid-solid coupling finite element model for productivity of gas production from hydrate reservoirs by depressurization

3.1.1 Basic parameters used in the simulation

The parameters used in fluid-solid coupling numerical simulation include basic property parameters, reservoir fluid and the basic properties of solid hydrate. All of the parameters input in the simulation of this study are from a natural gas hydrate reservoir in the Gulf of Mexico, as is shown in Table 3-1 and Table 3-2. And Figs. 3-1 and 3-2 show the gas/water relative permeability capillary pressure curve, respectively.

Water Density /Kg.m^{-3}	Water viscosity /Pa.s	Water Thermal Conductivity /W.m^{-1}.K^{-1}	Specific Heat of Water /J.Kg^{-1}.K^{-1}	Methane Thermal Conductivity /W.m^{-1}.K^{-1}	Specific Heat of Methane /J.Kg^{-1}.K^{-1}	Thermal Conductivity of Rock Skeleton /W.m^{-1}.K^{-1}
1000	0.001	0.6	Rely on temperature	0.00335	Rely on temperature	1.5
Specific Heat of Rock /J.Kg^{-1}.K^{-1}	Hydrate Density /Kg.m^{-3}	Hydrate Thermal Conductivity /W.m^{-1}.K^{-1}	Specific Heat of Solid Hydrate /J.Kg^{-1}.K^{-1}	Hydrate Decomposition Rate/mol. (m^2.Pa.s)$^{-1}$	Hydrate Decomposition Activating Energy /J.mol^{-1}	Methane Hydrate Number
835	910	0.393	1600	8060	77330	6.0

Table 3-1. Basic parameters for reservoir fluid, rock and gas hydrate

Depth of Sea Water/m	Distance to Sea Bed/m	Initial Hydrate Saturation	Initial Water Saturation	Absolute Porosity (Without Hydrate)	Absolute Porosity to X-axis /μm^2	Absolute Porosity to Y-axis /μm^2
1310	365	0.5	0.4	0.4	0.2	0.2
Initial Temperature /K	Initial Pressure/ MPa	Initial Elastic Modulus/MPa	Poission's Ratio	Cohesion /MPa	Angle of Internal Friction	Biot Factor
290	16.9	4845	0.40	2.0	30°	1.0
Vertical Stress/MPa	Maximum Horizontal Principal Stress /MPa	Minimum Horizontal Principal Stress /MPa	Sensitiveness Coefficient of Permeability c	Sensitiveness Coefficient of Permeability d	Sensitiveness Coefficient of Permeability e	Sensitiveness Coefficient of Porosity m
21.80	20.45	19.70	0.0038	-0.1096	1.3086	1.1455
Sensitiveness Coefficient of Porosity n	Sensitiveness Coefficient of Elastic Modulus A	Sensitiveness Coefficient of Elastic Modulus B	Sensitiveness Coefficient of Elastic Modulus C	Borehole Radius/m		
-0.0439	-0.0023	0.0716	0.7934	0.15		

Table 3-2. Basic parameters of finite element simulation for gas hydrate reservoir

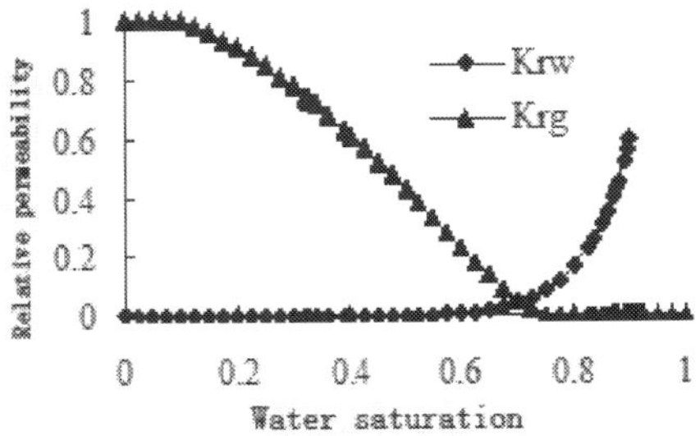

Fig. 3-1. Gas/water relative permeability curves

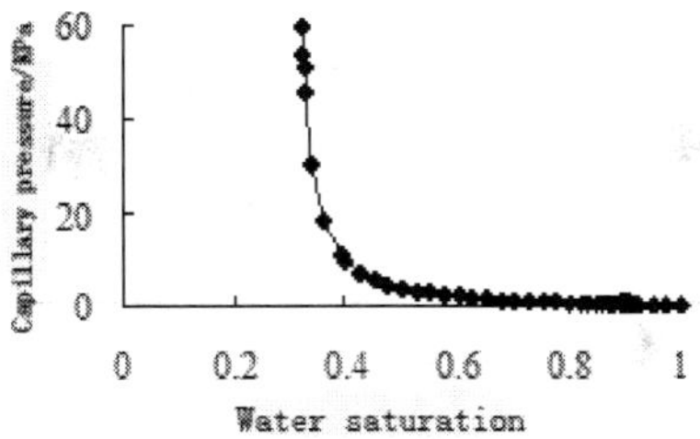

Fig. 3-2. Capillary pressure curve

3.1.2 Geometrical model and boundary conditions

To simplify the problem, the plane strain model is adopted considering the symmetry of the hydrate reservoir. Set the overall dimensions as 100m×100m for this finite element model. With respect to the significance of the near borehole formation to the whole capacity simulation, the model meshing is obtained with gradient grid technology, so the mesh density is higher in formation near borehole. The diagram of mechanical model and grid of the finite element model are shown in Figs. 3-3 and 3-4, respectively.

In the process of simulation the boundary conditions for the seepage field are as follows: Line *BC* and *CD* keep constant pressure boundary, Line *AB* and *DE* keep free pressure boundary, and at the position of the wellbore, namely Point *A*, the boundary keeps the bottomhole pressure.

The boundary conditions of the deformation field are listed as follows: The maximum effective horizontal principle stress (σ_H) is applied on the boundary line BC, and the minimum effective horizontal principle stress (σ_h) is applied on Line *CD*. On the Line AB the displacement is constant as zero along X-axis and free along Y-axis; And on the Line DE the displacement is zero along X-axis and free along Y-axis. At the Point A the displacement is fixed to both X-axis and Y-axis.

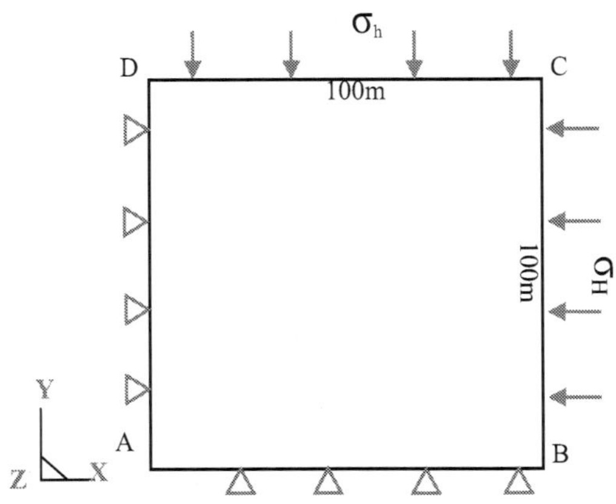

Fig. 3-3. Schematic diagram of mechanical model

Fig. 3-4. Grid of the FEM model

3.2 Fluid-solid coupling numerical simulation on productivity of gas production from hydrate reservoirs by depressurization

For the further studying the influential mechanism of fluid-solid coupling effects on the gas production from hydrate reservoirs by depressurization, this section also analyses the other simulation results of the two simplified models which is based on the gas-water two-phase non-isothermal fluid-solid coupling model established above.

The first model ignores physical parameters change caused by the fluid-solid coupling effect, but it considers the coupling effect between seepage field and deformation field, and it is named the simplified fluid-solid coupling model. Based on the first model, the second model further ignores the coupling effects of seepage field and solid field, so it is named the non-coupling normal model.

For the convenience of analysis, here the gas-water two-phase non-isothermal fluid-solid coupling model established above is renamed as the comprehensive fluid-solid coupling model. The basic characteristics of the three models are shown in Table 3-3.

3.2.1 Numerical simulation on productivity of non-coupling normal model for production by depressurization

Taking into account the non-coupling normal model with phase transition in Table 3-3, which considers the petrophysical properties change as a result of hydrates decomposition and neglects the coupling effect between the seepage field and the deformation field, the gas and water production rate are simulated. The results are shown Figs. 3-5 and 3-6, the unit of which is STCMD.

Num.	Model name	Factors in consideration	Contrast analysis method
1	Non-coupling normal model	Based on the seepage model of phases transition, consider hydrates decomposition and temperature variation, but neglect the fluid-solid coupling effect(bottomhole pressure is 13.9 MPa)	(1)By comparing the results of Models 1 and 2, we analyze the effect of matrix volumetric strain on production performance; (2)By comparing the results of Models 2 and 3, we analyze the effect of physical parameters on production performance.
2	Simplified fluid-solid coupling model	Based on the seepage model of phases transition, consider hydrates decomposition, temperature variation, the influence of rock skeleton volumetric strain on the seepage, but ignore reservoir properties and mechanical parameters variation (bottomhole pressure is13.9 MPa)	
3	Comprehensive fluid-solid coupling model	This is the gas-water two-phase non-isothermal fluid-solid coupling model for natural gas hydrate reservoir built in Section Two, which considers hydrates decomposition, reservoir physical and mechanical properties variation (bottomhole pressure is 13.9 MPa)	

Table 3-3. Essential characteristics of the three models for deliverability analysis

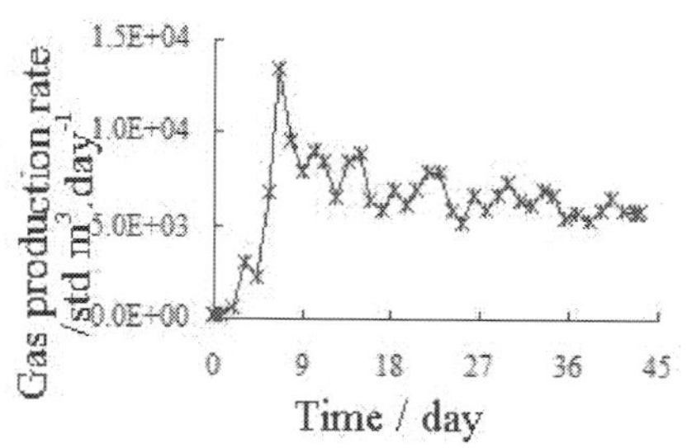

Fig. 3-5. Gas production rate and time

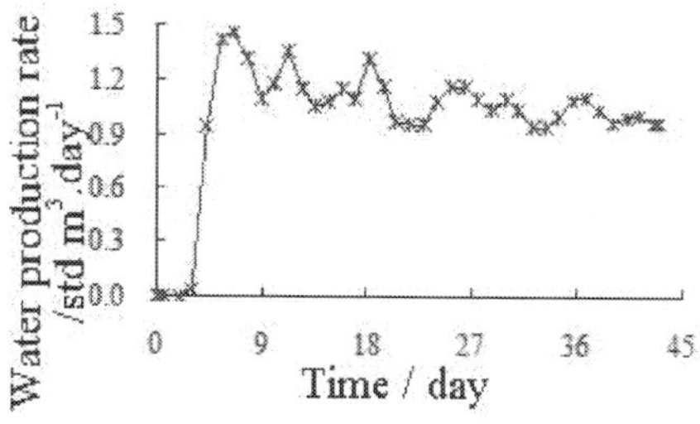

Fig. 3-6. Water production rate and time

Fig. 3-5 shows that the gas production rate curve fluctuates greatly because of the instability of gas-liquid two-fluid phase seepage. The variation tendency of gas production rate curve indicates that the gas production rate increases sharply at the early stage, and become stable lately. Only 43 days is set to exploit which is too short to finish hydrates decompose, but it can be indicated that gas production rate will decrease as hydrate saturation reducing, and finally no gas is produced. In general, the gas production rate curve can be divided into three stages, rapidly rising stage, stable stage and dropping stage. We can summarize the characteristic as follows:

1. The rising stage of gas production rate. As the pressure releases the gas production rate rises rapidly. This is because: a. the initial saturation of free gas is 0.1 and when the pressure decreases, the free gas could swell out quickly; b. the pressure of the reservoir prior to production is close to hydrate decomposition pressure at the initial temperature, so the pressure around wellbore will be below the equilibrium pressure of hydrates rapidly and they begin to decompose immediately.
2. The stable stage of gas production rate. Most of the gas produced after free gas output is from the hydrate decomposition, which is controlled by the rate of hydrate dissociation.
3. The dropping stage of gas production rate. The hydrate saturation become lower and lower as the decomposition front going ahead further, and the gas production becomes less correspondingly, and finally no gas could be produced. As only 43 days is set to simulate the gas production in Fig. 3-5, which is not long enough to lead to hydrate decomposition totally, so the obvious dropping stage of gas production is not reached yet in fact.

From Fig. 3-6 it can be seen that the water production rate curve is similar with the one of gas production. It can also be divided into three stages as follows:

1. Instead of gas, water is the only production in the original depressurization stage, and the time of water producing falls behind the gas producing time. This is because the existence of free gas in the reservoir has better flowing ability than the water when the pressure decreasing.
2. After the output of free gas, the water saturation of reservoirs increases gradually as hydrates decompose. The rate of water production decreases rapidly after reaching a peak under the influence of the relative permeability of water and gas. The reason for this phenomenon is as follows, the free water and produced water accumulate towards the wellbore as a result of the propelling of gas, and they come out rapidly when reach to some extent. Consequently, a water production peak takes place. However, due to the fact that both the gross of the free water and the water comes from the hydrate decomposition in the original stage are relatively limited, the rate of water production decreases speedily after reaching the peak.
3. The water producing rate remains steady level after falling. In this period, as the free water output is almost completely, the production is mainly the decomposing water from hydrates, which is controlled by the decomposition rate of hydrates. And the water production exhibits the similar character to that of gas. The decrease of the hydrate saturation and decomposition rate leads to the reduction of the water producing rate, and finally no water is generated.

Shuxia Lee and Yongmao Hao, who are from the Hydrate Research Center of China University of Petroleum, established hydrates synthesizing and exploiting equipment. They obtained the water and gas production rate characteristic curve through small scale hydrate exploiting experiment by depressurization[24,25]. Figs. 3-7 and 3-9 show the results.

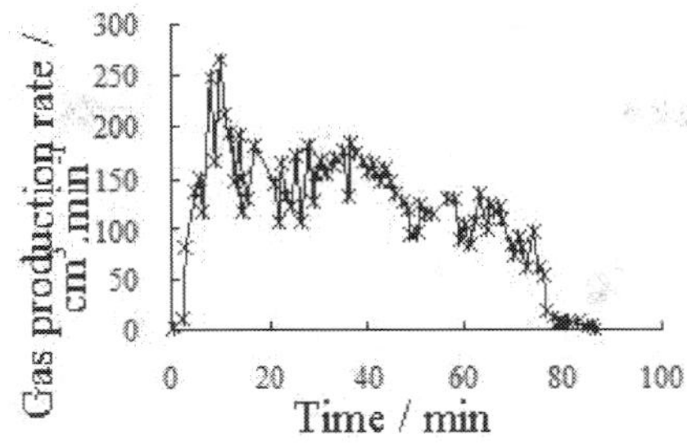

Fig. 3-7. Gas production rate from experiment

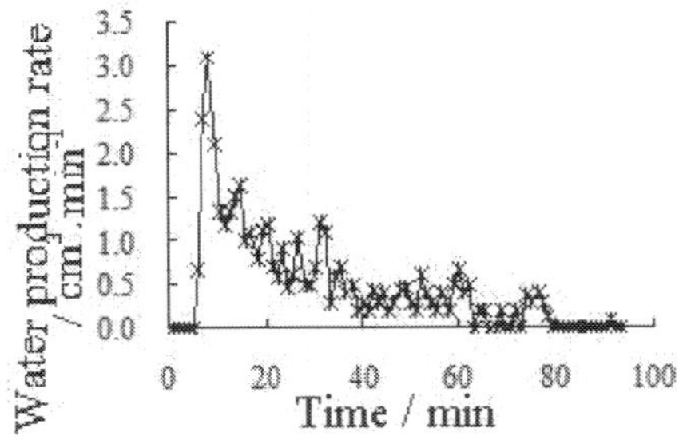

Fig. 3-8. Water production rate from experiment

The size of the pipe adopted in the simulation experiment by ShuXia Lee is Φ38x500mm and Φ38x800mm, which belongs to small one-dimensional model. However, in the numerical simulation of this study a plane strain model is used. The difference of the two models leads to no relative property of the gas and water production rate value. But it is significant to analyze the comparison of the change regulation. Because the model dimension is large and the time is limited in this research, the dropping stage of gas and water production rates have not been happened. Comparing Fig. 3-5 with Fig. 3-7, and Fig. 3-6 with Fig. 3-8, in general the characteristic of the gas and water production rate and the regulation variation match with results of physical simulation experiment. This shows the validity of the results of this simulation results.

The cumulative gas production is shown as Fig. 3-9. It can be seen that the cumulative gas production curve is well corresponding with the gas production rate curve. At the early stage of depressurization the gas production rate is bigger, and correspondingly the cumulative gas production rises quickly. When the gas production rate reaches the relatively steady stage, the cumulative gas production increases steadily and slowly. It can be inferred that, with gas production rate gradually decreasing at the later stage of depressurization production, correspondingly the increasing range of cumulative gas production will gradually decrease. Until the hydrates decomposition totally, the cumulative gas production finally stays at a certain level.

The cumulative water production is shown as Fig. 3-10. It is easy to see that, the cumulative water production curve is vary similar with the one of gas production, and it is also well corresponding with the water production rate.

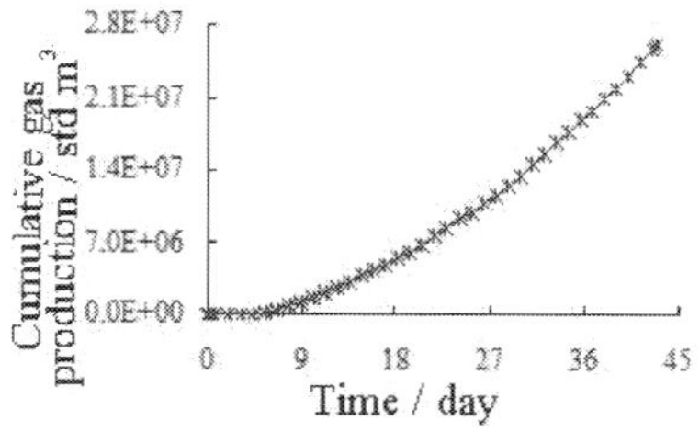

Fig. 3-9. Cumulative gas production

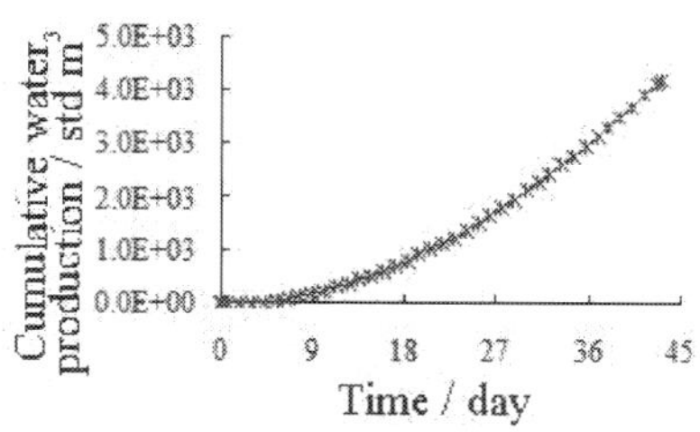

Fig. 3-10. Cumulative water production

3.2.2 Numerical simulation on productivity of simplified fluid-solid coupling model for production by depressurization

During the fluid-solid coupling effect of production by depressurization, the influence of reservoir deformation field to seepage field mainly reflects in two aspects. On the one hand, the deformation of reservoir skeleton will shrink the pore space, which will increase elastic drive energy. On the other hand, the change of effective stress state will cause petrophysical variations of stress sensitive reservoirs, such as porosity, permeability and so on. All of these will affect the production performance of hydrate reservoirs.

In this chapter, the simplified fluid-solid coupling model in Table 3-3 is adopted to simulate gas production from hydrate reservoirs by depressurization, in which the physical parameters alternation caused by fluid-solid coupling is neglected. By comparing with the simulation results of the non-coupling normal model above, we discuss the influence of fluid-solid coupling effects on the production behaviors of hydrate reservoirs.

Fig. 3-11 gives the gas production rate curves of both the non-coupling normal model and simplified fluid-solid coupling model. Fig. 3-13 gives the water production rate curves of the two models. The cumulative gas and water production curves of the two models are shown in Figs. 3-12 and 3-14.

From the comparison above, it can be seen that the results of the simplified fluid-solid coupling model and non-coupling normal model are similar in some aspects, but there is also some obvious difference, as follows:

1. Similar to the non-coupling normal model, the gas and water production curves of the simplified fluid-solid coupling model can be divided into three stages generally: rapidly rising stage, stable stage and gradually reduce stage. This is consistent with the results of experiments by Shuxia Lee and Yongmao Hao in China University of Petroleum.

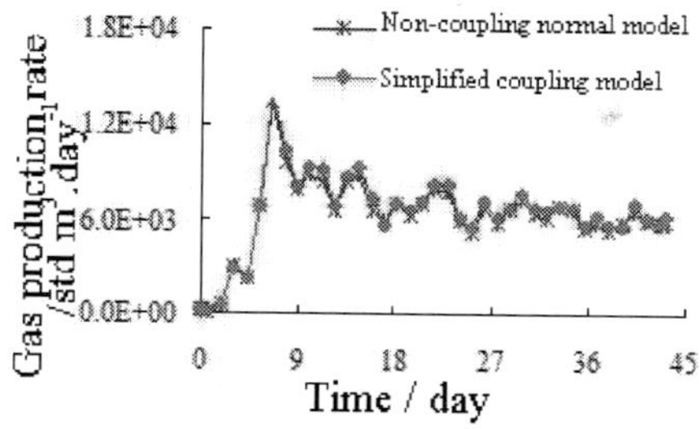

Fig. 3-11. Comparison of gas production rate

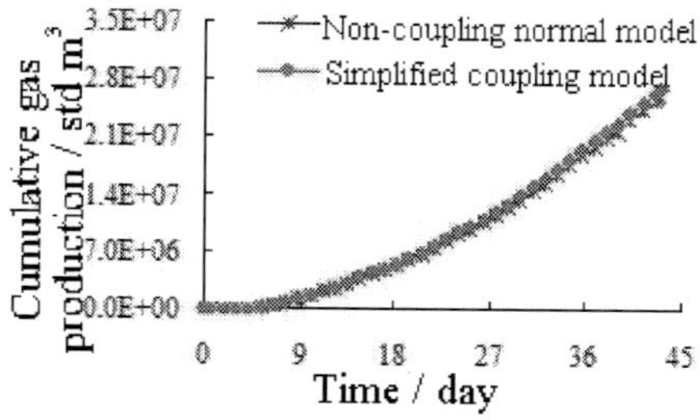

Fig. 3-12. Comparison of cumulative gas

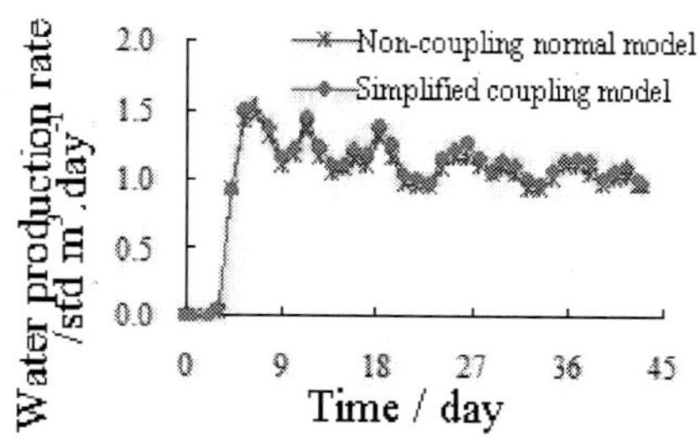

Fig. 3-13. Comparison of water production rate

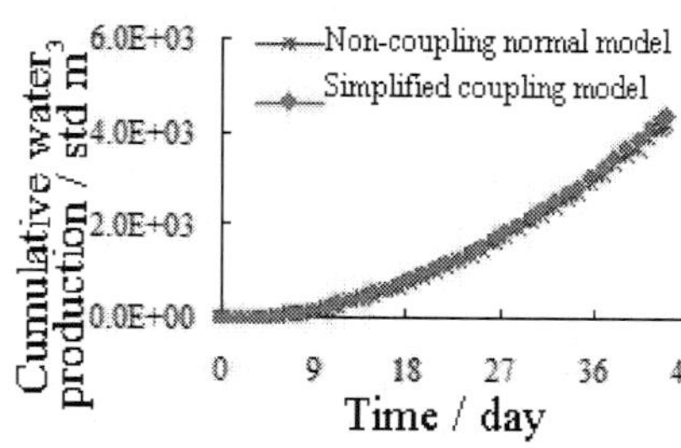

Fig. 3-14. Comparison of cumulative water

2. The major difference between two models lies in that both the average production rate and the accumulative production of gas obtained by the simplified fluid-solid coupling model are 3.45% higher than that of the non-coupling normal model. Similarly, both the average water production rate and the water accumulative production are 4.28% higher than that of non-coupling normal model. This is because reservoir skeleton volume strain can lead to reservoir porosity reducing and elastic drive energy increasing to some extent. This is flavor to raise gas and water production rate, accumulative gas and water production of hydrate reservoirs. The increasing of elastic drive energy has more effect to water production than gas production as a result of the poor liquidity of the gas with respect to the liquid.

3.2.3 Numerical simulation on productivity of comprehensive fluid-solid coupling model for production by depressurization

The analysis above shows that shrinkage of reservoir pore space can raise the elastic drive energy because of the fluid-solid coupling effect, which does help to increase the gas production rate and accumulative gas production. This is one side that the fluid-solid coupling affecting production performance. And on the other side, the change of the reservoir effective stress can lead to physical parameters, such as porosity and permeability, which can also influence production behaviors.

In this chapter, we adopt the comprehensive fluid-solid coupling model in Table 3-3 to simulate gas production from hydrate reservoirs by depressurization. Based on the simplified model, the comprehensive fluid-solid coupling model considers the change of physical parameters caused by the stress state variation. Thus, by comparing with the two models, we can study the influence of the physical parameters change of sensitive stress reservoirs, which is caused by the fluid-solid coupling effect, on the production performance of hydrate reservoirs.

Figs. 3-15 and 3-17 show the gas and water production rate curves respectively obtained from the comprehensive fluid-solid coupling model and the simplified model. The accumulative gas and water production curves of the two models are shown as Figs. 3-16 and 3-18.

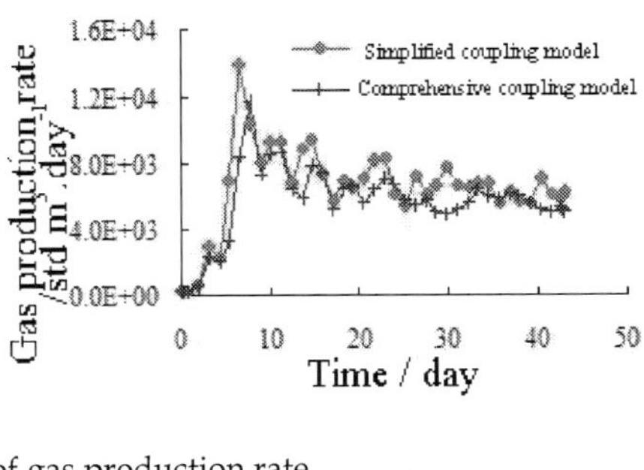

Fig. 3-15. Comparison of gas production rate

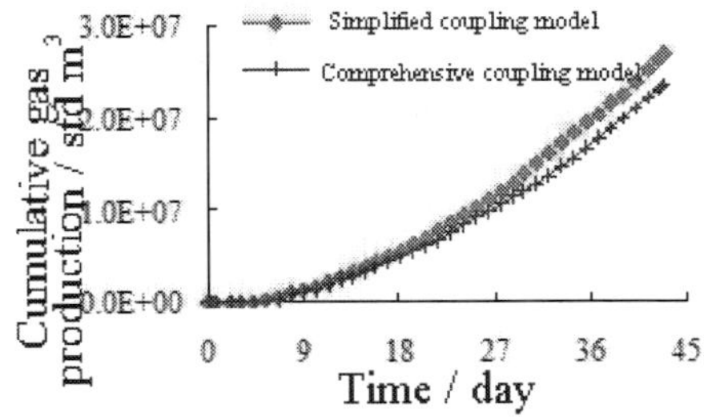

Fig. 3-16. Comparison of cumulative gas production

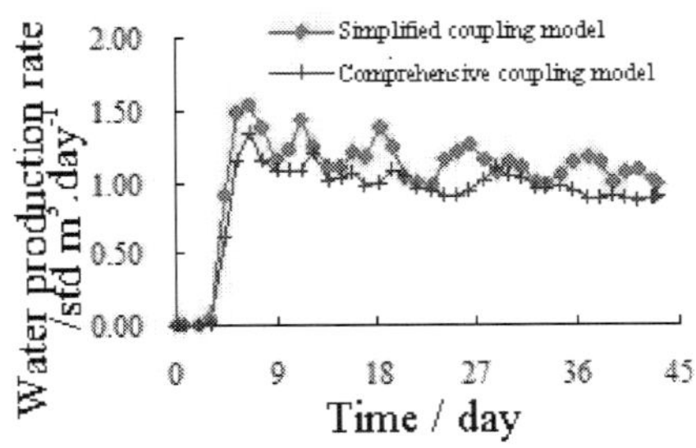

Fig. 3-17. Comparison of water production rate

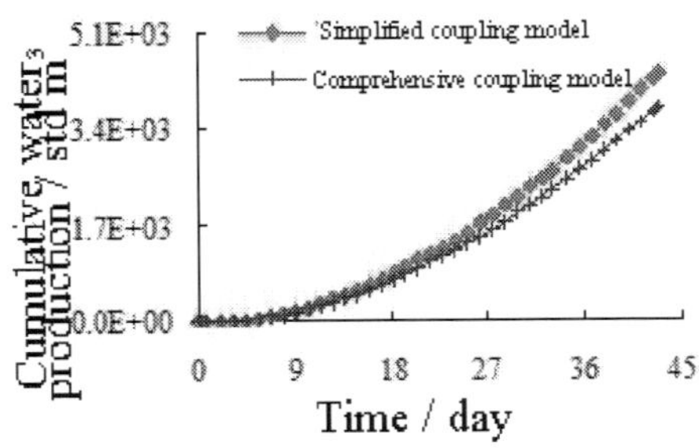

Fig. 3-18. Comparison of cumulative water production

The comparisons above show that the gas and water production rate curves of the comprehensive and simplified coupling models can also be divided into three stages, namely the rapidly rising stage, stable stage and gradually reduce stage, which is similar to that of the non-coupling normal model, and matches with the experiment results of Shuxia Lee. Further analysis shows that the numerical size of simulative result has significantly difference, as follows:

1. From Figs. 3-15 and 3-16 it can seen that the average gas production rate and accumulative production which are obtained by the simplified fluid-solid coupling model are obviously higher than that of the comprehensive fluid-solid coupling model. The average gas production rate in stable stage of the former model is about 1.144 times

greater than that of the later. Accordingly, the gas accumulative production is about 4.4% higher than that of the later.

2. Similarly, from Figs. 3-17 and 3-18 we can also see that the average water production rate and accumulative production simulated by the simplified fluid-solid coupling model is obviously higher than that of the comprehensive fluid-solid coupling model. The average water production rate in the stable stage of the former model is about 1.159 times greater than that of the later. Correspondingly, the accumulative water production is about 15.9% higher than that of the later.

Fig. 3-19 and Fig. 3-21 show the gas production rate and water production rate curve respectively, which are simulated by the comprehensive fluid-solid coupling model and the non-coupling normal model. The accumulative gas and water production curves respectively in two cases are shown in Figs. 3-20 and 3-22.

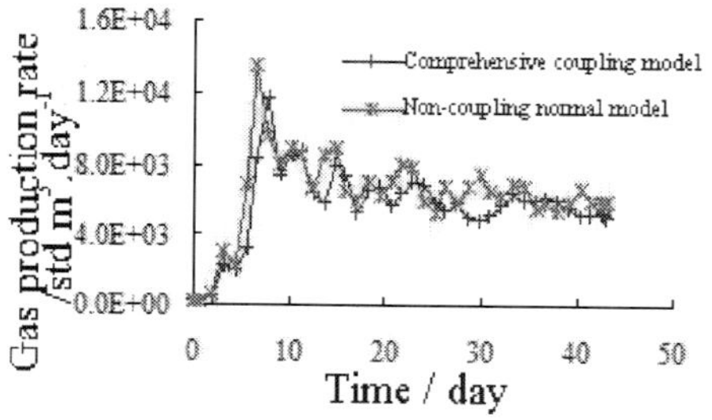

Fig. 3-19. Comparison of gas production rate

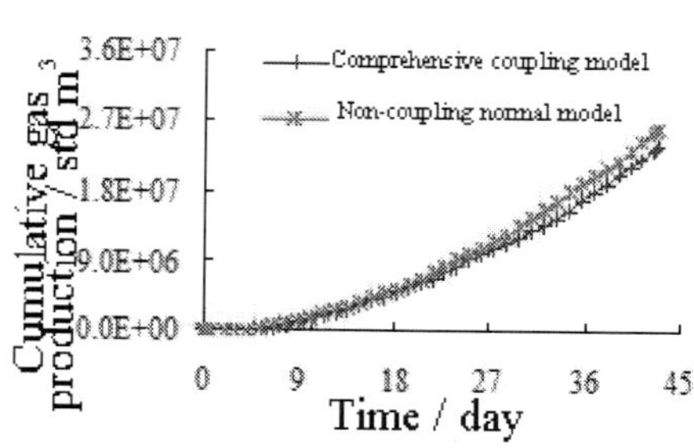

Fig. 3-20. Comparison of cumulative gas production

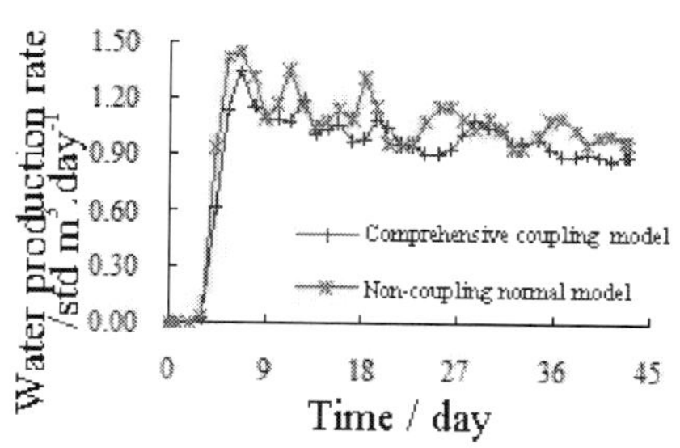

Fig. 3-21. Comparison of water production rate

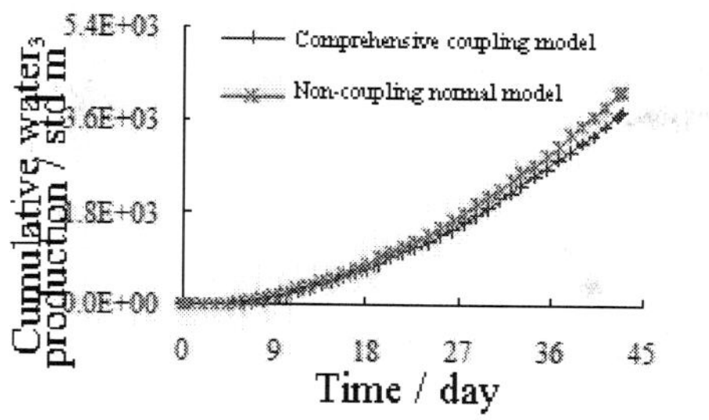

Fig. 3-22. Comparison of cumulative water production

The comparisons of Figs. 3-19 and 3-20 show that the average gas production rate and accumulative gas production of the non-coupling normal model are obviously higher than that of the comprehensive fluid-solid coupling model. The average gas production rate in the stable stage of the former model is about 1.106 times higher than that of the later. And accordingly, the accumulative gas production of the former is about 15.9% higher than that of the later. Similarly, from Fig. 3-21 and Fig. 3-22 it can be seen that the average water production rate and accumulative water production of non-coupling normal model is also obviously greater than that of comprehensive model. The average water production rate during the steady stage of the former model is about 1.111 times higher than that of the later. Correspondingly, the accumulative water production of the former is about 11.1% higher than that of the later.

In conclusion, by analyzing the production performances of the non-coupling normal model, the simplified coupling model and the comprehensive, we can find that the development index, namely the gas production rate, the accumulative gas production and so on, obtained from the simplified model is the highest in the three models. The results of the non- coupling normal model are slightly lower. But the development index simulated by the comprehensive coupling model is significantly lower than the two former models. The reason for this phenomenon is that fluid-solid coupling has different influencing mechanism to production performance.

During the production by depressurization, the fluid-solid coupling effect has two aspects. On one side, the shrinkage of rock matrix caused by fluid-solid coupling effect can raise the elastic drive energy, which is helpful for gas output. So the development index of simplified fluid-solid coupling model which only considers the skeleton volume strain is higher than that of the non-coupling normal model. This is the good aspect that fluid-solid coupling effect contributes to gas production. On the other side, the shrinkage of rock pore space can also cause the reduction of permeability and porosity, and accordingly the seepage captivity also decreases which causes the adverse effect for production. Relatively, the change of porosity and permeability which is caused by fluid-solid coupling plays the dominated role on affecting the production performance. But the increasing of elastic drive energy is limited. Therefore, the predictive value of production index simulated by the comprehensive fluid-solid coupling model which considers the skeleton volume strain and physical parameters change is obviously lower than the results of the simplified fluid-solid coupling model and non-coupling normal model.

The variation of skeleton volume strain and petrophysical properties caused by fluid-solid coupling in the reservoir is the objective existence. So, the overall effect of fluid-solid coupling causes the development index, namely the gas production rate, the accumulative gas production and so on, is obviously lower than that of the non-coupling normal model. Therefore, from the productivity point of view, the coupling effect between the seepage field and the deformation field must not be neglected in the exploitation of hydrate reservoirs. Although there are many numerical models for production from hydrate reservoir in previous research, they failed to totally consider the coupling effect between the seepage filed and rock deformation field.

Compared with the current numerical models, the fluid-solid coupling model, which this study established for the production from hydrate reservoirs production by depressurization, has the superiority as follows: fully consideration of hydrate decomposition effect, the coupling effect of seepage field and deformation field, the change of the temperature, and the dynamic variation of the physical and mechanical properties of formations under the influence of many factors.

In the next portion, based on the gas-water two-phase non-isothermal fluid-solid coupling model established in this research, we will carry out a sensitive analysis on the main factors which affect the productivity from hydrate reservoirs.

3.3 Influences analysis on productivity of gas production from hydrate reservoirs by depressurization

The factors that affect the production performance for hydrate reservoirs by depressurization are very more and the extent of them is various. In order to comprehensively evaluate the influence degree of all factors on the development index and provide the theoretical basis for the production optimization of hydrate reservoirs, we primarily study the influence of the reservoir absolute permeability, the bottomhole pressure and the reservoir temperature on hydrate reservoir production index, such as the accumulative gas production.

3.3.1 The effect of absolute permeability of formation

The absolute permeability of formation is a parameter reflecting the seepage ability of porous medium and it is an important factor affecting the production rate after gas hydrates decomposing. In this portion, we use the non-isothermal fluid-solid coupling model developed in this article to simulate the production behaviors with the condition of three reservoir absolute permeability, namely 100mD, 200mD and 300mD. In the simulation the other conditions of hydrate reservoir keep the same. The emphasis is to analyze the influence of formation absolute permeability on production index, such as the hydrate decomposing rate, the gas production rate and accumulative gas production.

Figs. 3-23 and 3-24 show the hydrate decomposition front and the distribution of hydrate saturation S_H under different permeability condition after the well producing for 43 days.

It is not difficult to find that the influence of absolute formation permeability is obvious to production performance of hydrate reservoir. The greater the formation absolute permeability is, the faster the propagate speed of the pressure drop is in the reservoir, the

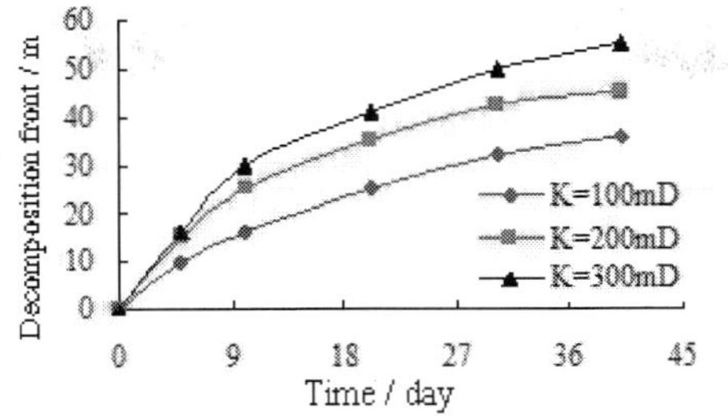

Fig. 3-23. Decomposition front for different permeability

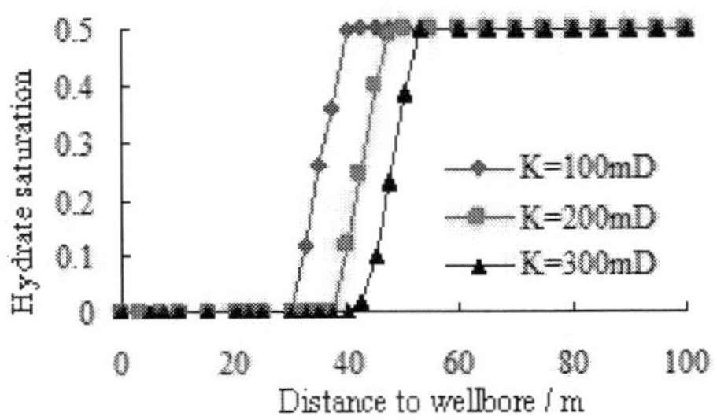

Fig. 3-24. S_H for different permeability

stronger the flow ability of reservoir fluid, the faster hydrate break down, and the faster decomposition front position promote. From Fig. 3-23 it can be seen that the distances between the decomposition front and the wellbore are 36m, 45.08m and 55.07m with respect to the formation permeability being 100mD, 200mD, 300mD after the well producing for 43 days.

Figs. 3-25 and 3-27 show the gas and water production rate of hydrate reservoir respectively under the condition of different permeability. The accumulative gas and water production curves with different formation permeability are shown as Figs. 3-26 and 3-28.

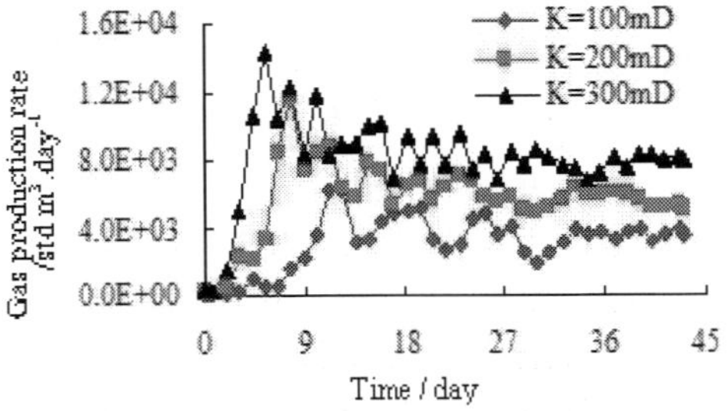

Fig. 3-25. Effect of permeability on gas rate

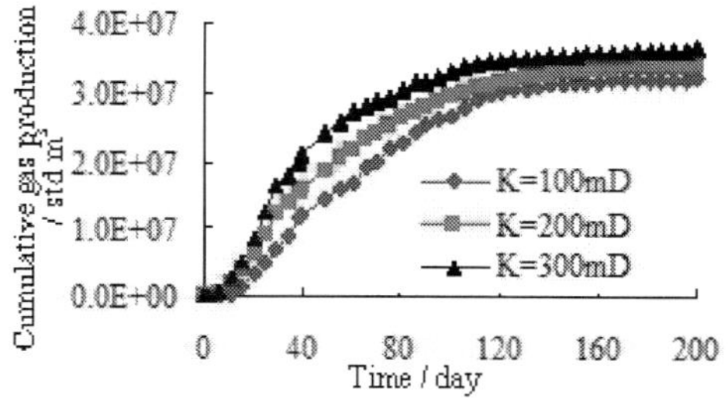

Fig. 3-26. Effect of permeability on cumulative gas

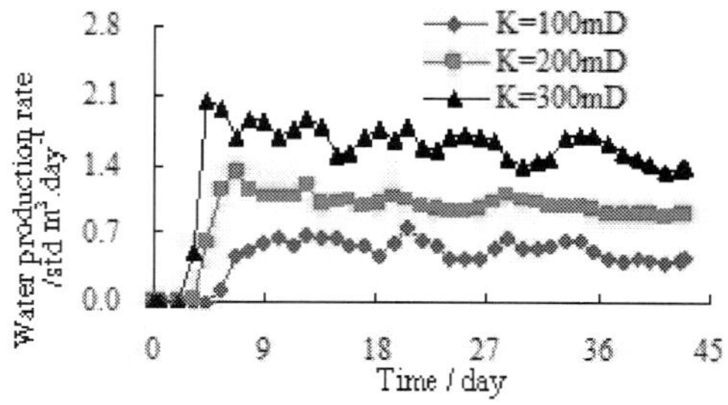

Fig. 3-27. Effect of permeability on water rate

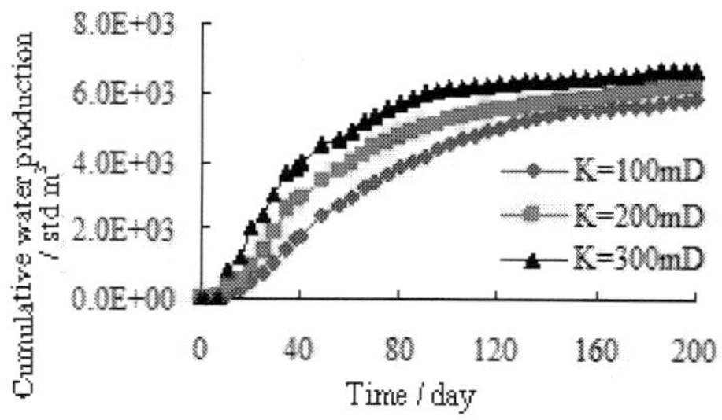

Fig. 3-28. Effect of permeability on cumulative water

From the results of the figures above, it is easy to find that despite of the different permeability of the formation, the gas and water production rate curves are similar, which can be divided into three stages on the whole, namely the rapidly rising stage, the stable stage and the gradually reducing stage. The greater the absolute permeability of reservoir is, the faster the propagate speed of pressure drop is, the faster hydrates break down, the greater the gas and water production rates are, and accordingly the bigger the accumulative gas and water productions are. After the well produces for about 200days, the accumulative gas production with the absolute permeability being 300mD is 7.05% higher than that of the formation with permeability being 200mD. The accumulative water production is 7.4% higher than that of the formation of which the permeability is 200mD. The accumulative gas and water production of the formation with the absolute permeability being 200mD is 5.9% and 6.34% respectively higher than that of formation whose absolute permeability is 100mD.

3.3.2 The effect of bottomhole pressure

The value of bottomhole pressure is an important parameter to control the gas production from hydrate reservoirs reasonably, and it not only affects the cumulative gas and water production of hydrate reservoirs, but also determines the propagate speed of hydrate dissociation front. Based on the non-isothermal fluid-solid coupling model above, this portion simulate the production performance under the conditions of the initial formation pressure 16.9MPa and three bottomhole pressure, namely 12.9MPa, 13.9MPa and 14.9MPa respectively. Other conditions in the progress keep the same. Emphatically the influence of bottomhole pressure on production index, such as the hydrate decomposing rate, the gas production rate, the accumulative gas production, and so on.

Figs. 3-29 and 3-30 show the hydrates dissociation front position and the hydrate saturation S_H distribution after 43 days production under the condition of different bottomhole pressures. Comparative analysis shows the lower bottomhole pressure leads to the faster pressure drop spread in reservoir, the greater pressure gradient, and the faster hydrate dissociation and decomposition front promoting rate. From Fig. 3-29, we can see that after 43 days production under the pressure being 12.9MPa, 13.9MPa and 14.9MPa respectively, the hydrate dissociation fronts move to the positions where the distances are 60.06m, 45.08m and 29.11m respectively far from the wellbore.

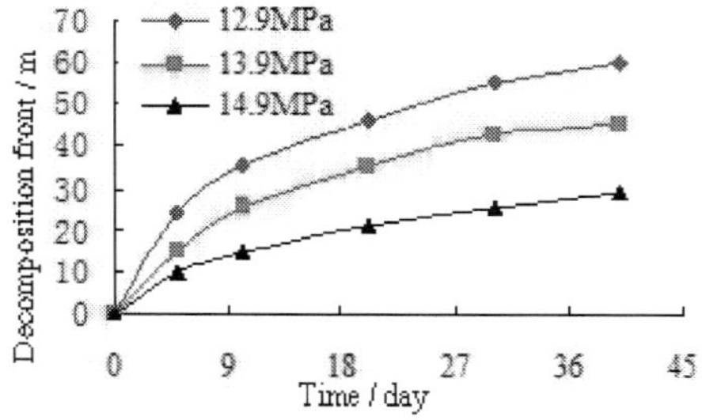

Fig. 3-29. Dissociation front with different pressure

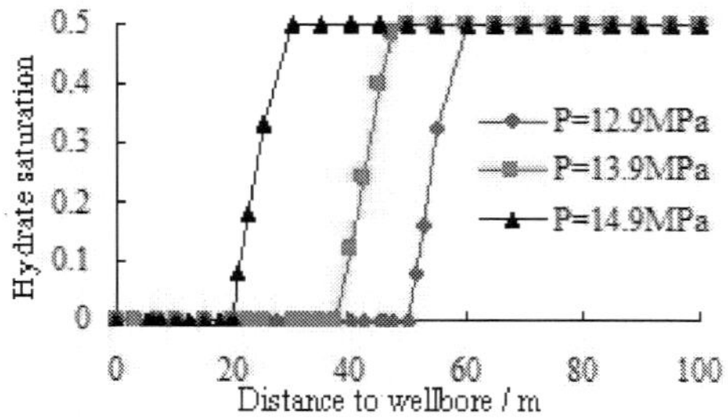

Fig. 3-30. S_H for different pressure

Figs. 3-31 and 3-33 show of the gas and water production rate of hydrate reservoirs respectively under different bottomhole pressures. And Figs. 3-32 and 3-34 give the accumulation gas and water production from hydrate reservoirs respectively.

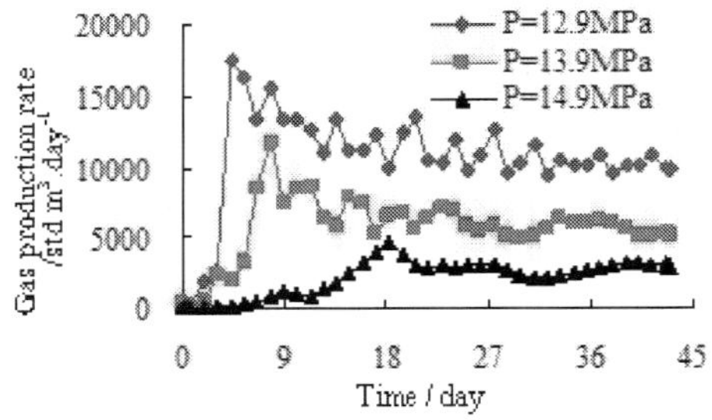

Fig. 3-31. Effect of pressure on gas production rate

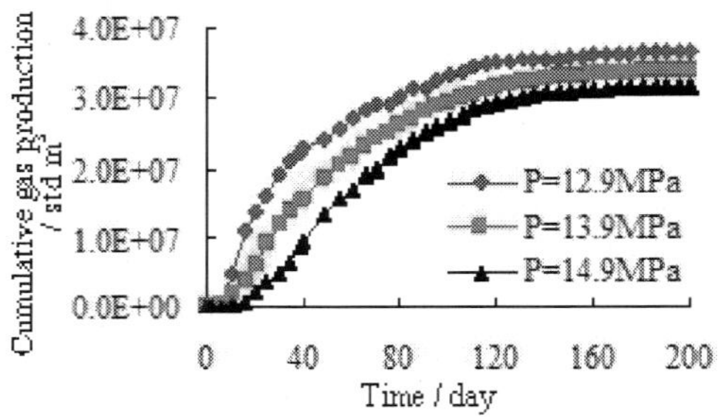

Fig. 3-32. Effect of pressure on cumulative gas

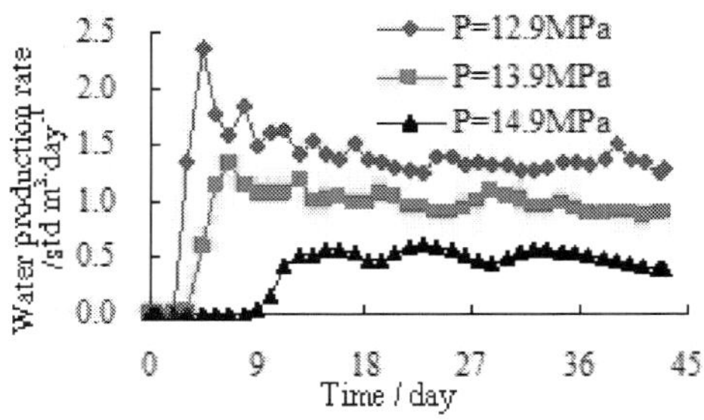

Fig. 3-33. Effect of pressure on water production rate

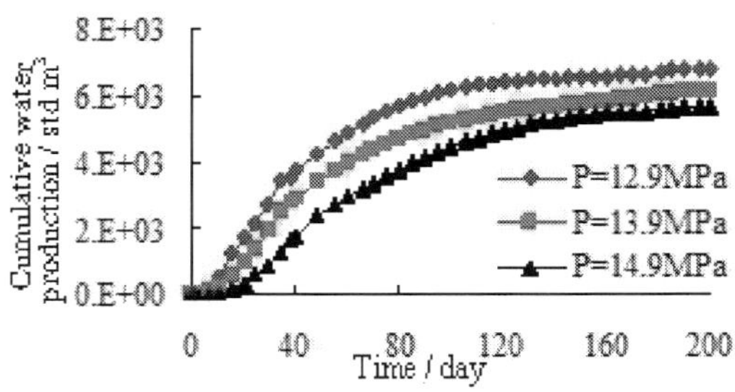

Fig. 3-34. Effect of pressure on cumulative water

Comparative analysis show that, similarly to the former, the gas and water production rate curves with different bottomhole pressures generally can be divided into three stages: rapidly rising stage, stable stage and gradually reducing stage. And the lower bottomhole pressure, correspondingly, leads to the faster hydrate dissociation, the bigger gas and water production rate and the greater gas and water accumulative production. After 200 days production, when bottomhole pressure is 12.9MPa, gas accumulative production is 7.9% higher than that with the bottomhole pressure being 13.9MPa, and accordingly, the water accumulative production increase 9.6%. When bottomhole pressure is 13.9MPa, the gas and water accumulative production increase 6.9% and 8.7% respectively compared with that being 14.9MPa.

It should be pointed out that low of the bottomhole pressure is contributive to improve the gas accumulation production and other development indexes. But at the same time, the lower bottomhole pressure can lead to the greater producing pressure difference and the worse stability in near well reservoir. Therefore, the stability of the reservoir near well bore should be considered in optimizing design of bottomhole pressure. To improve the production rate of hydrate reservoirs, the bottomhole pressure can not be reduced without restraint.

3.3.3 The influence of formation temperature

The formation temperature is a key parameter to affect the hydrate dissociation rate. Using the non-isothermal fluid-solid coupling model of the hydrate reservoir, this portion will simulate the production performance when formation temperatures are 289.5K, 290K and 290.5K respectively and other conditions are the same. We analyze emphatically the influence of formation temperature to the hydrate dissociation rate, gas production rate and cumulative gas production, etc.

Figs. 3-35 and 3-36 show the location of hydrate dissociation front and the distribution of the saturation of hydrate (SH) after 43 days production under different formation temperatures. Comparisons indicate that the higher formation temperature leads to the faster hydrate dissociation, the greater dissociation front promoting rate and the further forward distance during the same time. As Fig. 3-35 shows, when the formation temperatures are 289.5K, 290K and 290.5K respectively, the hydrate dissociation fronts are 34.105m, 45.08m and 56.08m respectively far away from borehole after 43 days production similarly.

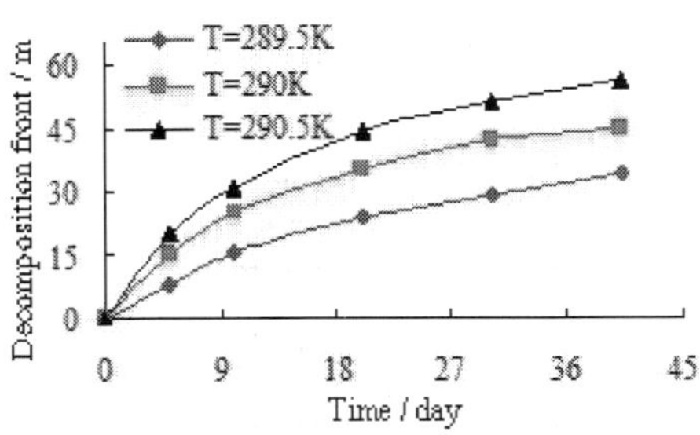

Fig. 3-35. Dissociation front with different temperature

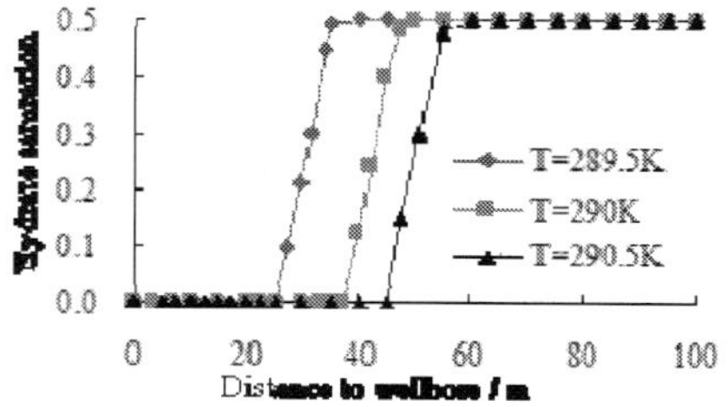

Fig. 3-36. S_H with different temperature

Figs. 3-37 and 3-39 show the gas and water production rate of the hydrate reservoir with different formation temperatures. And Figs. 3-38 and 3-40 give the gas and water accumulative production with different formation temperatures.

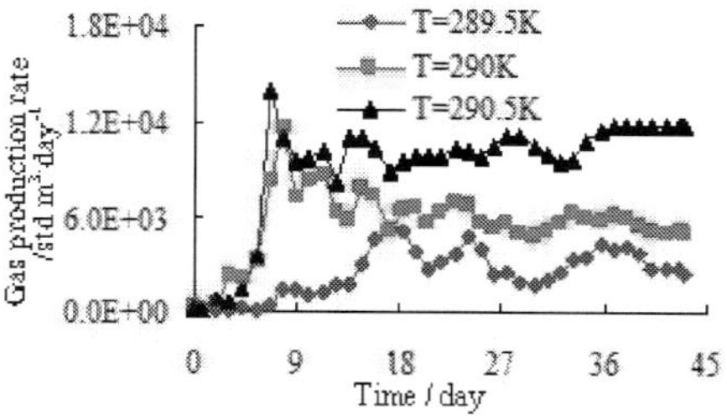

Fig. 3-37. Effect of temperature on gas rate

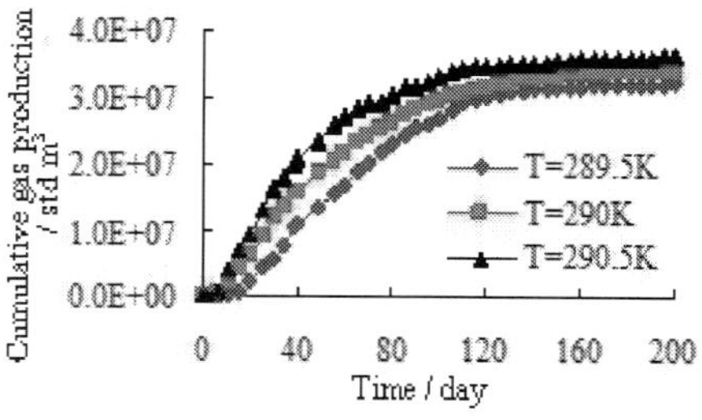

Fig. 3-38. Effect of temperature on cumulative gas

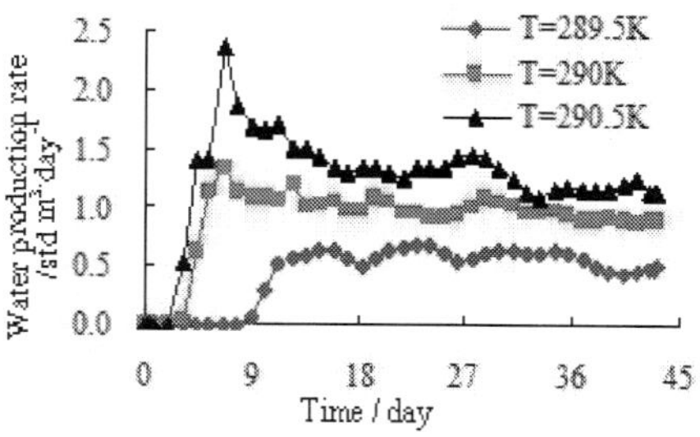

Fig. 3-39. Effect of temperature on water rate

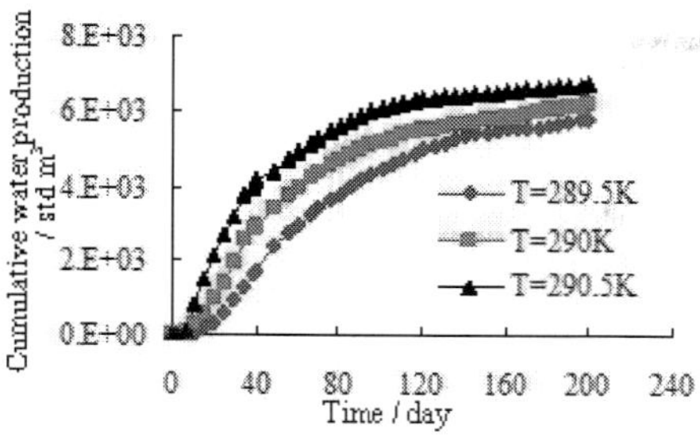

Fig. 3-40. Effect of temperature on cumulative water

Comparisons indicate that, as the same with the previous, the gas and water rate curves can be divided into three stages. They are the rapidly increasing stage, the stable stage and the gradually decreasing stage respectively. The higher formation temperature leads to the faster hydrate dissociation, the larger gas and water production rate. After 200 days production, the gas accumulative production with the formation temperature of 290.5 K is 7.0% higher than that of 290 K, and correspondingly, the water accumulative production increases 8.8%. And, the gas accumulative production with the formation temperature of 290 K is 5.9% higher than that of 289.5 K, correspondingly, the water accumulative production increase 6.9%.

4. Conclusions

Production from gas hydrate reservoirs by depressurization is a non-isothermal chemical and physical fluid-solid coupling process containing phase change. Based on the former study, this research closely held all kinds of the complicated physical and chemical mechanism during depressurization for gas production from hydrate reservoirs, and systematically studied the fluid-solid coupling numerical simulation of the stress state, the rule of physical parameters change and the productivity during the recovery process. Some conclusions are obtained as follows:

1. In view of the shortcomings of current mathematical models for gas production from hydrate reservoirs by depressurization and the comprehensive consideration of the hydrates decomposition thermodynamics, decomposition dynamics, mass and energy conservation in the decomposition process, gas-water two-phase flow in porous media, heat transfer and hydrates decomposition effect etc, we built up new phase change flow models of production by depressurization for gas hydrate reservoirs.
2. Based on phase change seepage models, according to the theory of the fluid-solid coupling seepage and considering the factors of reservoir permeability anisotropy etc, this study established the gas-water two-phase non-isothermal fluid-solid coupling mathematical model for gas production from hydrate reservoirs by depressurization. This model mainly includes the equation of fluid-solid coupling seepage, the equations of fluid-solid coupling solid deformation field, the equations of decomposition dynamics for gas hydrate reservoirs, the temperature field equation, the auxiliary equation and the initial and boundary conditions etc.
3. On the base of the previous stress sensitivity experiments for the unconsolidated sandstone, this study built up comprehensive dynamic models of physical property parameters for gas hydrate reservoirs, which include permeability, porosity and elastic modulus etc. This model reflects the relationship between reservoir physical property

parameters and the stress state, and at the same time it embodies the influence of hydrates decomposition effect. The factors considered in this model are full. And the model can lay a foundation for the fluid-solid coupling numerical simulation of depressurization for gas hydrate reservoirs.

4. The fluid-solid coupling numerical simulation was carried out, and the influence factors were analyzed for the productivity of hydrate reservoirs by depressurization. Results show that the comprehensive effect of the fluid-solid coupling model leads to the productivity from natural gas hydrate lower than that of the non-coupling model. The influence factors analysis reveals that the permeability and bottomhole pressure affect hydrate decomposition rate by changing the propagation rate of pressure drop, but the formation temperature affects hydrate decomposition rate by changing the hydrate phase balance pressure. The production index such as the gas production rate and the cumulative gas etc rises with the increase of reservoir absolute permeability and temperature, and with the decrease of bottomhole pressure.

5. Nomenclature

m_g = local mass rate of gas generation per unit volume of porous media
A_{dec} = the specific hydrate decomposition surface area in the porous media
M = the molar mass
f_e and f_g = the fugacities of methane gas at equilibrium pressure (P_e) and gas phase pressure (P_g), respectively
Φ_e and Φ_f = the fugacity coefficients of methane gas at P_e and P_g, respectively
K_d^0 = the temperature independent intrinsic rate constant for hydrate dissociation
ΔE = the activation energy
R = the universal gas constant
T = the temperature
φ = absolute porosity, namely the porosity of porous media when hydrate saturation is zero
S_H = the saturation of hydrate
A_{HS} = the specific surface area per unit hydrate volume
m_w = the mass rate of water generated from the hydrate decomposition
m_H = the mass rate of methane hydrate decomposition
ρ = the density
H = the caloric content
S = the saturation
$\vec{u}$ = the flow velocity
K_c = the effective heat conduction coefficient of formation with hydrate
Q_{in} = the supplement heat from outside
C_p = the specific heat capacity
$\vec{v}$ = the real velocity
$\vec{v}_{ra}$ = the real velocity of fluid relative to the rock
Q = the seepage flow
A_p = the interstitial surface area of seepage section, and S_a is the saturation of fluid.
A = the seepage sectional area
K_a = the permeability of fluid
∇P = the pressure gradient
μ = the viscosity
$\vec{v}_r$ = the seepage velocities relative to the rock skeleton

K_r = the relative permeability related with water saturation
σ_{ij}^T = total stress of all stress components
f_i = body force component.
σ_{ij} = the effective stress component of rock skeleton
$\overline{P}$ = the equivalent pore pressure
α = the Biot coefficient
δ_{ij} =the Kronecker symbol
$d\sigma_{ij}$ = the increment of effective stress
D_{ijkl} = the matrix tensor of elastoplastic coefficients
$d\varepsilon_{kl}$ = the increment of strain
$[D_{ep}]$ = the elastoplastic matrix
$[D_e]$ = the elastic matrix
$[D_p]$ = the plastic matrix.
K = the coefficients of heat conduction, W/m/K

6. Subscripts

g = gas
w = water
r = rock
H = hydrate
a = fluid

7. References

[1] Makogon Y.F., Holditch S.A., Makogon T.Y. Natural gas-hydrates – A potential energy source for the 21st Century. Journal of Petroleum Science and Engineering, 2007, V56 (2), 14–31.
[2] Collett T.S., Lee M.W. Reservoir characterization of marine and permafrost associated gas hydrate accumulations with downhole well logs. New York Academy of Sciences, 2000, V19 (4):51-64.
[3] Guangquan Lee. Fluid-solid coupled flow simulation and remaining oil distribution research in unconsolidated sandstone reservoirs. Phd thesis, Southwest Petroleum Institute, Canada; 2004. [in Chinese]
[4] Kim H C, Bishnoi P R, Heidemann R A. Kinetics of methane hydrate decomposition. Chemical Engineer Science, 1987, V42 (7):1645-1653.
[5] Fung L S K.A coupled geomechanical multiphase flow model for analysis of in-situ recovery in cohesiontess oil sands. JCPT, 1992, V31(6): 56-67.
[6] Zhongmin Yin, Qiang Wu, Jianjun Liu, Chunhe Yang, Xiating Feng. Numerical simulation on effect of pressure relief of water injection well on casing squeeze pressure. Chinese Journal of Rock Mechanics and Engineering, 2004, 23(14): 2390-2395. [in Chinese]
[7] Masuda Y., Naganawa S., Ando S · Numerical calculation of gas production performance from reservoirs containing natural gas hydrates. SPE Journal, 1997, V29 (3):201-210.
[8] Kambiz.Fundaments and applications of environmental and geophysical multiphase flows. Phd thesis, Clarkon University, Potsdam; 2006.
[9] Van Genuchten M.T. A closed form equation for predicting the hydraulic conductivity of unsaturated soils. Soil Sci. Soc. Am. J, 1980, V44 (10): 892-898.

[10] Parker J.C., Lenhard R.J., Kuppusamy T. A parametric model for constitutive properties governing multiphase flow in porous media. Water Resour. Res, 1987, V23 (4): 614-618.

[11] Kambiz Nazridoust,Goodarz Ahmadi. Computational modeling of methane hydrate dissociation in a sandstone core. Chemical Engineering Science, 2007, V62 (22): 6155-6177.

[12] Hong Huifang. Modeling of gas production from hydrates in porous media. Phd thesis, University of Calgary, Canada; 2003.

[13] Anjani K. Formation and dissociation of gas hydrates in porous media. Phd thesis, University of Calgary, Canada; 2005.

[14] National Institute of Standards and Technology · NIST Chemistry Web Book, http://webbook.nist.gov/ chemistry/fluid.html, 2008-08-16

[15] Masuda Y, Kurihara M, Ohuchi H. A field-scale simulation study on gas productivity of formations containing gas hydrates. Proceedings of the 4th International Conference on Gas Hydrates, Yokohama, 2002, 19-23.

[16] Kamath V.A., Holder G.D. Dissociation Heat Transfer characteristics of Methane Hydrates. AICHE J., 1987, V33 (10): 347-350.

[17] Ji C., Ahmadi G., Smith D.H. Natural gas production from hydrate dissociation: An axisymmetric model. Journal of Petroleum Science and Engineering, 2007, V58 (2):245-258.

[18] Ji C., Ahmadi G., Smith D.H. Constant rate natural gas production from a well in a hydrate reservoir. Energy Conversion and Management, 2003, V44(15), 2403-2423.

[19] Ahmadi G., Ji C., Smith D.H. Production of Natural Gas from methane hydrate by a constant downhole pressure well. Energy conversion and Management, 2007, V48 (7): 2053-2068.

[20] Yizhong Zhao. Fluid-solid coupling numerical simulation of deliverability of screen out fracturing in unconsolidated sandstone reservoir. Phd thesis, China University of Petroleum, China; 2008. [in Chinese]

[21] Tan C.P., Clennell M.B., Freij-Ayoub R. Mechanical and petrophysical characterization and wellbore stability management in gas hydrate-bearing sediments. Proceedings of the 40th Symposium on Rock Mechanics: Rock Mechanics for Energy, Mineral and infrastructure Development in the Northern Regions Alaska: 2005:810-833

[22] Freij-Ayoub R, Tan C.P, Clennell M.B, et al. A wellbore stability model for hydrate bearing sediments. Journal of Petroleum Science and Engineering, 2007, V57 (2), 209-220.

[23] Manoochehr Salehabadi, Jin Min,Yang Jinhai,et al. Finite Element Modelling of Casing in Gas Hydrate Bearing Sediments. 2008 SPE Europe/EAGE Annual Conference and Exhibition, Rome, Society of Petroleum Engineers, 2008: 1-13.

[24] Shuxia Li. The study of physical and numerical simulation on the exploitation of natural gas hydrates. Phd thesis, China University of Petroleum, China; 2006. [in Chinese]

[25] Yongmao Hao. The experiment and theory on the exploitation of natural gas hydrates. Phd thesis, China University of Petroleum, China; 2005. [in Chinese]

7

Natural Gas Hydrates

Mert Atilhan[1], Santiago Aparicio[2], Farid Benyahia[1] and Erhan Deniz[1]
[1]Qatar University, Chemical Engineering Department, Doha, Qatar,
[2]Univeristy of Burgos, Chemistry Department, Burgos, Spain

1. Introduction

Sir Humphry Davy witnessed the first chlorine hydrate crystallizing in 1811. Couple of century later his discovery, natural gas hydrates has begun to play an important role in energy business. From being a mere chemical curiosity, they have proven to be a nuisance for the natural gas industry. The problem of hydrate induced blockage in "wet gas" flow systems has been widely reported and became a major flow assurance issue in the energy sector[1]. The importance of pipeline blockage increased in the 70's when plugging of even the largest diameter pipelines from offshore, arctic fields or the wells from high-pressure underground storage facilities were reported. Studies over the past two decades showed that large gas hydrate plugs form most often after shut-in pipelines or wells begin to flow[2]. When a pipeline is shut-in, the fluid separates into the gas water and hydrocarbons as the temperature decreases[3].

Natural gas hydrates are non-stoichiometric, solid substances that consist of a low amount of gas molecules captured in a mesh cage system made up of water molecules. As seen in Figure (1), when the constituents of hydrates come into contact under high pressure and low temperature conditions a solid structure at different types of crystals with higher densities than typical fluid hydrocarbons[4] is formed. Hydrates are solid metastable compounds and their properties and stability depend upon temperature and pressure. Natural gas hydrates are dangerous compounds not only during construction stages but also during operation stages of process facilities such as platforms, pipelines and other engineering structures. Hydrates can easily form in pipelines and producing gas wells before the gas has been dehydrated. The prevention of hydrates requires substantial investments up to 10 to 15% of the production cost[5]. Flow assurance management has become one of the major critically important engineering practice in custody transfer of oil and natural gas and it is essential for successful and economic operation of oil and gas production systems[6]. Besides economical impact, an inherent problem associated with natural gas production and transportation with the presence of natural gas hydrates is the thread on operational safety. Natural gas hydrates may lead to safety hazards to production, transmission, and transportation systems. For both economical and operational safety perspective, understanding of formation kinetics, where and when natural gas hydrates form is necessary to better manage and mitigate this phenomena. Moreover, during the last decade several researchers pointed out that understanding the formation mechanism of natural gas hydrates may open new avenues in alternative ways of natural gas transportation. With the

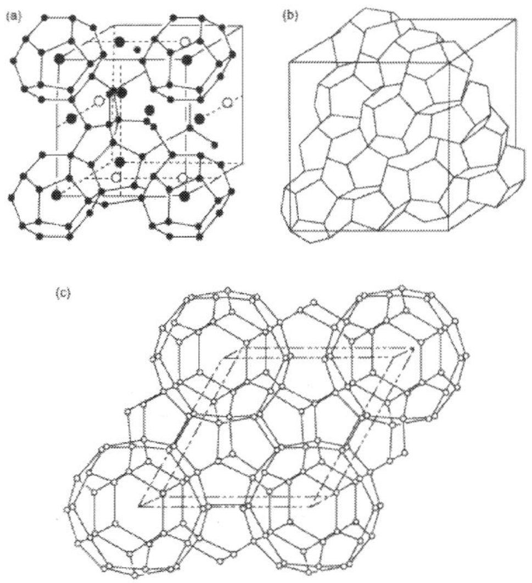

Fig. 1. Hydrate crystals (a) sI type, (b) sII type, (c) sH type[3].

increased demand for natural gas there is an additional incentive in exploring ways to monetize stranded gas that can not be economically developed due to low natural gas volumes to justify effective LNG facilities or lack of proper pipeline infrastructure[7, 8, 9].

Gas hydrates can form at the gas liquid interfaces along the entire length of the static pipeline. This can create small volumes of hydrate over time, but usually do not block the pipeline. However, when flow resumes, plugs can form at any point where the flow regime changes. Small-scale hydrate formation in the interface sometimes cannot be avoided in the pipeline. Moreover, under certain conditions, small-scale agglomerates are also observed in the bulk phase. Hydrate formation does not become a threat to pipe flow unless the agglomerates and hydrates formed at the interface start forming bridges. In such cases blockage occurs where the small accumulations of hydrates adhere to the walls and begin to bridge and reduce flow. This bridging can eventually shut down the entire pipeline or field until the hydrates have been removed[10]. Like hydrate formation, dehydration of hydrates in the pipelines is another major operational safety risk[6].

Hydrates form as a result of slow cooling of a fluid as in a pipeline or rapid cooling caused by depressurizing across valves or through turbo expanders. Studies have shown three conditions promote hydrate formation in gas pipelines and in petrochemical processes: *Coexistence of water, natural gas components and low temperatures and high pressures.* Other factors that favor hydrate formation can be listed as high fluid velocities, agitation, pressure, pulsations (or any source of fluid turbulence), the presence of CO_2 and H_2S[11].

Gas molecules ranging from C_1 to C_4 and including CO_2, N_2 and H_2S are typical hydrate components. The water needed for hydrate formation can come from free water produced from the reservoir or from water vapor condensed by cooling the hydrocarbon fluid. At low temperature conditions, onshore pipelines suffer from hydrate formation during the winter months. Offshore, below 900 m of water depth and at the ocean bottom, the temperature is remarkably uniform around 3.8 °C and the pipeline cools to this temperature within a few miles of the wellhead. This situation may lead hydrate formation[12].

2. Natural gas hydrates structure and physical properties

Natural gas hydrates (NGH) form in raw multiphase flow as a result of crystallization occurring around the guest molecules at certain operating temperature and pressure conditions[13, 14, 15]. The most widely observed guest molecules in natural gas mixture are methane, ethane, propane, i-butane, n-butane, nitrogen, carbon dioxide and hydrogen sulfide. However, among those, methane based NGH occurs the most naturally[16, 17, 18, 19]. NGH are composed of approximately 85-mol% guest molecule; therefore they have physical properties very close to ice. They have crystal structure. The density of NGH varies somewhat according to former molecule(s) and the formation conditions[20]. NGH are part of a larger family of compounds called "Clathrates", which are inorganic container compounds[21]. Although there are many container-compounds and hydrate formers in crystal structure, the focus in this paper is NGH formers and NGH structures.

In general, hydrates are classified by the arrangement of the water molecules in the crystal structure. All common natural gas hydrates belong to the three crystal structures: cubic structure I (sI), cubic structure II (sII), hexagonal structure (sH) as shown in Figure (1). Structure I is formed with guest molecules having diameters between 4.2 and 6 Å, such as methane, ethane, carbon dioxide, and hydrogen sulfide. Nitrogen and small molecules including hydrogen ($d < 4.2$ Å) form structure II as single guests. Larger (6 Å $< d <$ 7 Å) single guest molecules such as propane or iso-butane will form structure II. Still larger molecules (typically 7Å$<d<$9Å) such as iso-pentane or neohexane (2,2-dimethylbutane) can form structure H when accompanied by smaller molecules such as methane, hydrogen sulfide, or nitrogen[3].

The crystal structures of NGH consisting of water molecules are hydrogen-bonded in a solid lattice. The interaction or degree of bonding between individual water molecules and the guests is very weak, but the overall interaction of the guests with the host structure can be quite strong[22]. In the literatures, more than 130 compounds that are known to form clathrate hydrates with water molecules are mentioned and more emphasis observed to be given to sI and sII hydrates since these are by far the most common NGH structures[23, 24, 25, 26]. The sH structure NGH are well described by[27, 28, 29, 30, 31, 32, 33, 34]. Moreover very high-pressure hydrate phases are studied by[35, 36, 37, 38]. Crystal properties of NGH have been extensively studied by Sloan and Koh[3] and Jeffrey[24].

2.1 Crystal structure of sI, sII and sH type of NGH

In 1965, structure I type (sI) of hydrates was first observed by McMullan et al. and Jeffery [39] with definitive x-ray diffraction method on ethylene oxide hydrates. In Figure (1) sI structure, 46 water molecules present along with 8 polyhedra within the cubic structure. Moreover, Mak [40] observed structure II type (sII) hydrate crystal structure in the same year with definitive x-ray diffraction method. He showed that the sII type crystal consists of a face centered cubic lattice structure with a side dimension of 17.3 Å. Structure type (sII) type hydrate's typical crystal structural view is given in Figure (1b). Later, Jeffrey[24] pointed out that the sII type structure voids are formed by connecting 16 polyhedra and might accommodate larger guest molecules than it is commonly observed in NGH structures. On the other hand, first structure H type (sH) of hydrates were discovered in year 1883 by de Forcrand during his studies with binary hydrates with iso-butyl chloride or bromide as the guest molecule (now, they are known as sH formers). However it was not recognized at that time. The first sH type structure was reported by Ripmeester et al. [41, 42] with his study conducted via NMR spectroscopy and X-ray powder diffraction methods on different clathrates. An important feature of sH type hydrate crystals is that two sizes of molecules are required to stabilize the hydrate structure. This typically happens with a smaller molecule such as CH_4 or H_2S and a bigger molecule such as dimethlybutane. sH type hydrate's typical crystal structural view is given in Figure (1c). This type crystal structures observed to have unit cell length of 12.3 A. For sI, sII and sH type structures at usual pressures, only one guest molecule can be filled in the cavity. More information can be obtained about the crystal structure and cavity occupation by guest molecules in[43]. Scanning Electron Microscopy and other micro-imaging techniques are very powerful tools to study the porous structure of NGH's. Such techniques should accompany NGH crystal growth to understand better the relationship between thermodynamic conditions and structure. The conceptual crystal growth at the molecular level still remains a fuzzy area. Indeed the interaction between the contributory factors leading to crystallization oh NGH is still not understood: kinetics, mass transfer and heat transfer. A great deal of fundamental work is still required to disentangle this interdependency.

2.2 Physical properties of NGH

NGH are solid materials which have higher densities than hydrocarbon components forming natural gas mixtures. In the open literature detailed investigation on NGH physical properties tend to focus mainly on mechanical, elastic and thermal properties. Compression deformation measurements on NGH sediments are conducted by Parameswaran et al.[44] and Cameron et al.[45] showed that strength of NGH is approximately similar to that of ice. Later Stern et al.[46] did compression deformation measurements at constant applied stress (creeping test) on NGH (methane hydrate) and his results showed the same trend as Parameswaran's and Cameron's. However, in 2003 Durham et al.[47] showed that the impurities in previous studies effected the previous studies in creeping tests and they determined that NGH was 20 times more creep resistant than ice[48]. Elastic properties of NGH can be estimated accurately since they are function of crystal structures and crystal structures are well defined. Whalley first proposed that the elastic properties of NGH are similar to that of ice in 1980[49]. Later this theory was confirmed with first experimental studies conducted by Whiffen et al. in 1982, Pearson et al. in 1984 based on experiments on simple hydrates via Brillouin spectroscopy method, which was later followed by Kiefte et al.[50, 51, 52]. More recently in 2002, Shimizu et al.[53] performed in situ measurements on NGH via improved Brillouin spectroscopy technique and looked at the effect of of pressure on shear stress as well as compression velocities. This study showed that the shear velocities of NGH (mainly methane hydrates) are similar to that of ice[53]. First experimental studies on thermal properties of NGH were conducted by Stoll and Bryan in 1979 and they showed that the thermal conductivit of NGH as 0.393 $Wm^{-1}K^{-1}$ at 215.15 K which is 5 times less than that of ice (2.33 $Wm^{-1}K^{-1}$)[54]. Low thermal conductivity of NGH is confirmed with several studies later and a nice mapping of the thermal conductivity measurements from several experiments were recently published by Gupta[55, 56, 57].

3. The effect of NGH on the flow assurance

Flow assurance can be defined as an operation that provides a reliable and controlled flow of fluids from the reservoir to the sales point. Flow assurance operation deals with formation, depositions and blockages of gas hydrates, paraffin, asphaltenes, and scales that can reduce flow efficiency of oil and gas pipelines. Due to significant technical difficulties and challenges, providing safe and efficient flow assurance needs interdisciplinary focus on the issue and joined efforts of scientists, engineers and operation engineers[58]. It was mentioned by Guo et al.[59] that as a rule of thumb, methane caged NGH will form if the temperature is as high as 4.5 ^{0}C and pressures are as low as 11.7 bars. As seen in Figure (2), mild conditions are required for NGH formations. NGH predictions can be determined by using simulation software and computational methods. However, predicting hydrate formation requires more detailed experimental studies for each reservoir fluid since the operating conditions and compositions vary vastly. As a result of both theoretical and experimental investigations, five different NGH prevention methods have been implemented to provide flow assurance[6, 60]. These are:

i. Dehydration of wet gas and water removal (onshore or offshore)
ii. Avoid operation temperatures lower than the hydrate formation temperatures
iii. Avoid operation pressures higher than the hydrate formation pressures
iv. Injection of Thermodynamic Inhibitors (TI) such as methanol, glycol etc. to effectively decrease the hydrate formation temperature and inhibit or retard NGH crystal formation
v. Injection of Kinetic Inhibitors (KI) to prevent the aggregation of hydrate crystals

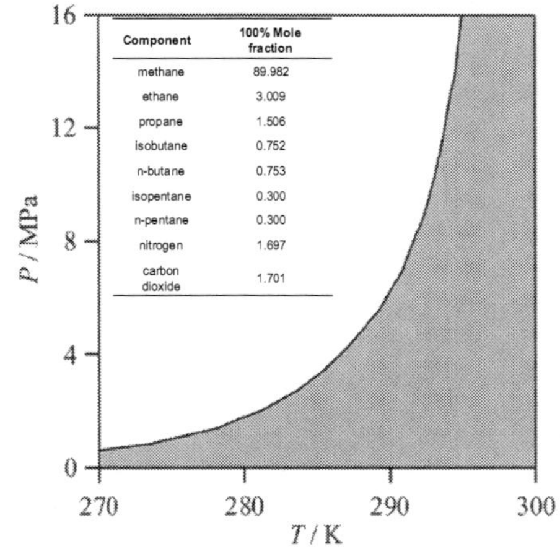

Component	100% Mole fraction
methane	89.982
ethane	3.009
propane	1.506
isobutane	0.752
n-butane	0.753
isopentane	0.300
n-pentane	0.300
nitrogen	1.697
carbon dioxide	1.701

Fig. 2. Calculated pressure – temperature diagram for hydrate formation for the a typical lean multicomponent gas mixture which composition is indicated within the figure. Gray area shows the hydrate free region and white are the region where hydrate formation is possible. Calculation according to the Baillie and Wichert Method[136].

Above options are applied separately or in matching combinations. Selection of above options depends on fixed and operating cost restrictions, technology availability and know-how, system characteristics and operation/process flexibility.

NGH plugging up a pipeline may cost the gas industry in excess of approximately $1 million each day that production is shut down[58] and once a hydrate plug has formed, it can take weeks or even months to dissociate it safely, or it will require complex operations such as pigging for plug removal[61]. The petroleum industry has tried to prevent hydrates from forming by drying and heating the oil as well as by adding TI such as glycol based antifreeze agents and methanol. Annual approximation of an operating expense is greater than $500M, which is devoted to hydrate prevention via TI injections[62]. Hydrates are a problem not only for pipelines but also for wells drilling: the use of water based drilling fluids for offshore operations may lead to the formation of hydrates that strongly affect deep-sea drilling operations. Experimental studies on kinetic mechanism and when & how NGH forms are conducted in two major mechanisms: analytical equipment based and purpose made engineered apparatuses[63]. Such figures and the negative environmental impacts of the use of TI's and KI's have motivated the academia and research centers all around the world to conduct fundamental research on NGH and their mitigation. It has been observed that over the last couple of decades, fundamental NGH research tendency has shifted from time independent TI studies to time dependent KI studies[64]. The replacement of traditional TI and KI by biodegradable equivalent may be the subject of future work to reduce the environmental impact.

3.1 NGH inhibition for subsea pipelines

In theory if all the water is removed from the natural gas stream, conditions for NGH formation will no longer be effective for the hydrate formation. Offshore dehydration may not be feasible for all the operations due to physical footprint constraints in the production facility at the offshore facilities. It may also not be necessary to have an offshore dehydration facility if the risers down in the sea bed do not have steep slopes, yet, temperature and pressure of the pipeline is as important as the riser slope. Having said that, dehydration facilities are not the most cost effective way of preventing NGH formation in subsea transport pipelines. Operationally, avoiding the pipeline conditions outside of the hydrate equilibrium loop (HEL) would also prevent NGH formation. To achieve these conditions, TI, KI and thermal methods can be applied to the pipeline. Manipulation the HEL can be achieved by injecting KI or TI to the pipeline. Some physical thermal methods are used in order to provide environmentally friendly HEL control without injecting chemicals to the pipeline. Heat conservation is the common practice for thermal methods and the application is made through insulation of the pipeline such as hot water jacket, concentric tube bundle or trace heating along the pipeline. Numerous alternative thermal method applications are available[65, 66, 67]. It is worth noting that under thermal methods, the so-called "memory effect" of NGH should be exploited to prevent hydrate formation again once the hydrate melted[3].

3.1.1 Thermodynamic inhibitors for subsea pipelines

Typical subsea pipelines do not have insulation and they require chemical inhibition not only for preventing NGH formation in the pipelines but also to prevent plugging during the start-up and shutdown conditions[68]. Both TIs and KIs are injected at the wellhead and selection of both inhibitors is based on the recovery cost and easiness of the injected inhibitor. Historically, methanol and glycols were considered for NGH prevention as TI agents. Basically, these chemicals make hydrogen bond with water molecules and prevent them to form ordered cages to entrap gas molecules . Methanol can be separated from the flow easily since it has lower density, surface tension and viscosity under separation conditions (below -25^{0}C). This brings lower cost in recovery of methanol in the system. Moreover, glycol based solutions are also used for plug prevention in the pipes. Typically ethylene glycol (or mono ethanol glycol known as MEG) is used due to lower cost, lower viscosity and lower solubility in liquid hydrocarbons. However, glycols must be added up to 100% of the weight of the water content present in the gas stream. This leads to economical issues since glycols are expensive inhibitors. Moreover, there is a need for a large footprint in onshore and offshore facilities. Difficult recovery and regeneration are other disadvantageous of plug prevention via glycols[69, 70, 71, 72]. There are some other electrolytes used as TIs in NGH prevention[73]. All the TI molecule in combination with the water molecule in the flow changes the HEL by changing the chemical potential of the hydration[74] which results in a shift of the HEL towards the lower temperature and higher pressure side. Hammerschmidt[1] gives an empirical formula and rule of thumb calculation for lowering the temperature for the NGH formation.

On the other hand, once the reservoir conditions are considered, the stability of hydrates in a reservoir also depends on the interactions between minerals, surrounding reservoir fluids and hydrate. As the reservoir depth increases, the level of salinity increases. Moreover, the formation of hydrate then will lead to increased salinity of the fluids surrounding the formed hydrate. This may result in liquid pockets of residual aqueous solution with increased salinity and non-uniform hydrates[75].

3.1.2 Low dosage hydrate inhibitors for subsea pipelines

An observation of made on an ocean fish that bond itself with proteins in order to avoid freezing in severe sub-zero conditions led discovery of low dosage hydrate inhibitors (LDHI) and LDHI have caught the attention of researchers working on NGH in academia and industry shortly after its discovery. Kinetic Inhibitors (KI) have been the most widely investigated sub-class LDHI in the past decade. By contrast to TIs, KIs limit or delay the growth of hydrate formation[76] by reducing the nucleation rate of hydrate and preventing the formation of critical nucleus. In other words, hydrate crystal growth is controlled with KIs. However, it is not economically viable to ensure complete coverage of all the hydrate nucleation and growth sites with KIs; and thus, it might be required to have huge amount of KI dosages to ensure the complete crystal growth inhibition.

First KI studies were conducted under low/moderate pressure test chambers in early 90s. The first generation KIs were mostly water-soluble polymers. For instance, PVP was the first KI to be found as an effective plug former inhibitor[77]. KIs are typically polymer solutions such as Vinyl Caprolactam/VP/Dimethylaminoethyl Methacrylate Copolymer (known as VC-713), Polyvinylpyrrolidone (PVP), Poly(VP/VC), PVCap (polyvinylcaprolactam), Polyvinlymethylacetamide etc. KIs are preferred to TI's because they are more efficient in plug formation inhibition and injection is made in much lower quantities[74]. Many others followed PVP; Later studies in KI investigation has been shifted to higher pressure chamber applications and performance of KIs are modified for: *low temperature, high pressure, inhibitor concentration* and *salinity*. KIs are time time-bound components and they can prevent the formation of NGHs for a limited period of time. There is rapid conversion of the remaining water into large accumulations of hydrates once this time period passes, which results in blockage. Use of KIs are not limited by the water cut, however, it is limited by the sub-cooling in the system that KIs will be used.

A KI is generally used in conjunction with surface-active agents. Surface-active agent along with KI polymer will cause emulsification to occur between the free water and gas/condensate phases in the pipe which then prevents the agglomeration of hydrate crystal well before the first crystal formation starts. In other words, KIs keep the particles small and well dispersed so that fluid viscosity remains low, allowing the hydrates to be transported along with the produced fluids[6].

On the other hand, in deep and ultra deep water cases where very extreme conditions exist (i.e. Gulf of Mexico, North Sea and West Africa-Nigeria), Anti Agglomerant (AA) another sub-class LDHI, are observed to perform better rather than classical type KIs and they are often known as hydrate growth inhibitors[78, 79]. The concept of AA is basically preventing the formation and accumulation of large hydrate crystals into a composite hydrate blockage through having a hydrophilic head that is incorporated within the hydrate crystals and hydrophobic tail that disperses the hydrates into a liquid hydrocarbon phase[80]. Anti-agglomerant disperse the very small formed hydrate nuclei whereas kinetic inhibitors (although their mechanism of actuation is not fully clarified) bind to the forming crystals thus hindering, and delaying, their growing. Unlike KIs, AAs are not limited by the sub-cooling of the system and they can continue to be effective at sub-cooled conditions; however, they require the presence of a condensate hydrocarbon phase in order to suspend the hydrate crystals. Moreover, at high water cut conditions, dispersed hydrate crystals might cause an increase in the viscosity of the condensate phase, which might result in blockage of the flow due to increased viscosity. Often, it is required to have demulsifier for old and water separation since AAs are based upon homogenously dispersion polar hydrate crystals in the apolar oil and condensate phase. AAs have recently been more recognized in industrial use especially in the Gul of Mexico region[79].

In addition to classic KI's and TI's, biological hydrate inhibitors (BI) have also been investigated as well. Inhibitor proteins or in other words anti-freeze proteins (AFPs) and antifreeze glycoproteins (AFGPs) inhibit the hydrate formation by binding the surface of hydrate crystal nuclei[81]. Recent studies have also shown that AFPs also prevent the hydrate growth after the hydrate crystal is formed[82].

3.2 Experimental studies for NGH

Experimenting the nucleation stage for NGH is perhaps the most challenging step and it is essential in understanding the process of crystallization of gas hydrates. Sloan[19], Makogon[5, 83], and Koh[84] have provided an extensive reviews on the kinetics of gas hydrate crystallization in literatures. Theoretical description of the rate of hydrate nucleation has been attempted by Kvamme[85] from the perspective of the nucleation aspects of the crystallization process are concerned and models have been published on crystallization theory for the prediction of gas hydrate formation by Sloan[19].

Kashchiev and Firoozabadi[86] have measured the rate of gas hydrate formation after nucleation and they developed gas consumption rate models based on the early stage of the crystallization theory, which is, the hydrate growth stage.

NGH have been known since 1930s to have the potential to plug the oil and gas transportation pipelines, in particular when working under high pressure. This phenomena has started be even more problematic since the present trend of the natural gas industry to work at higher pressures will increase the hydrates problem for natural gas transportation. Since 1930s, thermodynamic conditions that lead to NGH formation have been the center of focus of researchers in academia and industry. Experimental studies started in late 1950s by Katz et al.[87] and followed by Makogan, Berecz and Balla-Achs[88, 89]. Main focus of the experimental research on NGH was to determine the thermodynamic conditions to operate the pipeline outside the hydrate formation region or HEL.

Okutani et al[90] reported their high pressure hydrate dissociation curve determination apparatus with some experimental findings. Their method uses a high pressure, stainless steel isochoric cell without agitation to investigate methane hydrate equilibrium in water-hydrate-gas systems. They have investigated hydrate formation by monitoring pressure changes in the system under a constant cooling rate. Kim et al.[91] investigated the hydrate dissociation curve with a similar high-pressure cell by observing the dissociation temperature of the hydrates. Their design also included large observation windows. They formed hydrates by injecting methane and ethane samples through a water container with constant heating rate. They observed dissociation temperature with visual data. Seo et al [92] used a similar experiment and investigated three-phase hydrate equilibria in the methane+water+cyclic-ether and nitrogen+water+cyclic-ether systems using a temperature search method up to 120 bars. Sakaguchi et al.[93] described another interesting isochoric design. They investigated the formation and growth of hydrate crystals in a Pyrex cell visually using observations through CCD cameras connected to micrographic zoom lens at atmospheric conditions. Kang et al.[94] and Fleyfel et al [95] used a method known as "rocking cell". They used a visual rocking cell to observe hydrate formation in metastable methane, ethane, propane and water systems. They combined the constant pressure-rocking cell with NMR and observed the peak with NMR and visually confirm this.

Some researchers used x-ray diffraction, Raman spectroscopy and differential scanning calorimeter (DSC) methods to investigate hydrate formation and dissociation properties. Hester et al.[96] used Raman spectrometer to investigate hydrate compositions for the Hydrate Ridge in Oregon. Koh et al.[84, 97] studied hydrates using DSC. Using DSC, they quantified and compared the effect of various kinetic inhibitors. Le Parlouer et al.[98] investigated the thermodynamics properties and kinetics of hydrate formation within minerals. They modified a Setaram DSC 111 for pressures up to 400 bars with a temperature range of -45 °C to 120 °C. They have shown the dissociation temperature of hydrates with respect to methane pressure and heat flow versus temperature profiles at the dissociation peaks for different isobars. Dalmazzone et al.[99] used an approach similar to that of Le Parlouer for hydrate formation and kinetics investigation. They used a modified high pressure Setaram Micro-DSC for water-in-oil emulsions systems up to 40 MPa between 223 and 393 K.

In contrast to static high-pressure cell approaches, researchers also have investigated hydrate formation mechanism and kinetics using dynamic models and techniques. Gaillard et al.[100] investigated methane hydrate formation in a recalculating flow loop and perform kinetics modeling of hydrate formation. They esentially studied the kinetics of methane hydrate formation and inhibitor performance in a laboratory scale mini flow loop under pressures up to 75 bars. Lee et al. [101] used a laboratory scale mini flow loop to study the mechanism of hydrate plugging and examine inhibitors that prevented hydrate plugging in natural gas pipelines up to 8 MPa. Some sections of the loop also had visual access. Urdahl et al.[102] and Lund et al.[102, 103] used loop design in a relative motion flow loop and studied gas hydrate formation and inhibition in hydrocarbon gas-water-oil systems. Their 150 bar loop had a temperature range of -10 to 150 °C and was fully automated. It enabled visual detection of hydrate formation. They studied methane rich methane+ethane+propane+water systems in the presence of several different KI's. Mork[104] described one of the most detailed flow assurance and loop experiment for gas hydrates in the literature. He studied the rate of methane hydrate and natural gas hydrate formation in a 100 bar flow loop between 7 and 15 °C to understand the performance and scale-up of a reactor for continuous production of natural gas hydrates. He has shown that the rate of hydrate formation is strongly influenced by gas injection rate and that the pressure the effect of stirring rate is less significant.

Additionally, many patented experimental setups and procedures have been proposed and published for gas hydrate investigation. The most widely accepted designs are flow loops such as those of Guo et al.[59], Hatton[105], Larue et al.[106], Behar et al.[107], Mitchell and Talley et al.[108, 109].

Much of the available funding of NGH research was granted for industrial field experiments and academic testing of field characterization. This was aimed at establishing secure transportation conditions by avoiding plugging in pipelines. Results of the industrial and academic experiments may not be available in the open literature completely due to the confidential nature of the data to the companies and governments[4] funding such work.

It is highly desirable to initiate research work on NGH that is aimed at not only mitigating the effects of NGH, but also the risk and safety aspects. The impact of a NGH plug dislodged by pressure differential can be very dangerous and may cause the pipeline to rupture leading to casualty in operating personnel as reported by Sloan[3].

3.3 Instrumental analysis for NGH

NGH formation is an exothermic crystallization process. Characterization of NGH can be made by investigating the nucleation single crystals, crystal growth and agglomeration[110, 111]. The kinetics of hydrate growth is concerned with the rate at which the hydrate phase grows after the induction time that marks the onset of hydrate crystallization[112]. Mostly gas uptake rates at which hydrate interface advances are investigated operationally by measuring pressure, temperature and composition of the different fluid phases in the investigated system. On the other hand, molecular studies and analytical techniques are used for hydrate crystal structure, composition, and cage occupancy determination. For these purposes Powder X-ray diffraction (PXRD), Raman and Nuclear Magnetic Resonance (NMR) techniques have been used for solid phase structural analysis at molecular level for NGH[113, 114, 115, 116, 117, 118]. Characterization of NGH sH structure crystals (methane type) and several large organic molecules are recently done by Susilo et al. via NMR and

Raman techniques[119]. They showed that when a molecule is trapped and encaged in hydrate structure, Raman and NMR spectra gas/liquid signals are shifted and showed distinctive behavior during hydrate formation. The distinct methane behavior in sI, sII and sH type hydrate structures is observed as lower frequency when compared with free gaseous state components in Raman spectroscopy measurements[120]. Raman spectroscopy is used mostly on determining the cage occupancies and the hydrate compositions quantitatively after careful calibration of the observed peaks for the specific system under study[120, 121, 122]. On the other hand, the distinct methane behavior in all three-hydrate structures in NMR is characterized by lower field outputs in comparison to free gaseous state components[122, 123]. If accurate data acquisition is provided with both NMR and Raman spectroscopy, cage occupancies of NGHs are determined accurately and the results are comparable between both methods[124, 125].

4. NGH as a potential energy source: Storage and transport issues

It was previously indicated that hydrates are one of the major problems that oil and gas industries suffer from in operations. Since the discovery of the hydrates, majority of the research efforts has been spent on the determination of the hydrate crystal structure, kinetics of hydrate formation, thermodynamic behavior and mechanisms to avoid plug formations in the pipelines. However, NGH are also called as "white coal" and they have great potential due to its capacity to store huge amounts of methane in its cage structure. On average, NGH volume gains are estimated almost as 155 times smaller than the equivalent amount of natural gas at standard conditions[126]. Volumetric gain changes depend on the crystal structure of the NGH[83]. Estimation of the volumetric gain has been done by Berner[127] and for typical sII type crystal structure NGH with no impurities nor inclusions have an approximate 191 m^3 of gas per 1 m^3 of NGH. With 5 vol% impurities and 96% occupancy of cavities in the crystal structure this value is estimated as 174 m^3, and with 8 vol% impurities and 96% occupancy it is 160 m^3. Storing and transporting natural gas in NGH form will require natural gas stream with some water content treated at 5 to 15 ^{0}C under 10 to 25 bars to process 155 m^3 of natural gas in order to produce 1 m^3 of NGH. On the other hand, similar or slightly better volumetric gain is achieved by well established liquefied natural gas (LNG) and compressed natural gas (CNG) processes. However, for LNG processes -163 ^{0}C and for CNG processes pressure operations up to 200 bars is required[128]. Successful implementation of natural gas transport in the form of NGH may reduce the operational cost dramatically by avoiding very low temperature operation cost in LNG production and compressing requirement in CNG production. Conversely, stability of the end product is a major concern. It has been proved that the stability of the end NGH product stayed stable up to 2 years in solid hydrate form in cold climates such as Russia and Norway[129]. With all the potential in NGHs, it may become an attractive means of transportation of natural gas located in the small and unutilized gas fields. Transportation of gas as NGH requires export terminals with natural gas with water to produce NGH and receiving ports equipped with re-gasification facility to extract gas from the hydrate cage. Mitsui Engineering and Shipbuilding Corporation in Japan are working on engineering and feasibility investigation of such application with a capacity of 600 kg/day[7]. In today's world it has been estimated that about 70% of the total gas reserve is either too far from an existing pipeline or too small to justify a liquefaction facility

Another interesting aspect of NGH which attracted the scientific community to conduct research on is the large sediments and deposits of NGHs located on the ocean floor. Yet, there are estimations and speculations on the NGH depositions on the ocean floor, which is greater than the combined total energy resources that we have currently in all means (see Figure (3)). Englezos estimated that total amount of NGH in ocean floor is about 10^{16} m^3[130]. There are 50 proven locations of NGH sediments on earth where large quantities of NGHs present[128]. However, quantification of how much methane, ethane, propane, i-butane and n-butane is trapped in these sources is difficult to ascertain and has not been done for the majority of the sources. The difficulty of handling these deposits are: *to reach the source and bring them above the ocean floor* and *they dissociate very quickly with the pressure applied on those (even a mechanical arm used to carry them above sea level)*. In order to prevent quick dissociation of these deposits following solutions are proposed: *reduce the temperature of deposit in a separate chamber* and *increase the deposit density* [126]. On the other hand, another alternative method has been proposed by Fanklin, which is based on removing these deposits under controlled dissociation. This idea is proposed by applying: *increase temperature of the deposit with sensitive control, decrease the deposit weight, decrease the penetration rate and apply controlled pressure on the deposits*[131]. Such techniques remained as theories and have not gained importance so far[132].

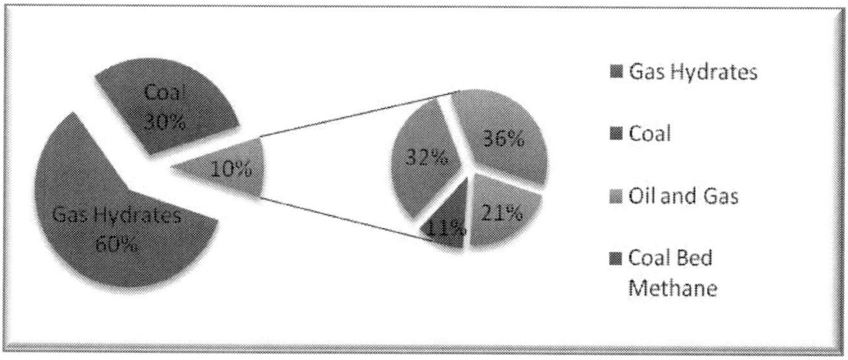

Fig. 3. Fossil energy resources. (Tboe: trillion barrels of oil equivalent).

Another biggest problem standing against the production of energy via NGH from the ocean bed deposits is the possibility that removal of the methane hydrates may trigger the uncontrolled release of methane gas to the ocean, which may increase the salinity of the oceans dramatically. This uncontrolled incident may result in global climate change since methane in the atmosphere is known to be the biggest contributor to the global warming[133, 134]. Utilizing NGH deposits in the ocean floor may power the planet in the future and the challenges stated above will become more important for the academia for further and detailed investigation in near future.

5. NGH safety and hazards

Gas processing plants and sites that experiences NGH problem considers the safety aspect of NGH has two different ways; in storing and pipeline transportation. Since the crystals matrix in hydrate systems has a high latent energy must melt before gases are released; explosive release of gas during accident is inherently inhibited in a hydrates system when compared with LNG and CNG. As transportation point of view, when ignited, NGH will burn slowly and will not explode. For this reason, once the wall of hydrates carrier is breached, the natural gas hydrates will not readily flow out of the carrier vessel, as is the case for LNG. On the other hand, from the view of hydrate plug formation in the transportation pipelines, there are bigger safety concerns exist especially during the removal of the hydrate plug. During the dissociation of hydrate plug in the pipeline, great pressure drop is occurred when the plug detaches from the pipe wall. This pressure difference might cause solid hydrate plug mass to reach to an approximate speed of 300 km/h within the pipeline. This phenomena will help to compress the downstream gas further which will result in blowouts, ruptures and damages in the pipeline[12].

Hydrates contain as much as 180 volumes of gas at standard temperature and pressure per volume of hydrate. Hydrates dissociation by heating causes rapid gas pressure increase in the system. Field engineers often call this phenomenon "hail-on-a-tin-roof". Small hydrate particles can contain considerable volume of gas. Attempts to blow the plug may rupture the pipe. In order to determine the best approach to remediation of hydrate blockage, the knowledge of the location and length of a hydrate blockage is very critical. During plug dissociation, there might be multiple plugs exist in the pipeline which threatens the pipeline both from safety and technical perspectives. As investigated by The Canadian Association of Petroleum Producers there are two key points to consider in that regard[135, 136]:

i. Always assume multiple hydrate plugs in the flow loop. There may be high pressure points between two plugs.
ii. Attempting to move hydrate plugs can rupture pipes and vessels in the flow loop.
iii. While heating a plug, the heating procedure should commence from the end of a plug rather than the middle section of the pipe.
iv. Single sided depressurization can potentially launch a plug like a high-speed bullet and result in ruptured pipes, damaged equipments and uncontrolled release of hydrocarbons to environment.
v. Actively heating a hydrate plug needs to be done in such a way that any released gas will not be trapped

6. Molecular simulations of natural gas hydrates

Compared to thermodynamic hydrate inhibitors like methanol and glycol, Low dosage kinetic inhibitors have several adventegous in terms of cost and environmental impact. They are active in low concentration, less than 1% of water phase, and generally they are not toxic. However, despite the studies have been done over the last years, the formation mechanism of hydrates, the kinetics of the process and the action mechanism of how LDKI's work are not fully understood. This lack of knowledge and the limitations of surface sensitive techniques prevents the development of more efficient LDKI's. Over the last two decades, Molecular simulations have been proven to become a powerful technique to study the molecular level details of solid/liquid interfaces and it provides valuable information about gas hydrates nucleation, growth mechanism together with a basic understanding of the action mechanism of LDKI's.

Liang et al. reviewed recently the studies that uses molecular simulations to analyze the properties of hydrates growing and inhibition[137]. Rodger et al.[138] analyzed the effect of an inhibitor based on quaternary ammonium zwitterions by molecular simulations methods and showed that its activity is comparable with the common kinetic inhibitor : PVP. Basically, molecular simulations allow to test new families of potential inhibitors before synthesizing the molecules or before doing expensive and time-consuming laboratory or field tests. There have been several computational studies that support these conclusions[139,140,141,142]. Kvamme et al. showed how computational experiments provide valuable information to select kinetic inhibitors and to understand the action mechanism of inhibiting hydrate formation[143].

Molecular simulations have also been used succesfully to analyze other aspects related with hydrates area in several recent works. Myshakin *et al.*[144] used molecular dynamics simulations to analyze the methane hydrate decomposition process, showing an Arrhenius type behavior and analyzing the regrowing process. Tanaka *et al.*[145] investigated the thermodynamic stability of the structure I and II hydrates using molecular simulations. Erfan-Nia *et al.*[146] have used molecular simulations to analyze the structure of hydrates formed from the methane+ethane mixture, showing that type I hydrates are formed in the whole composition range. Also, they used molecular simulations to show the methane storage capacity of structure II hydrates with the help of large guest molecules such as propane, i-butane, tetrahydrofuran etc[147].

7. Conclusion

This chapter reported the progress made to better understand natural gas hydrates and their mitigation in pipeline transportation. Whilst one acknowleges the benefits of techniques such as injection of inhibitors such as TI and KI to maintain gas production, the associated cost and negative environmental impact clearly signal to additional fundamental research in the area. Such work may include the use of more effective and environmentall friendly additives, a better understanding of the interaction between mass transfer, heat transfer and kinetics of NGH crystal growth and a better utilization of imaging techniques to support experimental work on NGH crystal growth. These studies must imperatively be conducted using properly designed reactors to produce hydrates under carefully controlled conditions. The field of "in-situ" hydrate formation detection is still far from being a reality and constitutes a desirable objective since this would make the continuous use of additives redundent and facilitate the use of selective local heating of pipeline where possible, or in the worst case, inject selectively the correct doses of KI's. This has been termed as "time dependent" growth model studies in the literature without providing any clues on how this might be achieved in practice.

The potential of NGH's as alternative state of natural gas during transportation still remains untapped even though it has been acknowleged given the abundance of natural gas in the form of hydrate in deep seas. This area of work would necessarily involve NGH stability studies.

8. References

[1] E.G. Hammerschmidt, Industrial and Engineering Chemistry 26 (1934) 851-855.

[2] A. Hunt, Fluid properties determine flow lone blockage potential, Pennwell, Tulsa, OK, USA, 1996.

[3] E.D. Sloan, C.A. Koh, Clathrate Hydrates of Natural Gases, Third ed., CRC Press, Boca Raton, FL, USA, 2007.
[4] E.D. Sloan Jr, Nature 426 (2003) 353-359.
[5] Y.F. Makogan, Hydrates of Hydrocarbons, PennWell Publishing Company, Tulsa, OK, USA, 1997.
[6] S. Mokhatab, R.J. Wilkens, K.J. Leontaritis, Energy Sources, Part A: Recovery, Utilization and Environmental Effects 29 (2007) 39-45.
[7] S. Saraf, Hydrocarbon Processing 86 (2007) 15.
[8] S. Thomas, R.A. Dawe, Energy 28 (2003) 1461-1477.
[9] H. Kanda, Economic study on natural gas transportation with natural gas hydrate (NGH) pellets, 23rd World Gas Conference, Amsterdam, 2006.
[10] T. Austvik, X. Li, L.H. Gjertsen, Annals of the New York Academy of Sciences 912 (2000) 294-303.
[11] J.J. Carroll, Pipeline and Gas Journal (2003).
[12] E.D. Sloan, J. Ben Bloys, Hydrate Engineering, Society of Petroleum Engineers Inc., Richardson, TX, USA, 2000.
[13] D.L. Katz, JPT, Journal of Petroleum Technology 35 (1983) 1205-1214.
[14] D.L. Katz, R.L. Lee, Natural Gas Engineering, McGraw-Hill, New York, 1990.
[15] S. Lee, J.S. Zhang, R. Mehta, T.K. Woo, J.W. Lee, Journal of Physical Chemistry C 111 (2007) 4734-4739.
[16] GPSA Engineering Data Book, Gas Processors Suppliers Association, Tulsa, OK, USA, 1998.
[17] R.E. Pellenbarg, M.D. Max, Journal of Chemical Education 78 (2001) 896.
[18] A. Demirbas,Energy Conversion and Management 51 (7), (2010) 1547-1561
[19] E.D. Sloan, Clathrate hydrates of natural gases, 2nd ed., Marcel Dekker, New York, 1998.
[20] Y. Bai, Q. Bai, Subsea Pipelines and Risers, Elsevier Science Ltd, 2005.
[21] M.T. Kirchner, R. Boese, W.E. Billups, L.R. Norman, Journal of the American Chemical Society 126 (2004) 9407-9412.
[22] W.B. Durham, L.A. Stern, S.H. Kirby, Ductile flow of methane hydrate, Natl. Res. Council Canada, Canada, 2003, pp. 373-380.
[23] D.W. Davidson, Y.P. Handa, C.I. Ratcliffe, J.S. Tse, B.M. Powell, Nature 311 (1984) 142-143.
[24] G.A. Jeffrey, Inclusion Compounds. in: J.L. Atwood, J.E.D. Davies, D.D. MacNichol, (Eds.), Academic Press, 1984.
[25] G.A. Jeffrey, T. Jordan, R.K. McMullan, Some new structures of organic polyhedral clathrate hydrates, 1966.
[26] G.A. Jeffrey, Y. Yeon, Acta Crystallographica, Section B (Structural Science) B42 (1986) 410-413.
[27] J.A. Ripmeester, C.I. Ratcliffe, Journal of Physical Chemistry 92 (1988) 337-339.
[28] J.A. Ripmeester, C.I. Ratcliffe, Journal of Physical Chemistry 94 (1990) 8773.
[29] J.A. Ripmeester, C.I. Ratcliffe, G. Enright, E. Brouwer, Thermodynamic and resonance studies of structural changes in crystals, Denmark, 1995, pp. 513-522.
[30] A.P. Mehta, A Thermodynamic Investigation of Structure-H-Type Clathrate Hydrates, PhD Dissertation, Colorado School of Mines, Golden, CO, 1996.

[31] M.M. Mooijer-van den Heuvel, C.J. Peters, J. De Swaan Arons, Fluid Phase Equilibria 172 (2000) 73-91.

[32] M.M. Mooijer-van den Heuvel, R. Witteman, C.J. Peters, Fluid Phase Equilibria 182 (2001) 97-110.

[33] K.A. Udachin, C.I. Ratcliffe, J.A. Ripmeester, Angewandte Chemie - International Edition 40 (2001) 1303-1305.

[34] K.A. Udachin, C.I. Ratcliffe, J.A. Ripmeester, Journal of Physical Chemistry B 111 (2007) 11366-11372.

[35] Y.A. Dyadin, K.A. Udachin, Journal of Structural Chemistry 28 (1987) 394-432.

[36] A.V. Kurnosov, A.G. Ogienko, S.V. Goryainov, E.G. Larionov, A.Y. Manakov, A.Y. Lihacheva, E.Y. Aladko, F.V. Zhurko, V.I. Voronin, I.F. Berger, A.I. Ancharov, Journal of Physical Chemistry B 110 (2006) 21788-21792.

[37] J.S. Loveday, R.J. Nelmes, M. Guthrie, Chemical Physics Letters 350 (2001) 459-465.

[38] J.S. Loveday, R.J. Nelmes, M. Guthrie, D.D. Klug, J.S. Tse, Physical Review Letters 87 (2001) 215501.

[39] R.K. McMullan, T.C.W. Mak, G.A. Jeffrey, Journal of Chemical Physics 44 (1966) 2338-2345.

[40] T.C.W. Mak, Journal of Chemical Physics 43 (1965) 2799-2805.

[41] D.W. Davidson, M.A. Desando, S.R. Gough, Y.P. Handa, C.I. Ratcliffe, J.A. Ripmeester, J.S. Tse, Nature 328 (1987) 418-419.

[42] D.W. Davidson, S.R. Gough, Y.P. Handa, C.I. Ratcliffe, J.A. Ripmeester, J.S. Tse, Some Structural Studies of Clathrate Hydrates, 1987, pp. 537-542.

[43] M.V. Stackelberg, H.R. Muller, Journal of Chemical Physics 19 (1951) 1319-1320.

[44] V.R. Parameswaran, M. Paradis, Y.P. Handa, Canadian Geotechnical Journal 26 (1989) 479-483.

[45] I. Cameron, Y.P. Handa, T.H.W. Baker, Canadian Geotechnical Journal 27 (1990) 255-258.

[46] L.A. Stern, S.H. Kirby, W.B. Durham, Science 273 (1996) 1843-1848.

[47] W.B. Durham, S.H. Kirby, L.A. Stern, Z. Wu, Journal of Geophysical Research 108 (2003) 2-1.

[48] W.B. Durham, L.A. Stern, S.H. Kirby, Canadian Journal of Physics 81 (2003) 373-380.

[49] E. Whalley, Journal of Geophysical Research 85 B 5 (1980) 2539-2542.

[50] H. Kiefte, M.J. Clouter, R.E. Gagnon, Journal of Physical Chemistry 89 (1985) 3103-3108.

[51] C. Pearson, J. Murphy, R. Hermes, Journal of Geophysical Research 91 (1986) 14132-14138.

[52] B.L. Whiffen, H. Kiefte, M.J. Clouter, Geophysical Research Letters 9 (1982) 645-648.

[53] H. Shimizu, T. Kumazaki, T. Kume, S. Sasaki, Physical Review B (Condensed Matter and Materials Physics) 65 (2002) 212102-212101.

[54] R.D. Stoll, G.M. Bryan, Journal of Geophysical Research 84 (1979) 1629-1634.

[55] A. Gupta, Methane Hydrate Dissociation Measurements and Modeling: The Role of Heat Transfer and Reaction Kinetics, Colorado School of Mines, PhD Dissertation, Golden, CO, 2007.

[56] A. Gupta, T.J. Kneafsey, G.J. Moridis, Y. Seol, M.B. Kowalsky, E.D. Sloan, Jr., Journal of Physical Chemistry B 110 (2006) 16384-16392.

[57] A. Gupta, J. Lachance, E.D. Sloan, Jr., C.A. Koh, Chemical Engineering Science 63 (2008) 5848-5853.

[58] D.B. Guo, D.S. Song, J. Chacko, D.A. Ghalambor, Offshore Pipelines, Gulf Professional Publishing, 2005.

[59] B. Guo, R.E. Bretz, R.L. Lee, Method and Apparatus for Generating, Transporting and Dissociating Gas Hydrates. in: U.S. Patent, (Ed.), New Mexico Tech Research Foundations, 1993.

[60] Y.F. Makogon, W.A. Dunlap, S.A. Holditch, Oceanic methane hydrate development: Reservoir character and extraction, Offshore Technol Conf, Richardson, TX, USA, 1997, pp. 8300.

[61] S.K. Kelkar, M.S. Selim, E.D. Sloan, Fluid Phase Equilibria 150 (1998) 371-382.

[62] Chemical & Engineering News (Cover Story), Volume 83, Number 24, , 2005, pp. 30-36.

[63] E.D. Sloan, F. Fleyfel, Fluid Phase Equilibria 76 (1992) 123-140.

[64] D.B. Robinson, Fluid Phase Equilibria 52 (1989) 1-14.

[65] A.B. Hansen, T.L. Clasen, R.M. Bass, JPT, Journal of Petroleum Technology 51 (1999) 61-62.

[66] R.K. Oram, Advances in deepwater pipeline insulation techniques and materials, Deepwater Pipeline Technology Congress, London, UK, 1995.

[67] J.K. Lervik, M. Ahlbeck, H. Raphael, T. Lauvdal, P. Holen, Direct electrical heating of pipelines as a method of preventing hydrate and wax plugs, ISOPE, Montreal, Can, 1998, pp. 39-45.

[68] P.K. Notz, S.B. Bumgardner, B.D. Schaneman, J.L. Todd, SPE Production and Facilities 11 (1996) 256-260.

[69] B. Edmonds, R.A.S. Moorwood, R. Szczepanski, Practical model for the effect of salinity on gas hydrate formation, Society of Petroleum Engineers (SPE), Stavanger, Norway, 1996, pp. 262-269.

[70] M.A. Kelland, T.M. Svartaas, L. Dybvik, New generation of gas hydrate inhibitors, Society of Petroleum Engineers (SPE), Dallas, TX, USA, 1995, pp. 529-537.

[71] M.A. Kelland, T.M. Svartaas, L. Dybvik, Studies on new gas hydrate inhibitors, Society of Petroleum Engineers (SPE), Aberdeen, Scotl, 1995, pp. 531-539.

[72] M.A. Kelland, T.M. Svartaas, L.A. Dybvik, Control of hydrate formation by surfactants and polymers, Society of Petroleum Engineers (SPE), New Orleans, LA, USA, 1994, pp. 431-438.

[73] M.D. Jager, C.J. Peters, E.D. Sloan, Fluid Phase Equilibria 193 (2002) 17-28.

[74] M. Wu, S. Wang, H. Liu, Journal of Natural Gas Chemistry 16 (2007) 81-85.

[75] J. Husebø, G. Ersland, A. Graue, B. Kvamme, Energy Procedia 1 (2009) 3731-3738.

[76] Y. Xu, X. Yang, J. Ding, G. Ye, Tianranqi Gongye/Natural Gas Industry 24 (2004) 135-138+118-119.

[77] J.P. Long, J.P. Lederhos, A. Sum, R.L. Christiansen, E.D. Sloan, Kinetic Inhibitors of Natural Gas Hydrates, 73rd Gas Processors Association Annual Convention, New Orleans, USA, 1997.

[78] L.M. Frostman, Anti-agglomerant hydrate inhibitors for prevention of hydrate plugs in deepwater systems, Soc Pet Eng (SPE), Dallas, TX, USA, 2000, pp. 573-579.

[79] L.M. Frostman, J.L. Przybylinski, Successful Applications of Anti-Agglomerant Hydrate Inhibitors, Society of Petroleum Engineers (SPE), Houston, TX, United states, 2001, pp. 259-268.

[80] A.P. Mehta, P.B. Hebert, E.R. Cadena, J.P. Weatherman, Fulfilling the Promise of Low Dosage Hydrate Inhibitors: Journey from Academic Curiosity to Successful Field Implementation, Offshore Technology Conference, 2002, pp. 565-571.

[81] M.A. Kelland, Energy and Fuels 20 (2006) 825-847.

[82] A.L. DeVries, Science 163 (1969) 1073.

[83] Y.F. Makogon, Journal of Natural Gas Science and Engineering 2 (2010) 49-59.

[84] C.A. Koh, Chemical Society Reviews 31 (2002) 157-167.

[85] B. Kvamme, Annals of the New York Academy of Sciences 715 (1994) 496-501.

[86] D. Kashchiev, A. Firoozabadi, Journal of Crystal Growth 250 (2003) 499-515.

[87] D.L. Katz, D. Cornell, R. Kobayashi, F.H. Poettmann, J.A. Vary, J.R. Elenbaas, C.F. Weinaug, Handbook Natural Gas Engineering, McGraw Hill, Newyork, NY, 1959.

[88] Y.F. Makogon, Hydrates of Natural Gas (Trasnlation by W.J. Cieslewicz), Pennwell, Tulsa, OK, 1981.

[89] E. Berecz, M. Balla-Achs, Gas hydrates, Elsevier, Amsterdam, Netherlands, 1983.

[90] K. Okutani, Y. Kuwabara, Y.H. Mori, Chemical Engineering Science 63 (2008) 183-194.

[91] Y.S. Kim, S.K. Ryu, S.O. Yang, C.S. Lee, Industrial & Engineering Chemistry Research 42 (2003) 2409-2414.

[92] Y.T. Seo, S.P. Kang, H. Lee, Fluid Phase Equilibria 189 (2001) 99-110.

[93] H. Sakaguchi, R. Ohmura, Y.H. Mori, Journal of Crystal Growth 247 (2003) 631-641.

[94] S.P. Kang, Y.T. Seo, H. Lee, B.J. Ryu, Journal of Chemical Thermodynamics 31 (1999) 763-772.

[95] F. Fleyfel, K.Y. Song, A. Kook, R. Martin, R. Kobayashi, Journal of Physical Chemistry 97 (1993) 6722-6725.

[96] K.C. Hester, R.M. Dunk, S.N. White, P.G. Brewer, E.T. Peltzer, E.D. Sloan, Geochimica Et Cosmochimica Acta 71 (2007) 2947-2959.

[97] C.A. Koh, R.E. Westacott, W. Zhang, K. Hirachand, J.L. Creek, A.K. Soper, Fluid Phase Equilibria 194 (2002) 143-151.

[98] P. Le Parlouer, C. Dalmazzone, B. Herzhaft, L. Rousseau, C. Mathonat, Journal of Thermal Analysis and Calorimetry 78 (2004) 165-172.

[99] D. Dalmazzone, N. Hamed, C. Dalmazzone, L. Rousseau, Journal of Thermal Analysis and Calorimetry 85 (2006) 361-368.

[100] C. Gaillard, J.P. Monfort, J.L. Peytavy, Oil & Gas Science and Technology-Revue De L Institut Francais Du Petrole 54 (1999) 365-374.

[101] J.H. Lee, Y.S. Baek, W.M. Sung, Journal of Industrial and Engineering Chemistry 8 (2002) 493-498.

[102] O. Urdahl, A. Lund, P. Mork, T.N. Nilsen, Chemical Engineering Science 50 (1995) 863-870.

[103] A. Lund, O. Urdahl, S.S. Kirkhorn, Chemical Engineering Science 51 (1996) 3449-3458.

[104] M. Mork, Formation Rate of Natural Gas Hydrate Reactor Experiments and Models, PhD Dissertation, Norwegian University of Science and Technology, 2002.

[105] G.J. Hatton, System and Method for Maintainging Multiphase Flow with Minimal Solids Degradation, United States Patents, Southwest Research Institute, 1996.

[106] J. Larue, J.C. Collin, A. Minkkinen, A. Rojey, Process and Apparatus for Transporting and Treating a Natural Gas, United States Patent, Institut Francais du Pertole, France, 1991.

[107] E. Behar, M. Cessou, C. Cohen, A. Rojey, M. Thomas, Device and Process for Studying the Behavior in Circulation of Multiphase Effluents, Applications to Effluents Forming Hydrates, United States Patent, Institut Francais du Pertole, France, 1995.
[108] G.F. Mitchell, L.D. Talley, Application of kinetic hydrate inhibitor in black-oil flowlines, Soc Pet Eng (SPE), Houston, TX, USA, 1999, pp. PI/.
[109] L.D. Talley, G.F. Mitchell, Proceedings of the Annual Offshore Technology Conference 3 (1999) 681-689.
[110] P. Raj Bishnoi, V. Natarajan, Fluid Phase Equilibria 117 (1996) 168-177.
[111] P. Englezos, Revue de l'Institute Francais du Petrole 51 (1996) 789-795.
[112] R. Kumar, P. Linga, I. Moudrakovski, J.A. Ripmeester, P. Englezos, AIChE Journal 54 (2008) 2132-2144.
[113] J.A. Ripmeester, C.I. Ratcliffe, Journal of Physical Chemistry 92 (1988) 337-339.
[114] C.A. Tulk, Y. Ba, D.D. Klug, G. McLaurin, J.A. Ripmeester, Journal of Chemical Physics 110 (1999) 6475-6483.
[115] E. Dendy Sloan, Journal of Chemical Thermodynamics 35 (2003) 41-53.
[116] R. Susilo, I.L. Moudrakovski, J.A. Ripmeester, P. Englezos, Journal of Physical Chemistry B 110 (2006) 25803-25809.
[117] R. Susilo, I.L. Moudrakovski, J.A. Ripmeester, P. Englezos, Journal of Physical Chemistry B 110 (2006) 25803-25829.
[118] R. Susilo, J.A. Ripmeester, P. Englezos, Chemical Engineering Science 62 (2007) 3930-3939.
[119] R. Susilo, J.A. Ripmeester, P. Englezos, Chemical Engineering Science 62 (2007) 3930-3939.
[120] A.K. Sum, R.C. Burruss, E.D. Sloan Jr, Journal of Physical Chemistry B 101 (1997) 7371-7377.
[121] T. Uchida, S. Takeya, Y. Kamata, I.Y. Ikeda, J. Nagao, T. Ebinuma, H. Narita, O. Zatsepina, B.A. Buffett, Journal of Physical Chemistry B 106 (2002) 12426-12431.
[122] S. Subramanian, R.A. Kini, S.F. Dec, E.D. Sloan Jr, Chemical Engineering Science 55 (2000) 1981-1999.
[123] T. Uchida, R. Okabe, K. Gohara, S. Mae, Y. Seo, H. Lee, S. Takeya, J. Nagao, T. Ebinuma, H. Narita, Canadian Journal of Physics 81 (2003) 359-366.
[124] T. Uchida, R. Ohmura, I.Y. Ikeda, J. Nagao, S. Takeya, A. Hori, Journal of Physical Chemistry B 110 (2006) 4583-4588.
[125] T. Uchida, S. Takeya, J. Nagao, T. Ebinuma, H. Narita, L.D. Wilson, C.A. Tulk, J.A. Ripmeester, Measurements of physical properties of gas hydrates and in situ observations of formation and decomposition processes via Raman spectroscopy and X-ray diffraction, Natl. Res. Council Canada, Canada, 2003, pp. 351-357.
[126] V. Lachet, E. Behar, Oil and Gas Science and Technology 55 (2000) 611-616.
[127] D. Berner, Marine transport of natural gas in hydrate form, Publ by Int Soc of Offshore and Polar Engineerns (ISOPE), San Francisco, CA, USA, 1992, pp. 636-643.
[128] M. Kelland, Marine Pollution Bulletin 29 (1994) 307-307.
[129] J.S. Gudmundsson, A. Børrehaug, Natural Gas Hydrate an Alternative to Liquified Natural Gas, 1996.
[130] P. Englezos, Industrial and Engineering Chemistry Research 32 (1993) 1251-1274.
[131] L.J. Franklin, In-Situ Hydrates - A Potential Gas Soure. in: J.L. Cox, (Ed.), Natural Gas Hydrates: Properties, Occurance and Recovery, Butterworth, Woburn, MA, 1983.

[132] S.R. Dallimore, T.S. Collett, T. Uchida, M. Weber, A. Chandra, T.H. Mroz, E.M. Caddel, T. Inoue, H. Takahashi, A.E. Taylor, The Mallik Gas Hydrate Field: Lessons learned from 30 years of gas hydrate investigation (2004)
[133] J.P. Kennett, K.G. Cannariato, I.L. Hendy, R.J. Behl, Science 288 (2000) 128-133.
[134] W.P. Dillon. in: J.P. Henriet, J. Mienert, (Eds.), Gas Hydrates: Relevance to World Margin Stability and Climate Change, Geol. Soc., London, 1998.
[135] R. King, CAPP Guidelines for the Prevention and Safe Handling of Hydrates, Canadian Assn. of Petroleum Producers, Calgary, 1994.
[136] C. Baillie, E. Wichert, Oil and Gas Journal 85 (1987) 37-39.
[137] Liang, S.; Kusalik, P. G. Chem.Phys. Lett. 2010, 494, 123.
[138] Storr, M.T.; Taylor, P.C.; Monfort, J.P.; Rodger, P.M. J. Am. Chem. Soc. 2004, 126, 1569
[139] Carver, T.J.; Drew, M.G.B.; Rodger, P.M. J. Chem. Soc. Faraday Trans. 1995, 912, 777.
[140] Kavamme, B.; Huseby, G.; Kristian, O. Mol. Phys. 1997, 90, 979.
[141] Lederhos, J.P.; Long, J.P.; Sum, A.; Christiansen, R.L.; Sloan, E.D. Chem. Eng. Sci. 1996, 51, 1221
[142] Storr, M.T.; Rodger, M. Ann. NY Acad. Sci. 2006, 912, 669.
[143] Kavamme, B.; Kuznetsova, T.; Aasoldsen, K. J.Mol.Graph and Mod. 2005, 23, 524.
[144] Myshakin, E. M.; Jiang, H.;_ Warcinski, R.P.; Joordan, K.D. J. Phys. Chem. A 2009, 113, 1913.
[145] Tanaka, H.; Nakatsuka, T.; Koga, K. J. Chem. Phys. 2004,121, 5488.
[146] Erfan-Niya, H.; Modarress, H.; Zaminpayma, E. Energ. Conv. Manag. 2011, 52, 523.
[147] Erfan-Niya, H.; Modarress, H.; Zaminpayma, J Incl Phenom Macrocycl Chem 2011, 70, 227-239.

8

Modelling and Simulation of Natural Gas Liquefaction Process

Alessandro Trigilio, Alexis Bouza and Sabrina Di Scipio
Thermodynamics and Transport Phenomena Department, Simón Bolívar University
Venezuela

1. Introduction

The global energy demand is increasing and the natural gas (NG) has obtained relevance as a clean fuel. At distances greater than 4000 km from the source of production, the most profitable way of transportation for the NG is as liquefied natural gas (LNG). Between years 2000-2010 the natural gas consumption increased 31.4% and LNG represents 30.5% of the global NG trade in 2010 (British Petroleum, 2011). With a world market in full development and a global perspective with a strong trend towards globalization and free trade, it is evident the importance of gas liquefaction processes.

LNG is obtained when the NG is cooled until its bubble point (or even below) at atmospheric pressure, which corresponds to -161°C. At this point, its density is 55% lower than the water and its volume is reduced 600 times. The volume reduction favours not only the transport (using LNG carriers) but also the storage in tanks (Institute for Energy, Law & Enterprise - University of Houston, 2003).

The number of installed plants for the NG liquefaction had a significant increase since 1960 (Haselden, 1977), as well as the operational improvements. The current installed plants are capable of processing more than 4 MTPA of NG.

In general, refrigeration cycles used to liquefy natural gas can be split into two main processes: 1) those that use pure refrigerants and 2) those that use mixtures of them. Each technology has its own properties and they have been consolidated as international patents, which seek to optimize energy consumption, capital investment and the physical space occupied by the required equipment.

Previous studies related with LNG processes using mixed refrigerant (MR) have been published. Del Nogal et al. (2005) optimized the mixed refrigerant composition focused on minimizing the power in the compression stages, but the complete model was not shown. Jensen & Skogestad (2006) proposed a control strategy for the process, based on a certain control and instrumentation philosophy, without reporting the process model. Jensen and Skogestad as well as Lee et al. (2002) optimized the refrigerant composition with the aim of minimizing the temperature gap between the natural gas and the mixed refrigerant in the cycle evaporator. Aspelund et al. (2010) optimized the refrigerant composition in three different ways: 1) using a fixed minimum internal temperature approach, 2) using a fixed exchanger area and 3) minimizing the cost for exchanger area, but again not model details were shown.

On the other hand, Haselden (1977) shown optimum operating conditions for cycles in cascade with three pure refrigerants, based on the operational experience of installed LNG plants. It is evident that there are very few attempts to model the processes for liquefying natural gas. Moreover, none of the above works show in detail a complete set of operating conditions for these systems.

Consequently, the purpose of this work is to model and simulate the natural gas liquefaction process in stationary state, using cycles with MR and cascade cycles with pure refrigerants. The specific objectives are:

- Calculate energetic properties and phase equilibria using the Peng-Robinson equation of state.
- Simulate the NG liquefaction process using mixed or pure refrigerants (considering traditional and optimized cascades), by developing Matlab® routines. The technical feasibility of operating conditions will be evaluated
- Based on the developed models, a sensitivity analysis of the variables involved will be made in order to determine its impact on operating conditions.

2. Refrigeration cycles

A refrigeration cycle uses changes in pressure and temperature on a mixed or pure compound (named refrigerant) in order to transfer heat from a cold zone (or fluid) to a hot zone (or fluid). In Fig. 1 is shown a simple throttle valve cycle, consisting in two heat exchangers, a throttle valve and a compressor.

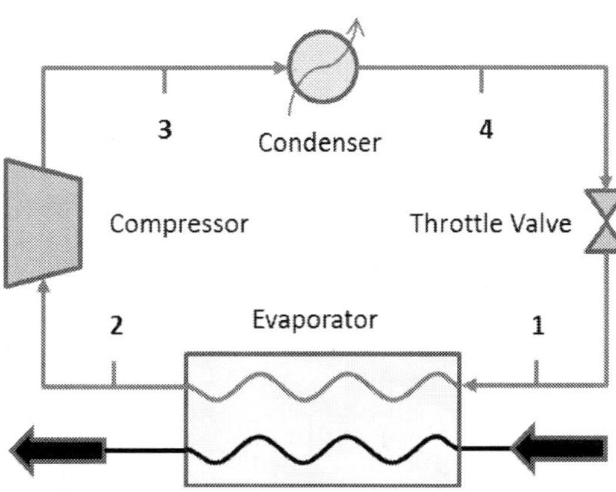

Fig. 1. Simple refrigeration cycle using a throttle valve.

The refrigerant is in a vapour-liquid equilibrium (stream 1) when enters the evaporator and exchanges heat with the hot source, until reaching saturation or even becoming overheated vapour (stream 2). Then, enters the compressor and suffers a pressure and temperature rise. The refrigerant in stream 3 exchanges heat with a cold source in the condenser, becoming liquid and entering to the throttle valve (stream 4), where it expands and its temperature falls to a minimum, starting the cycle again.

If the throttle valve is considered adiabatic, the First Law of Thermodynamics reveals that the expansion is isenthalpic. Moreover, if the compressor is adiabatic and the compression is reversible, then it is isentropic (Çengel & Boles, 2006). The existence of internal friction and the heat leaks to the surroundings result in a deviation from the isentropic behaviour, therefore is required the use of a polytropic efficiency (N_p).

2.1 Polytropic work in the compressor

The polytropic work (W_{poli}) in the compressor is used in order to determine the polytropic efficiency *(Np)* of the gas compression process, as shown in Eqs. (1) - (4) (Walas, 1990).

$$W_{poly} = \frac{k}{k-1} \cdot z \cdot R \cdot T \cdot \left[\left(\frac{P_2}{P_1} \right)^{\frac{n-1}{n}} - 1 \right] \tag{1}$$

$$n = \frac{k \cdot N_p}{1 - k \cdot (1 - N_p)} \tag{2}$$

$$k = \frac{C_p}{C_v} \tag{3}$$

$$C_v = C_p + R \tag{4}$$

where *Cp* is the heat capacity at constant pressure, *R* is the universal gas constant, *z* is the gas compressibility factor, P_1/P_2 is the pressure ratio and *T* is the inlet temperature.

2.2 Coefficient of performance

The Coefficient of Performance (*COP)* for a refrigeration cycle relates the amount of heat taken from the hot source (Q_{evap}) and the total work used (W_{comp}), as shown in Eq. (5) (Çengel & Boles, 2006).

$$COP = \frac{Q_{evap}}{W_{comp}} \tag{5}$$

A First Thermodynamic Law balance in the refrigerant relates the heats (including the condenser heat, Q_{cond}) and work in any refrigeration cycle, as seen in Eq. (6).

$$Q_{evap} = Q_{cond} - W_{comp} \tag{6}$$

Substitution of Eq. (6) in Eq. (5) gives Eq. (7), which is an alternative form for the COP.

$$COP = \frac{Q_{cond}}{W_{comp}} - 1 \tag{7}$$

In order to improve the performance of the cycle, it's important to maximize the COP. In that way it is removed as much heat as possible from the hot source, using the minimum work. It is important to note that, according to Eq. (7), the greater the COP greater the heat retired in the condenser for cooling the refrigerant, then more amount of utility stream is needed.

2.3 Refrigeration cycles in cascade

When there is a great gap between the temperature of the hot and cold zones (or fluids) and it is not possible to use only one refrigerant, the cascade cycles are used (Smith, 2005). This configuration uses more than one individual cycles with a common heat exchanger between cycles. The condenser of a cold cycle is the evaporator of the following hotter cycle, as shown in Fig. 2.

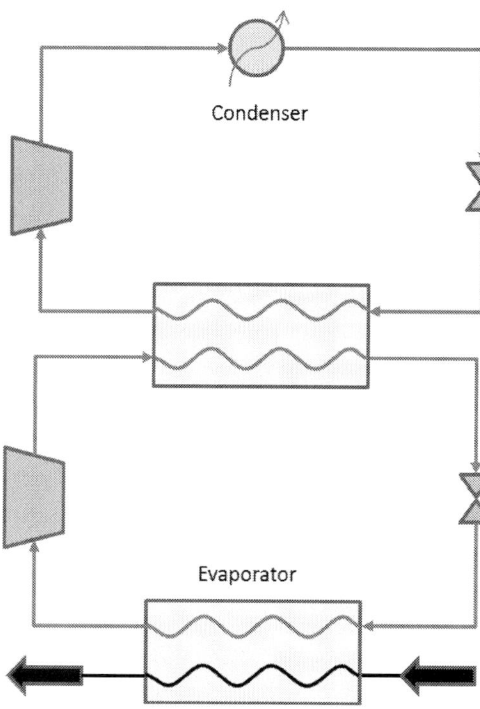

Fig. 2. Refrigeration cycle using a double cascade.

The advantages of using cascade cycles are: better distribution on the refrigerant load, decrease of the required work, and increase of the heat transferred in the non-common evaporator. Hence, the COP increases when this configuration is used.

These cycles are particularly useful in the liquefaction of natural gas because it is possible to cool it and achieve the desired temperatures using water as the external fluid (utility stream) in the condenser. Therefore, it is important to choose the right refrigerant for each cascade.

2.4 Refrigerant selection

In order to guarantee a minimum area in the heat exchangers, the refrigerant used usually is a pure compound, so the phase changes described previously occurs at a constant temperature and the drive force is maintained constant. In Fig. 3 is plotted the saturation curve for the liquid-vapor equilibria for some refrigerant commonly used in the liquefaction of natural gas (obtained from Lemon et al., 2005).

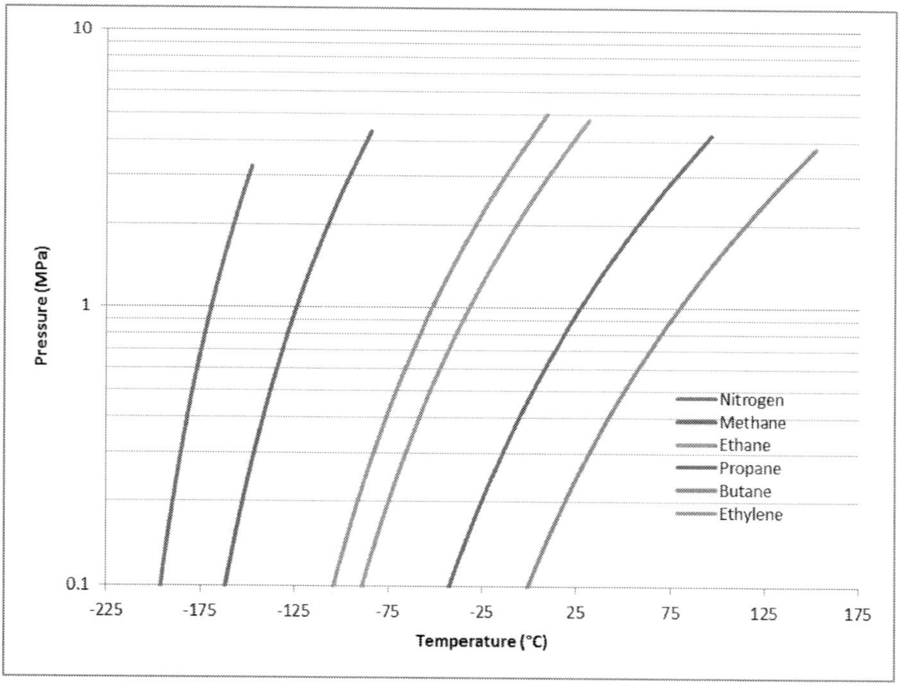

Fig. 3. Vapor pressure curves for refrigerants commonly used to produce LNG.

It should be noted that if the natural gas is between 50°C and -160 °C, these refrigerants can be used depending on the specifications of the process. For example, below 1 MPa and in the range of -125 °C to -160 °C, only methane can be used.

The natural gas is a mixture, mainly composed of methane and small quantities of ethane, propane and others, so the NG phase change occurs at non constant temperature. Despite the driving force, in the liquefaction cycles it is preferred to have a refrigerant cooling curve similar to the NG cooling curve (Smith, 2005). This reduces the energy consumption due to exergy considerations (Haselden, 1977). This can be achieved in two ways. The first is using more than one consecutive refrigeration cycle operating at different pressure levels, obtaining a discrete refrigerant temperature profile (as illustrated in Fig. 4). If more cycles are used, the whole process is more efficient, but implies larger amount of equipment required, and therefore a more complex system. The alternative way is to use a mixed refrigerant with a cooling curve as near as possible to the natural gas cooling curve, avoiding temperature cross between the two fluids.

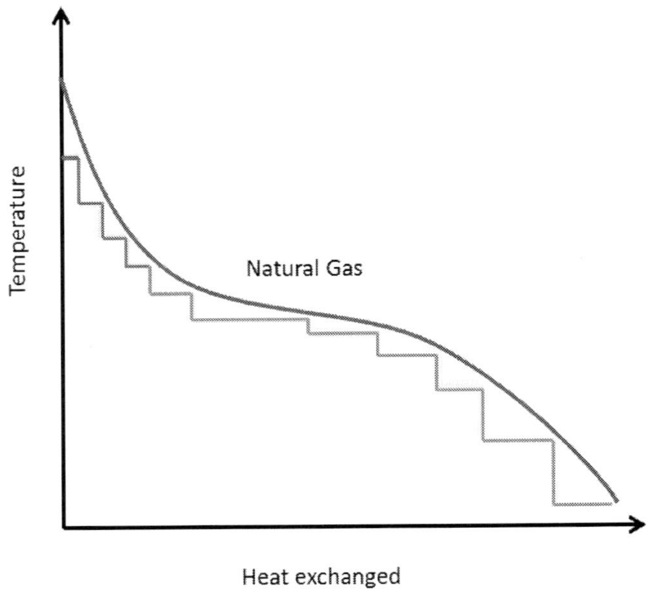

Fig. 4. Discrete refrigerant temperature profile using more than one cycle.

2.5 Natural gas liquefaction technologies

The most common technologies used for the natural gas liquefaction can be divided in two main groups according to the refrigerant used: those with pure refrigerant and those with mixed refrigerant. In this work two of the most used configurations have been selected (one of each group) for the modelling and simulation task.

2.5.1 Pure refrigerant cascade

Unlike the cycles explained in section 2.2, the cascade cycles used for the natural gas liquefaction usually have more than two cascade, and more than one common heat exchanger between them (as illustrated in Fig. 5). The main idea is to use the internal cycles to cool not only the natural gas, but also the refrigerants in the external cascade. Using cascades makes possible to work in a wide temperature range, so it can be employed cooling water as an external fluid in the evaporator. Also, the cascades improve the COP because of the temperature profile approach between the different refrigerants used and the natural gas, as seen in Fig. 4.

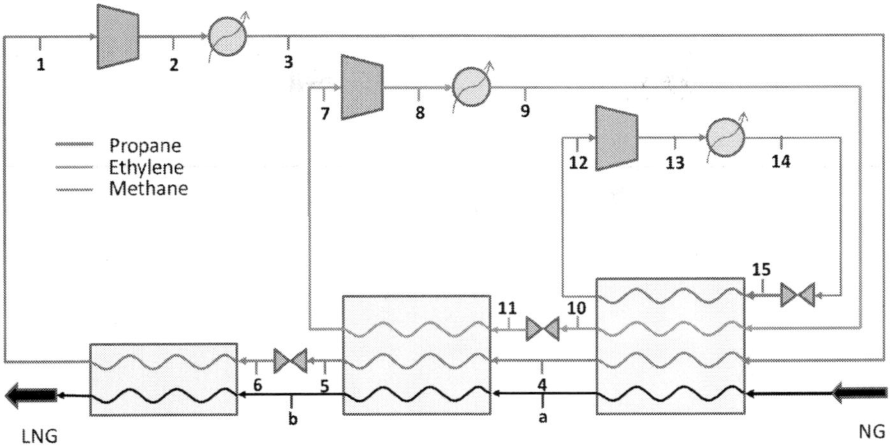

Fig. 5. Triple cascade liquefaction cycle with pure refrigerant.

2.5.2 Simple cycle with mixed refrigerant

As explained in section 2.4, the aim of using mixed refrigerants (MR) is for minimizing the approach between the cooling curves in the evaporator. One of the advantages of this cycles is the less amount of equipment required and, depending on the plant load, it is possible to use a single compressor if the power required is no more than 70 MW (Price & Mortko, 1996).

The cycle configuration is shown in Fig. 6. It is important to note the use of a double pass on the evaporator. According to Price & Mortko (1996), its purpose is to have best control over the

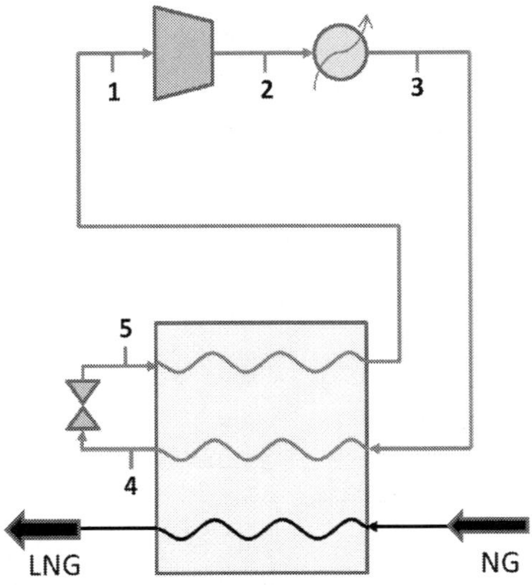

Fig. 6. Simple liquefaction cycle with mixed refrigerant.

temperature profile due to the possibility of using intermediate separators for the formed vapor, so the mixture composition can be regulated without interrupting the plant operation.

3. Modelling the liquefaction cycles

In order to model the stages described in Section 2, is necessary to use a thermodynamic model for the liquid-vapour equilibria and the energy properties, required for the work and heat calculations.

The Peng & Robinson (1976) equation of state was used. The vapour (y_i) and liquid (x_i) composition of each component (i) in phase equilibria can be related by Eq. (8). The thermodynamic model is used for the fugacity calculation in each phase ($\hat{\phi}_i^l, \hat{\phi}_i^v$), expressed in Eqs. (9) - (19). When performing iterative calculations, each side of the Eq. (8) should be satisfied.

$$k_i = \frac{y_i}{x_i} = \frac{\hat{\phi}_i^v}{\hat{\phi}_i^l} \tag{8}$$

$$\ln\hat{\phi}_i = \frac{b}{b_i}\cdot(Z-1) - \ln(Z-B) - \frac{A}{2\cdot\sqrt{2}}\cdot\left(\frac{2\cdot\sum_j x_j\cdot a_{ij}}{a} - \frac{b_i}{b}\right)\cdot\ln\left[\frac{Z+\left(1+\sqrt{2}\right)\cdot B}{Z-\left(1-\sqrt{2}\right)\cdot B}\right] \tag{9}$$

$$Z^3 - (1-B)\cdot Z^2 + \left(A - 3\cdot B^2 - 2\cdot B\right)\cdot Z - \left(A\cdot B - B^2 - B^3\right) = 0 \tag{10}$$

$$A = \frac{a\cdot P}{R^2\cdot T^2} \tag{11}$$

$$B = \frac{b\cdot P}{R\cdot T} \tag{12}$$

$$a = \sum_i\sum_j x_i\cdot x_j\cdot a_{ij} \tag{13}$$

$$a_{ij} = \left(1-\delta_{ij}\right)\cdot a_i^{1/2}\cdot a_j^{1/2} \tag{14}$$

$$a_i = 0.45724\cdot\alpha_i\cdot\frac{R^2\cdot Tc_i^2}{Pc_i} \tag{15}$$

$$\alpha_i^{1/2} = 1 + m_i\cdot\left[1-\left(\frac{T}{Tc_i}\right)^{1/2}\right] \tag{16}$$

$$m_i = 0.37464 + 1.54226\cdot\omega_i - 0.26992\cdot\omega_i^2 \tag{17}$$

$$b = \sum_i x_i \cdot b_i \tag{18}$$

$$b_i = 0.07780 \cdot \frac{R \cdot Tc_i}{Pc_i} \tag{19}$$

where Tc_i and Pc_i are the critical temperature and pressure of the component i, and ω_i is its accentric factor (Green & Perry, 2008).

If there are only hydrocarbons in the mixture, the binary interaction parameter (δ_{ij}) is zero. On the other hand, if nitrogen is present Nikos (1986) deduced temperature and pressure dependent correlations for this parameter (Eqs. (20) - (24)).

$$\delta_{ij}' = \eta_2 \cdot \left(\frac{T}{Tc_j} \right)^2 + \eta_1 \cdot \frac{T}{Tc_j} + \eta_0 \tag{20}$$

$$\eta_0 = 0.1751787 - 0.7043 \cdot \log \omega_j - 0.862066 \cdot \left(\log \omega_j \right)^2 \tag{21}$$

$$\eta_1 = -0.54474 + 1.328 \cdot \log \omega_j + 2.035767 \cdot \left(\log \omega_j \right)^2 \tag{22}$$

$$\eta_2 = 2.257079 + 7.869765 \cdot \log \omega_j + 13.50466 \cdot \left(\log \omega_j \right)^2 + 8.3864 \cdot \left(\log \omega_j \right)^3 \tag{23}$$

$$\delta_{ij} = \delta_{ij} \cdot \left(1.04 - 4.2 \cdot 10^{-5} \cdot P \right) \tag{24}$$

In order to determine the heats and works is necessary to calculate the enthalpy for each stream of the cycle. This can be done using Eq (25) taking as a reference state 0 J/mol when the temperature is 0 °C as an ideal gas. The residual enthalpy (h^{res}) for the Peng-Robinson equation of state is calculated using Eqs. (26) and (27).

$$h = h^{ref} + \sum_i x_i \cdot \int_{273.15\,K}^{T} C_p^{ig} \cdot dT - h^{res} \tag{25}$$

$$\frac{h^{res}}{R \cdot T} = (1 - Z) + \frac{a + D}{2 \cdot \sqrt{2} \cdot R \cdot T \cdot b} \cdot \ln \left[\frac{Z + \left(1 + \sqrt{2}\right) \cdot B}{Z + \left(1 - \sqrt{2}\right) \cdot B} \right] \tag{26}$$

$$D = \sum_i \sum_j y_i \cdot y_j \cdot \kappa_j \cdot \left(1 - \delta_{ij} \right) \cdot \sqrt{a_i} \cdot \sqrt{\frac{a_j}{\alpha_j} \cdot \frac{T}{Tc_j}} \tag{27}$$

The heat capacity at constant pressure (Cp^{ig}) is calculated using Eq. (28), and the coefficients where taken from Green & Perry (2008).

$$C_p^{ig} = C_1 + C_2 \cdot \left[\frac{C_3/T}{\sinh\left(C_3/T\right)} \right]^2 + C_4 \cdot \left[\frac{C_5/T}{\cosh\left(C_5/T\right)} \right]^2 \tag{28}$$

The liquefaction cycles where simulated using Matlab®. The conditions used for all the cycles were:

- The amount natural gas processed was 4 MTPA, equivalent to 26220 kmol/h.
- The throttle valve and the compressors were considered adiabatic.
- The pressure drops in the heat exchanger was set to 20 kPa on the cold side and 50 kPa on the hot side.
- The utility stream in the condensers was cooling water, so its temperature is 12 °C.
- The outlet temperature for the LNG from the evaporators is set to -153 °C (below its bubble point).
- It should be verified that the temperature at the compressor entrance is above the dew point.
- At the exit of common evaporators, all the streams have the same temperature (valid for streams a and b in Fig. 5 to 8).
- The LNG goes through a throttle valve after the evaporators, and its final outlet pressure is 180 kPa, as a liquid-vapor mixture with a vaporized fraction less than 1%.
- The initial conditions for the natural gas and its composition are shown in Table 1.

Methane (% mol)	0.924
Ethane (% mol)	0.056
Propane (% mol)	0.019
n-Butane (% mol)	0.001
GN Inlet Temperature (°C)	15
GN Inlet Pressure (kPa)	6000

Table 1. NG composition and inlet conditions.

3.1 Modelling of the cascade cycles

In order to verify the advantages of using cascade cycles, three cases were simulated. In general, the refrigerant must be selected to satisfy that the enthalpy at the condenser outlet (at the highest pressure and with the chosen utility) is such that the valve outlet temperature is cooler than the temperature desired for the natural gas.

In Fig. 7 is illustrated a simple cycle using methane in traditional cascade. According to the Figure 3, the methane cannot be used with an utility stream at 12 °C. In order to use this refrigerant two traditional cascades were added.

The second configuration is illustrated in Fig. 8. It is an intermediate between the one shown in Fig. 5 and Fig. 7. The principle is the same explained in Section 2.5.1, the internal refrigerants has to cool the natural gas and also the refrigerants in the external cycle. The third case is the complete cascade as in the Fig. 5.

In Table 2 initial operating conditions for the cascade cycles are shown. The number of variables specified where obtained after a degree of freedom analysis. The temperatures were set to guarantee the cooling of the NG according to the pressure levels.

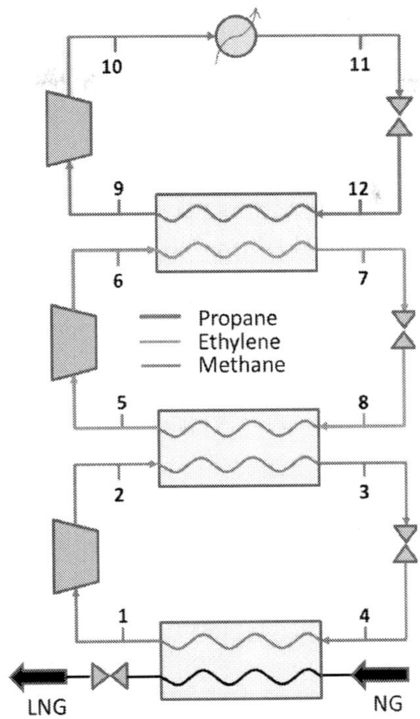

Fig. 7. Triple cascade in traditional configuration.

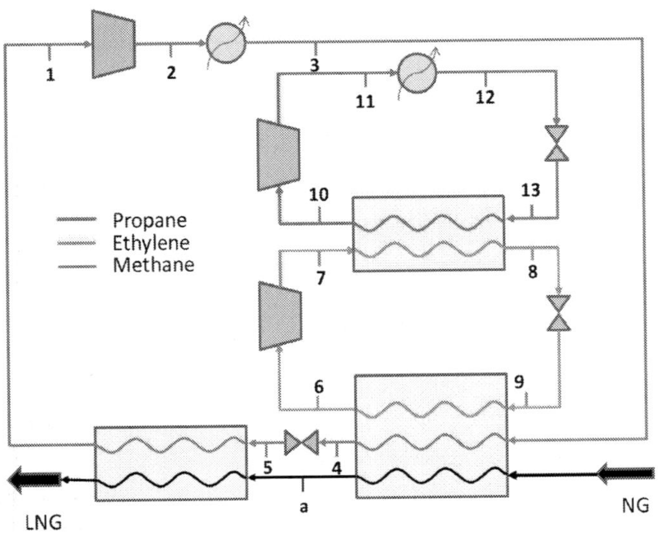

Fig. 8. Triple cascade in hybrid configuration.

Stream	Traditional Cascade	Hybrid Cascade	Triple Cascade
1	T=-155 °C	T=-155 °C	T=-155 °C
2	T=61 °C; P=4000 kPa	T=61 °C; P=4000 kPa	T=61 °C; P=4000 kPa
3	T=-94 °C	T=12 °C	T=12 °C
4	P=140 kPa	T=-94 °C	T=-30 °C
5	T=-96 °C	P=140 kPa	T=-94 °C
6	T=79 °C, P=1950 kPa	T=-96 °C	P=140 kPa
7	Vaporized fraction=0.2	T=79 °C, P=1950 kPa	T=-96 °C
8	P=140 kPa	Vaporized fraction=0.2	T=79 °C, P=2000 kPa
9	T=-32 °C	P=140 kPa	T=12 °C
10	T=38 °C, P=750 kPa	T=-32 °C	Vaporized fraction=0.2
11	T=12 °C	T=38 °C, P=750 kPa	P=140 kPa
12	P=120 kPa	T=12 °C	T=-32 °C
13	-	Vaporized fraction=0.2	T=38 °C, P=750 kPa
14	-	-	T=12 °C
15	-	-	Vaporized fraction=0.2

Table 2. Initial operating conditions for the cascade cycles.

The refrigerant load is unknown, and its value is calculated using an energy balance in the evaporators. For example, in the triple cascade (Fig. 5) the three energy balances are presented in Eqs. (29) - (31), as a function of enthalpies and flow rates (F). As can be seen in these equations, the inner cycle energy balance depends on the external cycle flow rate, so the cascades should be solved starting from the external cycle.

$$F_{methane} = \frac{F_{NG} \cdot (h_{LNG} - h_b)}{h_6 - h_1} \tag{29}$$

$$F_{ethylene} = \frac{F_{methane}(h_5 - h_4) + F_{NG}(h_b - h_a)}{h_{11} - h_7} \tag{30}$$

$$F_{propane} = \frac{F_{ethylene} \cdot (h_{10} - h_9) + F_{methane} \cdot (h_4 - h_3) + F_{NG} \cdot (h_a - h_{NG})}{h_{15} - h_{12}} \tag{31}$$

3.2 Modelling of simple cycles

After a freedom degree analysis, the conditions for the simple cycle were set as shown in Table 3. The pressure level is important because the temperature at the valve outlet (Fig. 6) should be less than the temperature at the inlet, and this value should be cooler than the LNG final temperature.

The composition of the mixed refrigerant was also a variable. A base case composition was used as well as others optimized compositions found in the literature, in order to compare the results, even if the optimizations were carried out at different conditions than the used in the present work. These compositions are presented in Table 1Table 4.

Stream	Operating Condition
1	T=-6 °C
2	P=3000 kPa, T=-1 °C
3	T=12 °C
4	T=-155 °C
5	P=340 kPa

Table 3. Simple cycle with mixed refrigerant conditions.

Component	Base Case	Lee et al. (2002)	Del Nogal et al. (2005)	Jensen & Skogestad (2006)
Methane	0.36	0.273	0.2712	0.2365
Ethane	0.28	0.356	0.3721	0.3949
Propane	0.11	0.052	0.0027	0.0000
Butane	0.15	0.209	0.2531	0.2914
Nitrogen	0.10	0.110	0.1008	0.0772

Table 4. Compositions used for the mixed refrigerant in the simple cycle.

4. Results and discussion

In the following sections the results obtained for the simulations will be discussed, as well as the results for the sensitivity analysis carried out.

4.1 Results for the cascade cycles

In Fig. 9 and Fig. 10 are presented graphically the results for the flow rate, heat exchanged, COP and work required for each cycle in each configuration used. Configuration 1

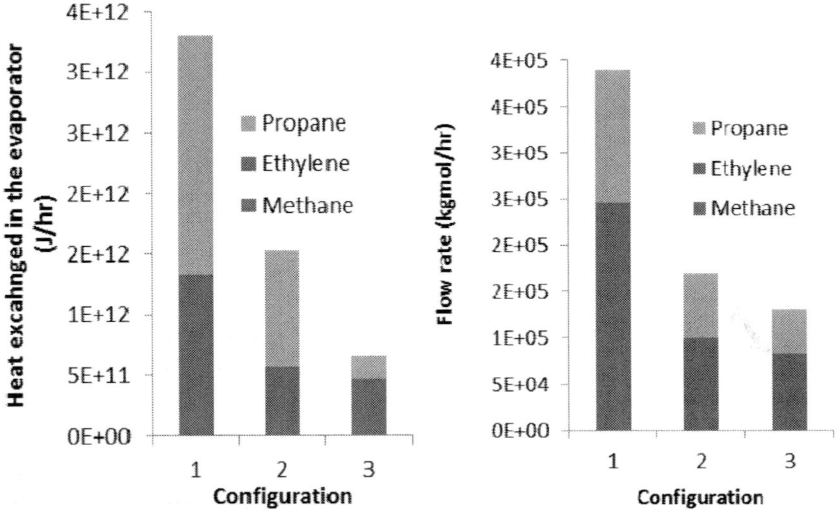

Fig. 9. Flow rate and heat exchanged results for the cascade cycles.

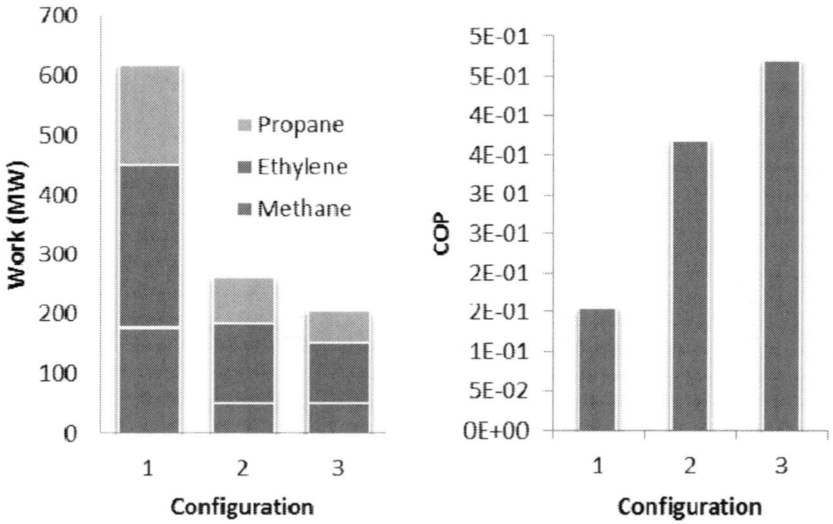

Fig. 10. Work required and COP of the cascades.

corresponds to the traditional cascade, configuration 2 to the hybrid cascade and configuration 3 to the triple cascade used for natural gas liquefaction.

It can be seen that, as commented previously, the use of the triple cascade is more convenient for the liquefaction of natural gas. It requires fewer amounts of refrigerants due to the less heat required to exchange in the evaporators. Moreover, the work required is less and as a consequence of these two effects, the COP of the cycle is improved greatly.

On the other hand, is important to note that other advantage of using the cascade cycles is the use of the propane for cooling the other refrigerats. Its heat capacity is greater, so it can exchange more energy using less flow rate.

In Table 5. Results for the traditional cascade are reported, along with the sensitivity analysis performed to the methane cycle. As one can expect, the compressor outlet pressure variation does not affects the valve outlet temperature because its outlet pressure corresponds to a saturation state, but it does affects the vaporized fraction after the valve. In the same way, the pressure variation at the valve outlet does not have consequences in the heat exchanged in the condenser.

The change in the flow rate of the lowest cycle modifies the energy balance in the common evaporator, so the flow rate of the other cycles changes in consequence. The work done in the compressor also varies due that it is presented as an extensive property.

Comparing the two parameters, the one with more impact in the operation was the outlet pressure of the lowest cycle. A variation of 12.5% in the pressure affects in more than 2% the COP value.

		Outlet Compressor Pressure at the lowest cycle		Outlet Valve Pressure at the lowest cycle	
	Base Case	**+12.5%**	**-12.5%**	**+14.3%**	**-14.3%**
Lowest Cycle					
Flow rate ratio (-)	3.51	-1.86%	+2.52%	+0.37%	-0.37%
W (MW)	176.20	-2.89%	+3.60%	+0.58%	-0.57%
Q Evaporator (J/mol)	3.76E+06	+1.90%	-2.46%	-0.37%	+0.37%
Q Condenser (J/mol)	1.07E+07	-0.01%	-0.19%		
T out valve (°C)	-157.49			+1.78 °C	-1.98 °C
Valve outlet vaporized fraction	0.55	-1.61%	+2.09%	-1.62%	+1.79%
Compressor Efficiency	0.84	+3.84%	-4.25%	-4.70%	+5.53%
COP (-)	0.55	+2.98%	-3.48%	-0.57%	+0.57%
Intermediate Cycle					
Flow rate ratio (-)	5.85	-1.87%	+2.33%	+0.37%	-0.37%
W (MW)	274.53	-1.87%	+2.33%	+0.37%	-0.37%
Q Evaporator (J/mol)	6.40E+06				
Q Condenser (J/mol)	1.28E+07				
T out valve (°C)	-98.29				
Valve outlet vaporized fraction	0.52				
Compressor Efficiency	0.90				
COP (-)	0.99				
Highest Cycle					
Flow rate ratio (-)	5.49	-1.87%	+2.33%	+0.37%	-0.37%
W (MW)	166.30	-1.87%	+2.33%	+0.37%	-0.37%
Q Evaporator (J/mol)	1.37E+07				
Q Condenser (J/mol)	1.78E+07				
T out valve (°C)	-34.55				
Valve outlet vaporized fraction	0.27				
Compressor Efficiency	0.93				
COP (-)	3.29				
Total Flow rate ratio (-)	14.85	-1.87%	+2.38%	+0.37%	-0.37%
Total COP (-)	0.1558	+2.21%	-2.62%	-0.43%	+0.43%

Table 5. Results for the traditional cascade.

The results for the hybrid cascade (as illustrated in Fig. 8) are presented in Table 6 and Table 7. Similar to the previous configurations, variation on the external cycles only affected the other cycles due to the change in the flow rates and thus in the evaporators energy balance.

The changes in the internal-lowest cycle only affected the internal-highest cycle. In particular it can be seen that variations in the compressor outlet pressure in the internal-lowest cycle has more impact than the variations due to the changes in the external cycle outlet compressor pressure. A modification of 20.5% in the highest pressure of the external cycle affects in around 4% the total COP.

External Cycle	Base Case	Outlet Compressor Pressure at the external cycle +12.5%	Outlet Compressor Pressure at the external cycle -12.5%	Outlet Valve Pressure at the external cycle +14.3%	Outlet Valve Pressure at the external cycle -14.3%
Flow rate ratio (-)	0.98	-1.91%	+2.61%	+0.37%	-0.37%
W (MW)	49.43	-2.94%	+3.69%	+0.58%	-0.57%
Q Evaporator 1 (J/mol)	3.75E+06	+1.95%	-2.54%	-0.37%	+0.37%
Q Evaporator 2 (J/mol)	-8.69E+06	-0.31%	+0.06%		
Q Condenser (J/mol)	-1.96E+06	+1.43%	-1.40%		
T out valve (°C)	-157.49			+1.78 °C	-1.98 °C
Valve oulet vaporized fraction	0.55	-1.65%	+2.15%	-1.62%	+1.79%
Compressor Efficiency	0.84	+3.84%	-4.25%	-4.70%	+5.53%
COP (-)	0.54	+3.03%	-3.56%	-0.57%	+0.57%
Internal Lowest Cycle					
Flow rate ratio (-)	2.80	-1.05%	+1.26%	+0.18%	-0.17%
W (MW)	131.73	-1.05%	+1.26%	+0.18%	-0.17%
Q Evaporator (J/mol)	6.44E+06				
Q Condenser (J/mol)	-1.29E+07				
T out valve (°C)	-98.29				
Valve outlet vaporized fraction	0.52				
Compressor Efficiency	0.90				
COP (-)	1.00				
Internal Highest Cycle					
Flow rate ratio (-)	2.64	-1.05%	+1.26%	+0.18%	-0.17%
W (MW)	79.79	-1.05%	+1.26%	+0.18%	-0.17%
Q Evaporator (J/mol)	1.37E+07				
Q Condenser (J/mol)	-1.78E+07				
T out valve (°C)	-34.55				
Valve outlet vaporized fraction	0.27				
Compressor Efficiency	0.93				
COP (-)	3.30				
Total Work (MW)	260.95	-1.41%	+1.72%	+0.25%	-0.25%
COP total (-)	0.3684	+1.43%	-1.69%	-0.25%	+0.25%

Table 6. Results for the hybrid cascade.

	Outlet Compressor Pressure at the internal-lowest cycle		Outlet Valve Pressure at the internal-lowest cycle	
Internal-Lowes Cycle	**+20.5%**	**-20.5%**	**+14.3%**	**-14.3%**
Flow rate ratio (-)	+8.90%	-8.26%	+0.31%	-0.31%
W (MW)	+6.27%	-6.09%	+0.62%	-0.61%
Q Evaporator (J/mol)	-8.17%	+9.01%	-0.31%	+0.31%
Q Condenser (J/mol)	-5.29%	+5.69%		
T out valve (°C)			-0.03 °C	+0.03 °C
Valve outlet vaporized fraction	+7.69%	-8.48%	-1.65%	+1.82%
Compressor Efficiency	+7.90%	-9.20%	-5.92%	+6.98%
COP (-)	-5.90%	+6.48%	-0.62%	+0.62%
Internal Highest Cycle				
Flow rate ratio (-)	+3.14%	-3.05%	+0.31%	-0.31%
W (MW)	+3.14%	-3.05%	+0.31%	-0.31%
COP (-)				
Total Work (MW)	+4.13%	-4.00%	+0.41%	-0.40%
COP total (-)	-3.96%	+4.17%	-0.41%	+0.41%

Table 7. Results for the hybrid cascade (continued).

The results for the triple cascade (as illustrated in Fig. 5) are shown in Table 8, Table 9, and Table 10. As in the previous cases, the hierarchy of the changes is from the external to the internal cycles, and the variations in the internal cycles did not affect the externals.

The sensitivity of the variations in the external cycle high pressure is slightly higher than in the intermediate cycle. A 12.5 % variation of the outlet compressor pressure in the external cycle changes more than 1.75 % the total COP of the system.

In Table 11 is reported the most relevant results for the sensitivity analysis. It is shown the results on the COP and on the total flow rate. The first gives an idea about the cycle performance, and the later about the equipment size required.

4.2 Results for the simple cycle using mixed refrigerant

In Table 12 are reported the sensitivity study for the simple cycle using mixed refrigerant. It can be seen that the high pressure of the cycle affects more the COP than the low pressure. In particular, a decrease of 10% in the high pressure produces a diminution of the COP by 8.48%.

On the other hand, on Table 13 are presented the results with other optimized compositions found in the literature. It is important to note that even if these compositions where optimized using particular conditions, using them results in an increase in the COP.

<table>
<tr><th></th><th></th><th></th><th colspan="2">Outlet Compressor Pressure at the external cycle</th><th colspan="2">Outlet Valve Pressure at the external cycle</th></tr>
<tr><th></th><th></th><th>Base Case</th><th>+12.5%</th><th>-12.5%</th><th>+14.3%</th><th>-14.3%</th></tr>
<tr><td rowspan="10">External Cycle</td><td>Flow rate ratio (-)</td><td>0.99</td><td>-1.96%</td><td>+2.69%</td><td>+0.37%</td><td>-0.37%</td></tr>
<tr><td>W (MW)</td><td>49.58</td><td>-2.99%</td><td>+3.78%</td><td>+0.58%</td><td>-0.57%</td></tr>
<tr><td>Q Evaporator 1 (J/mol)</td><td>3.75E+06</td><td>+2.00%</td><td>-2.62%</td><td>-0.37%</td><td>+0.37%</td></tr>
<tr><td>Q Evaporator 2 (J/mol)</td><td>-6.97E+06</td><td>-1.04%</td><td>+0.66%</td><td></td><td></td></tr>
<tr><td>Q Evaporator 3 (J/mol)</td><td>-1.72E+06</td><td>+2.72%</td><td>-2.55%</td><td></td><td></td></tr>
<tr><td>Q Condenser (J/mol)</td><td>-1.96E+06</td><td>+1.43%</td><td>-1.40%</td><td></td><td></td></tr>
<tr><td>T out valve (°C)</td><td>-157.49</td><td></td><td></td><td>+1.78 °C</td><td>-1.98 °C</td></tr>
<tr><td>Valve outlet vaporized fraction</td><td>0.55</td><td>-1.69%</td><td>+2.21%</td><td>-1.61%</td><td>+1.78%</td></tr>
<tr><td>Compressor Efficiency</td><td>0.84</td><td>+3.84%</td><td>-4.25%</td><td>-4.70%</td><td>+5.53%</td></tr>
<tr><td>COP (-)</td><td>0.54</td><td>+3.08%</td><td>-3.64%</td><td>-0.57%</td><td>+0.57%</td></tr>
<tr><td rowspan="9">Intermediate Cycle</td><td>Flow rate ratio (-)</td><td>2.16</td><td>-1.45%</td><td>+1.64%</td><td>+0.18%</td><td>-0.18%</td></tr>
<tr><td>W (MW)</td><td>101.13</td><td>-1.45%</td><td>+1.64%</td><td>+0.18%</td><td>-0.18%</td></tr>
<tr><td>Q Evaporator 2 (J/mol)</td><td>6.53E+06</td><td></td><td></td><td></td><td></td></tr>
<tr><td>Q Evaporator 3 (J/mol)</td><td>-9.52E+06</td><td></td><td></td><td></td><td></td></tr>
<tr><td>Q Condenser (J/mol)</td><td>-3.34E+06</td><td></td><td></td><td></td><td></td></tr>
<tr><td>T out valve (°C)</td><td>-98.29</td><td></td><td></td><td></td><td></td></tr>
<tr><td>Valve outlet vaporized fraction</td><td>0.52</td><td></td><td></td><td></td><td></td></tr>
<tr><td>Compressor Efficiency</td><td>0.91</td><td></td><td></td><td></td><td></td></tr>
<tr><td>COP (-)</td><td>1.02</td><td></td><td></td><td></td><td></td></tr>
<tr><td rowspan="9">Internal Cycle</td><td>Flow rate ratio (-)</td><td>1.80</td><td>-1.17%</td><td>+1.38%</td><td>+0.18%</td><td>-0.18%</td></tr>
<tr><td>W (MW)</td><td>54.37</td><td>-1.17%</td><td>+1.38%</td><td>+0.18%</td><td>-0.18%</td></tr>
<tr><td>Q Evaporator (J/mol)</td><td>1.37E+07</td><td></td><td></td><td></td><td></td></tr>
<tr><td>Q Condenser (J/mol)</td><td>-1.78E+07</td><td></td><td></td><td></td><td></td></tr>
<tr><td>T out valve (°C)</td><td>-34.55</td><td></td><td></td><td></td><td></td></tr>
<tr><td>Valve outlet vaporized fraction</td><td>0.27</td><td></td><td></td><td></td><td></td></tr>
<tr><td>Compressor Efficiency</td><td>0.93</td><td></td><td></td><td></td><td></td></tr>
<tr><td>COP (-)</td><td>3.29</td><td></td><td></td><td></td><td></td></tr>
<tr><td>Total COP (-)</td><td>0.4695</td><td>+1.78%</td><td>-2.05%</td><td>-0.28%</td><td>+0.27%</td></tr>
</table>

Table 8. Results for the triple cascade.

		Outlet Compressor Pressure at the intermediate cycle		Outlet Valve Pressure at the intermediate cycle	
	Intermediate Cycle	**+20%**	**-20%**	**+14.3%**	**-14.3%**
Intermediate Cycle	Flow rate ratio (-)	+8.76%	-8.15%	+0.29%	-0.29%
	W (MW)	+6.12%	-5.96%	+0.61%	-0.60%
	Q Evaporator 2 (J/mol)	-8.05%	+8.87%	-0.29%	+0.29%
	Q Evaporator 3 (J/mol)	-8.24%	+8.64%		
	Q Condenser (J/mol)	+3.08%	-2.67%		
	T out valve (°C)			+2.5 °C	-2.8 °C
	Valve outlet vaporized fraction	+7.69%	-8.48%	-1.65%	+1.82%
	Compressor Efficiency	+7.70%	-8.91%	-5.88%	+6.93%
	COP (-)	-5.77%	+6.34%	-0.60%	+0.60%
Internal Cycle	Flow rate ratio (-)	-0.18%	+0.18%	+0.25%	-0.24%
	W (MW)	-0.18%	+0.18%	+0.25%	-0.24%
	Total Work (MW)	+2.97%	-2.99%	+0.36%	-0.36%
	Total COP (-)	-2.89%	+3.08%	-0.36%	+0.36%

Table 9. Results for the triple cascade (continued).

	Outlet Compressor Pressure at the internal cycle		Outlet Valve Pressure at the internal cycle	
Internal Cycle	**+7.14%**	**-7.14%**	**+14.3%**	**-14.3%**
Flow rate ratio (-)	-0.001%	+0.001%	+0.23%	-0.23%
W (MW)	-1.56%	+1.53%	+1.00%	-0.98%
Q Evaporator (J/mol)	+0.001%	-0.001%	-0.23%	+0.23%
Q Condenser (J/mol)	-0.36%	+0.36%		
T out valve (°C)			+3.32 °C	-3.72 °C
Valve outlet vaporized fraction	-0.003%	+0.003%	-5.88%	+6.41%
Compressor Efficiency	+4.52%	-4.65%	-9.29%	+11.01%
COP (-)	+1.58%	-1.51%	-0.99%	+0.99%
Total Flow rate ratio (-)	-0.0003%	+0.0004%	+0.08%	-0.08%
Total Work (MW)	-0.41%	+0.41%	+0.27%	-0.26%
Total COP (-)	+0.41%	-0.40%	-0.26%	+0.26%

Table 10. Results for the triple cascade (continued).

Configuration	Variable	Variation	Sensitivity on the COP	Sensitivity on the Flow Rate
Traditional Cascade	Lowest cycle high pressure	-12.5%	-2.62%	+2.38%
Hybrid Cascade	Internal-lowest cycle high pressure	-20.5%	+4.17%	-4.86%
Triple Cascade	External cycle high pressure	-12.5%	-2.05%	+1.76%

Table 11. Summary of the sensitivity analisys for the cascades.

		High Pressure		Low Pressure	
	Base Case	+10%	-10%	+16.67%	-16.67%
Flow rate ratio (-)	2.69	-6.71%	+8.24%	+0.60%	-0.59%
W (J/mol)	7.26E+06	-1.29%	+1.27%	+0.40%	-0.40%
W (MW)	142.41	-8.09%	+9.62%	+1.00%	-0.98%
Q Evaporator 1 (J/mol)	-1.67E+07	-2.08%	+2.30%		
Q Evaporator 2 (J/mol)	2.16E+07	-0.05%	+0.05%	-0.13%	+0.13%
Q Condenser (J/mol)	-1.22E+07	+1.91%	-2.31%		
T out valve (°C)	-157.05	+0.11 °C	-0,11 °C	+0,93 °C	-1.04 °C
Valve outlet vaporized fraction	0.04	+1.61%	-1.64%	-25.99%	+27.82%
Compressor Efficiency	0.85	+4.81%	-5.41%	-6.97%	+8.14%
COP (-)	0.67	+7.49%	-8.78%	-0.99%	+0.99%

Table 12. Results for the simple cycle with mixed refrigerant.

	Lee et al. (2002)	Del Nogal et al. (2005)	Jensen & Skogestad (2006)
Flow rate ratio (-)	-25.56%	-32.19%	-20.31%
W (J/mol)	+3.15%	+19.62%	-11.67%
W (MW)	-23.22%	-18.88%	-29.61%
Q Evaporator 1 (J/mol)	-6.75%	-8.53%	+8.35%
Q Evaporator 2 (J/mol)	+2.56%	+4.16%	+12.23%
Q Condenser (J/mol)	+15.72%	+30.84%	+3.30%
T out valve (°C)	+0.46 °C	-0.01 °C	-0.03 °C
Valve outlet vaporized fraction	+11.83%	+33.19%	-6.96%
Compressor Efficiency	-0.01%	+0.03%	+0.05%
COP (-)	+30.24%	+23.28%	+42.06%

Table 13. Results for the simple cycle with mixed refrigerant (continued).

In order to use them it was necessary to adjust the high pressure to guarantee a valve outlet temperature of -157 °C. In all the cases the minimum temperature approach (MITA) of the temperature profiles inside the evaporator was checked in order to avoid temperature cross.

When using the compositions of Table 4, also the suction temperature of the compressor was changed, in order to have a MITA of at least 2 °C. The discharge temperature of the compressor was adjusted so the polytropic efficiency was 85%.

It must be noted that the amount the heat exchanged on each refrigerant pass in the evaporator must be maximized, in this way the difference in heat exchanged in each pass of is the heat retired from the natural gas, so the greater this quantity is, the less refrigerant would be used and the greater the COP.

This can be seen in Table 13, the mixture whit a greater heat transferred in each pass of the evaporator is the one proposed by Jensen & Skogestad (2006), which has the greater COP.

When comparing the results of the triple cascade with pure refrigerant and the simple cycle with mixed refrigerant can be seen that the later has the greatest COP and also the smallest and the smaller power consumption and refrigerant flow rate. Nevertheless, the advantage of the cascade cycle is its simpler operation, so the any decision on the technology selection should be studied carefully.

5. Conclusions

Starting with two of the most used technologies for natural gas liquefaction, it was possible to model and simulate the system using Matlab® routines and using the thermodynamic model of Peng-Robinson for calculating energetic properties of the fluids, which leaded to the calculation of heats and powers of the cycle.

It was also possibly to perform a sensitivity analysis of some of the variables of the cycle and it was obtained that the most affecting variables are the high pressure of the external cycle (for the cascades) and the high pressure of the simple cycle with mixed refrigerant.

Other compositions found in the literature were tested for the simple cycle. It was observed that every different composition could affect the performance, and we recommend performing an optimization using these operating conditions, in order to maximize the COP.

6. References

Aspelund, A.; Gundersen, T.; Myklebust, J.; Nowak, M. P. & Tomasgard, A. (2010). An Optimization-Simulation Model for a Simple LNG Process. *Computers & Chemical Engineering*, Vol. 34, No.10, (October 2010), pp. 1606-1617, ISSN: 0098-1354.

British Petroleum. (2011). Natural Gas. In: *BP Statistical Review of World Energy*, June 2011, Available from http://www.bp.com/statisticalreview

Çengel, Y. & Boles, M. (2006). *Thermodynamics: An Engineering Approach*, McGraw-Hill, 978-0070606593.

Del Nogal, F. L.; Kim, J.; Smith, R. & Perry, S. J. (2005). Improved Design of Mixed Refrigerant Cycles Using Mathematical Programming. *GPA Europe Meeting: GTL and LNG in Europe*, Amsterdam, Febrary, 2005.

Green, D. W. & Perry, R. H. (2008). *Perry's Chemical Engineering Handbook* (8th. ed.), McGraw-Hill, 0-07-154209-4.

Haselden, G. G. (1977). The Liquefaction of Natural Gas. *Contemporary Physics*, Vol. 18, No.5, 471-488, ISSN: 1366-5812.

Institute for Energy, Law & Enterprise. University of Houston. (2003). An Overview on Liquefied Natural Gas (LNG), Its Properties, the LNG Industry, Safety Considerations. Available from http://www.energy.uh.edu

Jensen, J. B. & Skogestad, S. (2006). Optimal Operation of a Simple LNG Process. *International Symposium on advanced control of chemical processes (ADCHEM)*, Gramado - Brazil, April 2-5, 2006.

Lee, G. C.; Smith, R. & Zhu, X. X. (2002). Optimal Synthesis of Mixed-Refrigerant Systems for Low-Temperature Processes. *Industrial Chemical Engineering Research*, Vol. 41, No.20, (August 2002), pp. 5016-5028, ISSN: 1520-5045.

Lemmon, E.W.; McLinden, M.O. & Friend, D.G. . (2005). Thermophysical Properties of Fluid Systems. In: *NIST Chemistry WebBook, NIST Standard Reference Database Number 69*, October 2011, Available from http://webbook.nist.gov

Nikos, V. (1986). Phase Behavior of Systems Compressing Noth Sea Reservoir Fluids and Injection Gases. *Journal of Petroleum Technology*, Vol. XXII, 1221-1233.

Peng, D.-Y. & Robinson, D. B. (1976). A New Two Constant Equation of State. *Industrial & Engineering Chemistry Fundamentals*, Vol. 15, No.1, (Febraury 1976), pp. 59-64.

Price, B. C. & Mortko, R. A. (1996). Prico - a Simple, Flexible Proven Approach to Natural Gas Liquefaction. *GASTECH, LNG, Natural Gas, LPG international conference*, Vienna, 1996.

Smith, R. (2005). *Chemical Process: Design and Integration* (1st. ed.), Wiley, 978-0471486817.

Walas, S. M. (1990). *Chemical Process Equipment Selection and Design*, Butterworth-Heinemann, 0-7506-9385-1.

9

Natural Gas Purification Technologies – Major Advances for CO_2 Separation and Future Directions

Biruh Shimekit and Hilmi Mukhtar*

Department of Chemical Engineering, Universiti of Teknologi PETRONAS, Malaysia

1. Introduction

In an effort to satisfy the rising global demand for energy and at the same time to combat the environmental impacts such as global greenhouse gas (GHG) emissions, it's worth searching for potential energy alternatives. As a base line for most approaches, the issues of producing sufficient quantity of energy with high quality, economical viability and environmental sustainability are the concern of the present. One of such vital components of the world's supply of energy that has fulfilled the aforementioned requirement is natural gas.

In addition to its primary importance as a fuel, natural gas is also a source of hydrocarbons for petrochemical feed stocks. Many researches have been undertaking on natural gas field as the presence of high component of methane in natural gas contributes for the production of other potential products such as syngas and high purity hydrogen.

Although natural gas is mostly considered as a "clean" fuel as compared to other fossil fuels, the natural gas found in reservoirs deposit is not necessarily "clean" and free of impurities. Natural gas consists primarily of methane as the prevailing element but it also contains considerable amounts of light and heavier hydrocarbons as well as contaminating compounds of CO_2, N_2, Hg, He, H_2S and etc. Thus, the impurities must be removed to meet the pipe-line quality standard specifications as a consumer fuel, enhance the calorific value of the natural gas, avoid pipelines and equipment corrosion and further overcome related process bottle necks.

In this chapter, the major advances, process advantages and limitations for the existing technologies in natural gas purification including absorption, adsorption, cryogenic and membrane processes are discussed. Moreover, special emphasis on the removal technologies of CO_2 from the natural gas is presented since CO_2 is the largest contaminant found in natural gas and the major contributor for the global GHG emissions. Comparisons in terms of advantages and disadvantages between these technologies are also described. Emerging concepts for new approaches in natural gas separation are highlighted. Finally, the future research and development directions of natural gas processing technologies are also presented.

* Corresponding Author

1.1 Natural gas consumption statistics

In consideration of the state of the art green technology requirements in promoting low-and zero-emission through the wise use of natural resources for the available reserves, natural gas becomes one of the most attractive, fastest and the premium growing fuel of world primary energy consumption (Economides and Wood 2009).

According to the technical report that was released on June 2011 by the International Energy Outlook 2010, the global utilization of natural gas raises to 44% from the reference year 2007 (108 trillion cubic feet) to the predicted year of 2035 (156 trillion cubic feet). In 2008, the need for natural gas was lowered due to the global economic recession that affects the global fuel markets, and in the following year, the worldwide natural gas utilization was reduced approximately by 1.1% (Energy Information and Administration 2011).

Nevertheless, the worldwide need for natural gas has revived as the global economies starts to make progress from the economic recession. Accordingly, the spending of natural gas rise on average by 1.8 percent per annum as of the year 2007 to the predicted year of 2020. However, far from the predicted year 2020 to 2035, the increase in natural gas utilization of natural gas is estimated to slow down to 0.9 percent per annum on average, due to the high prices on the natural gas product and its resources that will be introduced to the worldwide sell. Natural gas, also called "the prince of hydrocarbons" as it has many applications. The proportion of the natural gas consumed for energy production in major fields including industrial, commercial, residential, transportation and in generating electricity for the year 2009 is shown on Fig. 1.

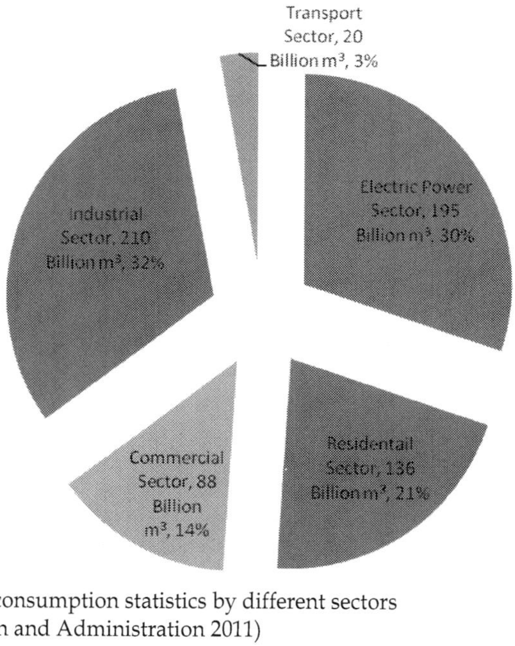

Fig. 1. Natural gas consumption statistics by different sectors (Energy Information and Administration 2011)

Again, according to the International Energy Outlook 2010 on its June 2011 release, natural gas was reported to be one of the main energy sources for industrial and electrical power generation applications. From the total worldwide natural gas consumption statistics for the year 2007, the industrial field shared almost 40 percent and it is predicted to maintain that share through 2035. In applications for electric power field, natural gas is usually preferred for new power plants. Moreover, when combusting natural gas, it usually emits less CO_2 than does other fuels such as petroleum or coal so it is expected that the use of natural gas will be encouraged by government decision and policy makers to at least reduce the impact of GHG (Energy Information and Administration 2011).

1.2 Theories of formation of natural gas

Like other non-renewable fossil fuels, natural gas is essentially formed from the decomposition of living matters such as plants, animals and micro-organisms that lived over millions of years ago and became an inanimate mixture of gases. Although there are various theories about the origin of fossil fuels, the most widely accepted theory states that fossil fuels are usually formed when organic matters are decayed and compressed under the earth's crust at high pressure and for a very long time. This kind of formation is technically referred to as thermogenic methane. This type of natural gas is formed by the decomposition process of the piled and compressed organic matters that are covered in mud, sediment and debris at high temperature beneath the crust of the earth. In another way, natural gas can also be formed by the action of tiny methane-producing microorganisms and it is technically termed as biogenic methane. In this case, methane formation usually takes place close to the earth's surface and the methane produced is usually dissipated into the atmosphere. However, in some cases, this methane can be trapped underground and recovered as natural gas. As a third theory, natural gas is formed by abiogenic processes where this process takes place at extremely underneath the earth's crust, where hydrogen-rich gases and carbon molecules are dominant. These gases may interact with minerals found in the underground in the absence of oxygen by the time the gases gradually rise towards the surface of the earth. In such processes reaction will take place and forms gaseous compounds such as nitrogen, carbon dioxide, oxygen, water and inert gases like argon. Hence, the condition will form methane deposits at very high pressure, similar to that of the thermogenic methane (NaturalGas.org 2010).

1.3 Composition of natural gas

The natural gas processed at the wells will have different range of composition depending on type, depth, and location of the underground reservoirs of porous sedimentary deposit and the geology of the area. Most often, oil and natural gas are found together in a reservoir. When the natural gas is produced from oil wells, it is categorized as associated with (dissolved in) crude oil or non-associated.

It is apparent that two gas wells producing from the same reservoir may have different compositions. Table 1 shows the composition of natural gas reservoirs in some part of the world. Further, the composition of the gas produced from a given reservoir may differ with time as the small hydrocarbon molecules (two to eight carbons) in addition to methane that existed in a gaseous state at underground pressures will become liquid (condense) at normal atmospheric pressure in the reservoir. Generally, they are called condensates or natural gas

Component	Reservoir				
	Groningen (Netherlands)	Laeq (France)	Uch (Pakistan)	Uthmaniyah (Saudi Arabia)	Ardjuna (Indonesia)
CH_4	81.3	69	27.3	55.5	65.7
C_2H_6	2.9	3	0.7	18	8.5
C_3H_8	0.4	0.9	0.3	9.8	14.5
C_4H_{10}	0.1	0.5	0.3	4.5	5.1
C_{5+}	0.1	0.5	-	1.6	0.8
N_2	14.3	1.5	25.2	0.2	1.3
H_2S	-	15.3	-	1.5	-
CO_2	0.9	9.3	46.2	8.9	4.1

Table 1. Composition of natural gas reservoirs in some parts of the world in volume percent basis (Rojey 1997; Bahadori, Mokhatab et al. 2007; Bakar and Ali 2010)

liquids (NGLs). This usually happens in a retrograde condensate reservoir (Beggs 1984; Bahadori, Mokhatab et al. 2007).

As one of the major contaminates in natural gas feeds, carbon dioxide must optimally be removed as it reduces the energy content of the gas and affect the selling price of the natural gas. Moreover, it becomes acidic and corrosive in the presence of water that has a potential to damage the pipeline and the equipment system. In addition, when the issue of transportation of the natural gas to a very far distance is a concern, the use of pipelines will be too expensive so that Liquefied Natural Gas (LNG), Gas to Liquid (GTL) and chemicals are considered to be an alternative option. In LNG processing plant, while cooling the natural gas to a very low temperature, the CO_2 can be frozen and block pipeline systems and cause transportation drawback. Hence, the presence of CO_2 in natural gas remains one of the challenging gas separation problems in process engineering for CO_2/CH_4 systems. Therefore, the removal of CO_2 from the natural gas through the purification processes is vital for an improvement in the quality of the product (Dortmundt and Doshi 1999).

As a gaseous fossil fuel, natural gas is relatively low in energy content per unit volume and emits lower quantities of green house gases (GHG) than other fossil fuels (Pascoli, Femia et al. 2001). However, when compared with other hydrocarbon energy sources, it is the most hydrogen-rich and has higher energy conversion efficiencies (Economides and Wood 2009).

Natural gas consists primarily of methane (70-90% of the total component) and other light and heavier hydrocarbons. The impurities present in natural gas need to be removed to meet the pipeline quality standard (NaturalGas.org 2010).

The allowable amounts of common impurities in U.S. for the delivery of the natural gas to the pipe line are given on Table 2 (Al-Juaied 2004; Baker 2004).

This standard may vary from pipeline to pipeline grids depends on the pipeline's system design. Generally, the following standards are used for natural gas specification in pipeline grid (Tobin J., Shambaugh P. et al. 2006). The natural gas should be (i) within a specific Btu

Components	U.S. pipe line specification
CO_2	<2 mol%
H_2S	<4 ppm
H_2O	<0.1 g/m^3 (<120 ppm)
C_{3+}	950-1050 Btu/scf dew point -20 °C
Total inerts (N_2,He, Ar, etc)	< 4 mol%

Table 2. U.S. pipeline composition specifications for natural gas delivery (Al-Juaied 2004; Baker 2004)

content range (1,305 Btu per cubic feet, +/- 50 Btu) (ii) delivered at a specified hydrocarbon dew point temperature level (below which any vaporized gas liquid in the mix will tend to condense at pipeline pressure (iii) contains no more than trace amounts of elements such as hydrogen sulfide, carbon dioxide, nitrogen, water vapor, and oxygen (iv) free of particulate solids and liquid water that could be detrimental to the pipeline or its ancillary operating equipment.

Comparatively carbon dioxide, that is produced from oil and coal, a GHG linked to global warming, has approximately higher production rate (1.4 to 1.75 times) than does from natural gas (Kidnay, Parrish et al. 2006).

1.4 Major processes in natural gas processing plant

The processing of wellhead natural gas into pipeline-quality dry natural gas can be quite complex and usually involves several processes. Most often, the number of gas treatment steps and the type of techniques used in the process of creating pipeline-quality natural gas depends on the source and makeup of the wellhead production stream. A typical natural gas processing plant whose simplified schematic representation shown in Fig. 2 consists

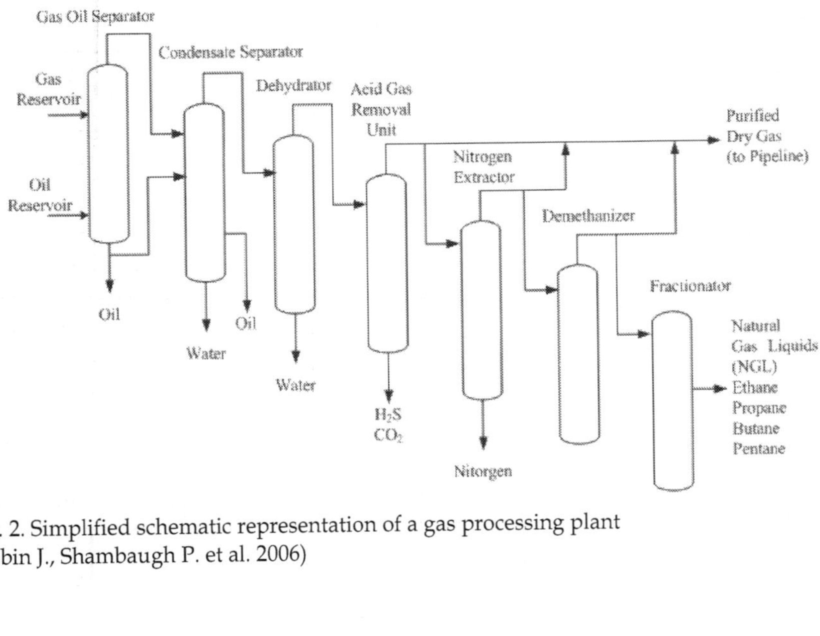

Fig. 2. Simplified schematic representation of a gas processing plant (Tobin J., Shambaugh P. et al. 2006)

mainly of the following processes: (i) gas oil separator (treatment unit); (ii) condensate separator; (iii) dehydrator; (iv) acid gas removal unit; (v) nitrogen extractor or Nitrogen rejection unit (NRU) and (vi) fractionator.

Brief descriptions about the major unit operations are provided as follows.

Gas-oil separators: are used to separate the gas stream and the crude oil at the top and bottom part of the cylindrical shell respectively by the action of pressure at the wellhead where gravity separates the gas hydrocarbons from the heavier oil.

Condensate separators: are used to remove condensates from the gas stream by mechanical separators at the wellhead. In condensate treatment section, two main operations, namely water washing and condensate stabilization are performed. Based on the quality of the associated water, the condensate may require water wash to remove salts and additives (Tobin J., Shambaugh P. et al. 2006).

Dehydrators: are used to remove water vapor using dehydration process so that the natural gas will be free from the formation of hydrates, corrosion problem and dew point. In this treatment, process of absorption using ethylene glycol is used to remove water and other particles from the feed stream. As another option, adsorption process can also be used for water removal using dry-bed dehydration towers (Tobin J., Shambaugh P. et al. 2006).

Acid gas removal units: are used to remove contaminates in the dry gas such as CO_2, H_2S, some remaining water vapor, inert gases such as helium, and oxygen. The use of alkanolamines or Benfield solution processes is mostly common to absorb CO_2 and H_2S from the feed gas (Tobin J., Shambaugh P. et al. 2006).

Nitrogen extractor or Nitrogen rejection unit: are used to remove nitrogen from the stream using two common ways. In the first type, nitrogen is cryogenically separated from the gas stream by the difference in their boiling point. In the second type, separation of methane from nitrogen takes place using physical absorption process. Usually regeneration is done by reducing the pressure. If there were trace amounts of inert gases like Helium then pressure swing adsorption unit can be used to extract them from the gas stream (Tobin J., Shambaugh P. et al. 2006).

Demethanizer: are used to separate methane from NGLs using cryogenic processing or absorption techniques. The demethanization process can take place in the plant or as nitrogen extraction process. As compared to absorption method, the cryogenic method is more efficient for the lighter liquids separation, such as ethane (Tobin J., Shambaugh P. et al. 2006).

Fractionators: are used to separate the NGLs present in the gas stream by varying the volatility of the hydrocarbons present in the stream. In fractionation, the NGLs after the demethanizer is subjected to rise through towers and heated to increase the temperature of the gas stream in stages, assisting the vapor and liquid phases thoroughly contacted, allowing the components to vaporize and condense easily and separate and flow into specific holding tanks (Tobin J., Shambaugh P. et al. 2006).

2. Acid gas removal technologies for natural gas purification

2.1 Overview on acid gas removal technologies for natural gas purification

Acid gas removal processes, also commonly known as gas sweetening processes are used to purify the natural gas from the acid gases such as CO_2 and H_2S. The technologies that are widely used to treat the natural gas include absorption processes, adsorption processes, cryogenic condensation and membranes. The technologies and their improvement have been developed over the years to treat certain types of gas with the aim of optimizing capital cost and operating cost, meet gas specifications and environmental purposes (Ebenezer and Gudmunsson 2006).

Comparative to the other natural gas separation techniques, the membrane process is a viable energy-saving alternate gas separation since it does not require any phase transformation. All the processes except membrane separation involves a change in the state of phase of the penetrant specious, where the desired penetrant is selectively transferred from gaseous state to liquid or solid state (Rojey 1997). Moreover, the necessary process equipment for membrane separation is very simple with no moving parts, compact, relatively easy to operate and control, and also easy to scale-up (Stern 1994).

2.2 Process selection factors

The processes that are used to remove acid gas are broad and the existing technologies are many that effective selection of process becomes a critical concern. This is because each of the processes has their own advantages and limitations relatives to others. Although common decisions in selecting an acid gas removal process can generally be simplified, factors such as nature and amount of contaminants in the feed gas, the amount of every contaminants present in feed gas and the targeted removal capacity, amount of hydrocarbon in the gas, pipeline specification, capital and operating cost, amount of gas to be processed, desired selectivity, conditions at which the feed gas is available for processing are the major factors that should also be considered (Dortmundt and Doshi 1999).

As highlighted on section 1.2.1, the aforementioned gas purification processes are briefly summarized in the following sections.

3. Absorption process

3.1 Basic separation principles for absorption process

Absorption process is one of the most important unit operations in natural gas purification process where a component of a gaseous phase is contacted with a liquid in which it is preferentially soluble.

Absorption is usually carried out in a countercurrent tower (column), through which liquid descends and gas ascends.

As per the needed surface area for gas-liquid contact, the absorption tower may be fitted with required trays, filled with packing, or fitted with sprays or other internals.

As overall selection criteria, the characteristics of a good solvent for absorption process fulfils the following main features: high gas solubility, high sclvent selectivity, high volatility, low effects on product and environment, high chemical stability, low cost and more availability, non corrosive, low viscosity, non flammable and low freezing point.

At equilibrium conditions, solubility of gases is referred as the quantity of gas dissolved in a given quantity of solvent. Further at equilibrium, the partial pressure (fugacity) of a component in the gas is equal to the fugacity of the same component in the liquid. This defines the equilibrium thermodynamic criterion for the relation of the concentration of a component in the gas and its corresponding concentration in the liquid (Meyers 2001).

As the important fundamental principles for physical absorption are solubility and mass transfer, the principles of reaction equilibria and reaction kinetics are for chemical absorption.

Based on the theoretical principles of mass transfer, the rate at which the equilibrium is attained when the gas (as a solute) is transferred into the solvent, no net transfer of material occurs between the phases. However, diffusion of materials between the phases will occur when a system is not in equilibrium. Thus, this ascertains that the driving force for diffusion of materials between the phases occur as far from the equilibrium (Meyers 2001).

When the gas (as a solute) is absorbed into a solution containing a reagent that reacts chemically with it, the amount of absorption (concentration profile) becomes dependent on the reaction kinetics and the concentration of the reacting reagent in the liquid (Meyers 2001).

The reverse process (which is also termed as stripping process, desorption, or regeneration), is employed when it is needed to remove the absorbed gases from the solvent for the purpose of recovery of the gas or the solvent or both. In the process of absorption, the selective solvent (absorbent) in a plate of packed column is contracted countercurrently with the gas to be processed (absorbate). As applied to gas purification process and based on the nature of interaction of absorbent and absorbate, absorption process is generally divided into physical and chemical absorption. In physical absorption, the desired gas component being absorbed (absorbate) is more soluble in the liquid solvent (absorbent) than other components in the gas phase but does not react chemically with the absorbent. Whereas in chemical absorption, the absorbate react chemically with the absorbent or a component within the absorbent. When the absorbent reaches its equilibrium level, regeneration step will take place by reducing the partial pressure in the gas phase and by thermal or chemical gradient for the case of physical and chemical absorption respectively (Kohl and Nielsen 1997; Scott 1998).

For cases where the removal of CO_2 is only needed in large quantities, or when only partial removal is required, the use of a hot carbonate solution or one of the physical solvents is preferred for economical advantages. The details on each of the absorption techniques for typical acid gases mainly for CO_2 separation are described in the next sections.

3.2 Physical absorption

Physical absorption processes are the type of absorption processes where the solvent interacts only physically with the dissolved gas. In this process, the solvent is used as an absorbent with thermodynamic properties such that the relative absorption of CO_2 is more favored over the other components of the gas mixture.

Mostly, physical solvent systems are used when the feed gas is characterized by high CO_2 partial pressure and low temperatures. Although heavy hydrocarbon restricts the wide use of physical solvent, its absorption capacity can be higher than chemical solvents. In addition, low CO_2 partial pressures as well as low outlet pressure of the product stream may also discourage the application of physical solvents (Ebenezer and Gudmunsson 2006).

Removal of CO_2 from the feed gas by physical solvent absorption is based on the solubility of CO_2 within the solvents. The partial pressure and the temperature of the feed gas are the two major factors that determine the solubility of CO_2. Although there are too many physical solvent processes for the removal of CO_2 from natural gas, not all the processes available are capable of removing CO_2 to meet LNG specification of 50-100 ppmv of <2.5% of CO_2 in the product stream (Ebenezer and Gudmunsson 2006).

In physical absorption, the interaction between CO_2 and the absorbent is weak relative to chemical solvents, decreasing the energy requirement for regeneration. Regeneration of solvents is assisted by either heating, pressure reduction or a combination of both.

Mostly, physical solvent scrubbing of CO_2 is well established. Selexol, a liquid glycol-based solvent, has been used for decades to process natural gas, both for bulk CO_2 removal and H_2S removal (Davison, Freund et al. 2001). Glycol is effective for capturing both CO_2 and H_2S at higher concentration. However, the CO_2 is released at near atmospheric pressure, requiring recompression for transportation and geologic storage. The Rectisol process, based on low temperature methanol, is another physical solvent process that has been used for removing CO_2. Glycerol carbonate is interesting because of its high selectivity for CO_2, but it has a relatively low capacity (Kovvali and Sirkar 2002).

3.3 Chemical absorption

In natural gas purification plant, chemical absorption processes are used to remove acid gases such as CO_2 in the gas stream by the action of exothermic reaction of the solvent with the gases.

Many alkanolamines are most widely used as the chemical solvent gas treating process for acid gas removal in the natural gas and petroleum processing industries. These processes use a solvent, either an alkanolamine or an alkali-salt (hot potassium carbonate processes) in an aqueous solution. The common amine based solvents used for the absorption process are monoethanolamine (MEA), diethanolamine (DEA) and methyldiethanolamine (MDEA) that reacts with the acid gas (CO_2 and H_2S) to form a complex or bond. H_2S, CO_2 and SO_2 are termed as acid gases since they dissociate to form a weak acidic solution when they come into contact with water or an aqueous medium. These amines are known as weak organic bases.

The basicity is provided by the amine function, and it provides reactivity to remove the acid gases. The hydroxyl groups serve to increase the solubility of amine in water. This effect reduces the vapor pressure of the amines so that less is lost out the top of the absorber or stripper (Glasscock 1990).

Table 3 shows the absorption capacity and some characteristics of commonly used amines for acid gases removal processes (Kohl and Nielsen 1997; Ritter and Ebner 2007).

A simplified schematic representation for the flow of a typical gas treating operation using amine solvents is shown in Fig. 3 (Kohl and Nielsen 1997; Al-Juaied 2004).

In the typical process, a sour gas (H_2S and/or CO_2) is introduced to absorber from its bottom at high-pressure and goes to rise up and counter currently contacts an aqueous alkanolamine solution that is routed to the top of the absorber. The resulting amine solution that contains high CO_2 is then transported to the heat exchangers to raise its temperature. Then, it is directed to the top of a regenerator (stripper) to countercurrently contact steam at high temperature and low pressure. Afterwards, the steam strips the CO_2 and H_2S from

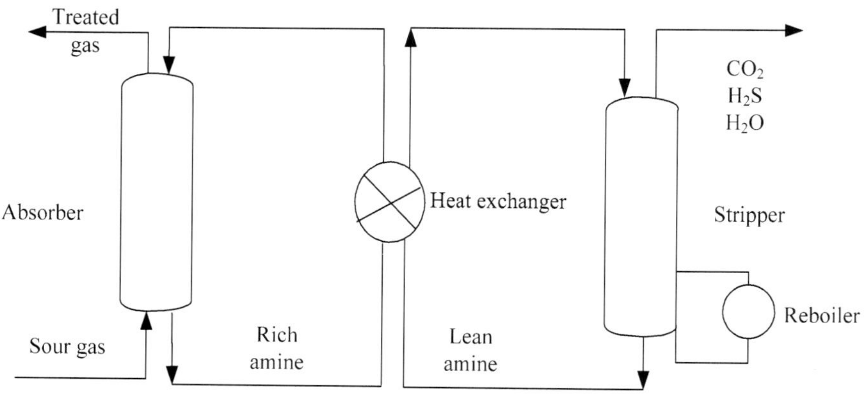

Fig. 3. Schematic representation for the flow of a typical gas treating operation using amine solvents (Kohl and Nielsen 1997; Al-Juaied 2004)

Name of amines	Vapour pressure (mmHg)	Relative acid gas capacity (%)	Remarks
Ethanolamine/ Monoethanolamine (MEA)	1.05	100	Good thermal stability, slow losses of alkanolamine but difficult to use MEA to meet pipeline specifications for H_2S.
Diethanolamine (DEA)	0.058	58	Lower capacity than MEA, reacts more slowly.
Triethanolamine (TEA)	0.0063	41	Low reactivity toward H_2S.
Hydroxyethanolamine/ Diglycolamine (DGA)	0.016	58	Same reactivity and capacity as DEA, with a lower vapour pressure and lower evaporation losses.
Diisopropanolamine (DIPA)	0.01	46	Selective for H_2S removal over CO_2 removal.
Methyldiethanolamine (MDEA)	0.0061	51	Selectively removes H_2S in the presence of CO_2, has good capacity, good reactivity, and very low vapour pressure, a preferred solvent for gas treating.

Table 3. Absorption capacity and some characteristics of commonly used amines for acid gases removal processes (Kohl and Nielsen 1997; Ritter and Ebner 2007)

solution and the lean alkanolamine solution is routed through the heat exchanger, where it is cooled, and reintroduced at the top of absorber (Al-Juaied 2004).

As the process proceeds, the complex or bond is subsequently reversed in the regenerator at elevated temperatures and reduced acid gas partial pressures releasing the acid gas and regenerating the solvent for reuse. These are well suited for low pressure applications where the acid gas partial pressures are low and low levels of acid gas are desired in the residue gas since their acid gas removal capacity is relatively high and insensitive to acid gas partial pressures as compared to physical solvents (Kohl and Nielsen 1997).

In most oil refineries sector, since stripped gas is mostly H_2S, much of which often comes from hydrodesulfurization (sulfur-removing process). The clause process is then used to convert the H_2S-rich stripped gas to elemental sulfur. In some plants, it is common to share a common regenerator unit for more than one amine absorber unit (Kohl and Nielsen 1997).

Recently, 2-amino-2-methyl-1-propanol (AMP) as another group of alkanolamines has received great deal of focus as it needs less energy for their regeneration. Moreover, blending of alkanolamines has also been researched as to obtain their potential synergy than the individual favorable features, e.g. high absorption capability coupled with thermal energy demand for regeneration (Rodriguez, Mello et al. 2011).

Generally, the chemical solvent processes are characterized by a relatively high heat of acid gas absorption and require a substantial amount of heat for regeneration.

As an overall, Table 4 shows the comparison of chemical and physical solvent absorption process for acid gas removal (Ritter and Ebner 2007).

	Chemical Absorption		**Physical Solvent Absorption**
	Alkanolamine	**Inorganic Carbonate**	
Type of absorbents	MEA,DEA,DGA,MDEA, DIPA	K_2CO_3, K_2CO_3-MEA, K_2CO_3-DEA	Selexol, Rectisol, Purisol
CO_2 absorption mechanism	Chemical reaction CO_2: $2RNH_2+CO_2+H_2O \leftrightarrow (RNH_3)_2CO_3$ $(RNH_3)_2CO_3+CO_2+H_2O \leftrightarrow 2RNH_3HCO_3$	Chemical reaction CO_2: $Na_2CO_3+CO_2+H_2O \leftrightarrow 2NaHCO_3$	Physical dissolution
Operating gauge pressure, mmHg	Insensitive to pressure	> 10,337.76	12,922.2-51,688
Operating Temp., °C	37.78-204.44	93.33-121.11	Ambient temperature
Absorbent Recovery	Reboiled stripper	Stripper	Flash, reboiler, or steam stripper
Swing variables (Temp. or Pressure)	Temperature principally	Both, but pressure principally	Pressure principally
Selectivity CO_2 vs. H_2S	Only MDEA selective for H_2S	May be selective for H_2S	Some selectivity for H_2S
Meets ppmv CO_2	Yes	Yes	Yes
Utility Cost	High	Medium	Low to medium

Table 4. Comparison of the major acid gas absorption processes (Ritter and Ebner 2007)

3.4 Limitation and challenges

The limitations of absorption process are briefly described as flows. The solvent used in absorption process such as amines and Benfield solution causes corrosion of the units (Polasek and Bullin 1984). When the solvent react with some corrosion inhibitors, it will cause erosion of the unit, high tendency for foaming and solid suspension thus reduce CO_2 solvent loading and require injection of antifoaming agents to reduce the surface tension of the solvent and to ensure better contact between the solvent and the CO_2, the regenerated solution leaving the stripper is at its saturated temperature and partially vaporize in the pump suction, resulting in vibration and excessive wear of the pump impellers. Moreover, since all of the solvents cannot be recycled back to the absorber column, the disposal of the solvents causes environmental hazards and thus showed the common disadvantages of using the absorption process (Bord, Cretier et al. 2004; Ebenezer and Gudmunsson 2006).

4. Adsorption process

4.1 Basic separation principles for adsorption process

In gas separation application, the process of adsorption is described as the adhesion or retention of selective components of feed gas stream brought into contact to the surface of certain solid adsorbent as the result of the force of field at the surface.

As the surface of an adsorbing material may exhibit different affinities for the various components of a fluid, it offers a straightforward means of purification (removal of undesirable components from a fluid mixture) as well as a potentially useful method of bulk separation (separation of a mixture into two or more streams of enhanced value). Bulk separation is defined as having the concentration of the adsorbed component above 10% by weight of the gas stream while for purification the concentration of adsorbed component is generally <2% by weight of the gas stream in the feed (Kerry 2007).

Although adsorption process is rarely applied for bulk separation of CO_2 from CH_4, there are kinetics-based adsorption processes that have been implemented in USA for the recovery of methane from landfill gas. These gases mainly comprises of methane (50-65%), carbon dioxide (35-50%), a trace amount of nitrogen and sulfur compounds. In this process, carbon molecular sieve is used as the adsorbent. In use of this process, it can be possible to recover more than 90% methane with 87-89% purity (Yang 1997).

One of the successful application for bulk separation of CO_2 from CH_4 is performed by using Engelhard molecular gate, a commercial brand name adsorbent developed by Engelhard Corporation.

The first application of molecular gate CO_2 removal system is at the Tidelands Oil Production Company operated facility in Long Beach, California. The feed source for this unit is hydrocarbon rich associated gas from enhanced oil recovery section. The feed more typically operates at 30-40% of CO_2 and the adsorbent able to reduce CO_2 level less than 2% (Ritter and Ebner 2007).

Depending on the nature and strength of the surface forces, adsorptive gas separation process can be divided into two types: physical adsorption and chemisorption.

Chemisorption can be considered as the formation of a chemical bond between the sorbate and the solid surface (covalent interaction of CO_2 and the surface of the adsorbent) that gives scope for much larger increases in adsorption capacity. Such interactions are strong, highly specific, and often not easily reversible. Chemisorption systems are sometimes used for removing trace concentrations of contaminants, but the difficulty of regeneration makes such systems unsuitable for most process applications (Meyers 2001).

In most operations, adsorption processes depend on physical adsorption. The forces of physical adsorption are weaker (a combination of Van der Waals forces and electrostatic forces) than the forces of chemisorption so the heats of physical adsorption are lower and the adsorbent is more easily regenerated as no covalent bonds are formed and heat is released upon adsorption. During capture, the chemical potential of the adsorbed CO_2 is lower than the chemical potential of CO_2 in the gas mixture. Several different types of force are involved. For nonpolar systems the major contribution is generally from dispersion-repulsion (van der Waals) forces, which are a fundamental property of all matter. When the surface is polar, depending on the nature of the sorbate molecule, there may also be important contributions from polarization, dipole, and quadrupole interactions. Selective adsorption of a polar species such as water or a quadrupolar species such as CO_2 from a mixture with other nonpolar species can therefore be accomplished by using a polar adsorbent. Indeed, adjustment of surface polarity is one of the main ways of tailoring adsorbent selectivity (Meyers 2001).

Physical adsorption at a surface is so fast, and the kinetics of physical adsorption are usually controlled by mass or heat transfer rather than by the intrinsic rate of the surface process (Meyers 2001).

Most of the separation processes is based upon equilibrium mechanism and the separation is accomplished by the adsorption equilibrium capacity difference of the adsorbent among the adsorbate.

The primary requirement for an economic adsorption separation process is an adsorbent with sufficient selectivity, capacity and service life. Adsorption selectivity may depend either on a difference in adsorption equilibrium or, less commonly, on a difference in kinetics. Kinetic selectivity is generally possible mainly with microporous or biporous adsorbents such as pelleted zeolites or carbon molecular sieves (Meyers 2001). These adsorbents offer three types of resistances to mass transfer: the external resistance of the fluid film, the diffusion resistance associated with transport through the macropores, and the intracrystalline or micropore diffusional resistance (Meyers 2001; Meyers 2011).

Thus the aforementioned resistances (one or their combined effects of resistance) will control the rate depending on the particular system and the conditions.

Since adsorption is a surface phenomenon, an adsorbent should have a high surface area to volume ratio. The main advantage of physical adsorption methods is its low energy requirement for the regeneration of the sorbent material with short period of time associated with the change in pressure.

The widely used adsorption processes includes the metal oxide (metal organic frame works), molecular sieves (zeolites, activated carbon) and promoted hydrotalcites based processes. Zeolite systems can produce nearly pure streams of CO_2, but have high energy expenses due to vacuum pumps and dehumidification equipment. As most effective adsorbent, the use of hydrotalcites at high temperatures (177-327°C) is widely for adsorption of CO_2 in or near combustion or gasification chambers. However, more study is still required to decrease the pressure difference requirement and enhance the capacity of current adsorbents (Hermann, Bosshar et al. 2005).

Commercial adsorbents that show ultraporosity have been used for the selective separation of gases, and included activated carbons, charcoal, activated clays, silica gel, activated alumina, and crystalline aluminosilicate zeolites.

The major advantages of using adsorption processes are simplicity of operation, the relative capability of the molecular sieve beds to withstand mechanical degradation and the possibility of simultaneous dehydration of gases and acid removal.

Once saturation of the adsorbent is reached, regeneration is carried out by either applying heat or by lowering pressure (concentration). Based on regeneration methods, adsorption process is most commonly divided into temperature swing adsorption (TSA), pressure swing adsorption (PSA) and displacement desorption.

Table 5 shows typical loading capacities of some commercial adsorbents towards acid gases with their corresponding regeneration methods. As an overall, the choice of adsorption methods depends on economic factor as well as technical considerations.

The next section briefly describes the aforementioned regeneration methods for adsorption processes.

Adsorbent	Adsorbate	T (°C)	P (mmHg)	Loading (mol/kg)	Regeneration Method
Activated carbon	CO_2	25-300	500	1.5-2.0@25 °C 0.1-0.2@300 °C	PSA
5A zeolite	CO_2	25-250	500	~ 3.0@25 °C 0.2@250 °C	PSA
Titanosilicates	CO_2	25-200	760-6x10^5	-	PSA
HTlc (K-promoted)	CO_2	300-400	200-700	0.4-0.7	PSA
Solid amine (supported PEI)	CO_2	75	760	1.5 - 3.0	PSA
Double-layer hydroxides	CO_2	375	230	1.5	PSA
Alumina (un-doped)	CO_2	400	500	0.06	PSA
Alumina (doped w/Li_2O)	CO_2	400	500	0.52	PSA
alumina(basic)	CO_2	300	500	0.3	PSA
Li zirconate	CO_2	500	760	3.4-4.5	TSA
CaO	CO_2	500	150	4-8@500 °C 7@500 °C	TSA

Table 5. Typical CO_2 loading capacities of some adsorbents at operating conditions (Ritter and Ebner 2007)

4.2 Thermal swing adsorption

The use of cyclic thermal swing processes (TSA) is widely applicable for the purification operations such as removal of CO_2 from natural gas or drying. In TSA, desorption is achieved by increasing the temperature of the adsorption bed by applying heat to the bed or more commonly by purging with a hot purge gas. At higher temperatures the adsorption equilibrium constant is reduced so that even quite strongly adsorbed species can be removed with a comparatively small purge gas volume. TSA is generally used for purification of the process such as drying or removal of CO_2 from natural gas (Meyers 2001).

Thermal swing adsorption is very reliable to remove minor component. The limitation in thermal swing adsorption process is the adsorption cycle time that is required to cool down the bed. Other obstacles are the high energy requirements and large heat loss (Mersmann, Kind et al. 2011).

In practice, in most two-bed purification processes, the desorption step controls the cycle, either directly or through the heat balance. Initial design of the regeneration cycle is commonly based on the assumption that during desorption the column approaches equilibrium. However, at the low concentrations that will prevail during the later steps of desorption, kinetic effects may be important, so more detailed understanding of the kinetics is necessary (Meyers 2001).

The rate at which the temperature is raised for regeneration is another important factor in the regeneration of molecular sieve driers. When this condition is too fast relative to the rate of moisture removal (drying), the probability of rapid desorption of moisture from the initial section of the bed will be there as it will be in contact with the hot desorbent gas, followed by condensation of liquid water in the cooler regions at some distance from the inlet. And the consequences will be serious for adsorbent life. To overcome the possibility of fluidizing the bed, the system will need to be operated in the down flow mode with up flow desorption since the gas velocity during adsorption is normally higher than that of desorption. Usually, the maximum up flow velocity is normally limited to 80% of the minimum fluidization velocity, while velocities as high as 1.8 times minimum fluidization can be tolerated in down flow (Meyers 2001).

When the adsorbent is regenerated by the TSA (when the equilibrium is favourable for this regeneration method) and unloaded by heating provided indirectly by an embedded heat transfer exchanger or directly by a hot regeneration gas or by both. The residual loading depends on the maximum temperature and a preconcentration of the possibly recircled gas and of the kinetic approach to equilibrium. Mostly the temperature swing adsorption is applied on molecular sieves which are loaded with H_2O, SO_2 and CO_2. In such conditions, the heats of adsorption are very high (ΔH>30kJ/mol) for polar adsorptives adsorbed on adsorbents with electrical charges like zeolites. Moreover, this TSA process is also applied on the regeneration of activated carbon, zeolites, and silica gel loaded with hydrocarbons present in natural gas when the heat of adsorption is higher than 30 kJ/mol (Mersmann, Kind et al. 2011).

4.3 Pressure swing adsorption

Pressure swing adsorption (PSA) is one of the most known industrial processes for gas separation. PSA is well known a technology for the removal of CO_2 from gaseous streams containing methane In such process, the removal of CO_2 from natural gas streams by adsorption processes are based on materials with selective adsorption to CO_2 by different equilibrium capacities or by differences in uptake rates (Cavenati, Carlos et al. 2006).

In PSA, regeneration is carried out by lowering the operating partial pressure to desorb the adsorbate (Kerry 2007). This can be obtained either by depressurization or by evacuation or by both. PSA is more suitable for bulk separation. Moreover, PSA is also used for drying of air and industrial gases. Air prepurification (purification of air prior to cryogenic distillation by removal of CO_2, water, and hydrocarbons) is also at developing stage for PSA (Yang and Wiley 2003).

As the most important decision in any adsorption-based technology is the adsorbent selection, zeolites as microporous materials that adsorb CO_2 strongly, are mostly used in PSA processes. Zeolites have good records for separation of CO_2 than activated carbons in the PSA (Cavenati, Carlos et al. 2006).

The PSA process is applied when there is a need for an essential reduction of the loading to be achieved by a certain pressure drop. This mainly depends on the loading as a function of the partial pressure at the operating temperature.

In a typical adsorption process, each bed in a PSA plant undergoes adsorption and regeneration cycle steps. These steps include 1) pressurization, 2) high pressure feed, 3) co-current depressurization, 4) countercurrent depressurization, 5) countercurrent purge (light reflux), and 6) several equalization (pressurization/depressurization) steps between two beds (Ritter and Ebner 2007).

While purge gas stripping depends on reducing the partial pressure by dilution with an inert purge gas, a pressure swing process desorption is achieved simply by reducing the total pressure. This generally requires a rather large purge volume, so such a process would normally be used only in special circumstances.

One of the important features of such processes is that the less strongly adsorbed species (the raffinate product) can be recovered at high purity but at relatively low fractional recovery, while the more strongly adsorbed species (the extract product) is always recovered in less pure form during the blowdown and purge steps. This type of process is therefore especially suitable for gaseous separations when the feed is inexpensive and the less strongly adsorbed species is the required product. All three major industrial applications of PSA (air drying, air separation, and hydrogen purification) fulfill these requirements (Meyers 2001).

Moreover, PSA systems are well suited to rapid cycling, making the process more preferable in obtaining relatively large throughput with relatively small adsorbent beds. However, the energy efficiency of such processes is not high, and since mechanical energy is usually more expensive than heat, PSA systems are generally not economical for large-scale operations. However, their process advantage lies in their compactness and simplicity, making them ideal for applications such as the production of medical oxygen in the home or in hospitals in remote areas. Nevertheless, with recent improvements in process efficiency, PSA processes are economically competitive with cryogenic distillation for oxygen production rates up to about 250 tons/day (Meyers 2001).

Engelhard Corporation is well known for its development of a typical PSA process to remove H_2O, CO_2, and heavier hydrocarbons from methane using their Molecular Gate adsorbent technology. Its adsorbents are comprised of titanium silicate molecular sieves that were originally developed to remove only N_2 from natural gas by kinetic separation. This class of materials was subsequently found to provide interestingly higher kinetic and adsorption selectivities for CO_2 and H_2O, exceeding those of more widely known aluminosilicate molecular sieves. A typical PSA system using the Molecular Gate adsorbent can separate CO_2 from CH_4 rich gas streams at feed absolute pressures between 5168.88 and 41,351.02 mmHg producing a product stream containing CH_4 at concentrations of > 90 vol%. This commercial technology is able to process gas streams containing up to 30 vol% CO_2 with favorable economics (Ritter and Ebner 2007).

The major advantages of PSA system are low capital and maintenance costs, high purity product, rapid shutdown and start-up characteristics, lack of corrosions problems, absence of heat requirement and pipe insulation and comparative straight forward operation.

Although the simple two-bed PSA cycle is widely used in small-scale units that achieve economic operations on a larger scale, it is necessary to improve the energy efficiency of the process. This can be accomplished by using multiple-bed systems in which blowdown and repressurization take place in several stages in such a way that the high-pressure enriched gas at the end of the adsorption step in one column is used to pressurize partially on the second column and so on. In contrast, PSA has a limitation as it requires high pressure and vacuum pressure that contribute to high operating cost (Meyers 2001).

4.4 Displacement desorption

Displacement desorption, is similar to purge gas stripping as the temperature and pressure are maintained constant, but instead of an inert purge, an adsorbable species is used to displace the adsorbed component from the bed similar to displacement chromatography. Displacement desorption is usually used when desorption by pressure swing or thermal swing fails to be practical.

Steam stripping process can be considered as a combination of thermal swing and displacement desorption as it is mostly used in the regeneration of solvent recovery system using an activated carbon adsorbent.

In a typical displacement desorption, the displacing component should be adsorbed somewhat less strongly than the preferentially adsorbed species so that the adsorption-desorption equilibrium can be shifted by varying the concentration of the desorbent. Such processes run more or less isothermally and offer a useful alternative to thermal swing processes for strongly adsorbed species when thermal swing would require temperatures high enough to cause cracking, coking, or rapid aging of the adsorbent (Meyers 2001).

4.5 Limitation and challenges

As compared to other methods, for adsorption to be economical, ideal adsorbents should have high transfer rates, high regenerability and high capacity that allow for thermal swing adsorption (TSA) or low pressure swing adsorption (PSA).

At the adsorption equilibria, the loading usually decreases with increasing temperature for a given partial pressure or concentration of the adsorptive in the fluid. In an isothermal system, the loading decreases with decreasing partial pressure or concentration. A further regeneration process is based on the replacement of the adsorbate by another adsorptive with a greater affinity to the adsorbent (Mersmann, Kind et al. 2011).

As an overall, the advantages and limitations of the aforementioned desorption techniques on the choice of the regeneration methods are summarised on Table 6 (Meyers 2001).

Method	Advantages	Limitations
Thermal swing	Good for strongly adsorbed species, since small change in temperature gives large change in adsorbate; desorbate can be recovered at high concentration; applicable to both gases and liquids	Thermal aging of adsorbent; heat loss means inefficiency in energy usage; unsuitable for rapid cycling, so adsorbent cannot be used with maximum efficiency; in liquid systems, high latent heat of interstitial liquid must be added.
Pressure swing	Good where weakly adsorbed species is required in high purity; rapid cycling, efficient use of adsorbent	Very low pressure may be required; mechanical energy more expensive than heat; desorbate recovered at low purity
Displacement desorption	Good for strongly held species; avoids risk of cracking reactions during regeneration; avoids thermal aging of adsorbent	Product separation and recovery needed. (choice of desorbent is crucial)

Table 6. Advantages and limitations of the regeneration method (Meyers 2001)

5. Cryogenic process

5.1 Basic separation principle for cryogenic process

Cryogenic separation (also known as low temperature distillation) uses a very low temperature (-73.30 °C) for purifying gas mixtures in the separation process (Ebenezer and Gudmunsson 2006).

The major industrial application of low-temperature processes involves the separation and purification of gases. Much of the commercial oxygen and nitrogen, and all the neon, argon, krypton, and xenon, are obtained by the distillation of liquid air (Meyers 2001).

Commercial helium is separated from helium-bearing natural gas by a well-established low-temperature process. Cryogenics has also been used commercially to separate hydrogen from various sources of impure hydrogen (Meyers 2001).

The cryogenic method is better at extraction of the lighter liquids, such as ethane, than is the alternative absorption method. Essentially, cryogenic processing consists of lowering the temperature of the gas stream to around -84.44 °C. While there are several ways to perform this function the turbo expander process is most effective, using external refrigerants to chill the gas stream. The quick drop in temperature that the expander is capable of producing condenses the hydrocarbons in the gas stream, but maintains methane in its gaseous form (Tobin J., Shambaugh P. et al. 2006).

While cryogenic separation is used commercially to liquefy and purify CO_2 from streams that have high CO_2 concentrations (typically greater than 50-70 percent), it has not been applied to large scale CO_2 capture from flue gas due to the low concentration of CO_2 that makes the application of this technique not economical. Cryogenic separation can separate CO_2 from other gases using pressure and temperature control resulting in solid or liquid CO_2 particulate matter and other contaminants are also removed in the process. Cryogenics use condensation of gases as the main principle. When CO_2 is cooled below its boiling point, it begins to condense and separate and turns into a liquid state. Differences in boiling points cause the gases to separate because each gas will turn to a liquid at a different point, but separation into pure components can also be influenced by the composition of the gas being cooled (Tobin J., Shambaugh P. et al. 2006).

The advantages of this process are the suitability to liquefy and purify the feed gas with high concentration of CO_2 and for producing a liquid CO_2 ready for transportation by pipeline and does not require compression since there is no additional chemicals used.

5.2 Limitation and challenges

The main disadvantage of cryogenic separation is that the process is highly energy intensive for regeneration and can significantly decrease the overall plant efficiency when applied to streams with low CO_2 concentration. Moreover, tendency for blockage of process equipment is high and some cryogenic fluids are flammable and toxic such as (acetylene, ethane) (Ebenezer and Gudmunsson 2006).

6. Membrane process

6.1 Basic principle of membrane process

In essence, gas separation membranes are thin films that selectively transport gases through the membrane based on differences in permeabilities of the species flowing through the membrane. The permeability of gases in a membrane is related as a function of membrane properties (physical and chemical structure), the nature of the permeant species (size, shape, and polarity), and the interaction between membrane and permeant species (Stern 1994; Burggraaf 1996; Shekhawat, Luebke et al. 2003).

The membrane properties and the nature of the permeant species determine the diffusional characteristics of a penetrant gas through a given membrane. The interaction between membrane and permeant, refers to the sorptivity or solubility of the gas in the membrane. The permeability coefficient (or permeability) of a penetrant is the product of the solubility coefficient or the sorptivity (thermodynamic parameter) and the diffusion coefficient (kinetic parameter). The permeability coefficient denotes the rate at which a penetrant traverses a membrane. The solubility (sorptivity coefficient) is a measurement of the amount of gas sorbed by the membrane when equilibrated with a given pressure of gas at a given temperature. The diffusion coefficient indicates how fast a penetrant is transported through the membrane in the absence of obstructive sorption.

The selectivity of the membrane to specific gas or liquid molecules is subject to the ability of the molecules to diffuse through the membrane. The permselectivity or ideal separation factor is simply the ratio of the pure gas permeability of the gases being separated.

Membranes utilized in separations need to possess both high selectivity and high permeation. The higher the permeability, the less membrane area is required for a given separation and therefore the lower the membrane cost. The higher the selectivity is the lower the losses of methane and therefore the higher the volume of the product that can be recovered (Porter 1990; Shekhawat, Luebke et al. 2003).

Gas transport through porous membranes takes place through a number of mechanisms, such as molecular sieving, Knudsen diffusion, surface diffusion, capillary condensation and micropore diffusion. Brief descriptions of these mechanisms are provided as follows.

Molecular diffusion: In molecular diffusion, the mean free path of the gas molecules is smaller than the pore size and diffusion occurs primarily through molecule-molecule collisions. Here, the driving force is the composition gradient. If a pressure gradient is applied in such pore regimes bulk (laminar) flow occurs, the transport is described by *Poiseuille flow* or viscous flow (De Lange, Keizer et al. 1995; Javaid 2005).

Knudsen diffusion: This mode of transport is significant when the mean free path of the gas molecules is greater than the pore size. In such situations, the collisions of the molecules with the pore wall are more frequent than the collisions among molecules. Separation selectivities with this mechanism are proportional to the ratio of the inverse square root of the molecular weights (De Lange, Keizer et al. 1995; Javaid 2005).

Surface diffusion: It takes place when the permeating species exhibit a strong affinity for the membrane surface and adsorb along the pore walls. In this mechanism, separation occurs due to the differences in the amount of adsorption of the permeating species. Surface diffusion often occurs in parallel with other transport mechanisms such as Knudsen diffusion (De Lange, Keizer et al. 1995; Javaid 2005).

The activation energy for the diffusion is strongly correlated with the heat of adsorption. Since it is assumed that diffusion takes place by molecules which jump from one site to another, the activation energy is a fraction of heat adsorption (Gilliland, Baddour et al. 1974). This implies that: a) strongly adsorbed molecules, and b) the total flux will decrease as the temperature is increased since the increased diffusivity is overruled by the decrease in surface concentration (De Lange, Keizer et al. 1995).

Capillary condensation: It is one form of surface flow where one of the gases is a condensable gas. Typically in mesopores and small macropores, at certain critical relative pressures as determined by the Kelvin equation, the pore gets completely filled by the condensed gas. Due to the formation of menisci at both ends of the pore, transport can take place through hydrodynamic flow generated by capillary pressure difference between the two ends. This mechanism of gas transport can be considered as the limiting case adsorption process when the pressure is increased. Theoretically capillary condensation can be used to achieve very high selectivities, as the formation of the liquid layer of the condensable gas will block and prevent the flow of the non-condensable gas (Javaid 2005).

Configurational or micropore diffusion: This type of diffusion may be considered as the limiting value of surface diffusion where the pore size becomes comparable to the molecular size. In this mechanism, diffusion is perceived as an "activated" process and separation is a strong function of molecular shape, molecular size, pore size, and the interactions between the pore wall and gas molecules. This type of mechanism is dominant in microporous inorganic membranes such as zeolite and carbon molecular sieves membranes (Javaid 2005).

In another dimension, membranes can be coupled with a solvent to capture desired gases such as CO_2 and the process is known as membrane gas absorption. In such process, the CO_2 diffuses between the pores in the membrane and is then absorbed by the solvent. The membrane maintains the surface area between gas and liquid phases. This type of membrane is used when the CO_2 has a low partial pressure, such as in flue gases, because the driving force for gas separation is small.

Unlike the gas separating membrane discussed earlier, the membrane gas absorption process does not undergo separation of the CO_2 from other feed gases, but rather function as a barrier between the liquid and gas with permeability through the pores.

As opposed to the traditional solvent absorption process whereby the solvent and the gas flow together and lead to transport problems such as foaming and channeling, the physical separation of the gas flow from the liquid flow through membrane absorption eliminates these problems.

Although a range of configurations exists either simply as gas separation devices or incorporating liquid absorption stages, the process has not yet been applied on a large scale and there are challenges related to the composition and temperature of the feed gases such as flue gases.

Currently, most researches are focussed on developing effective solvents and optimising the reaction time. The general advantages of the membranes process and mainly for the removal of CO_2 from natural gas includes: enhanced weight and space efficiency which make it more applicable for off shore environment, high adaptability to variation of CO_2 content in the feed gas and separation at low pressure and temperature, easy to combine with other separation process and use of other chemicals is not required, periodic removal and handling of spent solvent or adsorbent making the system more environmental friendly, relatively no moving parts make the process more flexible to operate, control and also easy to scale-up, low maintenance requirement, reduced energy consumption unless compression used, and low capital costs. However, it has some limitations as the separated CO_2 is at low pressure, it needs additional energy for compression of the feed gas to and meet pipeline pressure standard. Hence, economics remains a challenge while working towards pipeline specifications (Ebenezer and Gudmunsson 2006).

6.2 Membrane selection for natural gas separation

For more than two decades membranes have been known to constitute a mature technology that has been applied in natural gas processing. Currently membranes are used for CO_2 removal from natural gas at processing rates from million standard cubic feet per day (MMSCFD) to 250 MMSCFD. New units are also in design or construction stage to handle volumes up to 500 MMSCFD (Sridhar, Smitha et al. 2007).

The important criteria for selecting membrane materials for gas separation are based on the following key factors (a) intrinsic membrane permselectivity (b) ability of the membrane material to resist swelling induced plasticization (chemical resistance, which is quite rare but mostly fulfilled by inorganic membranes) and (c) ability to process the membrane material into a useful asymmetric morphology with good mechanical strength under adverse thermal and feed mixture conditions. The polymer membrane material should have good interaction and sorption capacity preferably with one of the components of the mixture for an effective separation. Molecular structure, specific nature and arrangement of chemical groups attached to the main chain are also some of the important factors, which affect the membrane properties and hence, their performance. Molecular weight distribution and membrane polarity are other parameters of interest for the development of novel membranes in natural gas separation.

Nowadays, the control of gas permeability and permselectivity of membrane material has attracted the attention of a large number of membrane researchers all over the world, from industrial sectors as well as academic laboratories.

In addition, knowledge and understanding of the scientific factors such as thermodynamics, mass transfer kinetics and surface science are equally important to control the complex morphologies that are needed for an efficient use of advanced materials in asymmetric and composite membranes. Furthermore, the behaviour of membranes with respect to various feed mixtures under practically realistic operating conditions using analytical techniques need also be understood.

In the field of removal of acid gases from natural gas, a feed stream constituting CH_4, CO_2 and H_2S must be dealt with an integration of three relevant areas such as material selection, membrane synthesis and system configuration.

A clear choice of membrane materials not only render high permeability ratios, but also yield good permeabilities. The specific membrane chemistry is also equally important which depends upon the type of separation to be achieved. Therefore, membrane material selection is an area of high significance.

With a few exceptions and based on the materials that the membranes were made from, gas separating membranes can be broadly categorized into the following major categories: polymeric, inorganic and mixed matrix. The following sections briefly explain about the aforementioned gas separating membranes and further address the current challenges in the area.

6.3 Polymeric membranes

Polymeric membranes are most commercially used for the bulk separation of gases in many processes such as recovery of nitrogen from air, separation of oxygen-nitrogen mixture and hydrocarbons in petrochemical industries and also purification of natural gas. With regards to the separation of CO_2 from gas streams, primarily for natural gas sweetening, these membranes selectively transmit CO_2 versus CH_4. The driving force for the separation is pressure gradient across the membrane. As such, compression is required for the feed gas in order to provide the driving force for permeation, and the separated CO_2 is at low pressure and it requires additional compression to meet pipeline pressure requirements (Pandey and Chauhan 2001; Sridhar, Smitha et al. 2007; Bernardo, Drioli et al. 2009).

The two types of polymeric membranes that are commercially available for gas separations are glassy and rubbery membranes. Glassy membranes are rigid and glass-like, and operate below their glass transition temperatures. On the other hand, rubbery membranes are flexible, soft and operate above their glass transition temperatures. Mostly, rubbery polymers show a high permeability, but a low selectivity, whereas glassy polymers exhibit a low permeability but a high selectivity. Glassy polymeric membranes dominate industrial membrane separations because of their high gas selectivities, along with good mechanical properties.

Based on structure, gas separation membranes can be classified as microporous membranes or nonporous dense membranes. Microporous membranes have a rigid, highly voided structure with randomly distributed, interconnected pores on the order of 0.01 to 10 μm in diameter. Nonporous, dense membranes consist of a dense film through which permeants are transported by diffusion under the driving force of a pressure, concentration, or electrical potential gradient. The separation of various components of a mixture is related directly to their relative transport rate within the membrane, which is determined by their diffusivity and solubility in the membrane material. Thus, nonporous, dense membranes can separate permeants of similar size if their concentration in the membrane material (that is, their solubility) differs significantly. Most gas separation processes use dense membranes to perform the separation. Usually these membranes have an anisotropic structure to improve the flux (Baker 2004).

In dense polymeric materials, solution-diffusion is widely accepted to be the main mechanism of transport. This mechanism is generally considered to be a three-step process. In the first step, the gas molecules are absorbed by the membrane surface on the upstream end. This is followed by the diffusion of the gas molecules through the polymer matrix. In the final step, the gas molecules evaporate on the downstream end. In glassy polymers, the sorption of gases becomes a complex process, which has been described by a combination of Henry's law and Langmuir expressions. This has been referred to as "dual mode sorption theory". Diffusion in glassy polymers is usually an activated process and Arrhenius relations may be used to express the permeability, diffusivity, and solubility coefficients (Kesting and Fritzsche 1993; Al-Juaied 2004).

Generally, polymeric membranes have been known for their excellent intrinsic transport properties, high processability and their low cost. However, a limit in the trade-off between permeability and selectivity as described by Robeson's "upper bound" (Fig. 4) has been seemingly reached by most polymeric membrane materials (Robeson 1991; Xiao, Low et al. 2009).

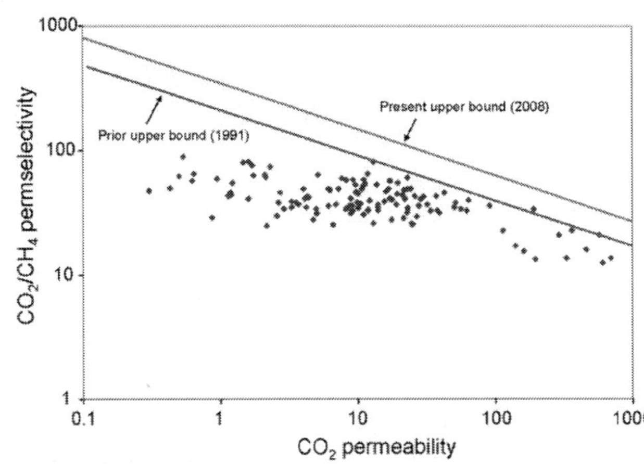

Fig. 4. Robeson's trade-off curve for carbon dioxide/natural gas pair (Xiao, Low et al. 2009)

Thus, the current research trend has focused on pushing the polymer performance above the upper bound and into the economically attractive region as showed by inorganic membranes.

6.4 Inorganic membranes

Inorganic membranes (also known as ceramic membranes) are used for gas separation due to their superior thermal, mechanical and chemical stability, good erosion resistance, insensitivity to bacterial action and a long operational life (Caro, Noack et al. 2000).

Microporous and dense membranes are the two types of inorganic membranes that are suitable for high-temperature gas separation applications.

Dense, nonporous inorganic membranes are made of polycrystalline ceramic material, in particular made of perovskites, palladium and its alloys, silver, nickel and stabilized solid electrolytes (zirconia). Mostly, they have been used or evaluated for separating gaseous components in laboratory work to characterize membrane properties. They are prepared as unsupported ones as well as thin films on porous supports. Application of dense membranes is primarily for highly selective separation of hydrogen (i.e.H_2 through Pd) and oxygen (i.e.O_2 through perovskites); transport occurs via charged particles. Dense membranes are impermeable to all gases except for a very limited number of gases that can permeate the material or can be incorporated into the structure of the membrane and transported through the material However, their applications are rarely been used in membrane separation processes because of their low permeability as they have relatively thick membranes which limited their application for practical separation processes as compared to porous inorganic membranes (Ismail and David 2001; Baker 2004).

Microporous inorganic membranes made of glass, metal, alumina, zirconia, zeolite and carbon membranes are commercially used as porous inorganic membranes. Other inorganic materials, such as cordierite, silicon carbide, silicon nitride, titania, mullite, tin oxide and mica also have been used to produce porous inorganic membranes. These membranes vary greatly in pore size, support material and configuration. A microporous ceramic membrane system generally consists of a macroporous ceramic support, some ceramic intermediate layers, and eventually a highly selective top layer. The support provides mechanical strength to the system. The intermediate layers bridge the gap between the large pores of the support and the small pores of the top layer. The top layer has separating capacities (Baker 2004).

Although inorganic membranes are more expensive than polymeric membranes, they possess advantages of temperature and wear resistance, well-defined stable pore structure, chemically inertness, and better selectivity than the polymeric membranes. However, due to the lack of technology to form continuous and defect-free membranes, high cost of production and handling issues e.g., brittleness, the commercial applications of inorganic membranes are still limited. For example, according to (Vu 2001; Baker 2004), a zeolite membrane module would cost around US$ 3000/m^2 of active membrane area compared to US$ 20/m^2 for existing gas-separation polymeric hollow-fiber membrane modules (Vu 2001; Baker 2004).

Later development has shown that there is a need to find cheap precursors and secure mechanical strength in order to make them an economically better choice. The development following the pyrolyzed carbon membranes was actually one step back, now trying to combine the excellent separation properties of the inorganic membranes with the more robust polymers (Pabby 2008).

Comparison in terms of the advantages and limitations of inorganic membranes over polymeric membranes are given on Table 7 (Caro, Noack et al. 2000).

Advantages for inorganic membranes	Limitation for inorganic membranes
High stability at high temperatures	High capital costs
High resistance to harsh environments	Brittleness (membrane cracking) due to extremely high sensitivity of membranes to temperature gradient
High resistance to high pressure drops	Low membrane surface per module volume
High resistance to microbiological degradation	Challenging in achieving high selectivities in large scale microporous membranes
Ease of cleanability after fouling	Generally low permeability of the highly selective (dense) membranes at medium temperatures
Ease of catalytic activation	Challenging for proper sealing of membrane to module at high temperatures

Table 7. Advantages and limitations of inorganic membranes over polymeric membranes (Caro, Noack et al. 2000)

Thus, the trend showed that the other alternative is the development of mixed matrix composite membranes which is a cost-effective membrane that combines homogeneously interpenetrating polymeric and inorganic particle matrices for high permeability and selectivity well above the upper-bound limit (Sridhar, Smitha et al. 2007).

6.5 Mixed matrix membranes

As discussed in previous sections, because of the economic competitiveness and the present challenges of aggressive environments for gas separating membranes, it is desirable to synthesize more durable membrane materials having higher selectivity and permeability with low cost and these have been an important goals of recent researches (Koros and Mahajan 2001).

In view of this situation, it remains a need for a new approach called mixed matrix membranes (MMMs) which is considered to be the most practical approach with the potential for future applications. It is also highly desirable to provide an alternative cost-effective membrane which combines homogeneously interpenetrating polymeric matrices for ease of processibility and inorganic particle for high permeability and selectivity well above the upper-bound limit (Shekhawat, Luebke et al. 2003).

The molecular-sieve type fillers used in mixed matrix membranes (MMMs) such as zeolite and carbon molecular sieve (CMS) are capable to discriminate between different molecules present in the feed mixture, usually on the basis of size and shape of molecules (Pal 2007; Shimekit, Mukhtar et al. 2009).

Gas transport through a mixed matrix membrane is considered as complex phenomena. Due to its heterogeneity, a number of theoretical permeation models have been used to predict the permeation properties of mixed matrix (heterogeneous) membranes as a function of the permeabilities of the continuous and dispersed phases (Hashemifard, Ismail et al. 2010).

Generally, permeation in MMMs occurs by a combination of diffusion through the polymer phase and diffusion through the permeable zeolite particles. The relative permeation rates through the two phases are determined by their intrinsic permeabilities (Baker 2004). The available models used in the prediction of the permeability of gases in mixed matrix membrane can be referred from (Shimekit, Mukhtar et al. 2011).

(Baker 2004; Shimekit, Mukhtar et al. 2011) states that CO_2 separation from natural gas are at the developing stages. Currently, significant growth in these applications, driven by the development of better membranes and process designs is occurring. For example, UNOCAL corporation offshore platform membrane systems and others such as cellulose acetate membranes have been widely used for the removal of carbon dioxide from natural gas for more than two decades mainly at small gas processing plants (less than 5 MMSCFD)(Baker 2002; Baker 2004).

Based on the current trend, the market for the separation of acid gases by membranes can be divided as follows: (1) Very small systems (less than 5 MMSCFD). At such flow rate, membrane units are very attractive and usually, the permeate gas is used or flared as fuel. Moreover, the system is a simple bank of membrane modules. (2) Small systems (5-40 MMSCFD). Two-stage membrane systems are used to reduce methane loss. In this gas flow range, amine and membrane systems compete and the choice depends on site specific factors. (3) Medium to large systems (more than 40 MMSCFD). In this range, membrane systems are too expensive to compete with amine plants. However, a number of large membrane systems have been installed on offshore platforms, at carbon dioxide oilfield flood operations, or where site-specific factors particularly favor membranes (Baker 2002).

6.6 Limitation and challenges

The application of membranes today for CO_2 separation in natural gas processing is mainly used for moderate-volume gas streams. For large-volume gas streams, membrane separation today cannot yet compete with the standard amine absorption as the flux and selectivity of the membranes are too low for processing large gas volumes.

Membranes are used in situations where the produced gas contains high levels of CO_2. However, a key sensitivity with these current membranes is that they must be protected from the heavier C_{5+} hydrocarbons present in wet natural gas streams. Exposure to these compounds immediately degrades performance and can cause irreversible damage to the membranes.

Thus, in order to fully exploit the use of membranes in natural gas purification, development of more selective, higher-flux and cost effective new membranes are still critical concerns. And hence, the outcome will make membrane processes much more competitive with other technologies such as amine absorption for large scale systems (Baker 2004).

7. Comparison of natural gas purification technologies for acid gas removal

In the previous sections the widely used natural gas purification technologies for the removal of acid gases from product natural gas streams have been discussed in detail. In the following section, an attempt is made to compare and summarize the aforementioned technologies briefly.

Absorption by physical solvents is mostly not recommended at low partial pressures as the compression of the gas for physical absorption is relatively no economical. In general, the economics of CO_2 separation is strongly influenced by the partial pressure of CO_2 in the feed natural gas. However, if the gas is available at high pressure, physical solvents might be a better choice than chemical solvents. While physical solvents can often be stripped of impurities by reducing the pressure without the application of heat, regeneration of chemical solvents is achieved by the application of heat. Mostly, physical solvents tend to be favored over chemical solvents when the concentration of acid gases or other impurities is very high. Unlike chemical solvents, physical solvents are non-corrosive, requiring only carbon steel construction. The concentration of heavy hydrocarbons in the feed gas also affects the choice of gas treating solvent. If the concentration of heavy hydrocarbons is high, a physical solvent may not be the best option due to higher co-absorption of hydrocarbons, particularly pentanes plus. Unlike synthesis gases, where there aren't appreciable quantities of hydrocarbons, natural gases can be a problem for physical solvents as the result of hydrocarbon co-absorption. This makes physical solvents particularly applicable to synthesis gas treating (Burr and Lyddon 2008).

Although adsorption technique is restricted to small gas stream and moderate pressure due to complexity of the design, the use of PSA technology is mostly used in shut-in natural gas wells that usually contained too much N_2. As a typical example, titanosilcate adsorbent (Engelhard Corporation) combined with a PSA process in a vacuum swing adsorption are used to remove N_2 and/or CO_2 from natural gas feed streams (Ritter and Ebner 2007).

When the natural gas well contains high CO_2 and H_2S, the use membrane process is most preferable at high pressure. CO_2 separation is accomplished by pressure-driven mass transfer through a permeable membrane where separation is due to the differences in permeation rate of different gas penetrants. Although the acid gas is usually recovered at low pressure, high purity product containing approximately 95% CO_2 can be achieved with one or two stages, depending upon feed gas pressure and percent recovery. Economic considerations may dictate additional capital and incremental energy requirements to increase feed pressure and/or utilize two-stage separation with recompression of gas from the first stage.

Cryogenic process, as compared to other methods of separating CO_2, it has the advantage that the CO_2 can be obtained at relatively high pressure. However, this advantage may be offset by the large refrigeration requirement. In such regards, the need of special materials is critical for cryogenic process.

The overall comparison of the widely used natural gas purification technologies with regards to acid gas removal are summarized in Table 8.

Although the choice of acid gas removal technology depends on the needs of the gas processor, the current market trend showed that membranes have also proven their usefulness to compete with absorption (amines) technology. Nevertheless, the absorption technology based on amine treatment is still an efficient method although the amine units are large and heavy. Table 9 compares amines and membranes for CO_2 removal systems (Baker 2004).

Process	Advantages	Disadvantages
Absorption	• Widely used technology for efficient (50-100) % removal of acid gases (CO_2 and H_2S).	• Not economical as high partial pressure is needed while using physical solvents. • Long time requirement for purifying acid gas as low partial pressure is needed while using chemical solvents.
Adsorption	• High purity of products can be achieved. • Ease of adsorbent relocation to remote fields when equipment size becomes a concern.	• Recovery of products is lower • Relatively single pure product
Membrane	• Simplicity, versatility, low capital investment and operation. • Stability at high pressure • High recovery of products • Good weight and space efficiency • Less environmental impact.	• Recompression of permeate • Moderate purity
Cryogenic	• Relatively higher recovery compared to other process • Relatively high purity products	• Highly energy intensive for regeneration • Not economical to scale down to very small size. • Unease of operation under different feed stream as it consists of highly integrated, enclosed system.

Table 8. Overall comparisons of natural gas purification technologies

Operating Issues	Amines	Membranes
User Comfort Level	Very familiar	Still considered new technology
Hydrocarbon Losses	Very low	Losses depend upon conditions
Meets Low CO_2 Spec.	Yes (ppm levels)	No (<2% economics are challenging)
Meets Low H_2S Spec.	Yes (<4 ppm)	Sometimes
Energy Consumption	Moderate to high	Low, unless compression used
Operating Cost	Moderate	Low to moderate
Maintenance Cost	Low to moderate	Low, unless compression used
Ease of Operation	Relatively complex	Relatively simple
Environmental Impact	Moderate	Low
Dehydration	Product gas saturated	Product gas dehydrated
Capital Cost Issues		
Delivery Time	Long for large systems	Modular construction is faster
On-Site Installation Time	Long	Short for skid-mounted equipment
Pretreatment Costs	Low	Low to moderate
Recycle Compression	Not used	Use depends upon conditions

Table 9. Comparison of Amines and Membranes for CO_2 Removal Systems (Baker 2004)

In the next sections, potential hybrid separation processes in natural gas processes are briefly presented.

7.1 Hybrid separation processes

In hybrid separation processes, an integration of one process is used with other separation process or processes in which the basic functioning of one process is joined with another physical or chemical process in a single unit operation (Bernardo, Drioli et al. 2009).

A properly designed hybrid process will balance the drawbacks of the specific process and favourably combine their advantages. It is anticipated that the result will be a better separation, contributing to a sustainable process improvement by allowing the reduction of investment and operational costs (Bernardo, Drioli et al. 2009).

As has been described in the previous sections and in spite of the greater potential of membranes for commercial application towards bulk removal of CO_2 in high-pressure natural gas streams with high CO_2 content, the technology faces some limitations as compared with other techniques especially when the CO_2 concentration is low. And it may be necessary to use another process (e.g., amine based or cryogenic system) to create a hybrid system (Tabe-Mohammadi 1999). Hence, this motivates researchers to look for an alternative technologies aimed at filling its technology gaps through the use of hybrid separation process.

Some researchers (Bhide, Voskericyan et al. 1998) also report that a membrane-amine hybrid system would be economical in comparison to an amine system alone or a membrane system alone for various conditions, such as low CO_2 compositions, and a recent study confirms the economic feasibility of another membrane-amine system design (Falk-Pedersen and Dannstrom 1997).

A study with a membrane-cryogenic distillation hybrid system also shows favourable economics as well (Vu 2001).

The integration of membranes with other separation processes such as PSA are also well-established in the chemical and petrochemical industries (Choe, Auvil et al. 1987; Doshi 1987; Doshi, Werner et al. 1989; Doshi and Dolan 1995; Feng, Pan et al. 1998). Membrane permeation can be an effective aid in the pressurization and high-pressure adsorption steps of a typical PSA process; the pressure difference available from the PSA can be used for operating the membrane incorporated into the blowdown step of the PSA cycle (Esteves and Mota 2007). Usually, the integration of membranes with PSA is considered in H_2 separation, while hybrid membranes + amine absorption are applied to the CO_2 separation. Cost comparison for membrane and absorption (diethanolamine) process showed that the membrane process is more economical for CO_2 feed concentrations in the range 5-40 mol % (Bhide, Voskericyan et al. 1998). It does ascertain that when the feed contains other gases such as H_2S, the operating cost for removing the additional gases also increases. In such cases, use of hybrid membrane processes (membranes for bulk separation of CO_2 and H_2S and gas-absorption processes for purification) is economically feasible (Bhide, Voskericyan et al. 1998). A hybrid system comprising of Cynara membranes and amine absorption is operating since 1994 in Mallet (Texas, U.S.A.) to perform the bulk removal of CO_2 from associated gas (90% CO_2 and heavy hydrocarbons), before downstream treating. The membrane system offered a 30% reduction in operating cost when compared with a methyl diethanolamine (MDEA) system and significantly reduced the size of the subsequent operations (Blizzard, Parro et al. 2005; Bernardo, Drioli et al. 2009).

The combination of membrane processes with another process (most commonly with amine units) for bulk removal of CO_2 offers an economical advantage over amine technologies. However, this process has also the limitation of being complex as it is a combination of two processes (Baker 2004).

8. Conclusions and future directions

In general, this chapter attempted to address a comprehensive assessment on the fundamental technologies that are widely used for natural gas purification (gas sweetening) processes namely: absorption, adsorption, membranes and cryogenics. More specifically, the removal of acid gases mainly of carbon dioxide (CO_2) from the natural gas has been given special emphasis aimed at increasing the heating value of natural gas, reducing pipe lines and equipment corrosion during transportation, storage and distribution and meeting environmental requirements. A comparative study on the aforementioned technologies was also made on the basis of their advantages and limitations. Furthermore, the economical competitive advantages of using hybrid separation processes over single gas separation processes havebeen discussed.

In the next section, potential research directions and future development opportunities of the promising natural gas purification processes are presented.

8.1 Research and development opportunities for natural gas separation

Research opportunities to improve and implement acid gas removal technologies in natural gas purification field have many drivers, including economical reasons, growing interest among policy makers and regulators, and concerns about GHG emissions.

Understanding of the current state of acid gas removal technologies, the direction of the researches and developments being undertaken and their likely outcomes is the key insight that energy company managers, engineers and planners are required to make informed strategic decisions. Thus, the following section describes the potential research and development activities with regards to the major acid gases separation process mainly on CO_2 separation using natural gas purification technologies and in accordance with what is set by the Chemical Industry Vision 2020 Technology Partnership (Ritter and Ebner 2007).

Absorbents and absorption process development

Although there are quite numerous absorbents available for CO_2 removal process in natural gas processing plant, in the near-future, development of new absorbents and absorption process will continue to be carried out as the need for enhanced capacity and improved heat of absorption will still be the interest of the processing industries for CO_2 removal. Moreover, researches and developments on new regeneration techniques for CO_2 removal from natural gas, as opposed to thermal or vacuum regeneration, need also be conducted at high pressure. Further, there is also a need to research and develop new absorbents that function at elevated temperature gases. More importantly, research and development on absorbents that are stable to contaminants such as amine environments need to be carried out. Moreover, preferential selective absorbents such as preferential absorption of CO_2 over other impurities such as sulfur in the acid gas removal process from natural gas or coal bed methane are some of the research directions with regards to absorption technologies (Ritter and Ebner 2007).

Adsorbents and adsorption process development

As a short-term adsorbent development research direction for CO_2 removal from natural gas, the focus of the current research interest are mostly on the development of high-capacity (~ 3-4 mol/kg) CO_2-selective adsorbents that can function at elevated pressures and operating temperatures. There are also related avenue of research that attempted to comp up with new designs for PSA cycle that have potentials to use the newly developed or the available CO_2 selective adsorbents (temperature swing adsorption or integrated PSA-TSA cycles). For example, development of new PSA cycles that used the idea of heavy reflux to show that CO_2 as a heavy product is more important than that of H_2 (light product). The issue of thermal energy management while designing the TSA and its hybrid cycle with PSA is also categorized in this research area to improve the overall efficiency of the removal of CO_2 from natural gas processing plant.

In the long-term, the research direction for the removal of CO_2 from natural gas using adsorption process will focus on fabrication of new, structurally modified adsorbents that can be used for the integration of gas-liquid absorption with PSA that is commonly known as rapid cycle PSA and improved design of the CO_2 removal process. Moreover, detail studies on how to reduce the cycling time in the rapid PSA process need to be conducted. Further studies on the adsorbent particle size, surface adsorption properties, and mass transfer limitations for enhanced removal of CO_2 from natural gas need to be addressed.

Once the novel adsorbent materials with enhanced adsorption capacity have been fabricated, the next studies will direct on development of temperature swing adsorption and/or integration of gas-liquid absorption with pressure swing adsorption using recently improved materials. Afterwards, further studies to quantify their effect on cycling time and bed sizes for a stressed TSA/PSA cycle will follow. The development of integration of gas-liquid absorption with pressure swing adsorption, studies for the improvement of CO_2 removal using the reaction processes that can improve the sorption for natural gas purification need equally be searched (Ritter and Ebner 2007).

Membranes and membrane process

In the near-term the research and development directions for the removal of CO_2 using membrane technologies will advance in areas such as: development of CO_2 permselective polymeric glassy or rubbery membranes that has a selectivity of CO_2 over H_2 with the range of 16-25, with high flux and high temperature stability, development of polymeric glassy or rubbery membranes for CO_2/CH_4 that can have selectivity of more than 50, high resistance for plasticization and stability towards heavy oil. More research is also need to be carried out on the development of facilitated transport membranes for enhanced CO_2 selectivity with higher resistance for adverse environments and service life. Although economical challenges are still there for the development of CO_2 permselective inorganic membranes, new inorganic membranes that can achieve higher selectivity (> 15-20%) under harsh environments need to be developed. Furthermore, enhanced permselective CO_2 surface flow membranes with higher CO_2 selectivity over H_2, N_2, and CH_4 need to be researched, development of CO_2 selective mixed-matrix membranes with higher selectivity, permeance and stability, development of CO_2 permselective hollow fiber membrane contactors with improved permeance and strength for natural gas purification are some of the typical future research directions.

In the near future, new membrane materials for CO_2 removal from natural gas will need to be given more focus so that membranes can be fabricated to have enhanced selectivity of CO_2 (> 100%), operating at high temperatures and pressures and with resistance to fouling and cracking or embrittlement.

Further in the long-term, the intensification of processes such as hybrid technologies using highly selective CO_2 adsorbent and membrane with high permeability for H_2 processes for the development of H_2 production (e.g., for steam methane reforming) will be the interest of researchers. Moreover, the extension of the results from the newly developed CO_2 adsorbent and membrane technologies will further be researched for their compatibility to be integrated into gasification combined cycle (IGCC) and other related technologies that targets for CO_2 sequestration as a long term prospect (Ritter and Ebner 2007).

In general, the potential of the hybrid processes over single processes for the removal of CO_2 from the natural gas in industrial scale need to be fully exploited through detail understanding of design methodologies and process flows. In this regards, modeling of the newly developed processes, new models for accurate prediction of permeation of gases in the newly developed processes such as (mixed matrix membranes and hybrid systems) along with molecular dynamics studies and module performance models can be recommended for further research direction.

9. Acknowledgments

The authors would like to thank Universiti Teknologi PETRONAS and the financial support provided by Ministry of Higher Education, Malaysia under Exploratory Research Grant Scheme grant number 15-8200-133.

10. References

Al-Juaied, M. A. (May 2004). Carbon dioxide removal from natural gas by membranes in the presence of heavy hydrocarbons and by Aqueous Diglycolamine®/Morpholine, PhD thesis, pp. 1-424, The University of Texas at Austin, Texas

Bahadori, A., S. Mokhatab, et al. (2007). Rapidly estimating natural gas compressibility factor. *Journal of Natural Gas Chemistry,* Vol. 16, No. 4, pp. 349-353

Bakar, W. A. W. A. and R. Ali. (2010). Natural Gas, In: *natural gas,* Primož Potočnik, pp. (1-616), Sciyo, Retrieved from http://www.intechopen.com/articles/show/title/natural-gas.

Baker, R. W. (2002). Future directions of membrane gas separation technology. *Industrial & Engineering Chemistry Research,* Vol. 41, No. 6, pp. 1393-1411

Baker, R. W. (2004). *Membrane Technology and Applications* (2nd edition), John Wiley & Sons, ISBN 0-470-85445-6, West Sussex

Beggs, H. D. (1984). *Gas Production Operations*, OGCI Publications, ISBN 0-930972-06-6, Oklahoma

Bernardo, P., E. Drioli, et al. (2009). Membrane gas separation: a review/state of the art. *Industrial & Engineering Chemistry Research*, Vol. 48, No. 10, pp. 4638-4663

Bhide, B. D., A. Voskericyan, et al. (1998). Hybrid processes for the removal of acid gases from natural gas. *Journal of Membrane Science*, Vol. 140, No. 1, pp. 27-49

Blizzard, G., D. Parro, et al. (2005). Mallet gas processing facility uses membranes to efficiently separate CO_2. *Oil and Gas Journal*, Vol. 103, No. 14, pp. 48-53

Bord, N., G. Cretier, et al. (2004). Determination of diethanolamine or N-methyldiethanolamine in high ammonium concentration matrices by capillary electrophoresis with indirect UV detection: application to the analysis of refinery process waters. *Analytical and Bioanalytical Chemistry*, Vol. 380, No. 2, pp. 325-332

Burggraaf, A. J. (1996). Important characteristics of inorganic membranes. *Membrane Science and Technology*, Vol. 4, pp. 21-34

Burr, B. and L. Lyddon (2008). A Comparison of Physical Solvents for Acid Gas Removal, In: *options for acid gas removal*, 11.08.11, Available from: http://www.bre.com/portals/0/technicalarticles/A%20Comparison%20of%20Physical%20Solvents%20for%20Acid%20Gas%20Removal%20REVISED.pdf

Caro, J., M. Noack, et al. (2000). Zeolite membranes-state of their development and perspective. *Microporous and Mesoporous Materials*, Vol. 38, No. 1, pp. 3-24

Cavenati, S., A. Carlos, et al. (2006). Removal of carbon dioxide from natural gas by vacuum pressure swing adsorption. *Energy & fuels*, Vol. 20, No. 6, pp. 2648-2659

Choe, J. S., S. R. Auvil, et al. (1987). Process for separating components of a gas stream, Google Patents 4701187

Davison, J., P. Freund, et al. (February 2001). Putting carbon back into the ground (1st edition), IEA Greenhouse Gas R&D Programme, ISBN 1 898373 28, Cheltenham

De Lange, R. S. A., K. Keizer, et al. (1995). Analysis and theory of gas transport in microporous sol-gel derived ceramic membranes. *Journal of Membrane Science, Vol.* 104, No. No. (1-2), pp. 81-100

Dortmundt, D. and K. Doshi (1999). Recent Developments in CO_2 Removal Membrane Technology, In: *design consideration*, 16.08.11, Available from: http://www.membrane-guide.com/download/CO2-removal-membranes.pdf

Doshi, K. J. (1987). Enhanced gas separation process, Google Patents, 4690695

Doshi, K. J. and W. B. Dolan (1995). Process for the rejection of CO_2 from natural gas, Google Patents, 5411721

Doshi, K. J., R. G. Werner, et al. (1989). Integrated membrane/PSA process and system, Google Patents, 4863492

Ebenezer, S. A. and J. S. Gudmunsson (December 2006). Removal of Carbon Dioxide from Natural Gas for LPG Production, In: *carbon dioxide removal processes*, 21.08.11, Available from: http://www.ipt.ntnu.no/~jsg/studenter/prosjekt/Salako2005.pdf

Economides, M. J. and D. A. Wood (2009). The state of natural gas. *Journal of Natural Gas Science and Engineering*, Vol. 1, No. 1-2, pp. 1-13

Energy Information, A. and U. S. E. I. Administration (2011). International Energy Outlook 2011, In: *Natural gas*, 22.08.11, Available from: http://205.254.135.7/forecasts/ieo/pdf/0484(2011).pdf

Esteves, I. A. A. C. and J. P. B. Mota (2007). Gas separation by a novel hybrid membrane/pressure swing adsorption process. *Industrial & Engineering Chemistry Research,* Vol. 46, No. 17, pp. 5723-5733

Falk-Pedersen, O. and H. Dannstrom (1997). Separation of carbon dioxide from offshore gas turbine exhaust. *Energy Conversion and Management,* Vol. 38, No. S81-S86

Feng, X., C. Y. Pan, et al. (1998). Integrated membrane/adsorption process for gas separation. *Chemical Engineering Science,* Vol. 53, No. 9, pp. 1689-1698

Gilliland, E. R., R. F. Baddour, et al. (1974). Diffusion on surfaces. I. Effect of concentration on the diffusivity of physically adsorbed gases. *Industrial & Engineering Chemistry Fundamentals,* Vol. 13, No. 2. pp. 95-100

Glasscock, D. A. (1990). Modeling and experimental study of carbon dioxide absorption into aqueous alkanolamines, PhD thesis, pp. (1-292), The University of Texas at Austin, Texas

Hashemifard, S. A., A. F. Ismail, et al. (2010). Prediction of gas permeability in mixed matrix membranes using theoretical models. *Journal of Membrane Science,* Vol. 347, No. 1-2, pp. 53-61

Hermann, W., P. Bosshar, et al. (2005). An Assessment of Carbon Capture Technology and Research Opportunities, In: *Physical Adsorption,* 25.08.11, Available from: http://gcep.stanford.edu/pdfs/assessments/carbon_capture_assessment.pdf

Ismail, A. F. and L. I. B. David (2001). A review on the latest development of carbon membranes for gas separation. *Journal of Membrane Science,* Vol. 193, No. 1, pp. 1-18

Javaid, A. (2005). Membranes for solubility-based gas separation applications. *Chemical Engineering Journal,* Vol. 112, No. 1-3, pp. 219-226

Kerry, F. G. (2007). *Industrial Gas Handbook: Gas Separation and Purification,* CRC, ISBN 978-0-8493-9005-0, New York

Kesting, R. E. and A. K. Fritzsche (1993). *Polymeric Gas Separation Membranes,* Wiley, ISBN 978-0-4715-6931-2, New York

Kidnay, A. J., W. Parrish, et al. (2006). *Fundamentals of Natural Gas Processing,* CRC, ISBN 978-0-8493-3406-1, New York

Kohl, A. L. and R. B. Nielsen (1997). Gas Purification (5th edition), Gulf Professional Publishing, ISBN 978-0-8841-5220-0, Texas

Koros, W. J. and R. Mahajan (2001). Pushing the limits on possibilities for large scale gas separation: which strategies? *Journal of Membrane Science,* Vol. 181, No. 1, pp. 141-141

Kovvali, A. S. and K. K. Sirkar (2002). Carbon dioxide separation with novel solvents as liquid membranes. *Industrial & Engineering Chemistry Research,* Vol. 41, No. 9, pp. 2287-2295

Mersmann, A., M. Kind, et al. (2011). *Thermal Separation Technology Principles, Methods, Process Design,* Springer, ISBN 978-3-642-12524-9, New York

Meyers, R. A. (2001). *Chemical Engineering. Encyclopedia of Physical Science and Technology* (3rd edition), Ramtech, Inc. California

NaturalGas.org. (2010). Overview of Natural Gas, In: *Background,* 22.06.11, Available from:http://www.naturalgas.org/overview/background.asp

Pabby, A. K. (2008). *Handbook of Membrane Separations: Chemical, Pharmaceutical, Food, and Biotechnological Applications* (1st edition), CRC, ISBN 978-0-849-39549-9, New York

Pal, R. (2007). New models for thermal conductivity of particulate composites. *Journal of Reinforced Plastics and Composites*, Vol. 26, No. 7, pp. 643

Pandey, P. and R. S. Chauhan (2001). Membranes for gas separation. *Progress in Polymer Science*, Vol. 26, No. 6, pp. 853-893

Pascoli, S. D., A. Femia, et al. (2001). Natural gas, cars and the environment. A (relatively) 'clean' and cheap fuel looking for users. *Ecological Economics*, Vol. 38, No. 2, pp. 179-189

Polasek, J. and J. Bullin (1984). Selecting amines for sweetening units. *Energy Progress*, Vol. 4, No. 3, pp. 146-149

Porter, M. C. (1990). *Handbook of Industrial Membrane Technology*, Noyes Publications, ISBN 08155-1205-8, New Jersey

Ritter, J. A. and A. D. Ebner (2007). Carbon Dioxide Separation Technology: R&D Needs For the Chemical and Petrochemical Industries, In: *Recommendation for future R&D*, 22.06.11, Available from:
http://www.chemicalvision2020.org/pdfs/CO2_Separation_Report_V2020_final.pdf

Robeson, L. M. (1991). Correlation of separation factor versus permeability for polymeric membranes. *Journal of Membrane Science*, Vol. 62, No. 2, pp.165-185

Rodriguez, H., L. Mello, et al. (2011). Absorption of Carbon Dioxide into Aqueous Solutions of Alkanolamines in a Wetted Wall Column with Film Promoter. *Chemical Engineering Transactions*, Vol. 25, No, 51, pp. 51-56

Rojey, A. (1997). *Natural Gas: Production, Processing, Transport*, Editions Technip, ISBN 2-7108-0693-2, Paris

Scott, K. (1998). *Handbook of Industrial Membranes*, Elsevier science publishers, ISBN 1-8561-7233-3, Oxford

Shekhawat, D., D. R. Luebke, et al. (2003). A Review of Carbon Dioxide Selective Membranes: A Topical Report, In: *CO_2 selective membranes*, 13.07.11, Available from: http://www.osti.gov/bridge/product.biblio.jsp?osti_id=819990

Shimekit, B., H. Mukhtar, et al. (2009). Ceramic membranes for the separation of carbon dioxide: A review. *Transaction of Indian Ceramic Society*, Vol. 68, No. 3, pp. 115-138

Shimekit, B., H. Mukhtar, et al. (2011). Prediction of the relative permeability of gases in mixed matrix membranes. *Journal of Membrane Science*, Vol. 373, No. 1-2, pp.152-159

Sridhar, S., B. Smitha, et al. (2007). Separation of carbon dioxide from natural gas mixtures through polymeric membranes-a review. *Separation & Purification Reviews*, Vol. 36, No. 2, pp, 113-174

Stern, A. (1994). Polymers for gas separations: the next decade. *Journal of Membrane Science*, Vol. 94, No. 1, pp. 1-65

Tabe-Mohammadi, A. (1999). A review of the applications of membrane separation technology in natural gas treatment. *Separation Science and Technology*, Vol. 34, No. 10, pp, 2095-2111

Tobin J., Shambaugh P., et al. (2006). The Crucial Link between Natural Gas Production and its Transportation to Market In: *Stages in the production of pipeline-quality natural gas and NGLs*. 13.07.11, Available from: http://lba.legis.state.ak.us/sga/doc_log/2006-01-01_eia_publication_natural_gas_processing.pdf

Vu, D. Q. (2001). Formation and Characterization of Asymmetric Carbon Molecular Sieve and Mixed Matrix Membranes for Natural Gas Purification, PhD thesis, pp. 1-362, The University of Texas at Austin, Texas

Xiao, Y., B. T. Low, et al. (2009). The strategies of molecular architecture and modification of polyimide-based membranes for CO_2 removal from natural gas-A review. *Progress in Polymer Science*, Vol. 34, No. 6, pp. 561-580

Yang, R. (1997). *Gas Separation by Adsorption Processes,* Imperial College Press, ISBN 9781860940477, Singapore

Yang, R. T. and J. Wiley (2003). *Adsorbents: fundamentals and applications*, John Wiley & Sons, ISBN 9780471297413, New Jersey

10

An Approach Integrating Chemistry and Toxicity for Monitoring the Offshore Platform Impacts

L. Manfra and C. Maggi
ISPRA Italian Institute for Environmental Protection and Research, Rome, Italy

1. Introduction

The world natural gas production grew substantially in 2010, by more than 6% (3275 billion cubic meters), the highest rise among fossil fuels and the strongest increase since 1994. (Enerdata, 2011). The gas production in European countries grew 3% in 2010 (314 billion cubic meters, with 8 billion cubic meters produced in Italy) (Assomineraria, 2011; Enerdata, 2011). More than 100 offshore gas platforms have been deployed since the 1960s in the northern and central parts of the Adriatic continental shelf, representing the highest concentration of fossil fuel extraction platforms in the Mediterranean area (Manoukian et al., 2010).

The offshore platform activities comprise different phases linked to exploitation of gas and oil reservoirs: a) the exploration phase to probe the position and the geological characteristics of well and then to install a steel platform; b) the production phase to extract oil and gas; c) the decommissioning phase, when the commercial life of well is finished (Oil Industry International Exploration & Production Forum/United Nations Environment Programme [E&P Forum/UNEP], 1997).

The environmental impacts linked to exploration and production phases are more remarkable than those related to decommissioning phase as often, when the platform life is ended, the submerged part of structure (the jacket) is left on the seabed avoiding the removal disturbs further the seafloor integrity.

Criteria for environmental monitoring in the vicinity of offshore oil and gas installations are according to international programmes (e.g. OSPAR Convention). The guideline goals are: a) to monitor selected chemical and physical variables of water column and sediment; b) to evaluate effects on biota; c) to identify spatial-temporal trend of the eventual alteration that could occurred. In order to give the best description of the environmental quality status an integrated approach may be applied by means physical analyses, chemical analyses, ecotoxicological assays, bioaccumulation studies, ecological investigations (macrozoobenthic community and fish assemblage studies) and acoustic investigations. The monitoring may be devised as flexible tool allowing modifications when the activities are in progress, if results of previous year activities request it (Trabucco et al., n.d.).

In this paper we would describe a key aspect of monitoring that allows to evaluate the environmental health status, potential load of pollution on environmental matrices and effects on biota due to offshore platform presence: integrated chemical and ecotoxicological approach.

The traditional chemical analyses allow to evaluate the behaviour of some environmental parameters (e.g. dissolved oxygen, chlorophyll, nutrients) and contaminants when a platform is installed and/or in production phase. Chemistry gives information on the presence, quantity and chemical form of substances linked to platform activities.

The integration of biological investigations allows an assessment of toxicity and bioavailability of contaminants, to understand the mechanisms of their toxic action and identification of the area of potential biological impact of platforms and/or their discharges.

The purposes of this paper are to a) provide an overview the main environmental impacts of installation and production activities, b) explain the importance of an integrated chemical and ecotoxicological approach for environmental monitoring plans, in particular for offshore platform monitoring, c) present the results collected, for many years, around several gas platforms located in Adriatic area (Mediterranean Sea).

2. Main environmental impacts of platforms and major production discharge

Environmental impacts may arise at all stages of oil and gas activities, including initial exploration, production and final decommissioning. The main pressures on the marine environment from oil and gas activities include the placement of installations on the seabed and operational and accidental discharges. In addition there could be concerns related to atmospheric emissions, low level naturally occurring radioactive material and noise (Commission protecting and conserving the North-East Atlantic [OSPAR], 2009). Finally, offshore platforms undergo subsidence, especially due to production activities (Setan & Othman, 2006).

The quantification and interpretation of the environmental effects of offshore platforms can be difficult and hard to generalize. Although the physical presence of the structure is a common characteristic to all the platforms, several local abiotic and biotic variables influence the effects of each platform in a unique and different way. The radius of impact can be different depending on the number of platforms placed in a restricted area, the structures' dimensions and biogeographic/climatic factors (Manoukian et al., 2010).

2.1 Impacts related to platform presence on the bottom

The placing a structure on the seabed may cause a physical impact on the seabed, that varies on a case by case basis depending on the particular sensitivities associated in the area. Physical impacts on the sea bottom will occur in connection with installing pipelines, cables, bottom rigs, templates, skids, and platforms including platform legs and anchoring. Due to the number and length of pipelines placed on or under the seabed the overall physical impact of pipelines is greater than those from the other installations or operations mentioned above (OSPAR, 2009).

First, the physical structure of the platform may determine some alterations in the environment: changes in local water flow, in the rate of erosion or sedimentation, modification in bottom morphology, variation in grain size and organic content in sediments, and then in the number of predators present.

Second, offshore platforms may cause disturbance of the seabed and mobilisation of sediment. The volume and distance that suspended sediments disperse depends on particle size, weight and current velocity. However, it is possible that the positioning of the structure results in a temporary increase in turbidity due to the material raised from the bottom. Third, the use of antifouling paints to combat bio-fouling and the use of corrosion protection systems (anodes) containing metals (e.g. Zn, Al and Cd) can lead to contamination phenomena in water, sediment and biota. Although there is a ban on using antifouling paints containing organotin compounds, their persistence is such that even today these compounds are found in sediments and tissues of organisms living in the vicinity of the platforms.

Fourth, physical structure may cause biotic modifications, affecting the benthic community composition and the responses of organisms exposed to eventual contaminations linked to installation phase. The scale of the impact also depends on the vulnerability of benthic communities in question, as well as the distance from the platform itself and the time elapsed since its installation (Trabucco et al., 2006). Historical monitoring has demonstrated that the impacts are largely transient, with re-colonisation of disturbed seabed habitats occurring within relatively short timescales. The creation of hard bottom substrate can, over time, give an opportunity for new benthic species to colonise the former sandy/mudflat areas. Pipelines, platform legs and subsea templates may act as shelter for fish and other mobile marine organisms, and provide a habitat for benthic organisms usually associated with hard substrates (OSPAR, 2009). Some studies highlighted the effects of platforms on benthic communities over the time. In most of these cases, the more relevant ecological alteration due to the presence of a gas platform was a mussel mound development affecting the surrounding seabed communities within a radius of a few meters (Manoukian et al., 2010 and references therein).

2.2 Impacts related to main waste fluid discharged from platforms

In the offshore oil and gas industry, produced water constitutes the main source of oil input in marine water basin. It is the principal aqueous waste resulting from production operations. It originates from water naturally present in geological formations (formation water) and water injected in the oil field (process water) to maintain reservoir pressure. Its disposal is carried out through the reinjection into the reservoir, the discharge into the ocean or the transport onshore. After the release into the sea, the produced water undergoes several different processes such as the dilution into the ambient fluid (which may occur more or less rapidly depending on local oceanographic conditions), the evaporation and biodegradation (that change the concentration and composition), the volatilization towards the atmosphere and the settling at the bottom (Stromgren et al., 1995 and references therein).

The environmental impact of this effluent is a function of (a) the loading and type of the contaminants present in produced water, (b) the partitioning of the chemical constituents in the environment, and (c) the nature of the receiving environment and its dispersion characteristics (Rabalais et al. 1992; E&P Forum/UNEP, 1997). Generally, the presence and, consequently, the effects of produced water substances on biota are dependent on a number of factors: bioavailability, bioaccumulation, biomagnification, toxicity and the capability of the organism to metabolise the substance. The bioavailability of the substance is dependent on the chemical form (speciation) in which it occurs. Rarely contaminants may occur in the water column or in solution. Many of the contaminants of concern in the marine environment have low water solubility and a high affinity for particles. In fact, some substances tend to adsorb on suspended solids both within produced water and in the water column; in this way, they can reach the seafloor and once there they may become incorporated in seabed sediments and become potentially available to filter feeders. Contaminants can be taken up by organisms either directly, by absorption from sea water, or by ingestion of particles and can be relayed to successively higher levels in the food chain via grazing and predation (OSPAR, 2009).

So on, the characteristics of some typical constituents of produced water (e.g. organic compounds, inorganic compounds and chemicals) and then their fate and effects are briefly described. Produced water contains organic compounds such as dispersed oil (in droplet form), dissolved oil, organic acids and phenols. In particular, volatile aromatic compounds as Benzene, Toluene, Ethylbenzene, Xylenes (BTEX) and low molecular weight aromatic hydrocarbons as Naphthalene, Phenanthrene and Dibenzothiophene and their alkyl homologues (NDP) are mainly in dissolved form, while high molecular weight Polycyclic Aromatic Hydrocarbons (PAHs) are present as dispersed oil. These compounds may be removed from seawater by means of volatilisation, adsorption to particles, sedimentation, biodegradation and photolysis. Many organic compounds interact whit suspended material and the particulate matter could affect their mobility. Exposure to marine organisms is low and may cause narcosis, alterations of permeability of cell membranes and developmental defects. PAHs are relatively insoluble and their potential for bioaccumulation increases with increasing molecular weight. They may be toxic in different way as narcosis, phototoxicity, biochemical activation, etc. Phenols are soluble in marine waters, highly volatile and degradable; they appear not particularly toxic to aquatic organisms (Cianelli et al., 2011 and references therein). Produced water also includes inorganic compounds as chlorinates, carbonates, sulphurs, sodium, potassium, calcium, ammonium and trace metals (copper, nickel, iron, chromium, manganese, lead, zinc, barium, arsenic, mercury and cadmium). Many important trace elements are transported through aquatic system as colloidal chemical compounds or are sorbed to solid particles. The chemical form of the metals depends upon pE: ions held by cation-exchanging clays, bound to hydrated oxides of iron and manganese or chelated by insoluble humic substances (Mansour, 1993). The most common trace elements in produced water are barium, iron, manganese, zinc and lead; co-precipitation reactions and adsorpion process favour partition of these metals on solid phases (Scott et al., 2007). Although less available than ones in solution, metals absorbed on small particles are more accessible than those in sediments. Among the factors involved in metal availability are the identity of the metal, its chemical form (type of binding, oxidation state), the nature of suspended matter, the type of organism and the physic-chemical condition of receiving water body. Some of heavy metals are essential at low levels but toxic at higher levels; lead, cadmium and mercury have notable toxicological and environmental significance. The most of heavy metals are particularly toxic in their chemically combined forms, the others, notably mercury, in the elemental forms (Sittig, 1991). Cadmium adversely effects several important enzymes; lead has a number of toxic effects, including inhibition of the synthesis of haemoglobin; arsenic is a metalloid which forms several toxic compounds; elemental mercury penetrates the blood-brain barrier, divalent ionic mercury damages the kidney and organometallic mercury compounds, as metilmercury, are very toxic (Maggi et al., 2009). The toxicity of heavy metals in suspended matter and their availability to organisms are very important in determining the environmental effects in aquatic system.

The reservoir fluids (oil, gas and water) are separated on the platform and chemicals may be added at various stages during this processing to aid oil-water separation and mitigate operational problems. Chemicals may include corrosion inhibitors, demulsiners, defoamers and biocides. Some of these production chemicals are appreciably toxic in standard toxicity tests, although the environmental fate and effects associated with their use will depend on a number of factors. These include: a) the fraction of the chemical released with the discharged water (as opposed to being entrained with the oil), which is determined largely by the oil/water partitioning characteristics; b) the mode of use (continuous low dosing or higher dose batch treatment); c) the chemical injection point within the process system (Henderson et al., 1999 and references therein).

3. Environmental monitoring plans and importance of an integrated chemical and ecotoxicological approach

In general, environmental monitoring programs collect data for one or more of the following purposes:

1. to establish a baseline; that is, gathering information on the basic site characteristics prior to development or to establish current conditions;
2. to establish long term trends in perturbed or unperturbed systems;
3. to estimate inherent variation within the environment, which can be compared with the variation observed in another specific area;
4. to make comparisons between different situations (for example, pre-development and post development; upstream and downstream; at different distances from a source) to detect changes; and
5. to make comparisons against a standard or target level.

What is the role of chemical analysis in evaluating environmental quality? And what the ecotoxicology? Why an integrated approach of these two disciplines is important?

3.1 Chemical analysis and bioassay in environmental quality assessment

The knowledge of chemical analysis today has become important not only for scientists in their research but in fact bears influence in our daily routine as well. Chemical analysis is a vital first step in environmental research. It is a body of procedures and techniques used to identify and quantify the chemical composition of a sample. Qualitative chemical analysis is used to identify a particular element, compound or substance present in a sample while quantitative analysis consists of determining the concentration of these substances. The primary advantage of measuring chemical concentrations in environmental media such as water, soil and biota is that they provide evidence of past or present exposure to environmental pollutants and of speciation. However chemical criteria usually do not include multiple chemical exposures and, for this reason, should be associated to biological investigations.

Biological criteria, on the other hand, are not only reflective of chemical exposure, but have the capacity to integrate many of the physical, chemical, and biological stressors that operate in ecosystems. In addition, many biological criteria are capable of integrating the effects of stressors on organisms both spatially and temporally, and are thus more suited for measuring and interpreting the possible effects of multiple stressors on both terrestrial and aquatic ecosystems (Adams & Tremblay, 2003). Biological tools for environmental assessment may be classified into three groups: 1) exposure and effect biomarkers that measure biological responses in local populations of non-migrating animals, 2) ecotoxicological bioassays that test the toxicity of environmental samples with standard biological models, 3) biological indices that measure diversity and related traits at community level. These three kinds of biological tools show, on one hand, increasing ecological relevance but on the other, decreasing early-warning value. In particular, community indices are not preventive, since effects may only be visible late in time, or may be impractical due to natural variability and cost-effectiveness limitations. Therefore, alternative rapid and sensitive biological tools, such as the effect biomarkers and bioassays developed in the last few years by the International Council for the Exploration of the sea and the Convention for the protection of the marine environment of the Northeast Atlantic, should be implemented (Beiras et al., Personal Communication).

Alterations at the molecular and cellular levels (biomarkers) can provide a sensitive indication of early changes, which often represent the first warning signals of environmental disturbance, even in the absence of acutely toxic responses. A multi-biomarker approach can be used in environmental prognosis to predict long-term toxicological effects or changes in the higher levels of biological organization.

Laboratory bioassays are a common procedure to evaluate toxicological endpoints at organism level, in a large number of test species from across several taxa, and across the main ecological or trophic positions (i.e. from bacteria to fish, and from decomposers to final consumers). (Piva et al., 2011 and references therein). It is very important give a toxicity judgment considering the response of more species. In fact, the differences in intrinsic properties of species determine their sensitivity to various chemicals. If this shape differs for different chemicals and the sensitivity of the various species to these chemicals differs strongly, one can expect, that in a sample (e.g. sediment) with a mixture of various contaminants and bioavailabilities the responses of organisms show very different response patterns, also depending on the endpoint – the physiological reaction that is measured in the test (Ahlf & Heise, 2005).

Environmental quality can no longer be assessed solely on the basis of a more or less comprehensive set of chemical analyses, nor by comparing individual determinations of chemical concentrations against arbitrary standards with no ecological basis. Pollution monitoring, either intended to identify geographical patterns of pollution or temporal trends, must nowadays be based on integrated approaches involving complementary chemical and biological methodologies (Beiras et al., Personal Communication). Toxicity tests permit to evaluate effects of contaminants alone or in mixtures; additionally, they also assess effects of those contaminants that have not been detected chemically (Ahlf & Heise, 2005). The interconnected roles of chemistry and ecotoxicology address questions relating to the presence of chemical pollutants, their bioavailability, and the onset of adverse effects at different levels of biological organization (Chapman, 2007). As example we report in Table 1 the advantages and limitations of chemical and biological criteria in environmental management.

Criteria	Advantages	Limitations
Chemical	• Provide specific chemical concentration • Provide chemical speciation • Well-proven technology • Short-term response • Generally reproducible • Fast turnaround results	• Not provide bioavailability • Not provide synergies • Not ecologically relevant
Biological	• Estimate to contaminant bioavailability • Relate directly to risk • Assess synergic effects • Tests in situ are realistic	• Variability a concern • Some criteria lack validation • Lab tests lack realism

(source: Volpi Ghirardini and Pellegrini, 2001; Adams and Tremblay, 2003)

Table 1. Some advantages and limitations of chemical and biological criteria in environmental management.

3.2 Chemical analysis and bioassay for offshore platform monitoring

A series of chemical and biological analysis tools may be employed to monitor environmental effects when a gas platform is installed and/or a produced water is discharged into the sea. In fact, concerns existed that contaminants associated to offshore activities (installation and/or production phase, with operational or accidental discharges) could be accumulated within sediments and marine organisms and result in significant negative effects on the marine environment.

Accumulation analyses (in sediment and biota), together biomarkers and toxicity tests in situ, may help the scientist to understand what happen to organisms present where there is an platform or a produced water discharge. Besides, laboratory tests may be carried out on species directly exposed to produced water or to environmental matrices, collected in gas platform and/or produced water discharge area.

The integrated chemical and ecotoxicological approach generally provides information on the following aspects in evaluation of potential effects of offshore platforms and produced water discharge:

1. Investigation in environmental media (water, sediment, biota) of contaminants potentially associated to the platform presence (e.g. trace metals as zinc, aluminium, cadmium or barium). In this context, chemical speciation allows to distinguish between

the anthropogenic impact connected with platform activities and the natural regional gradient of contaminant levels in water, sediment and biological tissues.

2. Assessment of toxicity and extent of toxic effects recorded near to platform area. Some chemical compounds used during installation may be dangerous for biota; however it is not always possible to distinguish the contribution of the toxicity caused by installing from that due to production phase.
3. Analysis of chemical composition of produced water and fate of its constituents in marine environment. If the concentration of some substances in environmental sample is found to be significantly higher than the background levels, then the bioavailability and ecotoxicity of these substances on selected species may be further considered.
4. Analysis of some produced water contaminants used as chemical tracers to describe the dispersion of effluent into the sea; in fact, some compounds (salts, nutrients, isotopes, ions) may be used as tracers. A substance may be conveniently used as a tracer if it has some proprieties: conservative, representative of produced water, easy to analyse and to monitor (Cianelli et al., 2008).
5. Bioaccumulation studies to evaluate produced water contaminant concentrations in edible tissue of fish and invertebrates collected near discharge;
6. assessment of short-term and long-term effects of produced water on marine organisms and estimation of potential biological impact area around discharge point. The toxicity studies are used as a complement to chemical measures for quantifying the potential toxic effects, together with the bioavailability of chemical substances and the possible synergies.

Details on chemical analysis, biomarkers and toxicity tests, with reference to sampling operations and laboratory analyses are reported in Trabucco et al. (n.d.).

4. A review of field studies: The Adriatic gas platforms (Mediterranean Sea)

The exploitation of gas fields in the Adriatic sea began in the 1960s and more than 100 platforms have been installed since then, which extract especially gas. These offshore structures are mainly distributed along the Northern and Central Adriatic coasts (about 90 platforms but also in the Ionian Sea and in the Strait of Sicily (De Biasi et al., 2006).

Adriatic is a shallow, landlocked semi-enclosed sea in the eastern Mediterranean, characterised by a wide shelf; with an average depth of few tens of meters in its northernmost section, and less than 200 meters in its middle portion, and a relevant seasonal variability in its circulation dynamics. It is a basin characterized by strong gradients in water properties due to important surface (air sea) and lateral (river runoff) fluxes. In the Adriatic Sea thermohaline stratification and current field may also crucially affect the impact of effluent discharges, such as produced water discharge (Cianelli et al., 2011).

The importance of assessing the impact of offshore activities on the marine environment has been widely recognized using different tools (chemical analysis, physical modelling, biological investigations etc.). In this context, we present results of some studies, carried out in the Adriatic Sea, that have adopted the chemical and ecotoxicological approach, both for monitoring of platforms with produced water discharge and platforms without discharge.

In case of release of produce waters into the sea, it is important consider that the most of Adriatic platforms produce gas and this means a higher content of low molecular-weight aromatic hydrocarbons and small water volumes compared to oil platforms. These may be important elements to take into account when they come to mitigating the impact of discharge.

Mariani et al. (2004) studied produced water collected from an off-shore gas platform located in the Central Adriatic Sea (Italy), including analytical determinations and acute toxicity bioassay by using the European sea bass (*Dicentrarchus labrax*) fish larvae. The chemical characterization, concerning inorganic and organic compounds, was carried out in the liquid phase, and also on the produced water solid fraction. The trace metals were found to be linked only to the solid fraction while the aromatic hydrocarbons and phenols were absent in all the matrices analysed. Examining and comparing the mortality time to unfiltered and to filtered produced water, they observed a shorter response time to the unfiltered sample, probably induced by chemical effect (contaminants associated with particle of produced water) or to particle mechanical effects (i.e. absorption through body surface and gills or oral ingestion/digestion).

Manfra et al. (2007) carried out a chemical and ecotoxicological assessment on the produced water and sediments around a gas platform. The chemical analysis has shown high concentrations of zinc and arsenic in sediment samples only in the nearest station to the platform. Zn was recorded in produced water particulate, but it is also associated with exploration and corrosion maintenance activities (i.e., galvanic anodes). As was not recorded in produced water and its concentration in sediment is not attributable to the discharge. The bioassays with the marine bacterium *Vibrio fischeri* and the sea urchin *Paracentrotus lividus*, have not generally shown sediment toxicity although the produced water toxicity, probably for the swift dilution process of produced water discharge into the sea.

Gorbi et al. (2007, 2008) developed a monitoring protocol with caged mussels to evaluate the potential ecotoxicological effects caused from the off-shore platforms. They transplanted native mussels from an Adriatic reference site to two different areas: a platform area with produced water discharge and another site without discharge. After the translocation period, trace metals were analysed in mussel tissues and early biological responses were detected at several cellular targets. The variations of chemical and biological responses were influenced by fluctuations of biological and environmental factors rather than the effects of platform activities. The effects of seasonality can be more pronounced in certain periods and significantly influence responses of caged organisms. The authors only observed an higher bioavailability for zinc and cadmium close to the platform without discharge, suggesting the influence of galvanic anodes for cathodic protection.

Fattorini et al. (2008) provided an extended 5 year data-set for basal concentrations of several trace metals in Adriatic mussels sampled on off-shore platforms with produced water discharge. While the majority of elements revealed seasonal changes mostly related to phytoplanctonic blooms and reproductive cycle, different fluctuations were observed for arsenic, not correlated with gonadic development, neither with other elements. The marked annual and geographical variations observed for this element suggested the influence of oceanographic or hydrological factors like salinity, of particular relevance for the Adriatic considering the elevated variability of environmental and meteorological conditions in this basin. The authors excluded the anthropogenic impact of exploitation activities, with the only exception of a certain enrichment of cadmium and zinc, probably associated to the use of anodic electrodes.

Tornambè et al. (2012) studied the toxicity of diethylene glycol (chemical additive largely used during oil and gas exploitation by offshore platforms) and its combined effects with produced waters by means acute bioassays with marine/estuarine species. They observed that diethylene glycol is toxic for marine/brackish organisms at concentrations not considered dangerous for the marine environment (orders of g/l)(Group of Experts on the Scientific Aspects of Marine Environmental Protection [GESAMP], 2002) (United Nations [UN], 2011) and normally not detected in Adriatic produced waters (2-13 mg/l) (Cianelli et al., 2008). Besides the results of toxicity tests showed that the presence of diethylene glycol in Adriatic produced waters may alter their toxicity; in fact the addition over 5 g/l has produced synergistic effects. The authors believe that safe threshold concentration of 3.5 g/l (established in Italy) could be considered acceptable for diethylene glycol in co-exposure with produced waters but it should be necessary to study the co-solvency mechanisms between water and glycols because they can act as co-solvents, greatly increasing the solubilization and transport of organic contaminants including BTEX (benzene, toluene, ethylbenzene and xylene), PAHs (polycyclic aromatic hydrocarbons) and alkanes (Sorensen et al. 2004).

Gorbi et al (2009) also studied the toxicity of diethylene glycol alone and in co-exposure with produced waters but using biomarkers. They investigated if diethylene glycol induces direct molecular/cellular effects in marine organisms, or indirectly modulate those of produced water. Sea bass (*Dicentrarchus labrax*) were exposed to diethylene glycol dosed alone or in combination with produced water. Biomarker results did not reveal marked effects of diethylene glycol dosed alone and with produced water, with the only exception of a slight genotoxic damage observed at an exposure dose higher than the maximum allowed limit (3.5 g/l) and of levels normally detected in Adriatic produced water.

Manfra et al. (2010) characterized three produced waters originated from Adriatic gas platforms by chemical analysis and toxicity tests on test-organisms belonging to different trophic levels such as bacteria, algae, crustaceans and fishes. Some differences among the produced waters were observed both for toxicity and chemical composition: the highest toxicity was recorded in the produced waters containing the highest concentrations of some metals (barium, manganese and zinc) and/or volatile aromatic compounds (BTEX). Finally, the authors observed that a filtration treatment of produced waters before their discharge into the sea reduces their toxicity and consequently it may decrease the ecological risk associated to the discharge. As already reported in Mariani et al. (2004), produced water particle presence induces toxicity due to chemical effect of contaminants and/or to particle mechanical effects.

5. Conclusion

The focal intend of this paper has been to explain the importance of use of an integrated chemical and ecotoxicological approach, in wide-ranging for environmental quality assessment and in particular for offshore platform monitoring. In fact, the combination of chemistry and ecotoxicology permits to evaluate the presence of chemical pollutants, their bioavailability, and the adverse effects. Only applying toxicity tests it is possible also see effects of contaminant mixtures and discover effects even if due to contaminants that have not been detected chemically.

Regarding field studies of the Adriatic Sea, quite limited contamination and toxicity were observed in presence of offshore gas installations and/or produced water discharge. Certainly, the principal environmental effects are recorded in sediments and biota, probably due to the nearness of platform anodes, to primary contaminants linked to produced water particles and also to natural regional gradient of some contaminants (e.g. arsenic levels). Although, the potential hazard area is limited (some meters) because the Adriatic platforms drain only small volumes of effluent into the sea and the dilution process is rapid in the near field (Cianelli et al. 2008, 2011). All results obtained for Adriatic platforms confirmed the utility of using ecotoxicology as an additional and complementary contribution to chemical analysis for monitoring offshore activities. Besides in these years Trabucco et al., n.d. provide an main background on each investigation that may be done to assess environmental impacts of offshore platforms, including chemical analysis and bioassay.

6. References

Adams, S.M. & Tremblay, L.A. (2003). Integration of chemical and biological tools in environmental management and regulation. *Australian Journal of Ecotoxicology,* Vol.9, pp.157-164

Ahlf, V. & Heise, S. (2005). Sediment Toxicity Assessment. *Journal Soils & Sediments,* Vol.5, No.1, pp.16-20

Assomineraria (2011). Le previsioni dell'upstream petrolifero per il 2011. Mediapoint & Communications. Gennaio 2011 (sito web www.assomineraria.org)

Beiras, R., Marigomez, I. & Lehtonen, K.K. Biological Tools for Marine and Estuarine Ecosystem Protection. Personal Communication rbeiras@uvigo.es, ionan.marigomez@ehu.es, kari.lehtonen@ymparisto.fi

Chapman, P.M. (2007). The interconnected roles of chemistry and biology (ecotoxicology and ecology) in evaluation of marine environmental quality. *Biologia Marina Mediterranea,* Vol.14, No.1, pp.10-16

Cianelli, D., Manfra, L., Zambianchi, E., Maggi, C., Cappiello, A., Famiglini, G., Mannozzi, M. & Cicero, A.M. (2008). Near-field dispersion of Produced Formation Water (PFW) in the Adriatic Sea: an integrated numerical and chemical approach. *Marine Environmental Research,* Vol.65, pp.325-337

Cianelli, D., Manfra, L., Zambianchi, E., Maggi, C. & Cicero, A.M. (2011). Modelling and Observation of Produced Formation Water (PFW) at Sea. In: *Fluid Waste Disposal,* Canton, K.W. (Ed.), 113-135, Nova Science Publishers, Inc., ISBN 978-1-60741-915-0, New York

De Biasi, A.M., Fabi, G., Pacciardi, L., Spinelli, O., Micheli, R., Puletti, M. & De Ranieri, S. (2006). Metalli pesanti e idrocarburi policiclici aromatici in prossimità di una piattaforma di estrazione: 3 anni di osservazioni (Adriatico settentrionale). *Biologia Marina Mediterranea,* Vol.13, No.2, pp.332-333

E&P Forum/UNEP (1997). Environmental management in oil and gas exploration and production. UNEP IE/PAC Technical Report 37. E&P Forum Report 2.72/254. ISBN 92-807-1639-5, pp.1-76

Enerdata (2011). Global Energy Statistical Yearbook 2011. http://yearbook.enerdata.net

Fattorini, D., Notti, A., Di Mento, R., Cicero, A.M., Gabellini, M., Russo, A. & Regoli, F. (2008). Seasonal, spatial and inter-annual variations of trace metals in mussels from the Adriatic sea: a regional gradient for arsenic and implications for monitoring the impact of off-shore activities. *Chemosphere,* Vol.72, pp.1524-1533

GESAMP (2002). (IMO/FAO/UNESCO-IOC/WMO/WHO/IAEA/UN/UNEP Joint Group of Experts on the Scientific Aspects of Marine Environmental Protection). Revised GESAMP Hazard Evaluation Procedure for Chemical Substances Carried by Ships. Rep. Stud. GESAMP No. 64, 126 pp.

Gorbi, S., Benedetti, M., Bocchetti, R., Virno Lamberti C., Moltedo, G., Di Mento, R., Fattorini, D., Pisanelli B., & Regoli, F. (2007). Use of caged mussel *Mytilus galloprovincialis* in an ecotoxicological approach to assess environmental impact in offshore activities. *Rapp. Comm. int. Mer Medit.,* Vol.38

Gorbi, S., Virno Lamberti, C., Notti, A., Benedetti, M., Fattorini, D., Moltedo, G. & Regoli, F., (2008). An ecotoxicological protocol with caged mussels, Mytilus galloprovincialis, for monitoring the impact of an offshore platform in the Adriatic sea. *Marine Environmental Research,* Vol.65, pp.34-49

Gorbi, S., Benedetti, M., Virno Lamberti, C., Pisanelli, B., Moltedo G. & Regoli, F. (2009). Biological effects of diethylene glycol (DEG) and produced waters (PWs) released from offshore activities: a multi-biomarker approach with the sea bass *Dicentrarchus labrax. Environmental Pollution,* Vol.157, pp.3166-3173.

Henderson, B., Grigson, S.J.W., Johnson, P. & Roddie, B.D. (1999). Potential impact of production chemical on the toxicity of produced water discharges from North Sea oil platforms. *Marine Pollution Bulletin,* Vol.38, No.12, pp.1141-1151

Maggi, C., Berducci, M.T., Bianchi, J., Giani, M. & Campanella, L. (2009). Methylmercury determination in sediment and marine organisms by Direct Mercury Analyser. *Analitica Chimica Acta,* Vol.641, pp.32-36

Manfra, L., Moltedo, G., Virno Lamberti, C., Maggi, C., Finoia, M.G., Gabellini, M., Giuliani, S., Onorati, F., Di Mento, R. & Cicero, A.M., (2007). Metal content and toxicity of produced formation water (PFW): study of the possible effects of the discharge on marine environment. *Archives of Environmental Contamination and Toxicology,* Vol.53, pp.183-190

Manfra, L., Maggi, C., Bianchi, J., Mannozzi, M., Faraponova, O., Onorati, F., Mariani, L., Tornambè, A., Virno Lamberti, C. & Magaletti, E. (2010). Toxicity evaluation of produced formation waters after filtration treatment. *Natural Science,* Vol.2, No.1, pp.33-40.

Manoukian, S.. Spagnolo, A., Scarcella, G., Punzo, E., Angelini, R. & Fabi, G. (2010). Effects of two offshore gas platforms on soft-bottom benthic communities (northwestern Adriatic Sea, Italy). *Marine Environmental Research,* Vol.70, pp. 402-410

Mansour, M. (1993). Fate and Prediction of Environmental Chemicals in Soils, Plants and Aquatic Systems. Lewis Publishers, Chelsea, MI

Mariani, L., Manfra, L., Maggi, C., Savorelli, F., Di Mento, R. & Cicero, A.M. (2004). Produced formation waters: a preliminary study on chemical characterisation and acute toxicity by using fish larvae (*Dicentrarchus labrax* L.,1758). *Fresenius Environmental Bulletin,* Vol.13, No.12a, pp.1427-1432.

OSPAR, 2009. Assessment of impacts of offshore oil and gas activities in the North-East Atlantic. Ospar Commission, London, No.453/2009, ISBN 978-1-906840-93-8, 40 pp.

Piva, F., Ciaprini, F., Onorati, F., Benedetti, M., Fattorini, D., Ausili, A. & Regoli, F. (2009). Assessing sediment hazard through a weight of evidence approach with bioindicator organisms: a practical model to elaborate data from sediment chemistry, bioavailability, biomarkers and ecotoxicological bioassays. *Chemosphere,* Vol.83, pp.475-485

Rabalais, N.N., McKee, B.A., Reed, D.J. & Means, J.C. (1992). Fate and effects of produced water discharges in coastal Louisiana, Gulf of Mexico, USA. In: *Produce Water.,* J.P. Ray, F.R. Engelhart, (Eds.), 355-369, Plenum Press, New York

Scott, K.A., Yeats, P., Wohlgeschaffen, G., Dalziel, J., Niven, S. & Lee K. (2007). Precipitation of heavy metals in produced water: Influence on contaminant transport and toxicity. *Marine Environmental Research,* Vol.63, pp.146-167

Setan, H. & Othman, R. (2006). Monitoring of Offshore Platform Subsidence Using Permanent GPS Stations. *Journal of Global Positioning Systems,* Vol.5, No.1-2, pp.17-21

Sittig, M. (1991). Handbook of toxic and hazardous chemicals and carcinogens. Ed. Park Ridge, N.J., Noyes Data Corporation, 3rd ed

Sorensen, J.A., Gallagher, J.R. & Hawthorne, S.B. (2004). Gas industry groundwater research program; final report for U.S. Department of Energy, National Energy Technology Laboratory. Cooperative Agreement No. DE-FC26-98FT40321; EERC Publication 2004-EERC-07-01; Energy & Environmental Research Center: Grand Forks, ND

Stromgren, T., Sorstrom, S.E., Schou, L., Kaarstad, I., Aunaas, T. & Brakstad, O.G. (1995). Acute toxic effects of produced water in relation to chemical composition and dispersion. *Marine Environmental Research,* Vol.40, pp. 147–169

Tornambè A., Manfra L., Mariani L., Faraponova O., Onorati F., Savorelli F., Cicero A.M., Virno Lamberti C., Magaletti E. (2012). Toxicity evaluation of diethylene glycol and its combined effects with produced waters of off-shore gas platforms in the Adriatic Sea (Italy): Bioassays with marine/estuarine species. *Marine Environmental Research* doi:10.1016/j.marenvres.2011.12.006

Trabucco, B., Cicero, A.M., Gabellini, M., Virno Lamberti, C., Di Mento, R., Bacci, T., Moltedo, G., Tommassetti, P., Panfili, M., Marusso, V. & Cornello, M. (2006). Study of the soft botom macrozoobenthic community around an offshore platform (Central Adriatic Sea). *Biologia Marina Mediterranea*, Vol.13, Vol.1, pp.659-662

Trabucco, B., Maggi, C., Manfra, L., Mannozzi, M., Nonnis, O., Cicero, A.M., Di Mento, R., Gabellini M. & Virno Lamberti, C. n.d. Monitoring of impacts of offshore platforms in the Adriatic Sea (Italy). *Natural Gas*, InTech ISBN 979-953-307-567-8

UN (2011). Globally Harmonized System of Classification and Labelling of Chemicals (GHS), fourth revised ed. United Nations, New York and Geneva

Volpi Ghirardini, A. & Pellegrini, D. (2001). Toxicity bioassays in water and sediment quality assessment of marine and transitional environments: suggestions for the choise, set up, evaluation and application methods. *Biologia Marina Mediterranea*, Vol.8, No.2, pp.1-16

Monitoring of Impacts of Offshore Platforms in the Adriatic Sea (Italy)

B. Trabucco et al.*
ISPRA Italian Institute for Environmental Protection and Research, Rome, Italy

1. Introduction

The oil and gas industry is truly global, with operations conducted in every corner of the world, since the worldwide increasing of energy demand has strongly raised the exploitation of non-renewable resources. The date of birth of the offshore industry is commonly regarded as 1947, when it was successfully completed by Kerr-McGee the first well in the Gulf of Mexico (Chakrabarti et al., 2005). Since 1950 about 7,000 platforms have been constructed and installed around the world, and today most of these structures, which represent more than 65% of the total, are located along the American coast of the Gulf of Mexico, the remainder is concentrated in the North Sea, Middle East, Africa, Australia, Asia and South America (Wilson III and Heath, 2008). Approximately 0.4% of world reserves of oil and gas are located in the Mediterranean basin (Pinder, 2001) and therefore this area represents one of the areas that will have greater expansion in offshore activities in the coming years (Maksound, 2004). In sixty years, Italy has produced 75% of the gas discovered. The natural gas production in 2010 was 8 billion cubic meters. The oil and gas production wells in the entire Italian territory are on the whole 1200 and the platforms at sea are 127, which extract especially gas (Paini, 2001; Assomineraria, 2004; 2011). These offshore structures are mainly distributed along the Northern and Central Adriatic coasts (about 90 platforms (De Biasi et al., 2006; Maggi et al., 2007), on depths between 10 and 120 meters, but also in the Ionian Sea and in the Strait of Sicily (www.eni.it).

The offshore activities comprise different phases linked to exploitation of gas and oil reservoirs: a) the exploration phase to probe the position and the geological characteristics of well and then to install a steel platform; b) the production phase to extract oil and gas; c) the decommissioning phase, when the commercial life of well is finished (Oil Industry International Exploration & Production Forum/United Nations Environment Programme [E&P Forum/UNEP], 1997).

Oil and gas exploration and production operations have the potential for a variety of impact on the environment, depending upon the stage of the process, the nature and sensitivity of the surrounding environment, pollution prevention, mitigation and control techniques. With regards to the aquatic environment, the principal problems are linked to

* C. Maggi, L. Manfra, O. Nonnis, R. Di Mento, M. Mannozzi, C. Virno Lamberti, A. M. Cicero and M. Gabellini
ISPRA Italian Institute for Environmental Protection and Research, Rome,Italy

the presence of the offshore structures and then to waste streams of drilling fluids, cuttings, well treatment chemicals and produced waters (Neff, 1987; Neff et al., 1992; Osenberg et al., 1992; Olsgard & Gray, 1995; Commission protecting and conserving the North-East Atlantic [OSPAR], 1999, 2009; Peso-Aguiar et al., 2000; Barros et al., 2001; Pinder, 2001; Cicero et al., 2003, 2004, Trabucco et al., 2006a, 2006b; Terlizzi et al., 2008; Manoukian et al., 2010). In particular, produced water is water produced along with oil and gas and so it may include a) naturally-occurring water layer present in oil and gas reservoirs, b) water that has been injected into the reservoirs to help force the oil to the surface and c) any chemicals added during the production and treatment process. The positioning of a permanent structure and the discharge of produced water, may generally modify environmental quality, causing effective changes of the physical-chemical characteristics of water column and sediment and also perturbations on the marine living communities and on the sea-bottom geomorphology. These impacts have been described in details in Manfra & Maggi (n.d.).

Monitoring programs have been developed worldwide in all the areas characterized by an intense extraction and production activity (e.g. North Sea, Gulf of Mexico and Adriatic Sea), also taking into consideration national and international sea protection policies and legislations. Among major international conventions, it's worth mentioning the Barcelona Convention (Barcellona Convention, 1979), with its Offshore Protocol, which represents a regional regulatory framework for the Mediterranean basin, but it doesn't supply technical tools to manage with environmental control activities. Good framework for the environmental monitoring of effluents resulting from offshore activities is provided by the OSPAR Convention (OSPAR, 1992; Stagg, 1998). This was born in 1992 for the protection of the marine environment, issued and signed by the countries of the North-East Atlantic area and by the European Economic Community. It worked out the guidelines for monitoring the environmental impact of offshore oil and gas activities, which represent a strategy adoptable to assess the impact resulting from the different phases of offshore activities (exploration, production and decommissioning) (OSPAR, 1999). Besides, some environmental protection measures are reported in the guidelines on environmental management in oil and gas exploration and production based on the collected experience gained by UNEP and the oil industry (E&P Forum/UNEP, 1997). At national level, the environmental compatibility of offshore platform construction and their eventual impacts are evaluated by an Environmental Impact Assessment (EIA) and a monitoring plan, respectively. Concerning waste streams, the Ministry Decree 28th July 1994 requires a) the produced water chemical composition, b) the produced water discharge control, imposing a threshold value (40 mg/l) for total mineral oil content, c) a toxicity assessment for drilling fluids, cuttings and chemicals. Then, the Legislative Decree 152/2006 permits produced water discharge exclusively if an environmental monitoring plan verifies the absence of risks for waters and marine ecosystems (MD, 1994; LD, 1999; LD, 2006).

Given the current (international and national) environmental regulatory, the application of environmental monitoring tools is required to avoid, minimize and mitigate the potential impacts arising from exploration and production of oil and gas (Maggi et al., 2007). In this paper we describe a multidisciplinary monitoring approach, applied from the Italian Institute for Environmental Protection and Research (ISPRA) which since 2000 assesses potential impacts due to offshore platform presence and produced water discharge in the Adriatic Sea.

2. An integrated approach to project monitoring plans

Monitoring of chemical-physical characteristics of water and sediment, together with ecotoxicological assays, bioaccumulation analyses and macrozoobenthic soft-bottom community structure investigations, let to provide the necessary information for assessing the actual spatial and temporal perturbations occurring in the marine ecosystem. The Adriatic Platform Monitoring (called next "APM ") matches information on physical, chemical and biotic variables in order to give the best description of the environmental quality status (Directive 2000/60/EC; Chapman, 1990, 1996), so as:

1. to monitor selected chemical and physical variables of water column and sediment;
2. to evaluate effects on biota;
3. to identify spatial-temporal trend of the eventual alteration that could occurred;
4. assessing mitigation measures;
5. defining guidelines supporting regulations and decisions.

The investigations on water column, sediment and biota, with geophysical analysis on the sea bottom allow to study, thus highlighting, the natural features of the marine environment and then the changes due to presence of the offshore structures. This in order to realize an ecological vision towards an ecosystem approach in the protection of the Mediterranean, one of the innovative features proposed to reduce land-based pollution in the Mediterranean Sea (Barcelona Convention, 1979). In particular, the APM provide to measure several different parameters (Cianelli et al., 2011):

- physical-chemical (temperature, salinity, density, oxygen, nutrient, current speed and direction, etc.)
- chemical (contaminant concentrations)
- geological (sediment grain size)
- geomorphological (bottom batimetry and morphology)
- bioassays (e.g. biomarkers and toxicity tests)
- ecological (benthic community structure)

The comparison of these parameters with the background level (or reference conditions) can provide an indication on the potential levels of concern. The APM was born with the characteristic of "flexibility"; in fact, results and data obtained during the occurring of the monitoring activities could be useful to arrange and modify the whole sampling strategy.

All aspects (sampling, storage, laboratory analysis and data analysis) related to the above parameters are described below, distinguishing between platform and produced water discharge monitoring.

3. Sampling strategy

In order to investigate the impact of the platform structure, a *radial* sampling design can be performed for all matrices, allocating stations around the platform according to fixed distance from it rather than using a randomized placement (Fig1). This choice seemed to be more appropriate to highlight environmental changes when the point source of disturbance is known (Ellis & Schneider, 1997; Cicero et al., 2004).

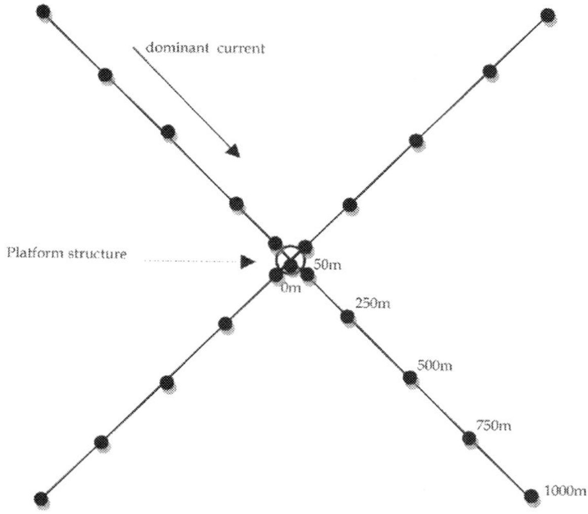

Fig. 1. Sampling strategy for platform monitoring.

To highlight the natural and anthropic features of the sea bed and provide detailed sea floor maps, geophysical investigations can be performed on a square shape area surrounding the platform (wide about 4 km2) by means of Multibeam Echosounder and Side Scan Sonar. The survey area can be completely covered by parallel navigation lines; transects have to ensure overlapping coverage.

With regards to the produced water impact, the water samples can be collected on a transect along the actual current from the point of discharge, instead sediment can be sampled on a transect oriented in the direction of the dominant current (Fig. 2 and 3).

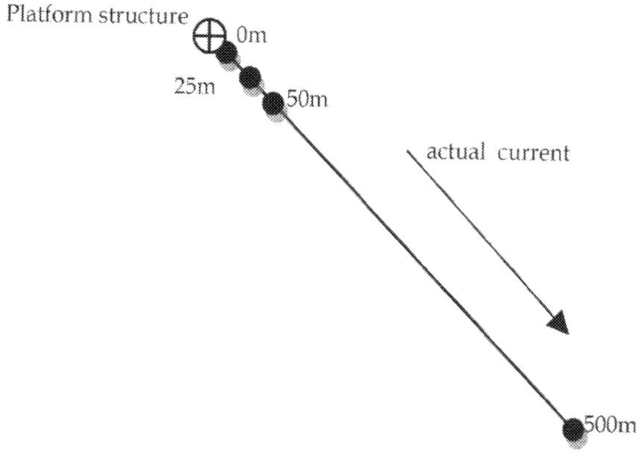

Fig. 2. Water sampling strategy for produced water monitoring.

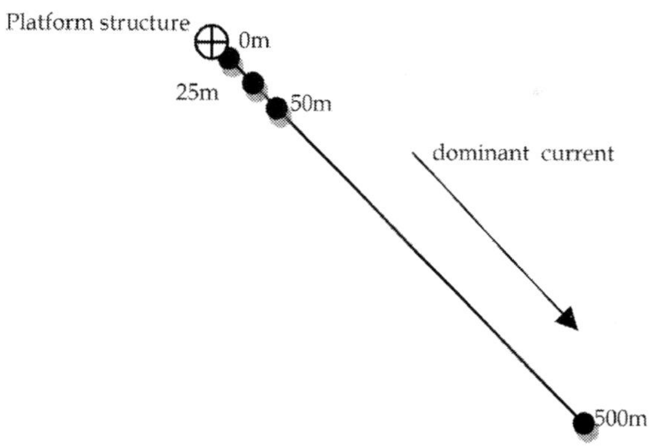

Fig. 3. Sediment sampling strategy for produced water monitoring.

Furthermore, in order to obtain an exhaustive rigorous environmental framework, all matrices should be sampled also in a control area, presenting the same geo-morphological characteristics of the investigated area, but not directly influenced by the off-shore activities.

Biological elements (e.g. mussels used for the bioaccumulation and biomarker analyses) can be taken on different depths, considering the level of the discharge, on the nearest leg to the discharged pipe along the main current (OSPAR, 1999) (Fig. 4).

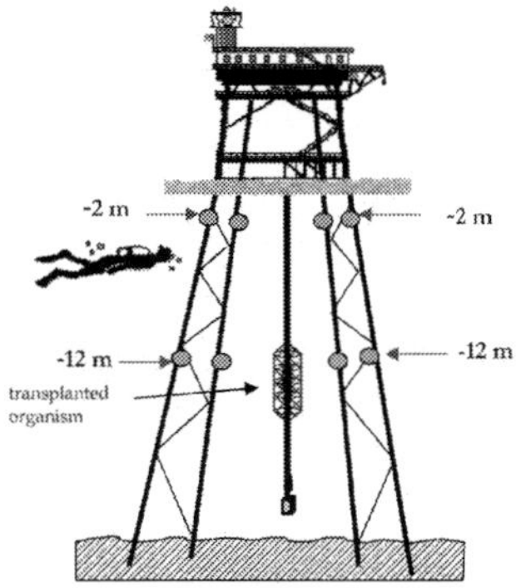

Fig. 4. Mussel sampling strategy for produced water monitoring.

Monitoring frequency can comprise one or two sampling campaigns before the installation of the platform or the beginning of the produced water discharge; two sampling cruises, during the first year of the platform life, or after the beginning of the produced water discharge, aimed at representing condition of maximum and minimum mixing of the water column. Then one survey, scheduled for each following year of activity, to monitor both types of impacts.

In particular, geophysical investigation have to be performed both before the installation of the platform and during the first year of the platform life. Then, these investigations can be carried out every two or three years.

The acquisition of meaningful data demands correct sampling and storage procedures. These may be quite diverse for different parameters in the same matrices or for the equal parameters in different matrices. In general separate samples must be collected for physical-chemical and biological analyses because the sampling and preservation techniques are quite different. Usually the shorter the time interval between sampling and analysis, the more accurate the analysis will be.

4. Matrix sampling

4.1 Water

Some physic-chemical properties can be measured in the water without sampling. Conductance, Temperature, and Density, generally, are monitored by means of a hydrological CTD profiles acquired using the multiparametric probe; the probe can also be equipped with additional sensors for fluorescence, transmittance, dissolved oxygen and pH.

In order to determine chemical compound concentrations, total suspended matter and toxicity, water samples can be collected by Niskin bottles. Consequently, some analyses can require the filtration treatment (e.g. nutrients or metal analysis), while others can be carried out on whole sample (e.g. PAH or TBT analysis).

The measure of water velocity during sampling cruises is fundamental to identify the directions of transect along the actual current from the point of discharge. The Acoustic Doppler Current Profile (ADCP) is, today, the instrument commonly used to collect the profiles of water velocity. It is based on a principle of sound waves called the Doppler effect and is useful to measure how fast water is moving across an entire water column.

4.2 Sediment

Bottom sediment deposits can be sampled by grab samplers or box corer, which are designed to recover sediment material from the top few centimeters of the seabed (IAEA, 2003). One of the most commonly grab samplers used for surface sediments is the Van Veen, which is capable of collecting sediment material up to about 20 cm in depth. The sampler has large access doors for convenient removal of the upper sample layers (USEPA, 1997). The box corer uses an open-ended rectangular box, found in various sizes. It is designed for a minimum of disturbance of the sediment surface by bow wave effects, which is important for quantitative investigations (IAEA, 2003; USEPA, 1997). Through box corer sediments may be sampled at different level (e.g. 0-3 and 8-10).

The samples for chemical, physical and ecotoxicological analyses can be carried out by Van Veen grab or box corer, depending on the number of sediment levels to collect. In any case, sampling device has to permit to collect a representative, undisturbed samples. The material, after homogenization, has to be stored in suitable vessels and at adequate temperature depending on different analyses.

The quantitative samples of benthic soft-bottom macrofauna can be performed on all stations identified for sediment analysis, using a grab or a box corer. At each platform and at each station, the number of replicas to be made is established to be sure to take representative volume (Picard, 1965). The material collected can be sieved on board with a mesh of 0.5 mm or 1 mm, then fixed in appropriate preservative.

4.3 Biota

Bioaccumulation and biomarker analyses may be carried out on organisms living in the surround of the offshore structure. Mussels (e.g. *Mytilus galloprovincialis*) can be collected on platform legs; otherwise native mussels, collected from a reference site and transplanted for 4-6 weeks in the investigated platform area, can be used. These organisms may be sampled at two different depths (e.g. superficial and 12 m, respectively) and, for each depth, different samples, constituted by the whole tissues, prepared and stored at -20°C, until processed for accumulation analysis. Specifically for biomarker analysis, after recovery, mussel tissues have to be frozen in liquid nitrogen and maintained at -80°C until biochemical and histological parameter analyses; haemolymph have to be withdrawn from adductor muscle and immediately analysed for lysosomal membrane stability and DNA damages (Gorbi et al., 2008).

4.4 Produced water

Produced water must be sampled directly onboard platform, after oil separation system and stored in glass or teflon vessels for different analysis (chemical analysis, toxicity tests and physical characterization).

5. Analysis

5.1 Physical-chemical analysis

The principal parameters to be monitored are those providing general physical-chemical characteristics of the water column (Fig. 5). Temperature, transparency and turbulence are, certainly, among main physical parameters affecting aquatic life: for example, temperature is able to regulate the rate of biological processes, transparency to determine the growth of algae and turbulence is an important factor in mixing and transport of nutrients. Nevertheless the salinity of water and the dissolved oxygen play a key role in determining the kinds of life forms present, while conductance reflects the total concentration of dissolved ionic material.

In addition, if platform discharges produced water it may be important to investigate others parameters in environmental matrices. These components must be chosen in light of the produced water composition (e.g. metals, total mineral oils, phenols, aromatic and aliphatic hydrocarbons) (Fig. 5). In fact, produced water is characterized by organic (aliphatic hydrocarbons, aromatic hydrocarbons, etc.) and inorganic components (metals, nutrients, etc.) and then, in some cases, by chemicals (anticorrosive agents, biocides, etc.) added to the water after its removal from the reservoir. The specific chemical composition of the produced water depends on the geological reservoir, the field operations and the exploitation level (Cianelli et al., 2011). Besides, the knowledge of the speed and direction of current (Fig. 5) near the produced water discharge point allows to collect water samples along the true direction of the produced water plume. For this reason, the profile collection of water velocity during sampling cruises is important to determine the directions of transect along the actual current from the point of discharge.

5.2 Chemical analysis

Chemical analysis is known to allow to identify and quantify different single compounds, potentially responsible for the environmental contamination; in this case it is possible to obtain a measure of pollution. The presence of offshore platforms may cause chemical alteration of aquatic ecosystem, especially if the platform discharges produced water. The corrosion protection systems of the installations (sacrificial anodes) release amounts of heavy metals in the form of ions; through absorption on particulate matter, these ions can reach the sediment or bioaccumulate in organisms living on and/or nearby the submerged structures. Moreover the produced water discharge can have effects on marine environment as a consequence of its chemical composition. In addition, it should be considered that around the platform, environmental effects are also determined by the release of organotin compounds (present in the paint used to protect the underwater components from biofouling), although today these paints are prohibited. Due to their physical and biological characteristics, sediment and biota are be able to accumulate various classes of chemicals, such as trace metals, polycyclic aromatic hydrocarbons (PAHs), aliphatic hydrocarbons, halogenated organic compounds, phosphate organic pesticides, etc.

So, the chemical parameters investigated in water, sediment and biota should be selected taking into account all kind of release by offshore structure or activities linked to the presence of platform. Therefore, the analytical determinations cover inorganic compounds (e.g. metals, sulfurs and oxides) and organic compounds (hydrocarbons with low and high molecular weight, organometallic compounds and organochlorinated compounds) (Fig. 5).

5.3 Sedimentological analysis

Grain size is the most fundamental physical property of sediment. Sediments consisting of grains of different size and mainly categorized as sand, silt or clay; they are classified according to the size and size distribution of their grains in the categories described in the Wentworth classification. The information on sediment grain size allows to study trends in surface processes related to the dynamic conditions of transportation and deposition. Moreover, sediments are the ultimate sink for contaminants in the marine environment. The relationships between physical properties and chemical characterization of sediment are a basic point of environmental impact assessment in order to determine whether or not contamination related to offshore activities has occurred. Grain size distributions (Fig. 5) in sediments depend on the distance from the platform, current speeds and water depths. In some platforms the sand content of sediments at the near-field stations was increased above background sediment by 35–60%. Also fragmented carbonate skeletal material was visible, derived from organisms encrusting the underwater parts of the platform.

5.4 Geomorphological analysis

Geomorphological investigation (Fig.5) performed by geophysical devices (e.g. Side Scan Sonar Multibeam echosounder) allow to study the morphological characteristics of the sea bottom, thus highlighting the natural features and the presence of anthropic offshore structures. This kind of investigation represents an essential tool for the effective management of the marine environment and to study accurately the impact of man activities on the seabed (Kenny et al., 2003, Collier & Brown, 2005). The acoustic investigations (Fig. 5) permit to monitor sea-bottom geomorphological modification, can also allow to map the alterations and supply the background information necessary for a correct planning of sampling activities and represent the basis for a long-term monitoring study. These data, together with the results of the multidisciplinary approach, provide more information on the potential environmental impact due to the presence of offshore platforms (Virno et al., 2004). In fact, the interference of the structure with the seabed is confined to a limited area on which you could detect changes in the morphology of the seabed due to changes in current fields generated by stakes driven into the seabed.

5.5 Biological analysis

A series of biological analysis tools may be employed to monitor environmental effects when a platform is installed and/or a produced water is discharged into the sea (Manfra & Maggi, n.d). Ecotoxicological analyses (biomarkers and toxicity tests) may be carried out to study produced water and environmental matrices (Fig. 5). Some aspects related to these investigations are briefly described.

5.5.1 Biomarkers

Alterations at the molecular and cellular levels (biomarkers) can provide a sensitive indication of early changes, which often represent the first warning signals of environmental disturbance, even in the absence of acutely toxic responses. There are some studies on biomarkers related to the monitoring protocol of Adriatic gas platforms (e.g. Gorbi et al. 2007, 2008; Fattorini et al., 2008). In these papers some analyses are reported as main biomarkers for monitoring impacts of offshore platforms in the Adriatic Sea. Induction of metallothioneins, peroxisomal proliferation and activity of acetylcholinesterase may be measured as markers for specific classes of chemicals. Oxyradical metabolism and appearance of oxidative-mediated toxicity reveal a more general onset of cellular disturbance. In addition to individual antioxidants (superoxide dismutase, catalase, glutathione S-transferases, glutathione reductase, Se-dependent and Se-independent glutathione peroxidases, and levels of total glutathione), the total oxyradical scavenging capacity (TOSC) allow a quantification of the overall capability to neutralize specific forms of intracellular reactive oxygen species (ROS; i.e. peroxyl and hydroxyl radicals). Cellular damages are evaluated as lysosomal destabilization (membrane stability, accumulation of lipofuscin and neutral lipids), lipid peroxidation products (malondialdehyde) and DNA integrity (strand breaks and micronuclei); the air survival test is finally applied to evaluate the overall physiological condition of mussels. Biomarker analyses have to carried out according to standardized procedures. Multivariate statistical analyses (principal component analysis, PCA) of biomarkers and accumulation data may be applied to discriminate between different sites and/or different sampling period.

5.5.2 Toxicity tests

Toxicity tests may be carried out on species directly exposed to produced water or to environmental matrices, collected in offshore platform area. APM proposes toxicity tests both for monitoring of effects connected to platform presence and to produced water discharge. In the first case, it may be better exposure to sediments (whole or aqueous phases as pore water or elutriate) while in the second case seawaters and produced waters, not only sediments, may give useful information on toxic effects. The toxicity assessment requires that almost three species, belonging to different trophic levels, are employed for toxicity testing. In APM, some species have to be planned in time: *Vibrio fischeri* (Bacteria), *Dunaliella tetiolecta* (Algae), *Brachionus plicatilis* (Rotifera), *Artemia franciscana* and *Tigriopus fulvus*

(Crustacea), *Paracentrotus lividus* (Echinodermata) and *Dicentrarchus labrax* (Osteichthyes). It is very important that the test battery permits to observe both sub-lethal and lethal "endpoints" (type of effect that can be measured in a toxicological test). In APM the following endpoints are studied: bacterial bioluminescence, algal growth, fertilization and development with echinodermata, and finally mortality with rotifera, crustacea and osteichthyes. Each bioassay has to be carried out according to standardized protocols. Statistical programs (e.g. Trimmed Spearman-Karber method and Probit analysis) allow us to analyze the relationship between a stimulus (dose) and a response (such as death or sub-lethal effects), calculating e.g. the concentration that induces the 50% effect (EC_{50}) or no observed effect concentration (NOEC). These values are interpreted according to toxicity scales to give a toxicity judgment (e.g. no toxic, low toxic, high toxic etc.).

5.6 Ecological analysis: Macrozoobenthic community structure

Offshore activities can induce changes in the characteristics of sediment near the offshore structures. The extent to which the biota is affected by these types of perturbations depends on complex interactions between environmental characteristics of the site and the characteristics of the platform. Several studies have proved that the presence and the activity of off-shore platforms might have some sort of impact on biological communities inhabiting the surrounding seabed.

In general, the artificial structures placed in the sea create a discontinuity, and provide a hard substrate that alters the initial state, providing the conditions for the establishment of new ecosystems (e.g. Trabucco et al., 2008; Terlizzi et al., 2008). In general, the presence of the structures can cause perturbations on the macrofauna assemblage structure more understandable and clear than those raised by the produced water discharge. In particular the variations in sediment physical features (sediment grain-size, sedimentation rates) might determine qualitative and quantitative changes in the structure of soft-bottom benthic communities living in the area around the installations (Davis et al., 1982; Frascari et al., 1991, 1992; Kingston, 1992; Olsgard & Gray, 1995; Barros et al., 2001; Spagnolo et al., 2002; Hernández Arana, 2005; Fontana et al., 2008; Terlizzi et al., 2008; Trabucco et al., 2006a, 2008; Manoukian et al., 2010). In order to evaluate possible change in the soft-bottom community structure due to the presence of the offshore platform, qualitative and quantitative analysis should be performed and ecological studies carried out (Fig. 5). The systematic recognition of organisms should be made at the species level at least for the most representative groups (Polychaeta, Mollusca, Crustacea, Echinodermata). The number of individuals can be count for each species. The data thus obtained may be used for the calculation of the following biological indices:

- Total Abundance (N);
- Total species richness (S);
- Index of Shannon-Weaver species diversity (H ', Shannon & Weaver, 1949).
- Index of evenness (J', Pielou, 1974).

The temporal and spatial evolution of the benthic community may also be evaluated through the application of multivariate analysis (nMDS, PCO, CA, CCA, etc..) and appropriate indices to assess the stress level of the communities found (eg index M-AMBI – Multivariate AZTI Biotic Index, Muxika et al., 2007).

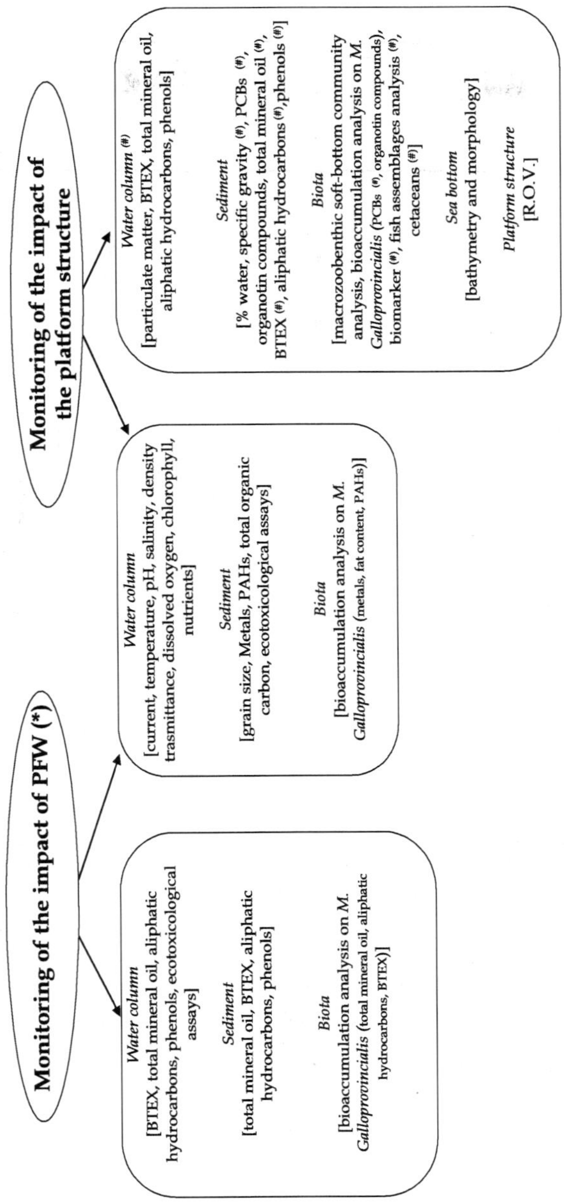

#optional investigation
(*) PFW characterization: BTEX, total mineral oil, aliphatic hydrocarbons, PAHs, phenols, metals, total nitrogen, sulphurs, sulphorate compounds, suspended matter, total organic carbon, ecotoxicological assays.

Fig. 5. Parameters to analyze for platform and produced water monitoring.

6. Conclusions

The APM is the result of more than ten years of experience, gained by ISPRA in the drawing up of monitoring projects to assess potential environmental impacts linked to offshore platforms, then the performance of the analytical methodologies and the carrying on the monitoring activities.

This approach has proved to be the best, based on a long-term multidisciplinary studies, allowing us to collect integrated data. In the framework of an ecosystem approach as recommended by the latest most innovative legislation in the field of environmental protection, APM provides environmental guidelines particularly useful for the public administration and decision makers in the safeguard of the marine ecosystem.

7. References

Assomineraria, (2004). Il Department of Energy e gli investimenti in USA in Italia. Da Assimon e notizie di Luglio 2004 (www.assomineraria.org)

Assomineraria, (2011). Le previsioni dell'upstream petrolifero per il 2011. Mediapoint & Communications. Gennaio 2011 (www.assomineraria.org)

Barcelona Convention (1979). Convention for the Protection Of The Mediterranean Sea Against Pollution, Signed 16 February 1976, in force 12 February 1978 (revised in Barcelona, Spain, on 10 June 1995 as the Convention for the Protection of the Marine Environment and the Coastal Region of the Mediterranean

Barros, F., Underwood, A.J. & Lindegarth, M. (2001). The influence of rocky reefs on structure of benthic macrofauna in nearby soft-sediments. *Estuarine, Coastal and Shelf Science*, Vol. 52, pp.191-199

Chakrabarti, S., Halkyard, J. & Capanoglu, C. (2005). Historical Development of Offshore Structure. In: Chakrabarti S. (Eds), Handbook of Offshore Engineering, Elsevier, London, pp.1-38

Chapman, P.M. (1990). The Sediment Quality Triad approach to determining pollution-induced degradation. Science Total Environment, Vol. 97, No. 8, pp.815-825

Chapman, P.M., Paine, M.D., Arthur, A.D., Taylor, L.A. (1996). A triad study of sediment quality associated with a major, relatively untreated marine sewage discharge. *Marine Pollution Bulletin*, Vol. 32, No. 1, pp. 47-64 - http://www.sciencedirect.com/science/article/pii/0025326X9500108Y - FN1

Cianelli, D., Manfra, L., Zambianchi, E., Maggi, C. & Cicero, A.M. (2011). Modelling and Observation of Produced Formation Water (PRODUCED WATER) at Sea. In: *Fluid Waste Disposal*, Canton, K.W. (Ed.), 113-135, Nova Science Publishers, Inc., ISBN 978-1-60741-915-0, New York

Cicero, AM., Di Mento, R., Gabellini, M., Maggi, C., Trabucco, B., Astori, M. & Ferraro, M. (2003). Monitoring of environmental impact resulting from offshore oil and gas installations in the Adriatic sea: preliminary evaluation. *Annali di Chimica, Journal of Analytical Environmental and Cultural Heritage Chemistry*, Vol. 93, pp.701-705

Cicero, A.M., Gabellini, M., Maggi, C., Nonnis, O., Trabucco, B. & Virno Lamberti, C. (2004). Methodological criteria for environmental monitoring of an offshore platform in the Central Adriatic Sea. *Rapport Commission international Mer Mediterranee*, Vol. 37, p.507

Collier, J.S. & Brown, C.J. (2005). Correlation of sidescan backscatter with grain size distribution of surficial seabed sediments. *Marinbe Geology*, Vol.214, pp.431-449

Directive 2000/60/EC of the European Parliament and the Council of 23.10.2000. A Framework for community action in the field of water policy. Official Journal of the European Community 22.12.2000, pp. 72

De Biasi, A.M., Fabi, G., Pacciardi, L., Gai, F., Ferrari, S., Puletti, M. & De Ranieri, S. (2006). Monitoraggio di una piattaforma di estrazione nel Mar Adriatico settentrionale: metalli pesanti. Atti Società Toscana Scienze naturali, Memorie, Serie B, Vol. 113, pp. 83-89

Ministry Decree 28th July 1994. Gazzetta Ufficiale 16th August 1994, n. 190

Legislative Decree 11st May 1999, n. 152. Gazzetta Ufficiale n. 124 29th May 1999

Legislative Decree 3rd April 2006, n. 152. Gazzetta Ufficiale n. 88 14th April 2006

E&P Forum/UNEP (1997). Environmental management in oil and gas exploration and production. UNEP IE/PAC Technical Report 37. E&P Forum Report 2.72/254. ISBN 92-807-1639-5, pp.1-76

Ellis, J.I & Schneider, D.C. (1997). Evaluation of a gradient sampling design for environmental impact assessment. *Environmental Monitoring and Assessment*, Vol. 48, pp.157-172

Fattorini, D., Notti, A., Di Mento, R., Cicero, A.M., Gabellini, M., Russo, A. & Regoli, F. (2008). Seasonal, spatial and inter-annual variations of trace metals in mussels from the Adriatic sea: a regional gradient for arsenic and implications for monitoring the impact of off-shore activities. *Chemosphere*, Vol.72, pp.1524-1533

Gorbi, S., Benedetti, M., Bocchetti, R., Virno Lamberti C., Moltedo, G., Di Mento, R., Fattorini, D., Pisanelli B. & Regoli, F. (2007). Use of caged mussel Mytilus galloprovincialis in an ecotoxicological approach to assess environmental impact in offshore activities. *Rapport Commission international Mer Mediterranee*, Vol.38

Gorbi, S., Virno Lamberti, C., Notti, A., Benedetti, M., Fattorini, D., Moltedo, G. & Regoli, F., (2008). An ecotoxicological protocol with caged mussels, Mytilus galloprovincialis, for monitoring the impact of an offshore platform in the Adriatic sea. *Marine Environmental Research*, Vol.65, pp.34-49

IAEA (2003). Collection and preparation of bottom sediment samples for analysis of radionuclides and trace elements. IAEA-TECDOC-1360

Kenny, A.J., Cato, I, Desprez, M., Fader, G, Schuttenhelm, R.T.E. & Side, J. (2003). An overview of seabed-mapping technologies in the context of marine habitat classification. *ICES Journal of Marine Science*, Vol.60, No. 2, pp. 411-418

Maggi, C., Trabucco, B., Mannozzi, M., Manfra, L., Gabellini, M., Di Mento, R., Nonnis, O., Virno Lamberti, C. & Cicero, A.M. (2007). A methodology approach to study the environmental impact of oil and gas offshore platforms. Rapport de la Commission Internationale pour l'Exploration Scientifique de la Mer Méditerranée 38, 688.

Maksound, J., (2004). International report- Mediterranean and Black seas to see more investment. *Offshore*, Vol. 64, p. 44

Manfra, L. & Maggi, C. n.d. An approach integrating chemistry and toxicity for monitoring the offshore platform impacts Natural Gas, InTech ISBN 979-953-307-567-8

Manoukian, S., Spagnolo, A., Scarcella, G., Punzo, E., Angelini, R. & Fabi, G. (2010). Effects of two offshore gas platforms on soft-bottom benthic communities (northwestern Adriatic Sea, Italy), *Marine Environmental Research*, doi: 10.1016/j.marenvres.2010.08.004

Muxika, I., Borja, A. & Bald, J., (2007). Using historical data, expert judgement and multivariate analysis in assessing reference conditions and benthic ecological status, according to the European Water Framework Directive. *Marine Pollution Bulletin*, Vol. 55, pp.16–29

Neff, J.M., (1987). Biological effects drilling fluids, drill cuttings and produced waters. In: Boesch D.F. and Rabalais N.N. (eds) Long-term environmental effects of offshore oil and gas development. Elsevier Applied Science, London, UK, pp.469-538

Neff, J.M., Sauer, T.C. & Maciolek, N. (1992). Composition, fate and effects of produced water discharges to nearshore marine waters In: Ray J.P. and Engelhart F.R. (eds) Produced Water Technological/Environmental Issues and Solutions. Plenum Press, New York, pp.371-385

Olsgard, F. & Gray, J.S. (1995). A comprehensive analysis of effects of offshore oil and gas exploration and production on the benthic communities of the Norwegian continental shelf. *Marine Ecology Progress Series*, Vol. 122, pp.277-306

Osenberg, C.W., Schmitt, R.J., Holbrook, S.J & Canestro, D., (1992). Spatial scale of ecological effects associated with an open coast discharge of produced water. In: Ray J.P. and Engelhart F.R. (eds) Produced Water Technological/Environmental Issues and Solutions. Plenum Press, New York, pp.387-402

OSPAR Convention (1992). Convention for the protection of the marine environment of the north-east atlantic.

OSPAR Commission. (1999). Guidelines for Monitoring the Environmental Impact of Offshore Oil and Gas Activities, Adopted at ASMO 2001, OSPAR Commission Agreement 01-10, www.ospar.org/

OSPAR (2009). Assessment of impacts of offshore oil and gas activities in the North-East Atlantic. Ospar Commission, London, No.453/2009, ISBN 978-1-906840-93-8, 40 pp.

Paini, G. (2001). Petrolio e gas in Italia: La proposta delle imprese. Convegno Annuale Settore Idrocarburi e Geotermia, Ischia. Assomineraria (sito web www.assomineraria.org)

Peso- Aguiar, M.C, Smith, D.H., Assis, R. C.F., Santa-Isabel, L.M., Piexinho, S., Gouveia, E.P., Almeida, T.C.A., Andrade, W.S., Carqueija, C.R.G., Kelmo, F., Carrozzo, G., Rodrigues, C.V., Carvalho, G.C. & Jesus, A.C.S. (2000). Effects of petroleum and its derivatives in benthic communities at Baia de Todos os Santos/Todos os Santos Bay, Bahia, Brazil. *Aquatic Ecosystem Health and Management*, Vol. 3, pp.459-470

Pielou, E.C. (1974). Population and community ecology, principles and methodos. Gordon and Breach Sci. Pubbl., New York. 424 pp

Picard, J., (1965). Recherches qualitatives sur le biocenoses marines des substrats meubles dragables de la region marseillaise. *Recuil Traveaux Station Marine Endoume*, Vol. 36, No.52,pp.1-160

Pinder, D., (2001). Offshore oil and gas: global resource knowledge and technological change. Ocean & Coastal Management, Vol. 44, pp.579-600.

Shannon, C.E. & Weaver, W. 1949. The mathematical theory of communication. Urbana IL: University of Illinois Press. 117 pp

Stagg, R.M. (1998). The development of an international programme for monitoring the biological effects of contaminants in the OSPAR convention area. *Marine Environmental Research*, Vol. 46, pp.307-313

Terlizzi, A., Bevilacqua, S., Scuderi, D., Fiorentino, D., Guarnieri, G., Giangrande, A., Licciano, M., Felline, S. & Fraschetti, S. (2008). Effects of offshore platforms on soft-bottom macro-benthic assemblages: A case study in a Mediterranean gas field. *Marine Pollution Bulletin*, Vol. 56, pp.1303–1309

Trabucco, B., Cicero, A.M., Gabellini, M., Virno Lamberti, C., Di Mento, R., Bacci, T., Moltedo, G., Tomassetti, P., Panfili, M., Marusso, V. & Cornello, M. (2006a). Studio del popolamento macrozoobentonico di fondo mobile in prossimità di una piattaforma off-shore (Adriatico Centrale). *Biologia Marina Mediterranea*, Vol. 13, No. 1, pp.659-662

Trabucco, B., Maggi, C., Virno Lamberti, C., Bacci, T., Marusso, V., Vani, D., Gabellini M. & Cicero, A.M. (2006b). Marine benthic assemblages around a gas platform (Central Adriatic sea). *Coastal Innovations and Initiatives - Proceedings Littoral*, pp.39-46

Trabucco, B., Bacci, T., Marusso, V., Lomiri, S., Vani, D., Marzialetti, S., Cicero, A.M., Di Mento, R., De Biasi, A.M. , Gabellini, M & Virno Lamberti, C.. (2008). Studio della macrofauna attorno alle piattaforme off-shore in Adriatico Centrale - Study of the macrofauna sorrounding off-shore platforms in the Central Adriatic sea . *Biologia Marina Meditteranea*, Vol. 15 No. 1, pp.141-143

U. S. Environmental Protection Agency (1997). Recommended guidelines for samplingmarine sediment, water column, and tissue in puget sound

Virno Lamberti , C., Nonnis, O., Cicero, M., Gabellini, M. & Ferraro, M. (2004) Acoustic investigations to assess the environmental impact due to the installation of a gas offshore platform in the central Adriatic, Sea (Italy). *Fresenius Environmental Bulletin*, Vol.7

Wilson III, C.A. & Heath, J.W. (2008). Rigs and offshore structures. In: Steele J.H., Thorpe S.A., Turekian K.K.(Eds), Encyclopaedia of Ocean Science, Academic Press, Orlando, pp.2414-2419

www.eni.it

Part 3

Natural Gas Marketing and Transportation

12

Natural Gas Market

Joseph Essandoh-Yeddu
University of Cape Coast, Energy Commission, Ghana

1. Introduction

The natural gas market is the collection of entities or players including buyers and sellers of the gas that compete among themselves or work together on different segments of the gas value and supply chain. It therefore encompasses all the players - producers, transporters, regulators, sellers and buyers and their activities in the market and industry.

Most natural gas trading shares some standard specifications which include the following:

- Specifying the buyer and seller.
- The price.
- The quantity of natural gas to be sold.
- The receipt and delivery point.
- The tenure of contract.
- Payment dates.
- Quality specifications for the gas.
- Arbitration in times of misunderstanding.

Physical contracts are usually negotiated directly between buyers and sellers but electronic bulletin boards and e-commerce trading sites are taking over the transactions in recent times.

Natural gas before the 1970s was once considered a mere byproduct of oil and thus not worth the significant capital investment required to find, gather, process and transport or distribute this resource. So for many places where oil was being prospected and natural gas was found, the wells would be ceiled for locations where local usage demand did not exist.

Driven by energy security and greenhouse gas emissions concerns, the wide spread and cleanliness of the gas has pushed it to the forefront of the fossil fuels such that it has become the clean fossil fuel of choice as well as a global commodity. This has changed the global landscape such that it has become almost equally valuable as the oil being prospected wherever it is found at present times. One important advantage that natural gas has over other fossil fuels besides, its relatively low carbon emissions, is that it leaves no solid residues and produces less other pollutants like sulphides/sulphates on combustion.

Worldwide, the natural gas industry has grown rapidly in recent years. IEA (2011) projects that the global energy share of natural gas would increase from 21% in the 2008-2010 to about 25% by 2035 (Figure 1).

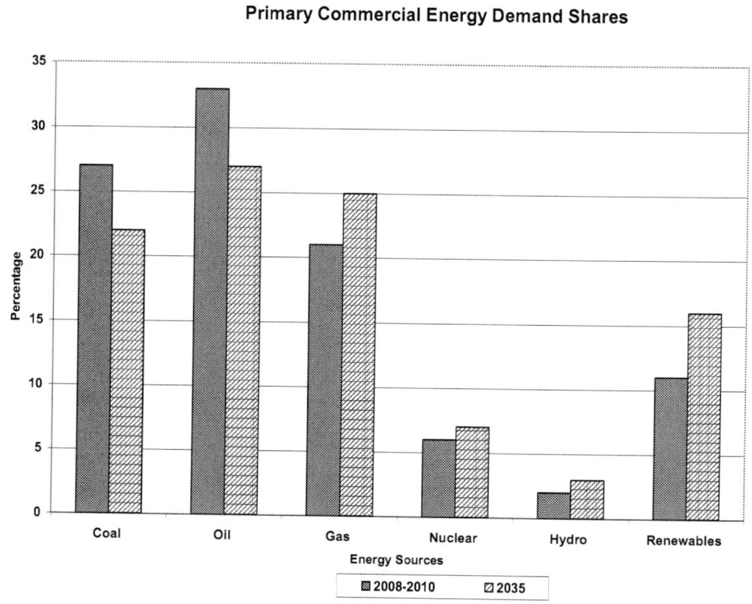

Fig. 1. Global Primary Commercial Energy Demand Shares.

Global marketed gas production has grown from an average of 1.2 trillion cubic metres[1] in the 1970s to about 3.3 trillion cubic metres in 2010. Russia and the United States with annual average production each of about 550 billion cubic metres since 2008 are the largest producers but also the largest consumers, with the United States consuming over 600 billion cubic metres and almost twice the consumption of Russia (BP, 2011).

Global average annual production had almost matched consumption since 2000; 3.05 trillion cubic metres (296 bcf/d)[2] and 3.03 trillion cubic metres (294 bcf/d) respectively. Mean annual growth rate was about 3.2% for both consumption and production.

This significant growth in gas consumption shot up the average price of the gas in the United States from about $6/MMbtu[3] in the 2000s to about $12/MMbtu in early 2008. The increasing price signals accelerated the pace of investment in unconventional gas production facilities, particularly in the United States (IEA, 2011).

The global financial and economic crisis in 2008 however led to a sudden and pronounced drop in consumption from the 3.03 trillion cubic metres in pre-2008 to about 2.94 trillion cubic metres by end of that year. Consumption growth rate between 2007-2008 was 2.4% but dropped to negative 2.2% in 2009. Average annual production of 3.05 trillion cubic metres from 2000-2008 correspondingly dropped to 2.97 trillion cubic metres in 2009.

[1] 1.09 billion tonnes of oil equivalent (BTOE) or 116.5 billion cubic feet of gas per day (bcf/d).

[2] Billion cubic feet per day

[3] MMbtu is million British thermal unit

The significant fall in gas demand had led to emergence of significant amount of over capacity in gas production in the United States, causing the average price of the gas in the country to fall to $4.00-$4.50/MMbtu range since the late 2008.

Consumption growth has however risen again; it jumped to 7.5% in 2010 (from negative 2.2% in 2009) dragging production to increase to 3.19 trillion cubic metres (almost 310 bcf/d) that year (from 2.97 trillion cubic metres in 2009).

Most of the world's proved natural gas reserves (about 72%) are located in two regions: the former Soviet Union (FSU) and the Middle East, although there is little production relative to the size of its reserves in the latter region. Both regions hold roughly about 56 trillion cubic metres (about 2,000 trillion cubic feet) of natural gas reserves as compared to the estimated world total reserves of 150 trillion cubic metres (5,300 trillion cubic feet) (CEE, 2006).

Despite the size of their reserves, North American region rather than the former Soviet Union and the Middle East has been the leading producer as well as consumer of the gas for decades. Nevertheless, production in the former Soviet Union region had increased significantly from 0.2 trillion cubic metres (19 bcf/d) in 1970 to 0.71 trillion cubic metres (69 bcf/d) in 2000, i.e. over 260 percent increment whilst the North American production increased only from 0.64 trillion cubic metres (62 bcf/d) to 0.74 trillion cubic metres (72 bcf/d); a mere 16% increase over the same period (CEE, 2006). The former is obviously catching up and this scenario is likely to influence the evolving gas market and industry into a new paradigm in the years to come.

2. Natural gas value chain

Generally, worldwide, natural gas market and industry comprises Upstream, Midstream and Downstream. Their various operational elements are elaborated below (Figure 2).

2.1 Upstream

Upstream activities cover exploration, field production, processing of the raw or associated gas and separation into various other molecules, gathering of the gas from feeder wells before it is pumped midstream.

Production involves extraction of discovered supplies from hydrocarbon fields either with crude oil (as associated/dissolved gas[4]) or separately (non-associated natural gas[5]).

Production time span for gas used for economic analysis is usually longer than oil and can sometimes go up to 30-35 years. During oil production, excess associated gas is disposed of by direct (venting) or by controlled burning (flaring). Under normal conditions, flaring would only take place in the start-up phase of production. Flaring of natural gas however, releases carbon dioxide, the most significant greenhouse gas into the atmosphere. Besides, the flaring gas is also a resource and commodity being wasted since it could be used for power production or as industrial feedstock or exported to earn foreign exchange, of course, all depending upon the quantities involved.

---

[4] Natural gas that occurs in crude oil reservoirs either as free gas (associated) or a gas in solution with crude oil (dissolved gas).

[5] Natural gas that is not in contact with no or significant quantities of crude oil in the reservoir.

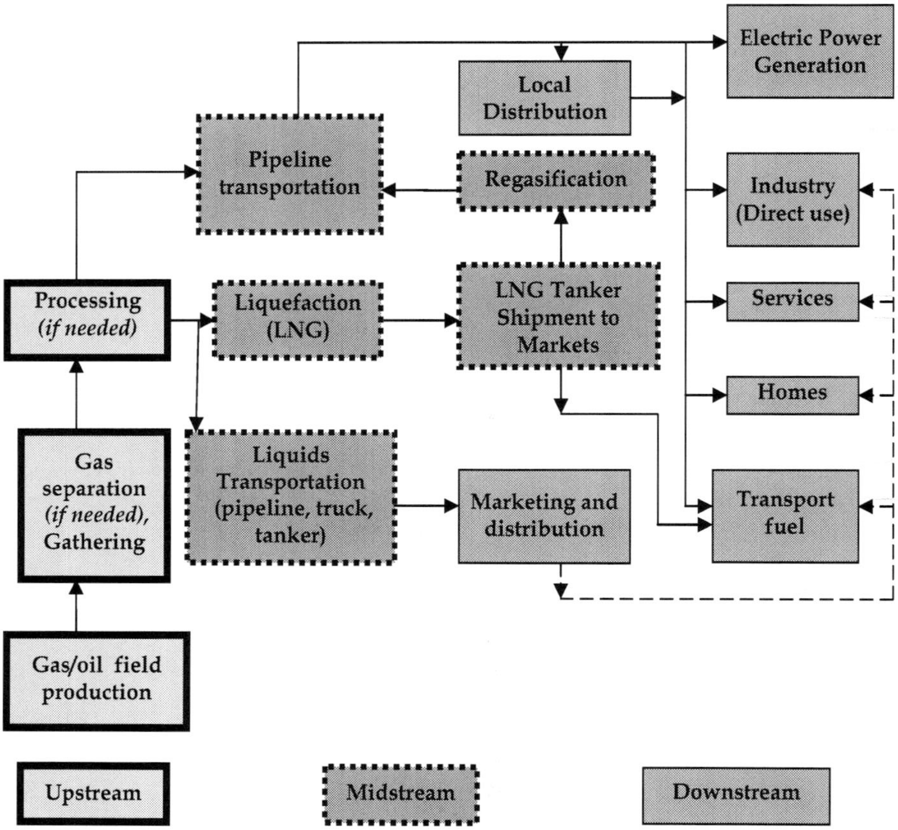

Fig. 2. Natural gas major value chain.

For where technically and economically possible and attractive therefore, the gas is re-injected into the geological formation for latter use and also for secondary recovery to enhance production. Re-injection also is a way of prolonging and increasing or maintaining the oil output. The longer however the re-injection, the likelihood of also losing some significant quantities of the gas in the geological formation, and in some extreme cases, if care is not taken, could damage the reservoir. So re-injection goes with some risk and for that matter gas flaring option technically tends to be more attractive in most cases to the field production operator.

Globally, between 10-15% of the gas produced annually is used for re-injection. Further 5-10% of the gas is flared. Owing to environmental concerns however, pressure to monetize 'flaring' gas is greatly being directed at locations where crude oil is being produced and the associated gas is being flared (IEA, 2011).

In 2010, out of the global production of about 3.2 trillion cubic metres, around 500 billion cubic metres of gas was used for re-injection About 250 billion cubic metres was lost through 'shrinkage' due to extraction of natural gas liquids including LPG[6] and for fuelling production facilities. Between 100-125 billion cubic metres was flared largely in developing countries and in large oil field operations as in Western Siberia of Russia. Much of the flaring however is expected to be reduced significantly by 2020 as most of such host countries are putting in place policies to compel operators to adopt 'zero flaring' of this potential fuel source (IEA, 2011).

Gathering involves collection of natural gas production from multiple wells connected by small diameter pipeline systems to a large pipeline for treatment.

Processing is the separation of heavier molecules and unwanted substances such as water. Out of the heavier molecules could be LPG or Natural Gas Liquids (NGLs) - ethane and larger molecules are stripped out as feedstock for petrochemical industry. LPG, a propane/butane mixture, could be obtained during the processing of **wet** natural gas[7] to obtain the **dry** natural gas[8]. If the gas stream contains impurities such as sulphide compounds then **treatment** is required. The gas is branded as **sour** if it has excessive sulphur compounds, giving it offensive odour. It is **sweet**, if it has much lower concentration of sulphur compounds, particularly, hydrogen sulphide.

2.2 Midstream

Midstream activities include pipeline transportation, liquefaction of the gas into LNG and storage of the gas. Some **upstream** activities however over lap with those of **midstream**. For instance, if the **gathering** involves delivery to a processing plant through a long-distance low pressure pipeline, it is usually placed under **midstream.** Some literature also consider **processing and treatment** of the raw gas, like gas-liquid extraction as a **midstream** activity.

Storage is usually in depleted underground reservoirs or caverns like those associated with salt domes. Storage can be located either near production or near demand. Re-enforced steel

---

[6] LPG is liquefied petroleum gas.

[7] A mixture of hydrocarbon compounds and small quantities of various non-hydrocarbons existing in the gaseous phase or in solution with crude oil in porous rock formations at reservoir conditions. When this mixture gas reaches the surface at normal temperature and pressure conditions, some of the hydrocarbon molecules become liquid. The principal hydrocarbons normally contained in the mixture are methane, ethane, propane, butane and pentane. Typical non-hydrocarbon gases that may be present in reservoir natural gas are water vapour, carbon dioxide, hydrogen sulphide, nitrogen and traces amounts of helium.

[8] Natural gas which remains after; (a) the liquefiable hydrocarbon portion has been removed from the gas stream, (b) any volumes of non-hydrocarbon gases have been removed where they occur in significant quantity to render the gas unmarketable. Dry natural gas is also called consumer-grade natural gas. Dry gas also indicates that the fluid does not contain enough of the heavier molecules to form liquid at surface temperature and pressure conditions.

dome engineered storage systems are also available where natural gas is stored in the form of LNG.

Pipeline transportation covers delivery of the gas from producing basins to local distribution networks and high-volume users via large diameter, high volume pipelines. Countries vary greatly with respect to allowable pipeline specifications for heat content. Some pipelines transport **dry** gas and some **wet** depending upon the standard set by the operators and the buyers.

Liquefaction, shipping and re-gasification known collectively as the LNG value chain, entails conversion of the gas to a liquid form via refrigeration to a cryogenic fluid (temperature -160°C[9]) for transportation from a producing country or region to a consuming country or region via ship.

2.3 Downstream

Midstream activities yield **downstream** activities. **Downstream** covers end-use conversion and distribution of the gas for power generation and to various sectors of the economy.

Distribution **downstream** involves retailing and final delivery of the gas via small diameter and low pressure local gas networks operated by local distribution companies (LDCs) - often called gas utilities. **End use conversion** covers direct use or conversion for use in other forms (petrochemicals, electric power or vehicle fuels).

Overall, there are also losses from **upstream** to **downstream**; for instance, about 20%, i.e. between 800-900 billion cubic metres of the gross gas volume produced globally in 2010 never reach the market (IEA, 2011).

2.4 Commercial and other practices governing the natural gas value chain

The following commercial elements and practices serve to bind the operating segments of the natural gas value chain and link suppliers, transporters and distributors with their customers. They are:

- Aggregation: Consolidation of supply obligations, purchase obligations or both as means of contractually - as opposed to physically - balancing supply and demand.
- Marketing: Purchase of gas supplies from multiple fields and resale to wholesale and retail markets.
- Retail marketing constitutes sales to final end users (typically residential, commercial, industrial, electric power and services).
- Capacity brokering: Trading of unused space on pipelines and in storage facilities.
- Information services: Creation, collection, processing, management and distribution of data related to all the other industry and market functions.
- Financing: Provision of capital funding for facility construction, market development and operation start-up.
- Risk management: Balancing of supply, demand and price risks.

[9] -256 °F

- Environmental issues: Impact of the activities of the various segments of the value chain on the ecosystem, comprising land, water bodies, and the atmosphere including the living beings within them.
- Health and Occupational Safety: impact of the various activities on the health and safety of workers in the various segments of the value chain.

2.5 Challenges for efficient and harmonious market along the value chain

The **first** challenge is to create a framework for efficient production of the gas upstream (Foss, 2005).

The **second** challenge is achieving efficiency in pipeline transportation to distribution of the gas to various consuming sectors.

The **third** challenge is developing transparent markets for natural gas supply and consumption. The evolving trend has been to separate infrastructure and product (often referred to as "unbundling") and to search for ways of providing competitive access to pipeline systems for multiple suppliers and users of natural gas (often termed "third party access" or "open access"). In these cases, pipelines become like toll roads, priced through tariff design, while the gas is priced in discrete competitive markets. Traditionally or previously, it was a "bundled" market.

When pipelines become subject to "third party access" regimes, the linking of suppliers to buyers, can be separated into competitive business activities. Such has triggered growth in marketing and trading (both of the physical product as well as financial derivatives) as separate businesses, particularly in North America.

Both the evolution of market-based policies for natural gas and international trade linkages have given rise to a **fourth challenge** which is, timely and accurate data and information on supply, demand and prices. Data and information must be accurate and timely to consistently attract the necessary long-term investment and minimize market disruptions and distortions. Unreliable forecasts create conflicts to the extent that they result in supply-demand imbalances which neither industry nor government at times has the flexibility to correct in a timely manner.

A **fifth** and increasingly **challenge** is dealing with integration, with respect to industry organization and international trade. Industry organization encompasses both vertical (meaning up and down the value chain) and horizontal (meaning over some geographic or international market region) integration. Integration of physical infrastructure across international boundaries has grown rapidly with increased demand for overland pipeline natural gas.

3. Recoverable reserves

Recoverable resources of conventional gas[10] worldwide are estimated to be about 400 trillion cubic metres based on current technology and price range and this is equivalent to more than 120 years of global consumption today (Table 1).

[10] i.e. including associated/dissolved gas

REGION	CONVENTIONAL		UNCONVENTIONAL					
			Tight Gas		Shale Gas		CBM	
	tcm	$/MMBTU	tcm	$/MMBTU	Tcm	$/MMBTU	tcm	$/MMBTU
Eastern Europe and Eurasia	136	2-6	11	3-7	N.A-	N.A	83	3-6
Middle East	116	2-7	9	4-8	14	N.A		N.A
Asia/Pacific	33	4-8	20	4-8	51	N.A	12	3-8
OECD North. America	45	3-9	16	3-7	55	3-7	21	3-8
Latin America	23	3-8	15	3-7	35	N.A	N.A	N.A
Africa	28	3-7	9	-	29	N.A	N.A	N.A
OECD Europe	22	4-9	N.A-	N.A-	16	N.A	N.A	N.A
Total	**404**	**2-9**	**84**	**3-8**	**204**	**3-7**	**118**	**3-8**

tcm: trillion cubic metres; MMBTU: million British thermal unit; CBM is coal-bed methane
Source: IEA, 2011

Table 1. Global recoverable resources of conventional gas.

Unconventional resources such as shale gas and coalbed methane (CBM) are now estimated to be as large as conventional resources **(Table 1)**. Production of unconventional gas is estimated to represent about 13% of global production as at 2010 **(Table 2)**.

REGION	**PROVED RESERVES as at 2010**		**PRODUCTION**		**CONSUMPTION**		**TRADE MOVEMENT**			
							LNG Exports		**Pipeline flows**	
	tcm	%	bcm	%	bcm	%	bcm	%	bcm	%
North America	9.9	*5.3*	826.1	*25.9*	846.1	*26.7*	20.0	*6.7*	123.6	*18.2*
S & C America	7.4	*4.0*	161.2	*5.0*	147.7	*4.7*	9.2	*3.1*	14.3	*2.1*
Europe & Eurasia	63.1	*33.7*	1,043.1	*32.7*	1,137.2	*35.9*	87.8	*29.5*	470.0	*69.4*
Africa	14.7	*7.9*	209.0	*6.5*	105.0	*3.3*	0	*0.0*	4.9	*0.7*
Middle East	75.8	*40.5*	460.7	*14.4*	365.5	*11.5*	2.9	*1.0*	31.5	*4.6*
Asia - Pacific	16.2	*8.7*	493.2	*15.4*	576.6	*18.2*	177.8	*59.7*	33.4	*4.9*
Total	**187.1**		**3,193.3**		**3,169**		**297.6**		**677.6**	

NB: tcm is trillion cubic metres; bcm is billion cubic metres
S & C America is South and Central America
Source: BP, 2011

Table 2. Proved Reserves as against Production and Consumption of natural gas in 2010.

Production cost of unconventional gas in North America ranges from $3-7/MMBtu. Other regions could be higher (**Table 1**).

Timely and successful development of new fields however depends on complex factors including industry's capability, policy choices of host countries and market demand.

4. Natural gas demand

IEA (2011) estimates that global gas demand could reach 5.1 trillion cubic metres by 2035 from about 3 trillion cubic metres in 2010. To meet the growing demand, global gas production must increase by 1.8 trillion cubic metres annually and unconventional resources would account for about 40% of it (IEA, 2011).

At the regional level, it seems clear that United States and Canada would continue to play minor role in net global gas trade in the short to medium term due to adequate gas and over-capacity production in North America. On the other hand, rapid growth in Asian gas demand during the same period would put pressure on supply (notably LNG) and consequently, prices and would stimulate greater supply investments within the Asian region and Greenfields elsewhere, particularly in Africa.

All regions are expected to see significant gas growth in the long term (IEA, 2011).

Most important factors determining the demand for natural gas in a country include the following:

- Level of economic activity of the country.
- Richness or suitability of the gas as feedstock for industry.
- Ease and access to supply.
- Competitiveness of the gas versus other fuel sources regarding pricing and environmental considerations.
- Host government policies involving incentives and commercial framework for investors.

Level of economic activity being driven by rising household incomes and increased commercial activities would boost demand for secondary and modern energy such as electricity and gas for cooking and heating. Natural gas use in residential, commercial and transport sectors of the economy requires construction of distribution networks which could be very expensive. Higher household income levels would therefore be necessary to cover cost of delivery and service. Growing industrial sector would require more power to support the growth.

Rich or wet gas if processed could yield other Petroleum Chemicals and for that matter add more value to upstream-midstream activities culminating in industrial growth.

Ease and access to supply of the gas if the price is competitive compared to alternative fuels would attract more customers for the gas. Natural gas has become environmentally **competitive fuel** of choice for environment-conscious economies and also economically competitive for carbon constrained economies because it emits less greenhouse gas per unit combustion compared to other fossil fuels. The global push to utilize relatively clean burning natural gas for power generation has both environmental benefits and power generation diversification and have triggered strong convergence between the natural gas and electric power value chains.

Power generation is the largest gas consuming sector and the biggest driver of gas demand today and it is expected to be so for the foreseeable future. Natural gas is used in both gas turbine and combustion/steam generating plants and prices of the gas in many parts of the world had remained strongly linked to oil, due to its interchangeability with other fossil fuels as industrial and power generation fuel.

Gas demand in the power sector is however said to be more sensitive to changes in the rate of growth of GDP than its use in the other applications. It was observed that from 1990-2008 a global average increase of 1% in GDP led to 1% demand in gas for power generation (IEA, 2011).

Besides power generation, transport stands out as having significant potential to drive demand upwards. Natural gas to supplement or substitute diesel as transportation fuel is one of the measures to reduce urban transport pollution. This is available in a number of European countries but also some developing countries like India.[11]

Natural gas vehicles (NGVs) are fuelled by CNG (compressed natural gas) and LNG (liquefied natural gas). The vehicular features are similar to the conventional ones using internal combustion engine except for the fuel injection system and the size of the fuel storage tank. These differences have given rise to additional costs of $2,000-10,000 higher than the standard vehicles for the natural gas fuelled vehicles.

LNG ships are currently diesel or light crude fuelled but most could be replaced by LNG in the long term and in the foreseeable future. There are approximately 120 LNG ships currently involved in worldwide trade.

Host government policies should be geared towards the ability to facilitate investment in pipeline networks. It means having commercial frameworks in place that not only attract investors but also protect affected public interests. For regional natural gas trade to occur, contiguous countries must have commercial frameworks that are similar enough to encourage market participants to develop efficient cross-border networks as well as dealing with risks. A good example for developing countries is the West African Gas Pipeline (WAGP) that carries Nigerian gas to Benin, Togo and Ghana (WAGPCO, 2011).

5. Natural gas supply

Most of the supply of natural gas is by pipeline for overland delivery and ship for transoceanic delivery as LNG.

Most of the imported gas would come from countries forming the Gas Exporting Countries Forum which has Russia, Oman, Malaysia, Algeria, Nigeria, Equatorial Guinea, Brunei, Iran, Turkmenistan, Indonesia, Qatar and Norway as the main members.

Globally, LNG trade is divided into two international trade blocks; West of Suez Canal and East of Suez Canal. There are old and new exporters as well as new and old importers for both sides of the Suez (Figure 3). For the new, the list is still evolving.

[11] Natural gas compressed to about 200 times at standard atmosphere.

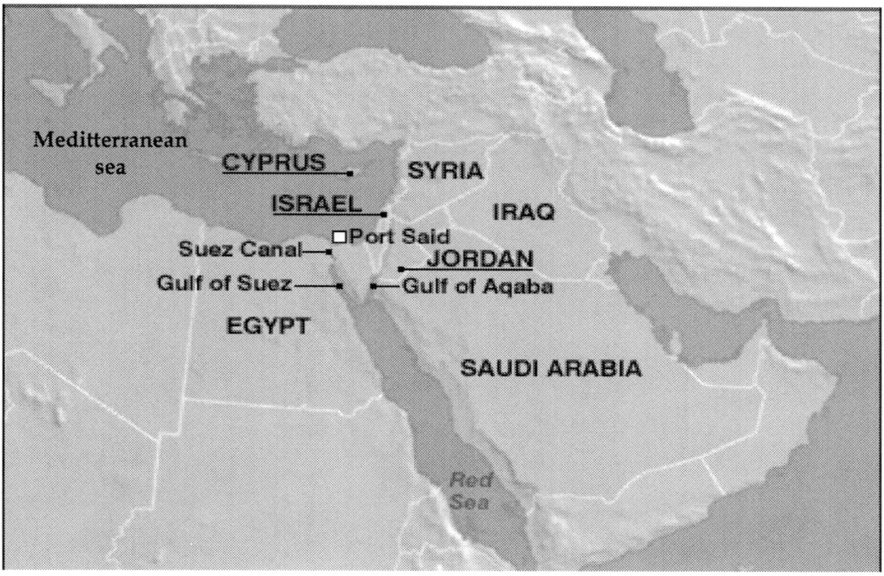

Fig. 3. Suez Canal in Egypt, linking Mediterranean sea to the Red sea.

For the countries involved as at the time of compiling the lists, the picture is as below (Tables 3 and 4):

5.1 East of Suez

Old Exporters	**Old Importers**
Indonesia	Japan
Malaysia	South Korea
Australia	Taiwan
Brunei	
New Exporters	**New Importers (started since 2005)**
Qatar	China
UAE	India
Yemen	
	Future
Future	Thailand
Iran	Pakistan

Table 3. LNG Exporters and importers east of Suez Canal.

5.2 West of Suez

Substantial LNG regasification was built in the United States early 2000s in anticipation of the country becoming a large importer of LNG. However the unexpected rise in domestic (local) gas production particularly from shale gas during the same period reduced imports to the country significantly.
The following sections would elaborate gas transportation options.

Old Exporters Algeria Nigeria Trinidad and Tobago	**Old Importers** Western Europe USA
New Exporters Equatorial Guinea Egypt Angola Norway Peru	**New Importers** Western Europe USA Brazil
Future North America	*Future* Ghana as hub for West Africa

Table 4. LNG Exporters and importers West of Suez Canal.

6. Natural gas transport

The prevailing global market structure is such that for conventional gas, most consumption takes place far away from gas producing centres, which means the gas has to be transported by overland pipelines and or as LNG via specialized ships depending upon the distance and location of the consuming market and this condition is prevalent among developing and emerging market countries that are gas rich. In some cases however, gas is stranded due to physical distances and technical complexities associated with transportation.

Within local or domestic markets, there thus must be sufficient demand and the capacity for potential consumers to pay in order to justify investment in transportation and distribution infrastructure. For most developing countries therefore, the gas is largely used in industry as feedstock or fuel, thus very little goes to the residential and the commercial sectors due to relatively low income per capita. In the developed countries on the other hand, residential, commercial and the industrial sectors compete for the gas almost equally.

6.1 Overland pipeline transportation

Overland pipelines are generally found to have lower cost per unit and higher capacity compared to shipment by rail or road and for that matter the most economical way to transport large quantities of natural gas[12] over land. Overland pipeline transportation and distribution have therefore being the dominant mode for terrestrial gas transport.

The oldest gas pipelines in the world are said to be located in the United States. The first long-distance natural gas trunklines to serve the Midwest Region from the prolific Hugoton Basin located in the Texas/Oklahoma Panhandle were built during the 1930s (U.S EIA, 2011).

[12] Also crude oil and refined products.

North America also has the world's longest gas pipeline network. The Canadian system is about 20,000 km whilst the United States has over 90,000 km of natural gas pipeline on more than 66 intrastate natural gas pipeline systems (including offshore-to-onshore and offshore Gulf of Mexico pipelines) delivering gas to local distribution companies, municipalities and to many large industrial and electric power facilities (U.S EIA, 2011).

Europe on the other hand has 28,000-29,000 km of gas pipeline network. Most of the natural gas being used in Europe is imported via pipelines from Russia, Central Asia, the Middle East and North Africa, with imports from Russia accounting for a quarter of the total supply Table 5.

Pipeline	Origin of Gas	Capacity bcf (bcm)*	Completed	Distance Km
Trans-Austria	LNG - Austria	1695 (48)	1960s[13]	380
Trans-Europa	Northsea to Germany	565 (16)[14]	1974	968
Transitgas	Switzerland	NA	1974[15]	293
Frigg U.K	Northsea to Scotland	NA	1977	362
MEGAL	Chech/Autrian to France	777 (22)	1980	1,115
FLAGS	Northsea to U.K	NA	1982	451
Trans-Mediterranean	Tunisia to Italy	1060 (30)	1983	2,475
West-Siberian [16]	Russia	1100 (31)	1984	4,500
STATpipe	Northsea-Norway	671 (19)	1985	890
Fulmar	Northsea to Scotland	160 (4.4)	1986	290
MIDAL	Northsea to Germany	459 (13)	1992	702
NOGAT	Northsea to Netherlands	NA	1992	62
STEGAL	Germany	NA	1992	314
CATS[17]	Northsea to U.K	618 (17.5)	1993	404
Zeepipel-III	Northsea to Belgium	918 (26)	1993	1,416
Rehden-Hamburg	Germany	NA	1994	132
Eorupipe1	Northsea	635 (18)	1995	660
Netra	Northsea to Germany	NA	1995	408
Maghreb-Europe	North Africa	424 (12)	1996	1,620
Yamal-Europe	Russia	1165 (33)	1997	4,196
WEDAL	Germany - Belgium	353 (10)	1998	319
JAGAL	Russia to Germany	847 (24)	1999	338
Europipe11	Norway to Germany	847 (24)	1999	658
Blue Stream	Russia	16 (0.47)	2003	1,207
Green Stream	Libya (North Africa)	283 (8)	2003-2004	520
South Caucasus	Caspian Sea	247 (7)	2006	700
BBL	Netherlands to U.K	668 (19.2)	2006	230
Franpipe	NA	706 (20)	1998	840
Langeled	Norway	2.5 (0.07)	2007	1,200
South Wales	NA	NA	2007	197
Gazela	Russia border to Czech	NA	2011	176
	Total			**28,023**

bcf is billion cubic feet; NA is information not available

Table 5. Major existing and operative gas pipelines in Europe.

[13] Extension completed in 2007
[14] Upgraded in 1998 and 2009
[15] Upgraded continued up to 2003
[16] Also called Urengoy-Pomary-Uzhgorod or Trans-Siberian pipeline to Europe
[17] Also called Central Area Transmission System

Australia has around 5,000 km gas pipeline network whilst Asia including Middle East has about 15,000 km pipeline network in operation as of 2011.

The longest gas pipeline in sub-Saharan Africa is the West African Gas Pipeline. The 678 km pipeline was completed in 2006 and transports natural gas from Nigeria to Ghana through Benin and Togo (WAPCO, 2011).

6.1.1 New pipeline projects

Relatively few pipelines are under construction and most are still planned as indicated in **Table 6**.

REGION	PIPELINE NAME	DELIVERY POINT	CAPA CITY bcm	STATUS	STATE DATE
Russia	Altai	China	30	Planned	2015
	Russian-Asian Pacific	Korea	10	Planned	2015-17
	Nord Stream	N.W. Europe	27.5	Under construction	2011 end
	Nord Stream 2	N.W. Europe	27.5	Planned	2012
	South Stream	S.E. Europe	63	Planned	2015 end
Caspian/ Middle East	Nabucco	S.E Europe	26-31	Planned	2017
	ITGI	S.E. Europe	12	Planned	2017
	TAP	Italy	10+10	Planned	2017
	IGAT 9	Europe	37	Planned	2020+
Caspian	CAGP	China	+30	Under construction	2012
	CAGP Expansion	China	+20	Planned	Post CAGP
	TAP1	Pakistan	30	Planned	2015+
Middle East/ Turkey	IPI	India	8	Planned	2015+
	Arab Gas Pipeline	Middle East / Turkey	10	Partially built	n.a
Asia Pacific	Myanmar-China	China	12	Under construction	2013
Africa	GALS1	Europe	8	Planned	2015

Start date are as reported by sponsors
CAGP is Central Asian Gas Pipeline; TAPI is Trans-Afghanistan Pipeline; IRI is Iran-Pakistani India; ITGI is Interconnection Turkey Greece Italy; TAP is Trans Adriatic Pipeline; IGAT is Iranian Gas Trunkline; GALSI is Gasdotto-Algeria-Sardegna-Italia. Source: IEA, 2011

Table 6. New pipelines planned and under-construction in the world.

6.2 Liquefied natural gas transportation

Although pipelines can be built under the sea, that process is economically and technically demanding, so the majority of gas at sea is transported by LNG tanker ships which is also seen as flexible compared to pipeline[18].

[18] LNG is natural gas that has been compressed 600 times its volume at standard atmospheric conditions (STP) and cooled to an extremely cold temperature (-260° F/ -162.2° C).

Around 75% of the inter-regional global trade is by LNG shipment as compared to pipeline.

LNG ships vary in size from 20,000 cubic metres to over 145,000 cubic metres cargo capacity but the majority of modern vessels are between 125,000 cubic metres and 140,000 cubic metres capacity (58,000 to 65,000 tonnes).

From LNG Liquefaction Plants, ships travel to demand markets to unload the LNG at specially designed terminals where it is pumped from the ship to insulated storage tanks and regasification Plant, where it is warmed up and converted back to gas before being delivered into the local gas pipeline network. Specially designed vehicle trucks are also used to deliver LNG to other storage facilities in different locations.

The liquefaction segment along the value chain is the most expensive and could constitute about 40% of the total LNG value chain cost. The regas segment is the least expensive depending upon the technology opted for (Table 7).

Production	**Liquefaction**	**Shipping**	**Regasification & Storage**
$0.5-1.0/MMBtu	$0.8-1.2/MMBtu	$0.4-1.0/MMBtu	$0.3-0.5/MMBtu
13–27%	22-40%	10-27%	8-15% *(15-30% for storage)*
Total value chain cost = $2.00-$3.70; with cost escalation = $2.60 - $4.80 *between 2006-2010*			

Source: Center for Energy Economics, UT-Austin, USA, 2008

Table 7. LNG Value Chain Costs.

LNG delivery from the vessels is accomplished through the following technologies:

i. Permanent LNG re-gasification plants.
ii. Floating Re-gasification plants using grounded LNG vessels which have retired from services.
iii. Temporary or stop-gap through "Energy Bridge Re-gasification Vessels" (EBRVs)

Permanent LNG discharge/re-gasification terminal: Development of permanent LNG re-gasification plant requires at least two years if funding for EPC[19] is readily available.

Energy Bridge Regasification Vessels: The energy bridge re-gasification is the LNG technology that delivers the gas in the shortest possible time; i.e. within a year. Energy Bridge Regasification Vessels, or EBRVs™, are purpose-built LNG tankers that incorporate onboard equipment for the vapourisation of LNG and delivery of high pressure natural gas. These vessels load in the same manner as standard LNG tankers at traditional liquefaction terminals, and also retain the flexibility to discharge the gas in two distinct ways. These are through:

- the EBRV's connection with subsea buoy in the hull of the ship; and
- a high pressure gas manifold located in front of the vessel's LNG loading arms.

Floating Re-gasification plants: Average lifetime of most LNG vessels is 25 years. This means LNG vessels built in the 1980s have become less competitive for transport services. Such an LNG ship is retired and reconfigured as floating LNG re-gasification facility.

[19] EPC is Engineering, Procurement and Construction.

Typical LNG ship has capacity of 120,000-125,000 liquid cubic metres (lcm). The larger the containment the greater the application for floating storage and regasification applications. Some 59 ships built worldwide before 1983 with containment between 122,000-133,000 lcm are due for retirement.

The LNG market and industry is expanding rapidly. LNG liquefaction capacity was 270 bcm in 2008 increasing to 370 bcm as of 2011 of which Qatar accounts for over 25%. This is expected to expand to 450 bcm in 2015 and doubled by 2020. New projects under construction are due to be on stream from 2014-2016 (Table 8).

COUNTRY	PLANT	CAPACITY		START DATE
		bcm	MTPA	
Algeria	Skikda (rebuild)	6.1	4.5	2013
	Gassi Touil	6.4	4.7	2013
Angola	Angola	7.1	5.2	2012
Australia	Pluto	6.5	4.8	2011
	Gorgon	20.4	15.0	2014
	Gladstone_LNG	10.6	7.8	2014
	Queensland Curtis	11.6	8.5	2015
	Donggi Senoro	2.7	2.0	2014
Papua New Guinea	PNG LNG	9.0	6.6	2014

Note: Start dates are as reported by project sponsors. MTPA is million tonnes per annum. Source: IEA, 2011.

Table 8. Liquefaction plants under construction by country.

500 bcm additional liquefaction capacity is being evaluated for 2015-2020 of which about 75% of it has been earmarked for construction in Australia, Russia, Nigeria and Iran (IEA, 2011).

6.3 Downstream gas delivery to customers

From the transmission pipelines, the gas reaches the consumers through the distribution network. Direct access to gas supply to maintain higher pipeline load factors improves efficiency of operations. However, occasionally, natural gas delivery commitments to downstream customers could be wholly or partially curtailed due to:

- force majeure conditions;
- compressor (station or unit) unavailability;
- line break; and
- supply unavailability for other reasons other than those mentioned above.

For short term (lasting up to 12 hours or 24 hours) "upset" conditions therefore, **linepack** could be used to address the shortfall. The usefulness of the **linepack** would depend on the duration, magnitude and location of this transient condition.

For long term supply disruptions (lasting days, weeks or months), however, storage facilities or an LNG storage system might be required. Depleted oil or gas reservoirs, aquifers or salt caverns are examples of such storage facilities.

If longer (over 6 months) disruptions are anticipated, alternative gas supply network is essential. In general, a mix of overland pipeline and LNG source, with an associated re-gasification unit, for a country or region is recommended.

Because of some of these challenges, domestic or local pipeline transportation of natural gas tends to be characterized by state intervention in various forms to manage market power and to protect the public interest as well.

The next question posed *is at what price is the gas delivered? How is the pricing of the gas computed and what kinds of contracts are made for the supply and delivery of the gas?*

7. Gas pricing, ownership and supply contracts

7.1 Pricing

Natural gas, particularly LNG is generally priced in CIF[20]. Formula 1 is the general pricing formula:

Total price = Gas head price + Transportation price + Other prices. (1)

Where Other prices could be distribution or profit margin, taxes and depreciation.

Some of the gas pricing mechanisms that have been introduced into the market are as follows:

- *Cost of Service*
- *Spot Price*
- *Netback Pricing*
- *"S-shaped" price curve*
- *Barter trade*
- *Price-Index provision*
- *Seasonal and weather-normalised rates*
- *Gas Swaps*
- *Production or Supply Payment*

7.1.1 Cost of service

The cost of service (COS)[21] of a project is defined as the minimum price required to provide for capital recovery, covering operating costs and paying taxes, royalties and production sharing, etc. The traditional gas market is based on long term contract and has used COS for pricing. COS is also basis for price regulation in most countries including approving appropriate return on investments.

[20] Unlike oil which is generally priced in FOB. CIF is abbreviation for Cost, Insurance and Freight, whilst FOB is abbreviation for Free on Board.

[21] Also called "Rate of Return"

The regulated company's revenue requirements are the total funds that the company may collect from customers and it is calculated by multiplying the company's rate base by an allowed rate of return (ROR) and adding this product to the company's operating costs (OC) as shown in formula 2.

$$\textbf{Revenue Requirement} = (\textbf{Rate base x ROR}) + \textbf{OC} - (\textbf{Taxes} + \textbf{Depreciation}) \quad (2)$$

The Rate base is the total value of the company's capital investments, which may include constructional work in progress. The allowed ROR constitutes a profit sufficient to pay interest on accumulated debt and to provide a negotiated acceptable return to investors.

A negotiated acceptable or what is termed as a fair return is determined through a comparable earning tests, where a company's earnings are measured against those of a firm facing comparable risks[22]. Operating costs include expenses on gas, labour, management, maintenance and commercials. Taxes and depreciation are also part of the company's revenue requirments.

Regulators are faced with the challenges of determining the price appropriate and acceptable to the seller to cover cost operations and future investments. Also, there is a challenge in allocating costs and acceptable prices/tariffs to the different customer classes. The regulator therefore first seeks to determine how much of an applicant's captial stock should be included in its Rate base, then attempts to determine which elements of test year costs and revenues should be allowed for regulatory purposes and whether to allow specific changes since the test year. The final step is to determine what the fair ROR is for the company.

7.1.2 Spot price

Spot price is the price of a commodity such as the gas on the spot and is dependent upon time and location. The spot price of gas say in Takoradi, Ghana at 01.00 GMT could be different from the spot price of gas of the same quality in Takoradi, Ghana at 02.00 GMT.

Spot market is characterized by:

1. Short purchase contract term, usually 18 months.
2. Best-effort delivery.
3. Negotiated price reflecting current market conditions.
4. Arranging transportation separately from the sales and usually provided on an interruptible basis.

There is an increasing frequency of spot sales. Participants in the **spot market** include:

1. Local distribution companies (LDCs) who buy the gas from the pipeline companies.
2. Marketers/Traders who buy the gas on the **spot market** to resell to LDCs and large customers. Some marketers do not sell gas themselves. Instead, they bring buyers and sellers together and help negotiate arrangements for transportation of the gas.

[22] A discounted cash flow approach, where a company's capital costs are estimated by analyzing conditions in the financial market, or other methods.

3. Pipeline operators at times purchase **spot market** gas and mix it in their supply portfolios in an effort to lower weighted average cost of their gas.
4. Industry, large commercial consumers and electric utilities may purchase **spot market** gas to increase supply security, but they have to arrange transportation of the gas to their vicinities.

As spot sales increase and proliferate, they are likely to undermine, if not topple the long term contract and price structure which had been an important feature of the traditional gas market.

7.1.3 Netback pricing

Netback pricing is retail price less all costs and expenses. The marketer after taking care of all associated costs, expenses and agreed profit margin, returns to the producer the balance. It encourages the marketer and the producer to increase market share and also receive a fixed profit margin.

Netback Pricing was first introduced by Saudi Arabia in 1985 and it was a revolution because the setting of the pricing was shifted from the producer to the consumer and it was the case until early 2000s when FUTURES market took over.

7.1.4 "S-shaped" price curve

This somehow operates on the principle of price ceiling and price floor. At below an agreed price floor of say \$6 per MMBtu, the buyer agrees to pay an additional premium for the gas. However above price ceiling of say \$10 per MMBtu, it is the reverse, the seller pays a premium to the buyer. The end result is a win-win situation where the buyer enjoys a discount at high prices and the seller is protected against low gas prices. The **S-shaped** pricing mechanism is likely to be used more in the future, particularly as civil society voice becomes louder to see fair share of profit for particularly host developing countries with the finite resource.

7.1.5 Barter trade

Barter trade is simply the exchange of gas for other commodities needed by the gas-exporting country. Existing barter agreements include Russian gas for Polish potatoes and Russian gas for Ukrainian consumer goods food and machinery.

7.1.6 Price-Index provision

Besides the base pricing described above, there is also the **indexing**. Meaning the pricing formula usually consists of two main parts, the base price and indexing. The major consideration for pricing in the case of gas is that is largely used as fuel and so the price is on energy basis and so indexed to price movement of other alternative fuels, such as crude oil, gasoline and fuel oil. Price-index provision ensures that the gas delivered is competitively priced compared to alternate fuels.

Besides price indexing and to cope with competition from alternative fuels and to avoid losing customers, some pipeline operators would provide special tariffs to keep multi-or dual fuel capable industrial customers from switching during period of high gas prices. They could offer incentives such as waiving transportation rate on the gas to reduce the tariff.

7.1.7 Seasonal and weather-normalised rates

Seasonal rates are where the price of natural gas is influenced by the seasonality of gas demand and supply. It is one of the new pricing mechanism for gas particularly in temperate regions. For weather-normalised rates, the price of natural gas is locked into an agreed weather and this can help reduce the effect of abnormal weather patterns on utility earnings.

Tariffs charged by pipelines and LDCs reflect gas sales and transportation volumes. The higher the volumes, the better the sales. However, residential demand is largely weather-sensitive in cold/ temperate climates. Meaning the warmer the weather the less the demand and vice versa. Abnormal weather patterns have therefore been the single most important factor for supplies to largely residential customers. To minimize the impact of weather on revenues, some gas utilities use the **weather-normalised rate** for customer billing; rates are increased when weather conditions are warmer to cover drop in demand; conversely, rates are reduced when weather is colder than normal.

7.1.8 Gas swaps

Besides, **weather-normalised rate,** an LDC may enter into agreements with a large industrial customer with dual-fuel capacity and high-load factors. In the summer when demand for gas is lower for the LDC, the industrial customers may utilize the LDC's excess supplies. Conversely, in winter periods when demand for gas is higher from the LDC, the industrial customer switch to their alternative fuels to enable the LDC meets their contracted volume supplies. Such **gas swap** contracts result in savings for both the LDCs and the customers.

7.1.9 Production or supply payment

In production or supply payment, the buyer makes upfront cash payment for gas to be produced or supplied over a long term in most cases, years. The major features of production payment are:

1. It represents a new capital source for the gas industry.
2. It is basically risk-free to producers. Further, the buyer only has recourse to the given field specified in the agreement.
3. The buyer is better off locking in firm title to reserves at a known price to back their sale commitments to guide against period of rising prices.

7.2 Ownership contracts

The different types of ownership or operational contracts regulating the gas market are:

- Concession

- Production Sharing Contract or Agreement, i.e. PSC or PSA
- Joint Venture
- Service Contract

Concession

Concession is an agreement that is royalty based. It means the host country government gets most of its share through royalty. Royalty is another name for Production Tax and it is related to gross revenue. Gross revenue less royalty equals NET revenue. Royalty also means, the government of host country gets its pay first. The advantage is that host country or government's risk is zero, because production loss or profit, the government takes the royalty. The next deductions from the net revenue include operating costs, depreciation, amortization and intangible drilling costs. Revenue remaining after royalty and the deductions is called **taxable income**. The remaining revenue after taxation is the **contractors take.**

Production - Sharing Contracts (PSC)

For purely Production Sharing Contract (PSC) or Agreement (PSA) as sometimes called, the operating company takes all the risks but manages the production, etc. Government in turn may take off taxes on all imported equipment and provide other tax incentives. After all the deductions by the operating company, there remains the NET revenue and this is what is shared between the operating company or entity and the government. The net revenue is called the profit hydrocarbon and in this case the **profit gas** and is split between the contractor and the government, according to the terms of the PSC negotiated. It means the company takes its money first before government/host country comes in, whilst in **concession** the government takes its money first.

The title of the hydrocarbons however, remains with the host country government (Johnston, D. 1994). With growing awareness of good governance, the state may also maintain the management control and, or would require the operator to submit annual work programmes for scrutiny and approval. Most PSC/PSAs are placing limit on cost recovery such that if the deductions amounts to more than the allowed limit, the balance would be carried forward and recovered later. In some cases, the host country would push royalty payment into the PSC/PSA, and may go by different names such as War Tax levy in Columbia (Johnston, D. 1994). In this case, the profit gas is Net Recovery less Cost Recovery. The operator's share of the profit gas may also be subject to taxation.

Joint venture

Joint Venture is a partnership where the parties share the risks and the rewards together. In **joint venture,** there is always a reference to WI (working interest) meaning - sharing of all costs and expenditures.

Investment in the gas industry in general is capital intensive and has long lead time and often involves financial risks. Investment through Joint Ventures therefore spreads the risks among parties and therefore reduces the share of responsibility of individual parties involved.

Service contract

Service contract means producing and may be selling in this case the gas on contract for the host country or government. In a typical **service contract** therefore, all the gas belongs to the host country or government. The operating company is paid for every per volume or energy of gas produced.

A kind of service contract is **Technical Assistance Contract** (TAC). TACs are often referred to as rehabilitation, redevelopment or enhanced oil recovery projects. They are associated with existing fields of production and sometimes, but to a lesser extent, abandoned fields. The contractor takes over operations including equipment and personnel if applicable. The assistance that includes capital provided by the contractor is principally based on special know-how such as steam or water flood expertise.

Rate of return / R- factor contract

Some countries have developed progressive taxes or sharing arrangements based on project rate of return (ROR). The effective government-take increases as the project ROR increases. The sliding-scale taxes and other attempts at flexibility may be based on profitability and production rates depending upon negotiation. To be truly progressive however, it should be based upon profitability.

Some contracts use what is called an R factor. **R** factors deal with all variables that affect project economics and it is expressed as:

$$Rfactor = \frac{Accrued.net.earnings}{Accrued.total.\exp enditures} \tag{3}$$

7.2.1 Some basic elements of operating contracts

The existing market is apparently being driven by a hybrid of **concession** and **PSC/PSA**.

PSC/PSA in the long term becomes **Service** contract, provided no new major investment is made in the production business, otherwise in most cases equipment purchased or imported under the contract become the property of the state at the end of the project. This is because, under most PSC/PSAs the contractor cedes ownership rights to the government for equipment, platforms, pipelines and facilities upon commissioning or startup. The government as owner is theoretically responsible for the cost of abandonment. Anticipated cost of abandonment is accumulated through a sinking fund that matures at the time of abandonment. The costs are recovered prior to abandonment so that funds are available when needed.

PSC/PSA is also being phased out in preference for **Joint Venture** contracts and that is the likely contract for the gas industry and market in the future, because host or producing (usually developing) nations demand quick cash revenue turnovers and these are assured under Joint-ventures. Most national oil companies (NOCs) are therefore moving into **Joint - Ventures**.

In summary, the characteristics of the contract types are as below (Table 9):

Contract type	Characteristic
Concession	Royalty for Government first.
PSC/PSA	Share of profit gas but company takes money first.
Joint Venture	Net Revenue interest.
Service Contract	All gas belongs to Government.
Rate of Return	Rate of return

Table 9. Summary characteristics of ownership contracts.

Many aspects of government/contractor relationship may be negotiated but some are normally determined by legislation. Elements not determined by legislation must be negotiated. Even though, it is usually better to have more aspects that are subject to negotiation, flexibility is the watchword.

Elements like r**oyalty is** taken right off the gross revenues and therefore contributes to their lack of popularity with the industry since they can cause production to become uneconomic prematurely. This works to the disadvantage of both the industry and government. One remedy that has become popular is to scale royalties and other fiscal elements to accommodate marginal situations. The most common approach is an incremental sliding scale based on average daily production. A sample sliding scale royalty could be as below (Table 10).

Average	Daily production	Royalty
First tranche	Up to 100 mmscf/d	5%
Second tranche	101-200 mmscf/d	10%
Third tranche	Above 200 mmscf/d	15%

Table 10. Sample sliding scale royalty.

Cost recovery may include the following items:

- *Tangible and intangible capital costs.*
- *Interest on financing (usually with limitations).*
- *Sunk costs*
- *General and Administrative Cost.*
- *Investment Credits and Uplifts*

Tangible vs. intangible capital costs: Sometimes a distinction is made between depreciation of fixed capital assets and amortization of intangible capital costs. Under some concession agreements, intangible exploration and development costs are not amortized. They are expensed in the year they are incurred and treated as ordinary operating expenses. Instances where intangible capital costs are written off immediately can be an important financial incentive. Amortizing intangible costs can take longer to recover, if not carefully negotiated.

Interest cost recovery: Sometimes interest expense is allowed as a deduction. Some contracts limit the amount of interest expense by using a theoretical capitalization structure such as a maximum 70% debt (Derman, A. and Johnston, D. 1999).

General and administrative costs: Many contracts allow the contractor to recover some office administrative and overhead expenses. Non operators are normally not allowed to recover such costs. Most unrecovered costs are carried forward and are available for recovery in subsequent periods. The same is true for unused deductions.

Sunk cost is applied to past costs that have not been recovered. Exploration sunk costs can have a significant impact on field development economics and can strongly affect the development decision. For this reason, many contracts may not allow pre-production costs to begin depreciation or amortization prior to the beginning of production.

Investment credits and uplifts allow the contractor to recover an additional percentage of capital costs through cost recovery. For example, an uplift of 20% on capital expenditures of $100 million would allow the contractor to recover $120 million. Uplifts can create incentives for the industry. Uplifts are the key of rate of return contracts.

Most contracts have a limit to the amount of revenues the contractor may claim for cost recovery but would allow unrecovered costs to be carried forward and recovered in succeeding years. In summary, the hierarchy of cost recovery can make a difference in cash flow calculations.

The basic elements of operating contracts besides royalty and cost recovery include work commitment, bonus payments, domestic obligation, ring fencing, commerciality, reinvestment obligations, tax and royalty holidays.

Work commitment refers to the obligations an exploration company incurs once a PSC/PSA is formalized and they are generally measured in kilometres of seismic data and the number of wells to be drilled in the exploration phase.

Cash bonuses are lump sums paid by the contractor to acquire a particular license. These cash bonuses are the main element in bidding rounds of very prospective acreage. **Production bonuses** are paid when production from a given contract area or field reaches a specified level.

Domestic obligation: Many contracts specify that a certain percentage of the contractor's profit oil be sold to the government. The sales price to the government is usually at a discount to world prices.

Ring fencing: Ordinarily all costs associated with a given block or license must be recovered from revenues generated within that block. The block is *ring fenced*. This element of a contract can have a huge impact on the recovery costs of exploration and development. From the government perspective, any consideration for costs to cross a ring fence means that the government may in effect subsidize unsuccessful operations. Allowing exploration costs to *cross the fence* may therefore be negotiated.

Commerciality deals with who determines whether or not a discovery is economically feasible and should be developed. Some regimes allow the contractor to decide whether or not to commence development operations. Other systems have a commerciality requirement where the contractor has to prove that the development of a discovery is economically beneficial for both the contractor and the government. The benchmark for obtaining commercial status for a discovery cannot be developed unless it is granted commercial status by the host government. The grant of the commercial status marks the end of the exploration phase and the beginning of the development phase of a contract.

Reinvestment obligations: Some contracts require the contractor to set aside a specified percentage of income for further exploratory work within the license.

Tax and royalty holidays: The purpose of tax and royalty holidays by the host country is to attract additional investment.

7.3 Supply contracts

There are four main types of physical trading supply contracts, namely, **swing, base-load, firm** and **futures contracts.**

Swing contract

Swing contracts are usually short term contracts and not longer than a month. It is also called 'interruptible'contracts. Under this type of contract, both the buyer and the seller agree that delivery of the gas can be interrupted on short notice; no legal commitment. They are the most flexible and are usually put in place when either the supply of gas from the seller, or demand for gas from the buyer are unreliable.

Base-load contract

Base-load contracts are similar to swing contracts in that neither the buyer nor seller is obligated to deliver or receive the exact quantities specified. However, it is agreed that both parties would attempt to deliver or receive the specified volume on a best-efforts basis. In addition, both parties generally agree not to terminate the agreement due to market price movements. There is however no legal recourse for either party if they believe the other party did not make the best effort to fulfill agreement. Such contracts rely instead on the relationship (being it personal or professional) between the buyer and the seller.

Firm contract

Firm contracts are different from swing and base-load contracts in that there is legal recourse available to either party, should the other party fails to meet its obligations under the agreement. These contracts are used primarily when both the supply and demand for the specified quantity of the gas are not likely to change.

Futures contract

Futures is one of the derivatives used in the financial markets for both commodities and securities.

Futures contract entitles the buyer of the gas through the contract to take delivery of it at an agreed location and at an agreed date specified in the contract in the future and it compels the seller to deliver the commodity to the buyer at the specified date in the future under the same conditions. Because the contract is tradable, i.e. can be bought and sold in open market, its value changes as the supply of and demand of these contracts changes.

A common feature for futures contracts is that they are standardized such that each futures contract represents the same quantity and quality, valued in the same pricing format, to be delivered and received at the same agreed delivery location and date. The only variable in a futures contract as to when it is bought and sold is *the price of the contract.*

The success of the natural gas market depends on factors including uncertainty in supply and demand, large trading volumes and price volatility. For this reason, futures traders base their price offers on the spot market prices and the markets also allow companies to hedge their price risks.

The major delivery locations in the United States are the Henry Hub in Louisiana and Waha Hub in West Texas. The high volume of trading activity and the high degree of volatility in prices at these two locations have made them points of choice for traders.

The New York Mercantile Exchange (NYMEX) introduced and began trading in natural gas futures with Henry Hub in 1990. The Kansas City Board of Trade (KSBT) also began trading in natural gas futures in 1995 with Waha Hub as the delivery point.

The International Petroleum Exchange (IPE) opened its gas futures for trading in 1997. More companies are now trading their gas futures through IPE. IPE quoted prices are increasing being used as guide by a number of firms in Europe and are gradually becoming a benchmark for gas market in Europe. IPE operates a 12-month range gas futures contract and it is likely to be extended to 15 months.

8. Gas market and industry structure

8.1 Traditional regulation

In the traditional regulatory environment, the main gas transmission entities have mostly been state owned. The state entity supplies gas to one or more distribution entities that are charged with the distribution of the gas to retail customers in specific concessional areas exclusively. In some cases, the transmission monopoly is legally separate from the distribution entity. Supply contracts between the transmission monopoly and the distribution entities are usually long term and are regulated by government agency. A distribution entity in turn has legal monopoly over the supply to retail customers in a concessional area. The gas supply to the public is also regulated by a government regulatory agency and it covers regulation of prices, return on investment, etc. Some regulatory agencies are fully or partially decentralized to the state, regional or district/local level.

The traditional gas marketing system also bound producers, pipelines, local distribution companies (LDCs) and consumers together with long-term contracts, with little room to respond to changes in the market place.

The government controlled price regulation and political pressures on price levels sometimes lead to operational inefficiencies. As demand grows however, pressure to remove subsidies would increase.

Bundled service

The traditional gas market is characterized by **bundled service** - production, transmission and distribution vertically connected and owned by one entity or a consortium. **Bundled service** certainly gives more monopoly to one company, but provides less complication for a new or emerging industry in a country.

8.2 Unbundling service: The new and future regulation

Unbundling is the process of separating natural gas services and supply into components with each component priced separately. Natural gas companies go through varying degrees of organizational unbundling based on market maturity, monopolistic power of the incumbent and the regulatory regime in place.

Unbundled service is usually appropriate for older, matured markets, typically with hundreds of customers. It is not uncommon to amend regulations to cater for unbundled service as the market gains maturity and more new customers (particularly the commercial and residential customers) come on stream. In the United States for instance, unbundling of the market and transportation services for interstate pipelines did not occur till 1993 (FERC Order 636), even though the networks had been in place for over 50 years. In South America, unbundling was born out of increased consumer groups and industry maturity.

The main objective is to prevent cross subsidy, abuse of market and rather maximize efficiency. The purpose of unbundling is to secure non-discriminatory treatment for companies seeking access to pipeline, by ensuring that a vertically integrated transport company does not discriminate in favour of its own gas supply business.

Unbundling ensures that cost is correctly allocated to transportation. This cost clarity provides a basis for establishing use-of-system charges. Unbundling enables customers to pick and create their own services package. Marketers can package the variety of options and sell these services to consumers without discrimination. This leads to the facilitation of competition, also allowing which components of the value chain could be offered for privatization should the need arise and consequently increased efficiency.

Unbundling would characterize the new and future gas market structure.

8.3 Privatization of the gas industry

Privatization allows governments to attract private capital and encourage private investments in its business portfolios. Objectives are to:

- Restructuring poorly run state-owned entities.
- Raising cash to relieve budgetary deficits.
- Raising foreign capital to repay foreign debt.
- Spreading ownership of operations.

Other advantages include attracting new technologies, increased competition and improved efficiency in operations.

8.3.1 Methods of privatization

i. Public offering of shares

Under this method, the government sells to the general public all or part of the shares it holds in a state-owned company. In both cases, widespread shareholding of the entity or enterprise is created.
When only a portion of the shares is sold, the result is a joint state-private ownership of the entity, or what is currently termed as Public Private Partnership.

ii. Private sale of shares

Under this method, the government sells all or part of its holdings in a state-owned company to a single purchaser or group of purchasers. The transaction can be direct acquisition or through a third party such as a broker.

iii. Asset acquisition

The transaction comprises sale of assets instead of shares. Assets can be sold individually to downsize the entity or, sold bundled to form a new corporate entity. The sale of asset can be by open competitive bidding or direct negotiation with the purchaser.

iv. Fragmentation

Fragmentation involves the breaking up or re-organisation of a state-owned entity into several separate entities or into a holding company with several subsidiaries. This method permits piecemeal privatization and allows other different methods of privatization to be applied to different component parts, thereby potentially maximizing the benefits of the overall process. Fragmentation also allows large state-owned entities into separate enterprises and eventually creating competition in the market.

v. Expanding state owned entities with private investment

Under this method, the government instead of disposing of any of its equity rather invites the private sector to buy into the venture. As with Public offering of shares, it results in a Private - Public Partnership joint venture.

vi. Management/Employee buy-out

Under this method, a group of managers or employees acquires a controlling interest in the entity or enterprise. The management/employee leverage use credit to finance the acquisition whilst the collective assets of the acquired enterprise is used as collateral.
This method provides means of transferring the ownership to management and employees and could be a solution for state-owned companies that are difficult to sell or very strategic to the economy, community or the nation. A strong cash-flow potential however is usually the pre-requisite for securing credit for the buy-out.

vii. Management contracts

Under this method, functions related to the entity's operations are contracted out to external usually private management group. There is usually no transfer of ownership and no divestiture of state assets. It has the potential to increase efficiency and effective use of state assets and it is seen as one of the feasible options for introducing private equity into state-owned enterprise especially in developing countries.

Examples of countries with privatization of gas sector

Argentina in 1985 embarked on privatization drive that led to the sale of its national gas transmission systems and a state-owned gas distribution company.

Belgium deregulated its gas sector in the year 2000 and allowed private participation in its local gas market.

Russian gas industry is dominated by GAZPROM, a state-owned company and it is known to be the largest in the world. In 1993, GAZPROM became a state-owned joint stock company and began privatization in 1994. Management/employees as well as local investors were allowed to buy shares whilst the government retained 40% of the shares. 9% of GAZPROM shares were set aside for foreign ownership.

9. The future market

In the evolving and future market and to the customer, cost of the gas and cost of transportation are the major considerations to determine the least-cost gas supply plan. This new marketing paradigm is putting a lot of pressure on gas producers, pipeline operators and LDCs to compete with each other eventually improving efficiency and potentially leading to decrease in cost of supply. New industry players including spot gas marketers and brokers of pipeline capacity create additional links between suppliers and customers. These developments are expected to intensify competition in the gas market and would have major implications for the industry in the future.

The focus of restructuring and regulatory reform in the evolving market therefore would be to reduce state regulations and introduce more competitive market whilst concurrently reforming the regulations to induce more efficiency in performance.

From the on-going restructuring worldwide, six lessons could be deduced which could also serve as guidelines for restructuring of the traditional gas industry. They are as follows:

i. Privatizing state-owned entities with the objective to create efficiency;
ii. Promoting competition in the supply of gas services by
   a. Opening up access to new suppliers; and
   b. Deregulating prices
iii. Ensuring that transmission access rules and associated prices are in most cases non-discriminatory to all applicants to support competition, except for foundation customers and where in special cases targeted at promoting local industry.
iv. Developing pricing arrangements that provide revenues to expand efficient investments maintain decent return on investments.
v. Providing sustainable return on investment for distribution companies but deemed fairly affordable for consumers.
vi. Putting in place regulatory and contractual mechanisms that ensure that market agreements are honoured.

9.1 Mergers and acquisitions in the gas industry

Merger in simple term means legally combining strengths but minimizing weaknesses. The end results include extending cooperative life. Acquisition simply means the entity being acquired has conceded weakness on its part.

Combining two corporate cultures could however cause serious challenges and could lead to failure if not managed well. Fundamental considerations for mergers and acquisition should therefore be:

1. Relate to the core business and that there is expertise to run the expanded business.
2. Supplement or extend current operations.
3. Not raise end-user tariff significantly, in most cases, it should lead to reduction in cost of doing business and eventually decrease end-user tariff.

9.1.1 Targets for mergers and acquisitions

Most attractive targets and successful ventures in the gas industry have been

- Parties involved are in a closely related business.

- The parties involved complement each other's weaknesses.
- Financing does not create unreasonably high debt-to-equity ratio.
- Employees of both parties receive a fair deal and most are happy with the merger or acquisition.

For natural gas pipelines and local distribution companies (LDCs), expansion through mergers and acquisitions offer numerous benefits including:

i. Reducing over all management costs.
ii. Creating large customer base.
iii. Opening access to new supply sources.
iv. Increase economies of scale.
v. Penetrating new markets and offering new services.
vi. Reducing or avoiding new investments by gaining access to new facilities.
vii. Cutting costs by eliminating duplicate services.
viii. Establishing name and recognition with customers.
ix. Flexibility in transporting large volumes of gas due to increase market share.
x. Expanded operational areas.

Unsuccessful mergers and acquisition on the other hand are characterized by the fact that the acquiring party:

- Pays too much for the acquisition. *Over-value of the acquired assets can cause acquisition failures. This can come about when the evaluator mistakenly over-values the entity being acquired.*
- Over estimate the market projections.
- Combine two different corporate cultures.
- Had limited access to information before merger or acquisition, in other words, lack of transparency.

9.2 National gas and international gas companies

National gas companies

National gas companies (NGCs) are currently and generally the same as the national oil companies (NOCs) of the host countries, since the gas market is still evolving compared to the oil market. NOCs usually represent the interest of their governments in the petroleum market. They act as gatekeepers and control access to the majority of resources for future oil and or gas production.

Many NOCs came into being during a period of relatively large-scale state intervention in their countries' economies, a process which only began to reverse in the 1980s-1990s (Stevens, 2003). It was envisaged in those times that market forces would not be sufficient to propel poor developing host countries out of their poverties. With most natural resources vested in the government, it was thought that only the state could marshall the resources required for the massive economic development. In recent times however, additional reasons have been (i) emergence of nationalism; (ii) hydrocarbon listed as a strategic resource; and (iii) commercially risky and technologically complex sector (Foss, 2005a; Foss, 2005b; Mommer, 2002).

The overall goal of an NOC therefore is to ensure the effective development of the hydrocarbon sector of the country and as well contribute to the country's socio-economic development.

The immediate objectives include:

- Earning revenue for the country;
- attracting new technology;
- ensuring infrastructure development;
- creating employment; and
- the latest addition, minimising damage to the environment.

NOCs differ from country to country based on the following:

- the level of government or state control or ownership.
- The extent of private shares.
- Their financial health.
- Access to international credit.
- Their operational experience and skills.

The more experience, skillful and higher the access to international credit, the more likely such an NOC would go international and become an IOC (International Oil Company).

International oil companies

The traditional examples of IOCs are Exxon-Mobil and Chevron of the United States, BP of United Kingdom, Royal Shell of the Netherlands, Elf and Total of France. The new entries include the PetroChina of China, Petrobras of Brazil, Petromas of Malaysia and Norsk Hydro of Norway.

The primary objectives of an IOC in a host country are:

- To look for good geology to find the resource, in this case gas.
- To expand operations.
- To maximize return on equity to shareholders, simply saying to make money.

What most IOCs consider before entering a host country after the presence of a good geology has been confirmed, include:

- Political stability.
- Respect for honouring contractual obligations, i.e. contract is not unilaterally changed in the middle of the course.
- Robust and transparent legal system, i.e. company believes that it shall receive a fair hearing during legal cases.
- Fulfilling bilateral cooperation between partners in the business and the host country.

Cooperate social responsibility goals have since 1990s been playing greater role in their host country operations.

9.3 Future of national and international gas/oil companies

More NOCs would move away from just exporting raw gas to adding value to the commodity. Nations with rich natural gas resources have aggressively added new petrochemicals capacity for the production of methanol and other industrial chemicals. It has been observed that a country like Trinidad and Tobago has been very successful in attracting foreign investments. In the mid-1970s, the country had a paradigm shift with respect to the focus of the hydrocarbon production - monetization of natural gas. Since 1975, the natural gas has been used to manufacture methanol and ammonia and exported to the United States.

A shift from "raw gas exporting" to "value-added" however would depend upon host Government policies and the type of instruments used to implement the policies, since they have impact on local gas consumption and fuel choices, directly or indirectly. For instance environmental policies to promote cleaner alternative fuels may encourage greater gas use locally if available through favourable taxation and financial incentives for development of infrastructure. Lower taxes on gas prices and CNG vehicles but high on diesel fuelled vehicle could help reduce local consumption of diesel and consequently local pollution in mega cities such as found in India.

With time and as nationalism sentiments grow, traditional IOCs such a Exxon-Mobil would be pushed out of most host developing countries whilst more NOCs would become IOCs. For instance, Petrobras of Brazil, Petromax of Malaysia and PetroChina of China which were formally NOCs have become the major new IOCs.

For IOCs therefore to exist in future, **first**, they must be at the cutting edge of technology where the NOCs are nowhere near. This can only be achieved through research and development (R&D). R&D investment however does not come cheap and does not yield immediate results; it may take between five to ten years.

IOCs that are short-sighted would not invest in R&D and by this posture, would be those which would cease to exist.

As oil and gas resources are becoming dear to explore and produce, the NOCs would look for IOCs with the requisite technologies. IOCs in the future therefore would be looked at as 'banks' of technologies and the knowhow. Therefore robust IOCs shall be the 'banks' with the skills.

Secondly, IOCs that takes on social responsibility and show respect to the environment are likely to survive for long in developing countries. This has become necessary because even though contractual commitments to host governments are honoured and the latter is supposed to take care of the social needs of its people, host governments usually renege on their commitments to the immediate communities. The IOCs may need not take on social responsibility programmes directly but could either set up a separate social enterprise or team up with a known one to implement the social and or environmental programmes.

The last but not the least, is local content. IOCs with programmes to train and employ local manpower are likely to have prolonged stay in developing countries as the latter exert pressure usually backed by civil society to increase local content share of the IOCs' operations.

10. Market options for developing countries

10.1 Policy and regulatory environment

Policy and regulatory environment would be a key driver for the growth of the natural gas market in any country. To ensure growth of the market, the policy and regulatory regime should first of all reflect fair returns to all stakeholders, including government and other stakeholders. The appropriate role of each stakeholder should be spelt out clearly without ambiguity (Energy Commission, 2007).

Secondly, the policy and regulatory regime should be transparent, predictable, clearly defined and open to all investors that meet laid down criteria. The nation's interest is best served by a well defined policy regime that is open to all eligible investors. The policy and regulatory regime should also be simple to administer and not imposing lengthy bureaucracy on private investors. Lengthy bureaucracy could breed corruption since civil and public officials at times take advantage of the long wait-time to promise 'short-cuts' to potential investors. Simplicity in administration is necessary to reduce the costs of compliance to both the investor and relevant government agencies. A policy regime that is loosely defined and subject to discretionary interpretation by public servants can prove costly and can lead to investor uncertainty, the uneven treatment of investment proposals, and can encourage counterproductive behaviour on the part of private sector interests.

10.2 Implementation models

In most developed countries, the Local Distribution Companies (LDCs) are typically owned and operated by municipalities and private enterprises with very little or no public/state sector involvement.

In the developing countries, however they seem to be two models: the **South American Model**, where there is equity participation by both the public and private sectors and the **South-East Asia Model** where most of the transmission and distribution systems are owned and operated by State-owned enterprises. Examples of the South American models are Bolivia (Transredes), Colombia (Promigas), Peru (Suez), Argentina (Metrogas), Chile (Metrogas). Examples of the South East Asia model are found in Pakistan, Bangladesh, India and Thailand.

The South-East Asia models require total funding from the state with all the associated risks. The shortcomings of this model are the inefficiencies in operation and conflicts between the regulatory and operating government entities.

The South American models (and indeed the developed world models), however, require little state involvement.

Since 1990s, most of the countries that subscribed to the South East Asia model are gradually transitioning to the South American Model, one that relies increasingly upon reduced government control and on a more market-responsive pricing climate to encourage foreign and private sector investments. This is expected to push faster the development of the gas sector.

10.3 National transmission and distribution system models

The national transmission and distribution systems in most developing countries have been designed based on (a) the public sector model and (b) the private-public partnership model (Energy Commission, 2007).

10.3.1 Public sector model

A public/state entity with complete operational autonomy, would on behalf of the Government, build and operate the infrastructure. The government would have 100% ownership of the assets. A consortium of LDC and EPCM (Engineering, Procurement, Construction, Management) team may install the facilities and transfer operational control of the network to local employees of the government entity, over a defined period of time under BOT[23] arrangement.

[23] Build Operate and Transfer

To be effective, this entity should operate at arm's length from the Government or sector ministry and should be managed by an expert and or, commercially oriented Board of Directors. Despite "best" efforts by most of the developing countries, the public sector models (Figure 4) have mostly been beset with operational failures around the world.

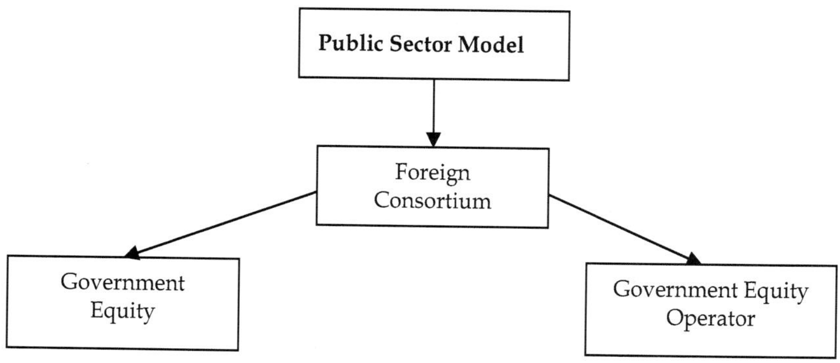

Fig. 4. Public Sector Model.

10.3.2 Public Private Partnership model

The **Public-Private Partnership** model (Figure 5) is usually a joint venture company, with majority private ownership, who builds, own and operate the transmission and distribution network.

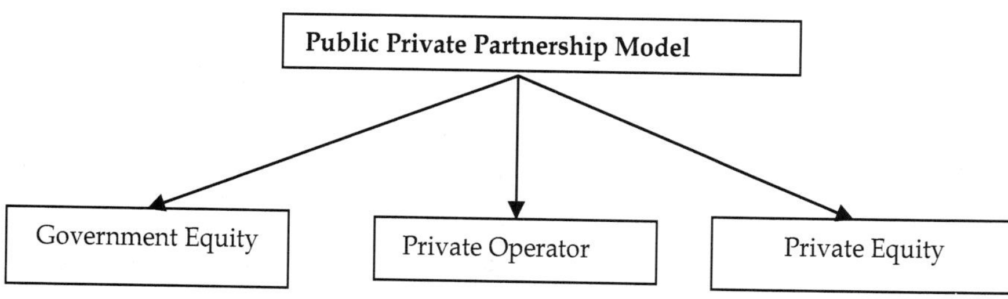

Fig. 5. Public Private Partnership Model.

The government typically may have minority or majority equity interest in the assets, however the minor the better since such usually limit the potential political influence. A consortium of LDC and EPCM team usually install the facilities and transfer operational control of the network to a private local company, over an agreed time period. The private consortium could be a partnership of local and foreign entities.

Table 11 summarizes the main characteristics between state or public ownership and public-private partnership (PPP).

ITEM	STATE/PUBLIC OWNERSHIP	PUBLIC-PRIVATE PARTNERSHIP
Set up	100% Government	Joint venture
Operation	Government entity	Private operating entity
Global Models	Asian Model: e.g. *India, Bangladesh, China, Thailand*	South American Model: e.g. *Bolivia, Colombia, Chile, Peru*
Observed Operational Efficiency	Inefficient, Government interference, subsidies, poor operational oversight	More efficient, commercially driven
System Growth	Expansion of low capacity system likely	Expansion only likely for high capacity load factor systems
Industry Rate of Maturity	Low rate of maturity	High rate of maturity

Table 11. Main characteristics between state/public ownership and public-private partnership.

Most foreign investors would prefer to enter into a joint venture arrangement with a State-owned entity to facilitate certain aspects of infrastructure development, such as land acquisition and landowner issues. A joint venture arrangement with the State could also reduce the risk associated with particular investments.

11. Conclusions

The global energy sector is beset with uncertainties in terms of supply security, development cost, greenhouse gas emissions and other environmental pollution. The flexibility of natural gas as a fuel, its lower carbon dioxide emission compared with other fossil fuels, its emerging global abundant reserves, relatively quick and lower development field cost make natural gas the most favourable fossil fuel in recent times.

The gas industry with its traditional structure of rigid regulations is being replaced by a regime that relies on market forces and is redefining the future gas industry. The structural and regulatory changes that are affecting the global natural gas transmission sector are designed to create competition, expand regional and international gas trade, and to reform the regulation of the transmission and distribution functions to allow for non-discriminatory access to pipelines. Concurrently, these structural changes are being accompanied by ownership changes in light of the global trend towards privatization. Amongst a series of developments, the most evident are the changes in asset structure of the industry, the changing role of gas pipelines, emergence of natural gas commodity markets, fuel switching capability and competition and new pricing mechanism.

In many instances, gas is stranded because of a lack of sufficient demand in locations where the gas is produced. Stranding of gas in locations far removed from big consuming markets calls for construction of pipelines or use of LNG vessels depending upon the distance and location of the consuming market. To complement the global expansion of overland pipeline, LNG investments are also fast growing. For nations that do not have large enough domestic demand relative to the size of their resource base, or that have not developed petrochemicals capacity for conversion of natural gas to other products, LNG is an important means of deriving value for their natural resource endowments through international trade.

Government policies and the type of instruments used to implement the policies however, have impact on gas consumption and fuel choices locally, directly or indirectly. Favourable policies would expand local demand and therefore could add value to the gas being produced by opening up local petrol-chemical industries. With time, there would be significant increase in local demand by the economy and consequently, reduce exports of raw gas. This is the likely market scenario for the future gas industry, particularly, in developing countries with abundant commercial gas resource.

Market and regulatory models for gas markets have been proposed for developing countries.

12. References

BP (2011). Statistical Review of World Energy 2011, www.bp.com

CEE (2006). International Energy Markets, *Economics of the Energy Industries*, Center for Energy Economics, Bureau of Economic Geology of the University of Texas at Austin, (1) pp. 1-19, 2006. www.beg.utexas.edu/energycom

Derman, A. and Johnston, D. (1999). Bonuses Enhance Upstream Fiscal System Analysis, *The Oil and Gas Journal*, Feburary 8, 1999.

Energy Commission (2007). *Natural gas Transmission and Distribution Infrastructure Plan for Ghana*, Energy Commission, Ghana August 2007. www.energycom.gov.gh

Foss, M. (2005a). The Struggle to Achieve Energy Sector Reform in Mexico. Prepared in 2004, for the U.S. Agency for International Development, The Nexus between Energy and Democracy, 2005.

Foss, M. (2005b). Global Natural Gas Issues and Challenges: A Commentary, *The Energy Journal*, Journal of the IAEE, January, 2005.

Johnston, D. (1994). Global petroleum fiscal systems compared by contractor take, *The Oil and Gas Journal*, December 12, pp. 47-50, 1994.

IEA (2011). World Energy Outlook, Special Report - ARE WE ENTERING A GOLDEN AGE OF GAS? International Energy Agency, OECD, 2011, www.iea.org

Mommer, B. (2002). *Global Oil and the Nation State*, Oxford University Press, New York, 2002.

Stevens, P., (2003). *National Oil Companies: Good or Bad? A literature Survey*, National Oil Companies Workshop Presentation, World Bank, Washington, D.C., USA, June 27, 2003.

U.S EIA (2011). Natural Gas: U.S. Natural Gas Pipelines, Energy Information Administration, www.eia.gov

WAGPCo (2011). www.wagpco.com

Major literature consulted in preparation of this chapter

Economics of the Energy Industries, Center for Energy Economics, Bureau of Economic Geology of the University of Texas at Austin, Copyright 2006 Revised Edition 2008. www.beg.utexas.edu/energycom

International Natural Gas Market, Lecture Notes for the International Petroleum Management Program, Institute for Petroleum Development, Austin, Texas 78746, U.SA, 2011 Edition. www.instituteforpetroleumdevelopment.com.

Guide to Natural Gas in Ghana, First Edition, prepared by the Resource Centre for Energy Economics and Regulation, ISSER, University of Ghana, August 2006.

World Energy Outlook, Special Report - ARE WE ENTERING A GOLDEN AGE OF GAS? International Energy Agency, OECD,2011 www.iea.org

13

The Gas Transportation in a Pipeline Network

Jolanta Szoplik
West Pomeranian University of Technology, Szczecin, Poland

1. Introduction

During the last few years we have been able to observe increasing development of different types of networks. Energetic, telecommunication, water or gas networks are only a few examples of systems whose main purpose is media transportation from the source (producer, manufacturer) to the target place of use. One of the basic features of network systems is their uniqueness with regard to the structure as well as transmission capability. In practice there are no two identical networks and their uniqueness leads to the necessity of use of individual approach to networks during designing and exploitation stage. Moreover networks have complex structure and any change or modification of their structure while in use is not an easy task. For these reasons flow improvement in current network and optimal exploitation of its capabilities is an important and actual issue.

In case of many real-time systems, examination of their performance in conditions different from currently existing is impossible. In such a situation a mathematical model of such system is constructed which is a kind of simplification of the reality (Kralik et al., 1988; Osiadacz, 1987 and Osiadacz, 2001). However, the model of the system can be used to simulate its behaviour in reaction on extortion. Simulation is an example of an experiment, performed with the help of an appropriate algorithm and a computer. An enormous advantage of computer simulation is its repeatability, and each simulation task in the same initial conditions can be repeated infinitely many times. In real conditions such an experiment is almost impossible to perform, as from technical point of view, it is extremely difficult to perform direct measurement of some values characterizing work of the system. However, conducting a series of simulations for different initial conditions allows to receive various solutions and to choose the best one out of those received. Such an example of using simulation results can be treated as quasi-optimisation which can be accepted as sufficient when the relations between elements of the system and extortion are not well known. The approach which uses mathematical model is a huge simplification of the real system.

A gas network can be an example of a system, where from the technical point of view, it is difficult to perform direct measurements of parameter values characterising flow in pipeline networks. Currently we observe evident increase in gas consumption by the municipal receivers as well as the industry, and this is the reason why gas transportation network from the place of extraction directly to the recipient results in pipelines system becoming more and more complex. Gas in the network is transported under appropriate pressure. Depending on its level we can distinguish three different types of networks: high, middle or low pressure networks (Osiadacz, 1987; Kralik et al., 1988). The analysis of network flow simulation results enables us to estimate such network bandwidth reserves, define possibilities of network expansion directions and possibility of connecting new recipients, calculate maximal value of the gas flow rate in the pipeline of the network or gas parameters (e.g. pressure) in output points. Overpressure layout, velocity of the gas and gas flow rate in each pipeline of the given gas network can be defined on the basis of network gas flow simulation.

In the literature, there are many examples of application various mathematical methods and IT tools, to facilitate solving complex simulation tasks, optimisation or steering in regard to gas flow in the gas pipeline or gas pipelines network.

The subject of several scientific papers was the analysis of flow issues and leakage detection in network systems. The authors stated, that these problems can play an important role in the management of pipeline system and to cause reduce the loss of leakage. Gonzalez et al. (2009) focused on modeling and simulation of gas pipeline network and presented two models derived from the set of partial differential equations and two numerical schemes for integration of such models. Fukushima et al. (2000) proposed and successfully implemented the leak detection system based on a dynamic simulation with wave equation using real operational data on their one of the longest gas pipeline in Japan. Brkić (2009) for construction and calculation of looped gas distribution pipeline network of composite structure with known node gas consumption proposed to use an improvement of Hardy Cross method procedure. Liu et al. (2005) presented an adaptive particle filter to tackle the leak detection and location in gas pipelines. Reddy et al. (2006) used dynamic simulation models (transfer function model) of gas pipeline network for on-line leak detection and identification and they stated, that proposed method was 25 times faster than the explicit finite - difference approach.

The problems of heat exchange between gas flowing through the pipe and the ground analysed Osiadacz & Chaczykowski (2001), Chaczykowski (2010), Ke &Ti (2000), Tao & Ti (1988). Ke & Ti, 2000 and Tao & Ti, 1988 proposed for transient analysis of isothermal gas flow in the pipeline network a new mathematical model based on electrical analogy. These papers treated mainly of gas flow in high pressure network that have much simpler structure than low pressure networks.

However, applying in the research Artificial Neuronal Network (ANN), does not require earlier knowledge of exact relations between particular values, and it is only sufficient to know input and output data. Examples of use of Artificial Neuronal Networks for modelling, steering or optimisation of fluid flow or transport supporting appliances can be found in literature. Zahedi et al. (2009) propose applying the ANN method to predict hydrates forming temperature during gas transportation with gas pipeline. According to the authors results received using the ANN are more adequate than analogous results received with traditional methods. Carvalho et al. (2006) applied the ANN to detect and then to classify faults on side of the gas pipeline to appropriate group. The authors proved that precision of the ANN to detect faults on gas pipeline side is approximately 94%, whereas correctness of recognition of fault type (corrosion or other) is approximately 92%. The ability of the ANN to recognise type of corrosion (internal, external) is clearly smaller (approximately 72%). Silva at al. (2007) in their work introduced proposal of applying the

ANN for analysis of influence of pipeline corrosion defects on chosen parameters characterising the flow. The authors however noticed that there is necessity of verification of work effects with real data. Based on the results from magnetic flux leakage signals, Hwang et al. (2000) presented a new approach for training, hierarchical wavelet basis function neural network for the three-dimensional characterization of defects on the pipeline. Nguyen et al. (2006, 2008) use the Neural Network for forecasting the demand of the hourly gas stream for customer. Experimental data obtain by ANN are used in the Genetic Algorithm to search the optimal combination of compressor scheduling.

The purpose of this chapter is to perform the analysis of results of the steady-state simulation calculation for gas flow in the low pressure gas pipelines network, based on which, two methods of steering the gas stream pressure entering the network will be described. Under discussion will be variability of gas consumption from the network by different recipients groups in seasonal and daily cycle. Air temperature will be important parameter with significant influence on gas consumption by consumers. There will be an algorithm developed to steer gas pressure in form of dependence of gas pressure on stream volume feeding the network, as well as the network characteristic with marked network nodes, where the pressure is lowest. It was proved, that steering gas pressure feeding the network, due to keeping lower gas pressure in the network, can significantly lower the network exploitation costs.

In the present chapter fragment of real low pressure network in Szczecin city (Poland) was used and calculations were performed for real data presenting hourly gas streams leaving 108 nodes of the network in successive hours of four chosen days with various air temperature.

2. The elements of gas pipeline network system

Gas network consists of connected and cooperating with each other objects that transport and distribute natural gas. Fig. 1 presents main objects that compose to the gas network and their main classification. Gas pipelines with equipments are used to transfer and distribute gas fuel to consumers. Polish classification of gas pipeline networks and short characteristics (gas overpressure (p), gas speed (w) and type of pipes material) are presented in Fig. 2. Taking into account the gas stream overpressure, there are four types of gas pipelines. Another classification is, when as classification criteria, the type of pipes material in the network. In this case there are gas pipelines made of polyethylene (PE) or steel. Fig. 3 presents all previously mentioned gas network objects.

Gas station is a set of appliances in the gas network fulfilling separate or simultaneous functions of pressure reduction, stream measurement and gas characteristics or stream division. The gas compressor station raises the gas stream pressure in order to overcome the pressure lost resulting from the frictional gas flow in the pipeline of network.

Turbines in compressor stations can be fed with gas from the gas pipeline or electrically. Gas storehouses are natural containers constructed in rock mass, underground mining excavations or salt caverns, and used to storage gas in periods of lower demand (months of high air temperature) and additional network feed during periods of higher gas demand (months of low air temperature).

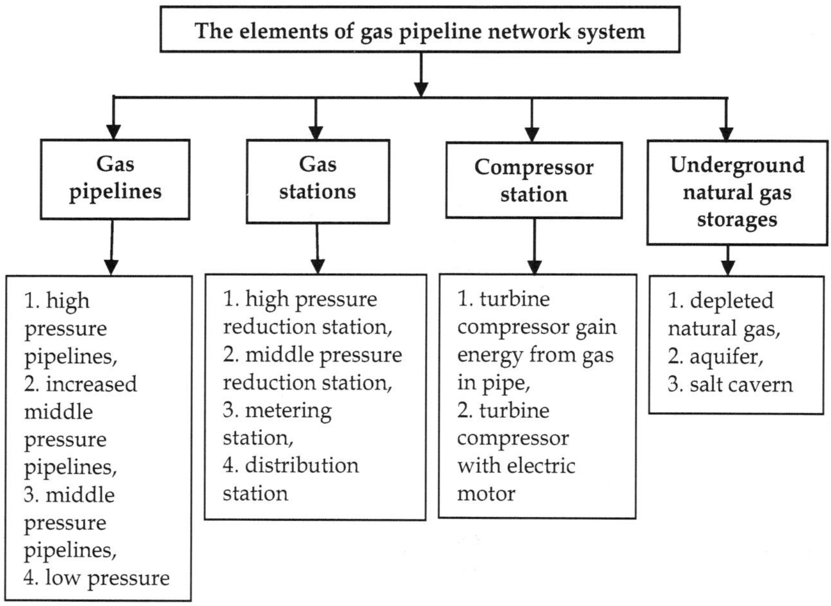

Fig. 1. Technical units composed the gas pipeline network.

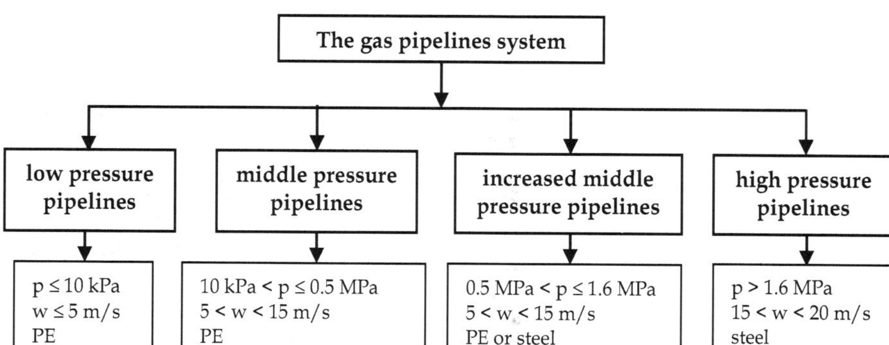

Fig. 2. The main types of gas pipeline networks.

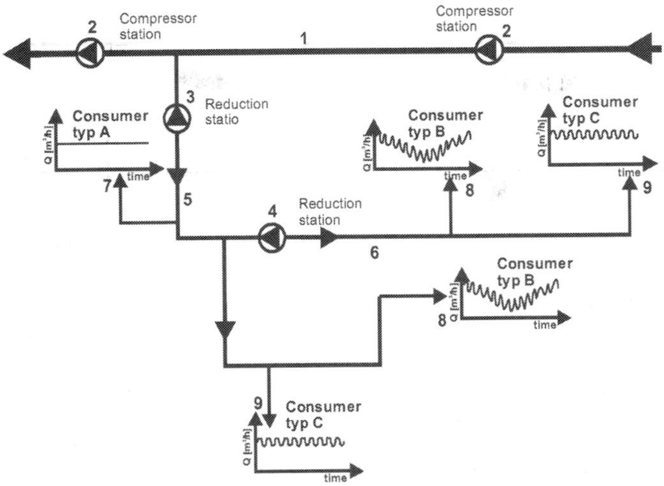

Fig. 3. Fragment of gas network with marked objects and gas use characteristics by different groups of consumer.

Fig. 3 shown that, large amount of natural gas is transported on long distances through the high pressure pipelines network (1). Compressor stations (2) are located on gas pipelines in distance of 100-150 km, that increase initial gas pressure lost during the flow. Whereas natural gas delivered to consumers is characterised with clearly lower overpressure that is received after two stage pressure reduction in reduction stations. High pressure gas reduction station (3) lowers gas pressure from high to medium level. However, in middle pressure gas reduction station (4) is second stage of gas pressure reduction to low level. Gas pipelines network between gas reduction stations (3) and (4) is called (middle pressure gas network) (5), whereas from the gas reduction station (4) distribution of gas with low pressure gas network (6) begins directly to municipal consumers (7), (8), (9).

3. The variety of the gas demand in the year

Natural gas transported with network presented in Fig. 3, is delivered to industrial (7) and municipal (8) and (9) consumers. In the industry the natural gas is used as raw material and main source of methane, whereas in households gas is used to prepare meals and heat water and accommodation. In Fig. 3, three types of gas consumers are marked and presented their gas usage characteristic within a year. The A type (7) is an industrial consumer, who consumes from the network the same gas amount regardless of time of the year and time of the day. The B type (8) and C type (9) are municipal consumers that use gas in households. The B type consumer is characterised with variable gas consumption within daily and seasonal cycle, which means, that gas is mainly used to house heating. The C type consumer uses gas only to prepare meals or possibly to heat water, therefore this kind of variation can be accepted as only daily variation.

Gas consumption by the B type consumers mainly depends on weather and calendar factors. Fig. 4 presents exemplar diagram of gas consumption variation by large group of B and C

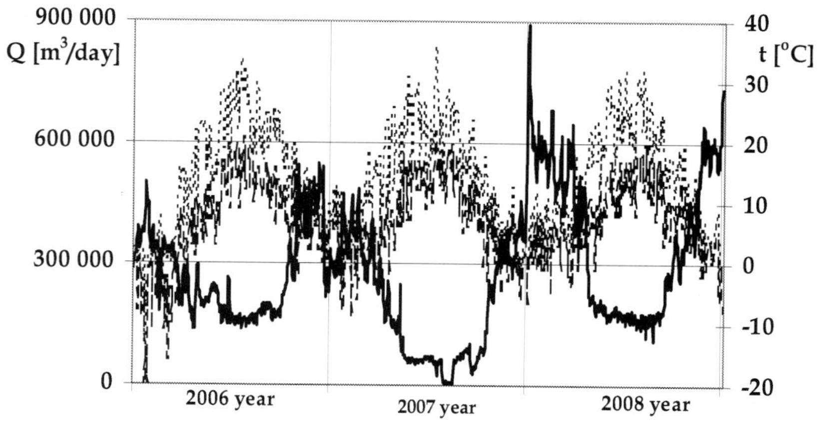

Fig. 4. The change of gas stream Q in days of years 2006, 2007 and 2008; (…) - t_{max} the day temperature, (--) t_{min} the night temperature, (—) Q daily volumetric gas flow; (Szoplik J., 2010a).

type consumers together, for one of Polish cities in successive days of years 2006, 2007 and 2008 depending on air temperature. The increase of air temperature during summer months (July, August) influences on visible gas consumption decrease and vice versa, low air temperatures (January, February, November and December) cause, that large gas amounts leave the network and the gas is used mainly to heating. Variation of gas consumption from the network results with various network loads.

4. Mathematical model for gas network

The computer programs used for simulation of gas pipeline distribution transport system are based on the gas network mathematical model. The detailed form of the mathematical model for such a system depends on the assumptions of flow as well as the conditions of the network operation. There is no universal formula to describe the flow of gas in a pipeline. The different equations are used depending on the working pressure of the network and the assumptions made with regard to the conditions of the network operation (Osiadacz, 1987; Osiadacz, 2001; Kralik et al., 1988). Three equations describing the gas flow through a pipeline network are derived from the equation of continuity, the equation of motion and the equation of energy (Osiadacz & Chaczykowski, 2001; Ke & Ti, 2000):

$$\frac{\partial \rho}{\partial t} + \frac{\partial (\rho w)}{\partial x} = 0 \tag{1}$$

$$\frac{\partial (\rho w)}{\partial t} + \frac{\partial (\rho w^2)}{\partial x} + \frac{\partial p}{\partial x} + 2\rho w^2 \left(\frac{\lambda}{D}\right) + g\rho \sin \alpha = 0 \tag{2}$$

$$\frac{\partial}{\partial t}\left[(\rho A dx)\left(c_v T + \frac{w^2}{2} + gz\right)\right] + \frac{\partial}{\partial x}\left[(\rho w A dx)\left(c_v T + \frac{p}{\rho} + \frac{w^2}{2} + gz\right)\right] - q\rho A dx = 0 \tag{3}$$

where: ρ is the density of gas, w is the gas flow velocity, λ is the Finning friction coefficient, D is inner diameter of the pipe, α is the angle between the horizon and the direction x, A is the cross-section area of the pipe, c_v is the specific heat at constant volume, q is the specific heat related per unit mass and T is the temperature of gas.

Stationary gas flow in a pipeline does not vary in time and is a special case of dynamic behaviour of gas flow in a pipeline. In this case, the variables in the equations of continuity, motion and the energy are only a function of coordinates and this system is described by the set of nonlinear algebraic equations (Osiadacz, 1987 and Osiadacz, 2001). For the low-pressure gas pipeline network, when the change of the gas pressure and the dynamics of the flow can be neglected, it is suitable to use the steady-state simulation. The more complex the pipeline network is, the more complex the mathematical model of such system is therefore in order to solve this problem in most cases computers are used. In case of gas networks the structure of the pipeline network can be presented by means of the graph theory, which allows simple representation of the structure in terms of the properties of its elements incidence.

A graph (Kralik et al, 1988; Osiadacz, 1987 and Osiadacz, 2001) consists of a set of nodes and a set of branches. The nodes are presented by points whereas the branches by line segments connecting two points. When a branch has a pair of nodes in a certain order the graph is called direct graph. Sometimes the node is connected to itself and this closed path of nodes and branches is called a loop of the graph. In case of the gas pipeline network the boundary nodes are the points where the elements are connected to the whole network and this is the node that belongs to a single element. The second type of node is an internal node of the network which is common to exactly two elements (simple node) or to at least three elements (crossing node). Gas enters the network at node called supplier node and leaves the network at boundary nodes. A branch in gas pipeline network is a pipe with constant diameter and roughness. Taking into account the position of a branch in the network there are boundary and internal branches.

The interconnection of the pipe in the network can also be presented by the branch-nodal incidence matrix $A = [a_{i,j}]_{n \times m}$. The number of rows ($n$) is equal to the number of nodes, but the number of columns (m) is equal to the number of branches (Osiadacz, 1987; Osiadacz, 2001 and Kralik et al., 1988). Each element $a_{i,j}$ of the matrix A is equal to 0 or (+1) or (-1).

$$a_{ij} = \begin{cases} +1- \text{ when branch } (j) \text{ enters node } (i), \\ -1- \text{ when branch } (j) \text{ leaves node } (i), \\ 0- \text{ when branch } (j) \text{ is not connected to node } (i) \end{cases}$$

The incidence of the loops and branches describes the matrix of branch-loop $B = [b_{i,j}]_{kxm}$. The rows (k) in matrix B correspond to the loops while the columns (m) correspond to the branches in the network. The elements $b_{i,j}$ are defined as:

$$b_{ij} = \begin{cases} +1- \text{ when branch } (j) \text{ has the same direction as loop } (i), \\ -1- \text{ when branch } (j) \text{ has opposite direction to loop } (i), \\ 0- \text{ when branch } (j) \text{ is not in loop } (i) \end{cases}$$

Osiadacz (1987); Osiadacz (2001) and Kralik et al. (1988) propose that for the simulation of the steady-state pipeline networks the analogy between fluid and electrical network can be successfully applied. The aim of the simulation of the gas flow is to estimate the values of the flow rates in each pipe of the network and the pressure at each node of the network. The calculated value of flow and pressure must satisfy the flow equation and together with the value of the loads and off-takes gas flow must satisfy the first and second Kirchhoff's laws.

The matrix form for the first and second Kirchhoff's law (Osiadacz, 1987; Osiadacz, 2001 and Kralik et al., 1988) represent the following equations:

$$A_1 \cdot Q = q \quad \text{(I Kirchhoff's law)} \tag{4}$$

$$B \cdot \Delta p = 0 \quad \text{(II Kirchhoff's law)} \tag{5}$$

where:

$A_1 = [a_{ij}]_{(n-n1) \times m}$ - reduced nodal-branch incidence matrix,
$B = [b_{ij}]_{k \times m}$ - loop-branch incidence matrix,
$Q^T = [Q_1, Q_2, \dots Q_m]$ - vector of flows in the branches,
$q^T = [q_1, q_2, \dots q_{(n-n1)}]$ - vector of stream at the output nodes,
$\Delta p^T = [\Delta p_1, \Delta p_2, \dots \Delta p_m]$ - vector of pressure drops in the branches,

Equations (4) and (5) complete with one of the following forms of flow equation (equation 6 or 7) are the matrix form of the mathematical model and describe the gas flow in the network.

$$Q = \Psi(\Delta p) \tag{6}$$

$$\Delta p = \Phi(Q) \tag{7}$$

where: $\psi(\Delta p)$ is the vector of pressure drop functions in the branches and $\Phi(Q)$ is the vector of flow functions

There are two methods which are most frequently used to carry out the simulation of the gas flow in the network. In the nodal method, the equations (4, 5 and 6) are resolved. Initial approximations are made to the nodal pressure and are corrected in the next iterations until the final solution is reached. The loop method is the other way to resolve the mathematical model for the gas flow. Initial approximations are made to the branch flow and next iterations are corrected until the final solution is reached. The loop method provides the solution of the equations (4, 5 and 7).

5. The subject of the analysis in the study

The subject of the analysis was the real low pressure pipeline network which consisted of 319 pipelines of various diameters (from 0.050 m to 0.25 m). The whole length of the pipeline in this network was equal to 4151 m and the overall amount of gas accumulated in the network was equal to 51 m^3. The low pressure gas pipeline network operated with overpressure in the range of 1.7 kPa to 2.5 kPa. The operating temperature was 283 K, the relative density of the gas was equal to 0.6. The velocity of the gas was always lower than 5 m/s in each pipe of the network.

The graphic representation of the network analysed in the study consisted of 316 nodes and 319 branches (Szoplik J., 2010b, 2010c). Taking into account the position of nodes in the network is possible to distinguish two types of nodes: boundary and internal nodes. There was 1 supplier node (Fig. 5, point Z1 – gas reduction station), where gas entered the network, 108 nodes, where gas left network (boundary or output nodes) and 207 internal nodes. There were also one input, 108 output and 210 internal branches in the graph of the network.

Boundary branches containing input or output nodes are called input and output branches respectively, but the elements which are neither input nor output elements of the network are called internal elements. The number of loops in this pipeline network was 3. Fig. 5 is the graphic representation of the network used in the study. The detailed data for the graph, diameter and length of each type of the pipe in the network presented in Fig. 5 are collected, respectively, in Table 1 and Table 2. There are only 13 nodes distinctly marked at the graph presented in Fig. 5 (A2; A51; A61, A64, A65, A70, A71, A75, A92, A90, A80, A146, A147). The flow rate at these nodes is divided or the diameter of the pipe is varied.

No	Description	Number
1	Total number of nodes	316
2	- supplier node	1
3	- boundary (output) nodes	108
4	- internal nodes	207
5	Total number of branches	319
6	- output branches	108
7	- internal branches	210
8	Number of loops	3

Table 1. The characteristics of the nodes and branches for the graph presented in Fig. 5.

The mathematical model of gas flow in the pipeline network consisting of equations (4), (5) and (7) was performed with computer program GasNet, used to steady-state simulation of gas flow by means of loop method. Calculations were performed for real fragment of the low pressure gas pipeline network in the city Szczecin and were based on real data characterising hourly values of gas streams which left the network at each of 108 output nodes for different values of gas stream overpressure entering the network in gas reduction station Z1 (Fig. 5). The drop pressure in pipelines constituting network was calculated as a difference of absolute pressure in two adjacent nodes. The friction factor λ was determined according to the guidelines concluded in the norm PN-76/M-34034.

No	D_{nom} [mm]	D_{in} [mm]	s [mm]	L [m]
1	250	204.6	22.7	18.7
2	225	184.0	20.5	687.6
3	180	147.2	16.4	1280.7
4	160	130.8	14.6	80.2
5	125	102.2	11.4	517.1
6	90	73.6	8.2	813.5
7	63	51.4	5.8	735.8
8	50	40.8	4.6	17.8

Table 2. The diameter D, length L and thickness s of the wall for the pipes of the network presented in Fig. 5.

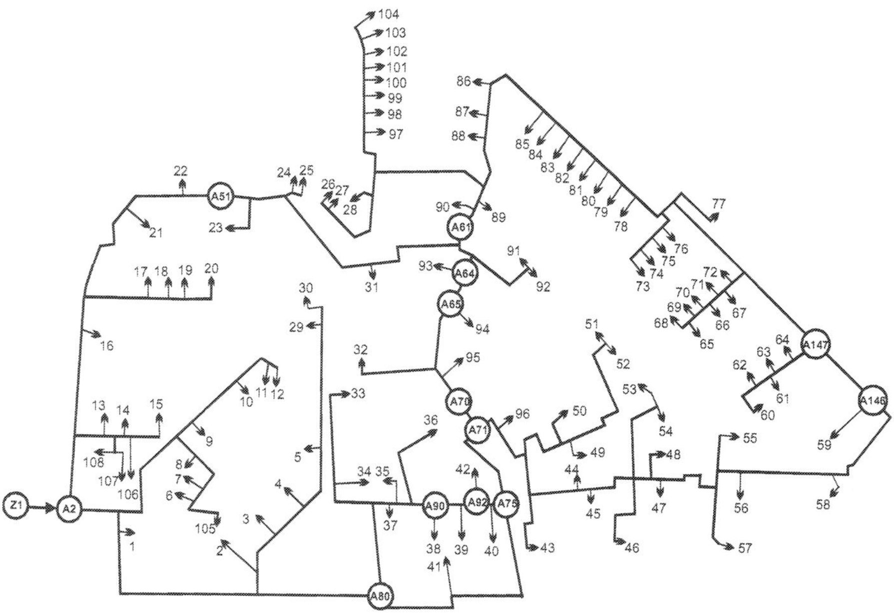

Fig. 5. The graph of the gas pipeline network; Z1 – middle pressure gas reduction station; (Szoplik J., 2010b).

In the case of the low pressure gas pipeline network the drop pressure in each branch was calculated as the difference of the pressure at two adjacent nodes. The friction factor λ in the laminar region ($Re \leq 2300$) is defined by the Hagen-Poliseuille:

$$\lambda = \frac{64\eta}{wD_{in}\rho} = \frac{64}{\text{Re}} \tag{8}$$

For the turbulent gas flow the friction factor depends on the Reynolds number as well as the relative roughness of the pipe wall ($e = k/D_{in}$). The friction factor in transitional region ($Re >$ 4000 for $e \leq e_{bound}$), depends only on Reynolds number and is described by the implicit relationship Prandtl-Karman:

$$\lambda = \left[2\log\frac{\sqrt{\lambda}\,\text{Re}}{2.51}\right]^{-2} \tag{9}$$

Calebrook-White equation is used to calculate the friction factor for the flow in fully turbulent region ($Re > 4000$ and $e > e_{bound}$) when factor λ depends also on the Reynolds number and relative roughness e:

$$\lambda = \left[-2\log\left(\frac{2.51}{\text{Re}\sqrt{\text{Re}}} + \frac{e}{3.72}\right)\right]^{-2} \tag{10}$$

The boundary relative roughness e_{bound} describes one of the relationships below:

- Filonienko-Altsul

$$e_{bound} = \frac{18 \log \mathrm{Re} - 16.4}{\mathrm{Re}} \tag{11}$$

- Blasius

$$e_{bound} = 17.85 \mathrm{Re}^{-0.875} \tag{12}$$

- Altsul-Ljacer

$$e_{bound} = \frac{23}{\mathrm{Re}} \tag{13}$$

5.1 The characteristic of input data

Steady-state gas flow simulation calculations are conducted based on properly prepared input data in form of 108 values of hourly gas streams consumed from the network by consumers in consumption nodes.

Amount of daily gas stream leaving particular node of the network depends inter alia on air average temperature. Based on archive data covering years 2006-2008, linear relations of daily gas consumption in function of air average temperature were developed. Details of defining linear models of gas consumption value from the temperature are described in literature (Szoplik J., 2010a). General form of linear model to define daily gas stream Q_d [m^3/day] consumed by consumers from particular node is presented by equation

$$Q_d = a(18\text{-}t) + b \tag{14}$$

where: a, b - model constant, individually determined for each node of the network based on real data of gas consumption by consumers, t - average air temperature [°C].

Fig. 6. presents results describing gas stream volume variation Q_d received within twenty four hours in four selected nodes of the network from the Fig. 5 depending on average air temperature t = +18, +8, -4, -16 °C (set according to equation (14)). Comparing results presented in Fig. 6 one can see, that the increase of daily gas stream Q_d leaving the network, caused by temperature increase, is not equal in all nodes of the network, as the number of gas consumers and gas receivers installed at consumers, consuming gas from a particular node of the network is different.

However amount of hourly gas stream Q [m^3/h] leaving particular node of the network is defined based on characteristic of percentage gas consumption in particular hours of the day. Such characteristics were developed based on real data describing gas stream flow through reduction and measurement station in successive hours of the day in various days of the year. Fig. 7 presents gas stream size change during different hours of the day with average temperature of -4 °C in four selected nodes of the network from the Fig. 5. However Fig. 8 presents the results in form of hourly stream size Q of the gas leaving the network in node 55 during various hours of four exemplar days, differ with air temperature.

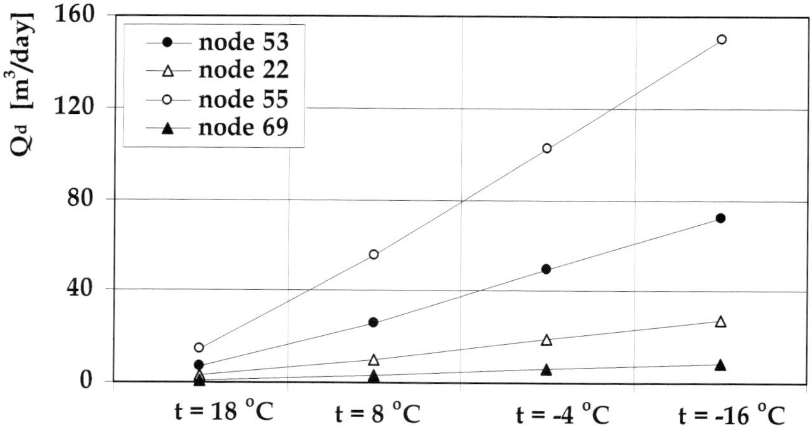

Fig. 6. The effect of air temperature t on the volumetric daily gas stream size Q_d in a given node of the network.

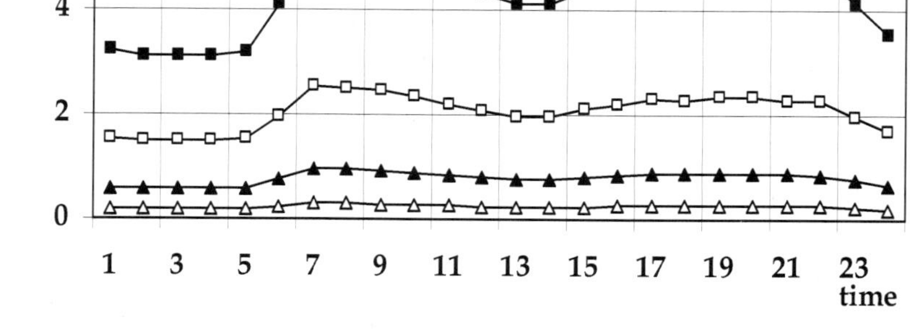

Fig. 7. Change of the gas stream size Q leaving the network in four exemplar nodes of the network in particular hours of the day; air temperature t = -4 °C.

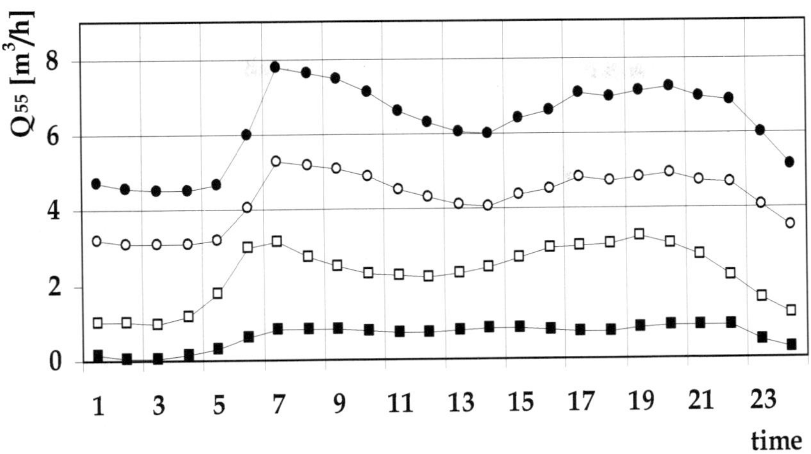

Fig. 8. Change of the gas stream Q_{55} leaving the network in node 55 in particular hours of the day; various values of air temperature; (•) t = -16 °C, (○) t = -4 °C, (□) t = 8 °C, (■) t ≥ 18 °C.

Analysing results presented in Fig. 7 and Fig. 8 one can notice, that size of the hourly gas stream Q depend on air temperature and hour of the day and type of the node in the network. It can also be noticed, that in the same node of the network (i.e. node 55) maximal gas stream size Q_{max} noted on the day with average temperature of -16 °C is nine times larger than that the day with temperature of 18 °C. There are also clear differences between maximal and minimal stream values Q during a day (for the same node). In percentage, larger diversity is noticed during summer days than during winter days. During winter time (t = -16 °C) maximal gas stream size Q_{max} is larger than minimal stream Q_{min} by approximately 70%, however it is two times larger for summer time (t ≥ 18 °C).

6. The results and discussion

Change of the gas stream leaving the network in 108 nodes during season and twenty four hour cycle is reflected on appropriate diversity of the network load. With the GasNet 3.8.1 programme, a steady-state simulation of gas flow in network, for each set of input data (108 values of gas streams) was conducted. Details of mathematical model of gas flow in the network and algorithm, according to which calculations were performed in the GasNet programme, were described in papers (Szoplik J., 2010b, 2010c). For each of four, exemplar days different with average temperature conducted 24 simulations. In total, performed 96 simulation calculations, are presented in this chapter.

In practice, it is accepted, that the network works correctly, when each collection node gas stream of appropriate size is delivered with pressure higher than $p \geq 1700$ Pa. Fulfilling above two conditions is possible due to correct work of the reduction gas station (point Z1, Fig. 5). Gas stream overpressure entered into the network through the middle pressure gas reduction station Z1 (Fig. 5) can be variable in range from 1700 to 2500 Pa. From the practical point of view, the overpressure of the gas entered to the network should be possible lowest, as this will allow to maintain respectively low gas pressure in the network, that in case of leak or the network failure will allow to minimize gas losses. Such effect can be achieved by steering appropriately gas stream pressure feeding the network in point Z1.

Steering gas stream pressure inputted into the network can be done with several methods. One of the methods is to connect pressure of gas stream p_{Z1} and stream size Q_{feed}. For that purpose it is necessary to develop relationship p_{z1} = f(Q_{feed}). Another method is to make dependent entry stream pressure on gas pressure in selected point of the network. In this case it is necessary to develop gas network characteristic and appointing network nodes, where pressure is the lowest, as those nodes shall be most exposed to possible entry pressure changes.

6.1 Pressure steering algorithm

During gas flow in the network simulation, for each entry data set (gas stream leaving network), overpressure gas stream entered into the network was multiple time changed, and it allowed to indicate (from range of 1700 ÷ 2500 Pa) the lowest gas overpressure feeding the network. During research a clear dependence was noticed of the feeding overpressure stream on the size of the stream entering the network. The stream overpressure p_{z1} dependence on the stream size Q_{feed}, being sum of streams received in 108 nodes of the network is presented on Fig. 9. The figure presents 96 points, that characterise minimal gas overpressure, that at given network load Q_{feed}, ensure gas delivery to all 108 gas consumption nodes under overpressure higher than 1700 Pa. Data on the Fig. 9 presenting results received for each of 24 hours and 4 days of various air temperature were described using the equation

$$p_{z1} = 5 \cdot 10^{-4} Q_{feed}^2 + 8.23 \cdot 10^{-2} Q_{feed} + 1703 \tag{15}$$

Equation (15) allows to estimate the value of minimal gas stream overpressure p_{z1} [Pa] feeding the network depending on size of that stream Q_{feed} [m³/h] with average relative error of ± 2 %. The equation (15) includes influence of both time of the day and air temperature on the size of the stream overpressure feeding the network. Air temperature influences on the decrease of gas streams leaving the network in 108 consumption nodes, and hence there is smaller total gas stream entering the network Q_{feed}. Influence of the time of the day on feeding pressure value is considered also in form of gas stream size leaving the network. Higher gas consumption from the network can be noticed during day hours, whereas lower during night hours. Exact entry overpressure value change p_{z1} in further hours of the exemplar winter day (t = -16 or -4 °C), summer day (t ≥ 18 °C) or autumn or spring day (t = 8 °C) presents in Fig. 10. Analysing results presented in Fig. 10 one can see, that value of the stream minimal overpressure feeding the network depends on the time of the day and achieves higher values in hours of maximal gas consumption from the network.

The equation (15) called pressure steering algorithm can be used to program the reducer in gas reduction station (point Z1 on Fig. 5). However, determination of such an algorithm requires many time consuming calculations.

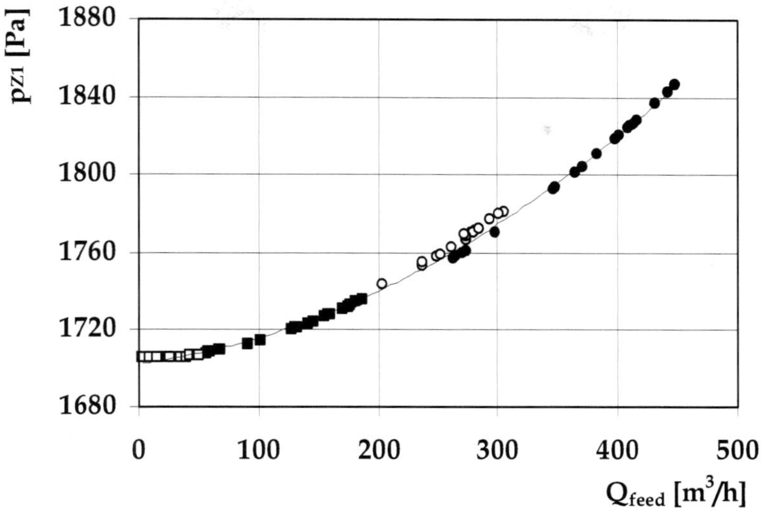

Fig. 9. The effect of gas stream size feeding the network Q_{feed} on the gas stream overpressure p_{Z1}; (•) t = -16 °C, (○) t = -4 °C, (■) t = 8 °C, (□) t ≥ 18 °C.

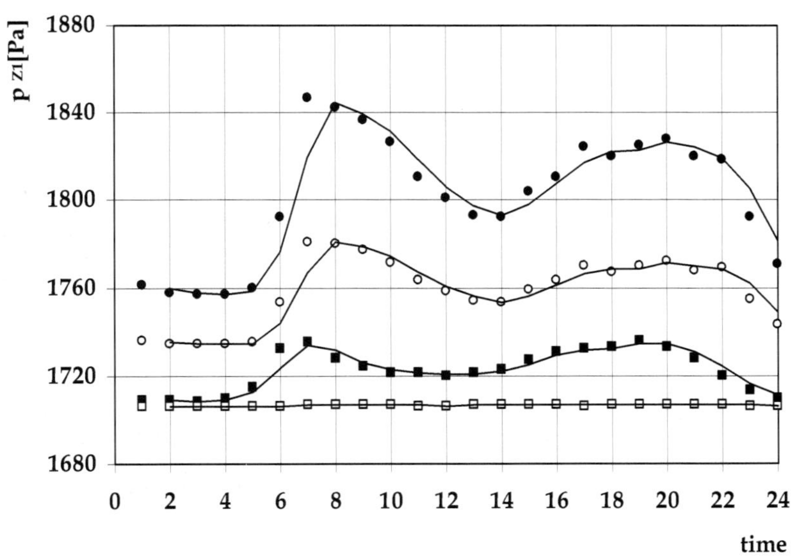

Fig. 10. The change of the gas stream overpressure p_{Z1} feeding the network in particular hours of the day; various values of air temperature; (•) t = -16 °C, (○) t = -4 °C, (■) t = 8 °C, (□) t ≥ 18 °C.

6.2 The characteristic of gas pipeline network

Based on gas flow in the network steady-state simulation results allowed to define also the network characteristic, that is presented in form of map of overpressure and gas streams layout in all gas pipeline of the network. Analysed data allowed to determine areas of the network, that are most sensitive to possible entry pressure stream fluctuation. Results presented in further part of the chapter were received for minimal value of the network feeding overpressure gas p_{z1}, but allowing to deliver gas to each node of the network under overpressure of $p \geq 1700$ Pa. Therefore on presented maps there are no areas with too low gas pressure. However based on detailed calculation results, received during simulation, one can at the same time define nodes of the network that due to their location in the network and characteristic of gas partition, are sensitive on gas pressure rapid decrease. Such nodes (called typical nodes) were marked on Figs. 11 - 14 with dashed line and their location in the network changes and depend on gas streams leaving the network. The reason of typical nodes location change in the network is irregular location of gas collection nodes, what causes uneven network load. In this case, even small gas pressure decrease at the network entry, below minimal value, will cause that gas streams consumed by the recipients focused in these nodes, will characterise with overpressure lower than minimal accepted (p = 1700 Pa), and this may cause damage or faulty running of appliances feeding with gas.

Joining typical node or nodes of the network with gas reduction station (point Z1) can be used in second mentioned gas pressure steering method. Gas pressure drop in typical node will be a signal to the reduction station to increase gas pressure entered to the network and vice versa.

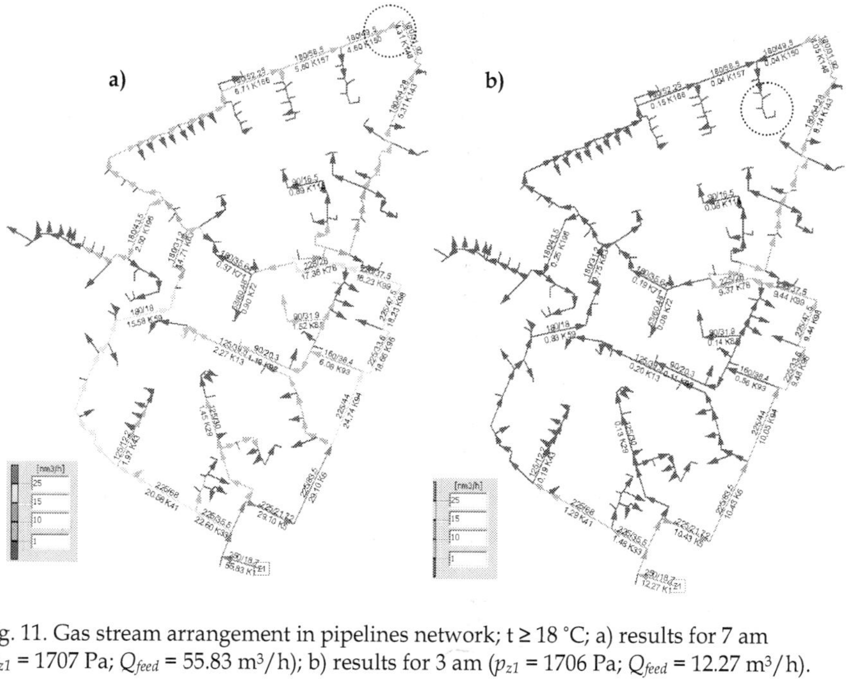

Fig. 11. Gas stream arrangement in pipelines network; t ≥ 18 °C; a) results for 7 am (p_{z1} = 1707 Pa; Q_{feed} = 55.83 m^3/h); b) results for 3 am (p_{z1} = 1706 Pa; Q_{feed} = 12.27 m^3/h).

The simulation results in form of gas streams layout in the gas pipeline network received for the hour of minimal and maximal gas consumption on the day of average air temperature higher than t >18 °C are presented in Fig. 11. However, analogous data, received however for a winter day with average air temperature t = -4 °C presents Fig. 12. Comparing results presented in Figs. 11 and 12 one can see, that clearly higher load of the network is during winter in hours of top gas consumption (at 7 am), whereas lowest is in summer during night hours (at 3 am). In summer time in definitely larger part of gas pipelines, the stream of flowing gas is lower than 1 m^3/h, whereas in winter is much higher and is approx. 8 m^3/h.

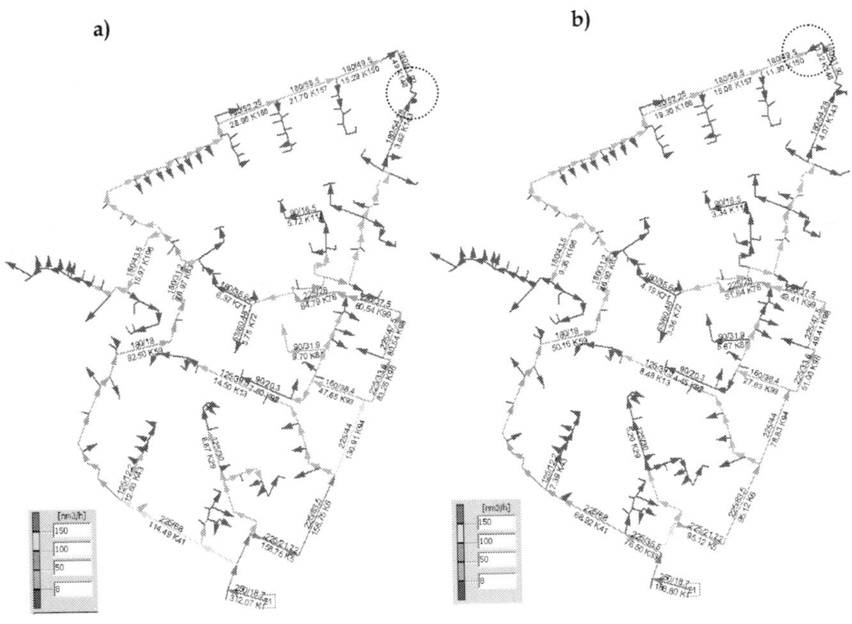

Fig. 12. Gas stream arrangement in pipelines network; t = -4 °C; a) results for 7 am (p_{z1} = 1781 Pa; Q_{feed} = 312.07 m^3/h); b) results for 3 am (p_{z1} = 1734 Pa; Q_{feed} = 186.80 m^3/h).

The network load increase caused by higher gas consumption from the network requires its feeding with stream of higher overpressure. Considering results presented on Fig. 13 and Fig. 14 gas overpressure layout can be analysed in all the network gas pipelines depending on the network load and feeding overpressure value. No significant differences were noticed in the overpressure values feeding the network in summer time (Fig. 13a and Fig. 13b). However in winter time (Fig. 14a and Fig. 14b) differences between gas overpressure values entered the network within hours of top consumption and during the night hours are clearly higher.

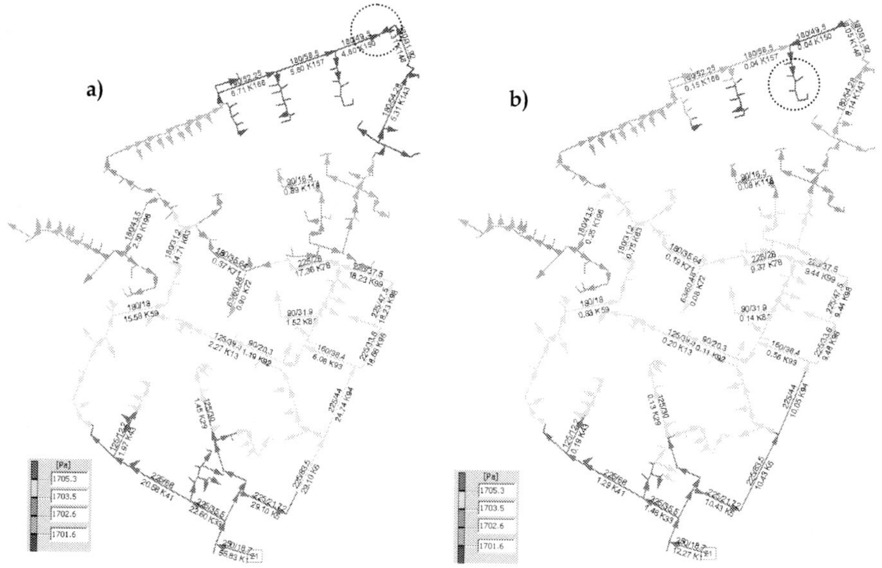

Fig. 13. Gas stream overpressure arrangement in pipelines network; $t \geq 18$ °C; a) results for 7 am (p_{z1} = 1707 Pa; Q_{feed} = 55.83 m^3/h); b) results for 3 am (p_{z1} = 1706 Pa; Q_{feed} = 12.27 m^3/h).

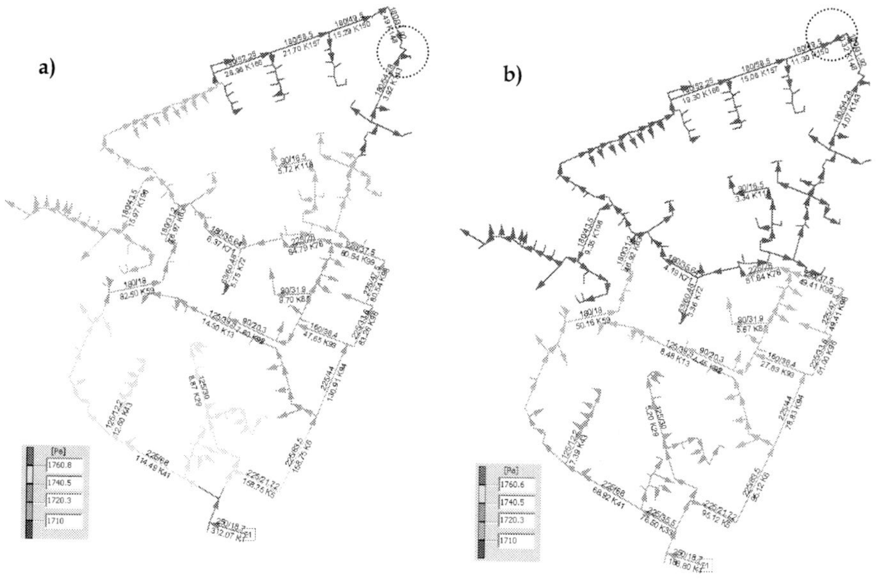

Fig. 14. Gas stream overpressure arrangement in pipelines network; t = -4 °C; a) results for 7 am (p_{z1} = 1781 Pa; Q_{feed} = 312.07 m^3/h); b) results for 3 am (p_{z1} = 1734 Pa; Q_{feed} = 186.80 m^3/h).

7. Conclusions

Natural gas transportation and distribution from production place is in gas pipelines that vary with gas overpressure value, gas speed in pipe, network pipeline materials and the network structure.

Gas transportation to consumers is a difficult task, because characteristic feature of gas flow in the network is irregularity of the network load caused with irregularity of gas consumption by consumers from the network, that shift in seasonal and daily cycle.

Gas transportation through network depends inter alia on the work quality of the gas network objects (compressor stations, reduction stations and gas pipelines). Large number of works dedicated to work optimisation of objects proves importance of this issue.

Assuming, that gas loss resulting from the network leak are proportional to gas pressure in the pipeline, steering gas pressure feeding the network allows to lower significantly the network exploitation costs due to lower gas loss, caused with the network leak or with the network gas pipeline damage.

8. Acknowledgment

The author would like to thank to the President and workers of the Gas Company in Szczecin, for access and preparation of data to necessary for preparation of the present work and for providing precious guidelines, enabling correct interpretation of the calculation results.

9. References

Brkić D. (2009). An improvement of Hardy Cross method on looped spatial natural gas distribution networks. *Applied Energy*, 86, pp. 1290-1300.

Carvalho A.A., Rebello J.M.A., Sagrilo L.V.S., Camerini C.S. & Miranda I.V.J. (2006). MFL signals and artificial neural networks applied to detection and classification of pipe weld defects. *NDT&E international*, 39, pp. 661-667.

Chaczykowski M. (2010). Transient flow in natural gas pipeline - The effect of pipeline thermal model. *Applied Mathematical Modelling*, 34, pp. 1051-1067.

Fukushima K., Maeshima R., Kinoshita A., Shiraishi H. & Koshijima I. (2000). Gas pipeline leak detection system using the online simulation method. *Computers & Chemical Engineering*, 24, pp. 453-456.

Herran-Gonzalez A., De La Cruz J.M., De Andres-Toro B., Risco-Martin J.L. (2009). Modeling and simulation of a gas distribution pipeline network. Applied Mathematical Modelling, 33, pp. 1584-1600.

Hwang K., Mandayan S., Udpa S.S., Udpa L., Lord W. & Atzal M. (2000). Characterization of gas pipeline inspection signals using wavelet basis function neural networks. *NDT&E international*, 33, pp. 531-545.

Ke S.L. & Ti H.C. (2000). Transient analysis of isothermal gas flow in pipeline network. *Chem. Eng. Jour.*, 76, pp. 169-177.

Kralik J., Stiegler P., Vostry Z. & Zavorka J. (1988). *Dynamic Modeling of Large-Scale Networks with Application to Gas Distribution*. Elsevier, Amsterdam

Liu M., Zang S. & Zhou D. (2005). Fast leak detection and location of gas pipeline based on an adaptive particle filter. *Int. J. Appl. Math. Comput. Sci.*, 15, pp. 541-550.

Nguyen H.H. & Chan Ch.W. (2006). Application ofartificial intelligence for optimization of compressor scheduling. *Engineering Applications of Artificial Intelligence*, 19, pp. 113-126.

Nguyen H.H., Uraikul V., Chan C.W. & Tontiwachwuthikul P. A. (2008). Comparison of automation for optimization of compressor scheduling. *Advances in Engineering Software*, 39; pp. 178-188.

Osiadacz A.J. & Chaczykowski M. (2001). Comparison of isothermal and non-isothermal pipeline gas flow models. *Chem. Eng. Jour.*, 81, pp. 41-51.

Osiadacz A.J. (1987). *Simulation and analysis of gas network.* E&FN Spon., London.

Osiadacz A.J. (2001). *Steady- state simulation of gas network*, BIG, Warszawa (in Polish).

Reddy H.P., Narasihman S. & Bhallamudi S.M. (2006). Simulation and state estimation of transient flow in gas pipeline networks using a transfer function model. *Ind. Eng. Chem. Res.*, 45, pp. 3853-3863.

Silva R.C.C., Guerreiro J.N.C. & Loula A.F.D. (2007). A study of pipe interacting corrosion defects using the FEM and neural networks. *Advances in Engineering Software*, 38, pp. 868-875.

Szoplik J. (2010a). Quantitative analysis of the heterogeneity for gas flow in the pipeline system. *Gaz, Woda i Technika Sanitarna*, 2010, 1, 2-6 (in Polish).

Szoplik J. (2010b). The application of the graph theory to the analysis of gas flow in a pipeline network. *37th International Conference of SSCHE*, proceedings on CD ROM, 24-28.05.2010, Tatranske Matliare, Slovakia

Szoplik J. (2010c). The steady-state simulations for gas flow in a pipeline network. *Chem. Eng. Trans.*, 21, pp. 1459-1464.

Tao W.Q. & Ti H.C. (1998). Transient analysis of gas pipeline network. *Chem. Eng. Jour.*, 69, pp. 47-52.

Zahedi G., Karami Z. & Yaghoobi H. (2009). Prediction of hydrate formation temperature by both statistical models and artificial neural network approaches. *Energy Conversion and Management*, 50, pp. 2052-2059.

14

Phase Behavior Prediction and Modeling of LNG Systems with EoSs – What is Easy and What is Difficult?

Blanca E. García-Flores[1], Daimler N. Justo-García[2],
Roumiana P. Stateva[3] and Fernando García-Sánchez[1]
[1]*Laboratory of Thermodynamics, Research Program in Molecular Engineering, Mexican Petroleum Institute, Mexico, D.F.,*
[2]*Department of Chemical and Petroleum Engineering, ESIQIE, National Polytechnic Institute, Mexico, D.F.,*
[3]*Institute of Chemical Engineering, Bulgarian Academy of Sciences, Sofia,*
[1,2]*México,*
[3]*Bulgaria*

1. Introduction

In the past decades, natural gas processing has increasingly encompassed low temperatures with the view to recover ethane and heavy components from natural gas. Liquefied natural gas (LNG) is a useful method for storing the gas in a small space at peak-shaving facilities of natural gas distribution companies. The increased density of LNG facilities allows transportation of natural gas via large ocean vessels from gas fields situated far from their potential markets. Also, in the past, natural gas has been an important source of chemical feed stocks like ethane, propane, butane, etc. In order to increase the recovery of these stocks, the gas processes have been shifting to lower temperatures where the recoveries are improved. In all of the above processes and applications of natural gas, knowledge of the LNG systems phase behavior and thermodynamic properties is required for the successful design and operation of LNG plants.

The question of how to predict and model phase equilibria behavior for natural gas systems is far from new and liquid-vapor equilibria problems have been successfully solved for many years now. At present, interest has shifted to systems containing not only species in the simple paraffinic homologous series, but also water, carbon dioxide, hydrogen sulfide, hydrogen, nitrogen– to mention a few.

Not unrelated to this growing variety of species in cryogenic process streams is the occurrence of multiphase phenomena. The situation of interest in natural gas processing is often a methane-rich stream for which there is the question whether a second solvent can alter dramatically the pattern of the phase behavior of the customarily liquid-vapor mixture and cause problems.

The active interest in the use of nitrogen gas to pressurize oil reservoirs to enhance recovery has resulted in natural gas process streams, rich in nitrogen, which are likely to display complex phase behavior. Investigators have studied experimentally ternary prototype LNG systems, containing nitrogen as a second solvent, and a lot of excellent data have been published. However, in order to help further understanding the possible occurrence of multiphase equilibria in LNG process systems, it is necessary to acquire knowledge of their phase behavior and of the variety of critical end point boundaries through an ability to predict, model and calculate them.

There are several aspects of separation/refinement of natural and synthetic gases where multiphase equilibria play a role. The occurrence of liquid-liquid-vapor (LLV) behavior during the recovery of natural gas by low temperature distillation, especially from nitrogen-rich gas mixtures, is just one such example. LLV behavior can also occur in the processing of gases containing high quantities of carbon monoxide that result from coal gasification, as well as in high pressure absorption processes for the removal of either desirable or undesirable components from natural and synthetic gas mixtures. In these latter processes, methanol can be used as the absorber to separate out feed stocks (ethane, ethylene, propane, propylene, etc.), harmful components such as hydrogen sulfide, carbon disulfide, and carbonyl sulfide, or simply unwanted species such as carbon dioxide. Obviously, the formation of a second liquid phase can upset the expected performance of these processes.

In the cryogenic processing of natural gas mixtures, species such as CO_2 and the heavier hydrocarbons can form solids and foul the gas processing equipment. For example, the formation of a solid phase can coat heat exchangers and foul expansion devices, leading to process shutdown and/or costly repairs. Knowledge of the precipitation conditions for gas streams is essential in minimizing downtime for cleanup and repairs. However, the appearance of a solid phase in a process is not always a liability. Off-gas (primary nitrogen) from power plants with light water reactors or from fuel processing plants can contain radioactive isotopes of krypton and xenon, which may be removed by solid precipitation. The phenomenological aspects of LLV/solid-liquid-vapor (SLV) behavior are also interesting because there is a need for a better understanding of the physical nature of the thermodynamic phase space, especially the non-analyticities of critical point region.

The success of the design and operation of separation processes in the oil and gas industry at low temperatures is critically dependent on accurate descriptions of the thermodynamic properties and phase behavior of the concerned multicomponent hydrocarbon mixtures with inorganic gases. Consequently, it is important to apply appropriate models within a thermodynamic modeling framework to predict, describe and validate the complex phase behavior of LNG mixtures. In this case, equations of state (EoS) are usually the primary choice.

The aim in this chapter is to examine and analyze the challenges and difficulties encountered when modeling the complex phase behavior of LNG systems, and to compare the capabilities of two numerical techniques advocated for phase behavior predictions and calculations of complex multicomponent systems. The chapter is organized as follows. In Section 2 the thermodynamics and topography of the phase behavior of LNG systems are presented. Section 3 describes two computational techniques, advocated by us, to predict and model complex phase behavior of nonideal mixtures, and their application to LNG systems. Section 4 is focused to the description of two numerical methods to directly calculate critical end points (K- and L-points). In Section 5 two representatives of the equations of state (EoSs) type thermodynamic models, namely SRK EoS (Soave, 1972) and PC-SAFT EoS (Gross and Sadowski, 2001) are used to represent the equilibrium fluid phases. Results and discussion of the multiphase behavior modeling for selected ternary systems studied are presented in Section 6. Finally, conclusions derived from the work are given in Section 7.

2. Thermodynamics and topography of the phase behavior of LNG systems

Though only a limited number of immiscible binary systems (methane–*n*-hexane, methane–*n*-heptane, to name the most prominent ones) are relevant to natural gas processing, LLV behavior can and does occur under certain conditions in ternary and higher LNG systems even when none of the constituent binaries themselves exhibit such behavior. It is also known that the addition of nitrogen to miscible LNG systems can induce immiscibility and this necessarily affects the process design for these systems.

The qualitative classification of natural gas systems and the topography of the multiphase equilibrium behavior of the systems in the thermodynamic phase space and the nature of the phase boundaries will be addressed in this section.

Kohn and Luks (1976, 1977, 1978), who carried out extensive experimental studies on the solubility of hydrocarbons in LNG and NGL (natural gas liquids) mixtures, qualitatively classify natural gas systems (or any system) as one of four types (Kohn and Luks, 1976). These types of binary systems show, for instance, that a first type system, methane-*n*-octane (Kohn and Bradish, 1964), has a solid-liquid-vapor (S-L-V) locus which starts at the triple point of the solute and terminates at a Q-point (S_1-S_2-L-V) near the triple point of the solvent with an S-V gap in the locus. The S-L-V branches terminate with a K-point where the liquid becomes critical with vapor in the presence of a solid. On the contrary, methane-*n*-heptane (Kohn, 1961), a second type system, has a Q-point (S-L_1-L_2-V) in the central portion of the SLV locus, at which point one loses the L_1 phase and gains the L_2 phase (solvent lean phase). These systems also have a L_1-L_2-V locus running from the Q-point to a K-point where the liquid L_2 becomes critical with vapor in the presence of L_1. The other two types of systems have no discontinuities between the triple points of the solvents; however, one of these types, methane-*n*-hexane (Shim and Kohn, 1962), a typical third type system, has a L_1-L_2-V locus which is terminated by critical points, points where the two liquid phases, or a liquid and vapor, become critically identical. A typical system of the fourth type is ethane-*n*-octane (Kohn et al., 1976). The topographical evolution of multiphase equilibria behavior is discussed in detail by Luks and Kohn (1978). Ternary ($N = 3$) or more complex systems exhibit similar types of phenomena, but the loci and exact points of a binary become $N-1$ and $N-2$ dimensional spaces in the more complex systems.

It is known that systems rich in methane can exhibit L_1-L_2-V behavior of the second and third type variety (e.g., methane-*n*-heptane and methane-*n*-hexane, respectively). However, investigations into ternary S-L-V and more complex systems revealed that L_1-L_2-V behavior can occur in systems whose binary pairs exhibit no immiscibility (Green et al., 1976; Orozco et al., 1977). Hottovy et al. (1981, 1982) observed that systems of the first type (methane-*n*-octane) could be modified to behave like a second type system by adding a second solvent (e.g., methane-ethane-*n*-octane, methane-propane-*n*-octane, methane-*n*-butane-*n*-octane, and methane-carbon dioxide-*n*-octane). Merrill et al. (1983) reported the phase behavior of the ternary systems methane-*n*-pentane-*n*-octane and methane-*n*-hexane-*n*-octane, which also exhibit L_1-L_2-V immiscibility. Additionally, these authors studied the systems methane-*n*-hexane-carbon dioxide, in which the carbon dioxide is added to the pair methane-*n*-hexane (third type) to induce L_1-L_2-V immiscibility.

In five of these seven ternary systems, methane-*n*-octane-(ethane, propane, *n*-butane, *n*-pentane, and carbon dioxide), the immiscibility region is bounded by loci of type K, type Q and LCST (lower critical solution temperature) points, whereas for the system methane-*n*-hexane-*n*-octane the region of L_1-L_2-V immiscibility is bounded, apart from the K, Q, and LCST points, by the L_1-L_2-V locus of the methane-*n*-hexane binary system. In the case of the system methane-*n*-hexane-carbon dioxide, the immiscibility region is bounded by a locus of K and LCST points and by the L_1-L_2-V locus of the methane-*n*-hexane binary system.

It should be pointed out that the onset and evolution of LLV behavior in mixtures is related to the evolution of SLV behavior in those same systems. Thus, in natural gas it is often of interest to predict whether a methane-rich stream with one of the solutes (such as an *n*-paraffin of carbon number four or higher; or benzene, or carbon dioxide) will form a solid phase. The reason behind that is that the presence of a second solvent can considerably change the solubility of the solid solute, if the solute is a hydrocarbon; this also occurs to a lesser extent if the solute is carbon dioxide.

On the other hand, the presence of nitrogen as a second solvent reduces the solubility of solids in methane, ethane, and mixtures thereof. Furthermore, the addition of nitrogen to miscible LNG systems can induce immiscibility. The number of nitrogen binary systems relevant to LNG that exhibit LLV immiscibility is few – nitrogen-ethane and nitrogen-propane, being among the most prominent ones. However, LLV phenomenon has been observed at certain conditions in many ternary and higher realistic nitrogen-rich LNG systems since LLV behavior can and does occur in multicomponent systems even when for none of the constituent binaries themselves an LLV locus is reported.

Another interesting example is the ternary mixture methane + ethane + *n*-octane; the species in the constituent binary pairs (methane + ethane) and (ethane + *n*-octane) are too similar in molecular nature to be LLV immiscible, while the pair methane + *n*-octane is too dissimilar to be immiscible. If a multicomponent mixture is considered to be solute plus solvent in the pseudo-component sense, it can be readily seen why the ternary mixture has a region of LLV behavior. The methane + ethane form a solvent background in which *n*-octane is immiscible.

The type of the LLV region displayed by a system depends on whether it contains an immiscible binary or not. For a ternary system with no constituent binary LLV behavior present the three-phase region is a "triangular" surface in the thermodynamic phase space with two degrees of freedom, while its boundaries have one. It is bounded from above by a K-point locus; from below by a LCST locus and at low temperatures by a Q-point locus. The systems methane + *n*-butane + nitrogen (Merrill et al., 1984a) and methane + *n*-pentane + nitrogen (Merrill et al., 1984b) belong to this class.

A K-point occurs when a liquid phase and a vapor phase become critical in the presence of a heavier noncritical equilibrium liquid phase, whereas an L-point occurs when two liquid phases becomes critical in the presence of a noncritical equilibrium vapor phase. These points, also called upper critical end points (UCEPs) and lower critical end points (LCEPs), are different depending on their location in the pressure-temperature space. That is, an UCEP is always located at a higher temperature and pressure than the LCEP, and when there exists only one critical end point, so that a K-point is always an UCEP point. On the contrary, an L-point can be an UCEP or an LCEP, depending on the global phase behavior of the system.

The system methane + ethane + *n*-octane does not exhibit immiscibility in any of its binary pairs. Although immiscibility has been reported in binary systems of methane + *n*-hexane and methane + *n*-heptane, solutes such as *n*-octane and higher normal paraffins crystallize as temperature decreases before any immiscibility occurs. On the other hand, with ethane as solvent, solutes beginning with *n*-C_{19} and higher paraffins demonstrate LLV behavior. Apparently, the addition of modest amounts of ethane to methane creates a solvent mixture exhibiting immiscibility with *n*-octane.

3. Modeling of the phase behavior of LNG systems

The success of the design and operation of separation processes in the oil and gas industry at low temperatures is critically dependent on the accurate descriptions of the thermodynamic properties and phase behavior of the concerned multicomponent hydrocarbon mixtures with inorganic gases. Phase-split calculations and phase stability analysis in natural gas systems simulation can take up as much as 50 % of the CPU time. In complicated problems it may take even more. Thus, it is important to develop a reliable thermodynamic modeling framework (TMF) that will be able to predict, describe and validate robustly and efficiently the complex phase behavior of LNG mixtures.

The TMF has three main elements: a library of thermodynamic parameters pertaining to pure-substances and binary interactions, thermodynamic models for mixture properties, and algorithms for solving the equilibrium relations. Reliable pure-component data for the main constituents of LNG systems are available experimentally; an equation of state is usually the primary choice for the thermodynamic model. Thus, the focal point of a TMF for phase behavior calculations of LNG systems is the availability of robust methods for thermodynamic stability analysis, and of reliable efficient and effective flash routines for three phase split calculations.

3.1 Computational technique 1

This first technique uses an efficient computational procedure for solving the isothermal multiphase problem by assuming that the system is initially monophasic. A stability test allows verifying whether the system is stable or not. In the latter case, it provides an estimation of the composition of an additional phase; the number of phases is then increased by one, and equilibrium is achieved by minimizing the Gibbs energy. This approach, advocated as a stagewise procedure (Michelsen, 1982b; Nghiem and Li, 1984), is continued until a stable solution is found.

In this technique, the stability analysis of a homogeneous system of composition $\mathbf{z}$, based on the minimization of the distance separating the Gibbs energy from the tangent plane at $\mathbf{z}$, is considered (Baker et al., 1982; Michelsen, 1982a). In terms of fugacity coefficients, φ_i, this criterion for stability can be written as (Michelsen, 1982a)

$$F(\boldsymbol{\xi}) = 1 + \sum_{i=1}^{N} \xi_i \left[\ln \xi_i + \ln \varphi_i(\boldsymbol{\xi}) - h_i - 1 \right] \geq 0 \quad \forall \boldsymbol{\xi} > 0 \tag{1}$$

where ξ_i are mole numbers with corresponding mole fractions as $y_i = \xi_i \Big/ \sum_{j=1}^{N} \xi_j$, and

$$h_i = \ln z_i + \ln \phi_i(\mathbf{z}) \qquad i = 1,\ldots,N \tag{2}$$

Equation 1 requires that the tangent plane, at no point, lies above the Gibbs energy surface and this is achieved when $F(\boldsymbol{\xi})$ is positive in all its minima. Consequently, a minimum of $F(\boldsymbol{\xi})$ should be considered in the interior of the permissible region $\sum_{i=1}^{N} y_i = 1$, $\forall \mathbf{y} \geq 0$. To test condition 1 for all trial compositions is not physically possible; it is thus sufficient to test the stability at all stationary points of $F(\boldsymbol{\xi})$ since this function is not negative at all stationary points. Here, the quasi-Newton BFGS minimization method (Fletcher, 1980) was applied to eq 1 for determining the stability of a given system of composition $\mathbf{z}$ at specified temperature and pressure.

Once instability is detected with the solution at $p-1$ phases, the equilibrium calculation is solved by minimization of the following function

$$\underset{n_i^{(\phi)}}{Min}\ \Delta g = \sum_{\phi=1}^{p} \sum_{i=1}^{N} n_i^{(\phi)} \ln \left(\frac{x_i^{(\phi)} \varphi_i^{(\phi)} P}{P^{\circ}} \right) \tag{3}$$

subject to the inequality constraints given by

$$\sum_{\phi=1}^{p-1} n_i^{(\phi)} \leq z_i \qquad i = 1,\ldots,N \tag{4}$$

and

$$n_i^{(\phi)} \geq 0 \qquad i = 1,\ldots,N\ ;\ \phi = 1,\ldots,p-1 \tag{5}$$

where z_i is the mole fraction of the component i in the system, $n_i^{(\phi)}$ ($i = 1,\ldots,N$; $\phi = 1,\ldots,p-1$) is the mole number of component i in phase ϕ per mole of feed, $x_i^{(\phi)}$ is the mole fraction of component i in phase ϕ, T is the temperature, P is the pressure, and P° is the pressure at the standard state of 1 atm (101.325 kPa). In eq 3 the variables $n_i^{(p)}$, $x_i^{(p)}$, and $\varphi_i^{(p)}$ are considered functions of $n_i^{(\phi)}$.

Equation 3 is solved using an unconstrained minimization algorithm by keeping the variables $n_i^{(\phi)}$ inside the convex constraint domain given by eqs 4 and 5 during the search for the solution. In this case, a hybrid approach to minimize eq 3 is used starting with the steepest-descent method in conjunction with a robust initialization supplied from the stability test to ensure a certain progress from initializations, and ending with the quasi-Newton BFGS method which ensures the property of strict descent of the Gibbs energy surface. A detailed description of this approach for solving the isothermal multiphase problem can be found elsewhere (Justo-García et al., 2008a).

3.2 Computational technique 2

The second approach to calculate multiphase equilibria used here applies a rigorous thermodynamic stability analysis and a simple and effective method for identifying the phase configuration at equilibrium with the minimum Gibbs energy. The rigorous stability analysis is exercised once and on the initial system only. It is based on the well-known tangent plane criterion (Baker et al., 1982; Michelsen, 1982a) but uses a different objective function (Stateva and Tsvetkov, 1994; Wakeham and Stateva, 2004). The key point is to locate all zeros ($\mathbf{y}^*$) of a function $\Phi(\mathbf{y})$ given as

$$\Phi(\mathbf{y}) = \sum_{i=1}^{N} \left[k_{i+1}(\mathbf{y}) - k_i(\mathbf{y})\right]^2 \tag{6}$$

where

$$k_i(\mathbf{y}) = \ln \varphi_i(\mathbf{y}) + \ln y_i - h_i \qquad i = 1,\ldots,N \tag{7}$$

with

$$h_i = \ln z_i + \ln \varphi_i(\mathbf{z}) \qquad i = 1,\ldots,N \tag{8}$$

and assuming $k_{N+1}(\mathbf{y}) = k_1(\mathbf{y})$. Therefore, from eqs 6–8, it follows that $\min \Phi(\mathbf{y}) = 0$ when $k_1(\mathbf{y}^*) = k_2(\mathbf{y}^*) = \ldots = k_N(\mathbf{y}^*)$. The zeros of $\Phi(\mathbf{y})$ conform to points on the Gibbs energy hypersurface, where the local tangent hyperplane is parallel to that at $\mathbf{z}$. To each zero $\mathbf{y}^*$, a number k^* (equal for each y_i^*, $i = 1,\ldots,N$ of a zero of the function) corresponds, such that

$$k_i^* = \ln y_i^* + \ln \varphi(\mathbf{y}^*) - h_i \qquad i = 1,\ldots,N \tag{9}$$

Furthermore, the number k^*, which geometrically is the distance between two such hyperplanes, can be either positive or negative. A positive k^* corresponds to a zero, which represents a more stable state of the system, in comparison to the initial one; a negative k^*, a more unstable one. When all calculated k^* are positive, the initial system is stable; otherwise, it is unstable.

It is widely acknowledged that the task to locate all zeros of the tangent-plane distance function (TPDF), $\Phi(\mathbf{y})$ in this particular case, is extremely challenging because a search over the entire composition space is required. The search is further complicated by the existence of a number of trivial solutions, corresponding with the number of equilibrium phases present (Zhang et al., 2011). The specific form of $\Phi(\mathbf{y})$ (its zeros are its minima) and the fact that it is easily differentiated analytically, allows the application of a non-linear minimisation technique for locating its stationary points, and in their works Stateva and Tsvetkov (1994) and Wakeham and Stateva (2004) used the BFGS method with a line-search (Fletcher, 1980). The implementation of any non-linear minimization technique requires a set of "good" initial estimates, and the BFGS method is no exception. All details of the organization and implementation of the initialization strategy employed by the stability analysis procedure are given elsewhere (Stateva and Tsvetkov 1994) and will not be discussed here.

Thus, as discussed by Wakeham and Stateva (2004), a method has been created which leads, in practice, to an "extensive" search in the multidimensional composition space. It has proved to be extremely reliable in locating *almost all* zeros of $\Phi(\mathbf{y})$ at a reasonable computational cost. The term "almost all" zeros is used because there is no theoretically-based guarantee that the scheme will always find them all. If, however, a zero is missed, the method is self-recovering. Furthermore, the TPDF is minimized *once only*, which is a distinct difference from the approach that stage-wise methods generally adopt. Since stability analysis on its own cannot determine unequivocally which is the stable phase configuration of a system (identified as unstable at the given temperature and pressure), it is suggested to run a sequence of two-phase flash calculations to determine the correct number of the phases at equilibrium, and the distribution of the components among the phases.

4. Calculation of K- and L-points

Ternary systems which exhibit $L_1 - L_2 - V$ behavior but don't exhibit such behavior in their constituent binaries have the immiscibility region bounded by a K ($L_1 - L_2 = V$)-point locus, a LCST ($L_1 = L_2 - V$) locus, and a Q ($S - L_1 - L_2 - V$)-point locus. Ternary systems which have immiscibility in a constituent binary can have boundaries similar to those mentioned above, besides the intrusion of the binary $L_1 - L_2 - V$ locus on the ternary $L_1 - L_2 - V$ region. The K-point and LCST loci can intersect at a tricritical point where the three phases become critical; i.e., $L_1 = L_2 = V$.

Needless to say that it is costly and time-consuming to determine the K and LCST loci in ternary systems experimentally; thus, the availability of appropriate algorithms and numerical routines in the third element of the TMF that will allow the prediction and reliable location in the thermodynamic phase space of such points, is indispensable in the study of the complex phase behavior of LNG model systems. Among the several such algorithms published in the open literature we have chosen to outline briefly and implement those of Gregorowics and de Loos (1996) and Mushrif and Phoenix (2006). In our choice we have been guided by the fact that the above algorithms can successfully predict the K- and L-points of binary and ternary systems. We will thus test their robustness and efficiency in the locating critical end points in LNG model systems.

4.1 Gregorowicz and de Loos' algorithm

In a study on the modeling of the three-phase LLV region for ternary hydrocarbon mixtures with the SRK EoS, Gregorowicz and de Loos (1996) proposed a procedure for finding K- and L-points of ternary systems, based on the solution of thermodynamics conditions for the K- and L-point using the Newton iteration technique and starting points carefully chosen. They applied their procedure to calculate the K- and L-point loci for two ternary systems, namely, C_2 + C_3 + C_{20} and C_1 + C_2 + C_{20}, in which the constituent binary C_2 + C_{20} exhibit immiscibility. Consequently, the extension of the three-phase LLV region of these systems is bounded by the binary $L_1 - L_2 - V$ locus of the system C_2 + C_{20} and the ternary K-point and L-point loci. Briefly, the strategy followed by these authors to find the K- and L-points of the two ternary systems was the following: (1) calculation of the critical line, the K-point, the three phase line, and the L-point for the system C_2 + C_{20} using thermodynamic conditions, and (2) calculation of the K- and L-point loci for the ternary systems by using as starting points the obtained coordinates of the K- and L-point calculations for the binary system C_2 + C_{20}. In this case, to obtain a K- and L-point for a ternary system, the following set of six nonlinear equations,

$$D = \begin{vmatrix} A_{VV} & A_{Vx_1} & A_{Vx_2} \\ A_{Vx_1} & A_{x_1x_1} & A_{x_1x_2} \\ A_{Vx_2} & A_{x_1x_2} & A_{x_2x_2} \end{vmatrix} = 0 \tag{10}$$

$$D^* = \begin{vmatrix} D_V & D_{x_1} & D_{x_2} \\ A_{Vx_1} & A_{x_1x_1} & A_{x_1x_2} \\ A_{Vx_2} & A_{x_1x_2} & A_{x_2x_2} \end{vmatrix} = 0 \tag{11}$$

$$\mu_i^c - \mu_i^\alpha = 0 \qquad i = 1,2,3 \tag{12}$$

$$P^c - P^\alpha = 0 \tag{13}$$

in seven variables, $T, V^c, V^\alpha, x_1^c, x_2^c, x_1^\alpha, and\ x_2^\alpha$ have to be solved, where α designates V for the L-point or L_2 for the K-point, D and D^* are the two determinants that must be satisfied at a critical point, and μ_i is the chemical potential of component i.

The critical criteria given by eqs 10 and 11 are based on the Helmholtz energy, which can be expressed as (Baker and Luks, 1980),

$$A = \int_V^\infty \left(P - \frac{nRT}{V} \right) dV - RT \sum_{i=1}^{N} n_i \ln\left(\frac{V}{n_i RT} \right) + \sum_{i=1}^{N} n_i (U_i^0 - TS_i^0) \tag{14}$$

where n_i is the mole number of component i and V is the system volume. Derivatives of the Helmholtz energy are denoted by a subscript in eqs 10 and 11 (e.g., A_V, A_{x_i}) indicating the differentiation variable (volume V or mole fraction x_i of component i). Details to obtain the elements of determinants D and D^* are given in Baker and Luks (1980).

It is worth mentioning that Gregorowicz and de Loos (1996) calculated the ternary K- and L-points at chosen values of the temperature, which it is important when the experiments are carried out isothermally.

4.2 Mushrif and Phoenix's algorithm

The second approach to calculate K- and L-points was proposed by Mushrif and Phoenix (2006). This approach utilizes an efficient critical point solver and a standard phase stability test within a nested-loop structure to directly locate K- and L-points. The algorithm consists of two nested inner loops to calculate a critical-point temperature and volume at fixed composition $\mathbf{z}$. An outer loop uses the critical point as a test phase, searches for an incipient phase at a trial composition $\hat{n}$, and updates the critical composition to iteratively decrease the tangent plane distance of the incipient phase to zero. The Newton-Raphson method with numerical derivatives was used in both the inner and outer loops.

This algorithm is similar to that proposed by Gauter et al. (1999) to calculate critical end points (CEPs) for ternary systems of carbon dioxide as the near-critical solvent and two low-volatile solutes (1-pentanol or 1-hexanol + *n*-tridecane) with the PR (Peng and Robinson, 1976) EoS. These authors used the approach of Heidemann and Khalil (1980) to follow a critical line in steps of the mole fraction of the carbon dioxide, searching for a CEP along this line; i.e., searching for the occurrence of an additional phase in zero amount by using the formulation of Michelsen (1982a).

The principal difference between the algorithm of Mushrif and Phoenix and that of Gauter et al. is that the former directly calculate the K- and L-points without following the critical lines.

In the algorithm proposed by Mushrif and Phoenix (2006), the critical criteria are based on the tangent-plane criteria developed by Michelsen and Heidemann (1988) at specified temperature and pressure. Michelsen and Heidemann (1988) also formulated the critical-point criteria in terms of the tangent plane distance based on the Helmholtz energy at fixed temperature and volume. The condition for the stability of a mixture, with respect to a trial phase $\hat{\mathbf{n}}$, is

$$F = \sum_{i=1}^{N} \hat{n}_i \ln\left(f_i / f_{i_0}\right) - V(P - P_0)/RT \tag{15}$$

where F is the tangent plane distance from the Gibbs energy surface to the hyperplane tangent of the Gibbs energy surface at the composition $\mathbf{z}$.

In eq 15 the sign of F will determine the stability of the test phase; i.e., if $F > 0$ the test phase is stable, if $F = 0$ the test phase is in equilibrium with some alternate phase, and if $F < 0$ the test phase is unstable. Michelsen (1984) developed the criteria for critical points by expanding the tangent plane distance function in a Taylor series around the test point as

$$F = bs^2 + cs^3 + ds^4 + O(s^5) \tag{16}$$

such that $F(0) = 0$ and $(dF/ds)_{s=0} = 0$ hold at the test point; s being a parameter that defines the distance in composition space from the test point at $s = 0$. As the sign of the tangent plane distance function F determines the stability of the test phase, it is necessary to find the minimum of this function using scaled mole numbers as $X_i = (n_i - z_i)/z_i^{1/2}$, where z_i are mole fractions in the test phase and n_i are mole fractions in any alternate phase. At the test point $(\hat{\mathbf{n}} = \mathbf{z})$, $\mathbf{X} = 0$ and $F = 0$. Expressions for the first g_i and second derivatives B_{ij} of F with respect to $\mathbf{X}$ are given in Michelsen (1984).

Function F is minimized by varying $\mathbf{X}$ under the constraint that $\mathbf{u}^T\mathbf{X} = s$, where $\mathbf{u}$ is a vector of unit length. By applying the method of Lagrange multipliers, coefficient b can be expressed as $b = (1/2)\mathbf{u}^T\mathbf{B}\mathbf{u}$, regardless of the choice of vector $\mathbf{u}$. The least possible value of coefficient b is obtained by choosing $\mathbf{u}$ as the eigenvector of $\mathbf{B}$ corresponding to the smallest eigenvalue $\lambda_{\min}$; i.e., $\mathbf{B}\mathbf{u} = \lambda_{\min}\mathbf{u}$.

At trial conditions of temperature and volume, matrix $\mathbf{B}$ is calculated and $(\lambda_{\min}, \mathbf{u})$ are determined by inverse iteration (Wilkinson, 1965), then $b = \lambda_{\min}/2$. If $b = 0$, the system is at the limit of intrinsic stability. At a critical point coefficients b and c in eq 16 are zero for a given eigenvector $\mathbf{u}$ of $\mathbf{B}$ corresponding to the smallest eigenvalue $\lambda_{\min}$. For the evaluation of coefficient c, Michelsen (1984) showed that this can be determined efficiently from information already available of u_i and g_i.

The solution procedure to calculate a critical point is as follows: since coefficient b is to be a zero eigenvalue of $\mathbf{B}$, then it must be a singular matrix with a zero determinant; i.e., $\det(\mathbf{B}) = 0$, with a vector $\mathbf{u}$ satisfying $\mathbf{Bu} = 0$ ($\mathbf{u}^T\mathbf{u} = 1$)

The criterion of $b = 0$ is met when the matrix $\mathbf{B}$ is singular and is used to find the critical temperature at a fixed composition and volume. The determinant of matrix $\mathbf{B}$ is calculated through a LU decomposition of $\mathbf{B}$ ($\mathbf{LU} = \mathbf{B}$ where $\mathbf{L}$ is lower triangular and $\mathbf{U}$ is upper triangular); i.e., the determinant of $\mathbf{B}$ is the product of the diagonal elements of the LU decomposed matrix. Once the iteration to find the stability limit has converged, the vector $\mathbf{u}$ is determined by inverse iteration technique.

The implementation of the equation-of-state approach for calculating critical points using this procedure requires that temperature T and volume V are iterated in a nested way. That is, based on an initial guess of V, the temperature is determined in a inner loop until the determinant of the matrix $\mathbf{B}$ becomes equal to zero; then the convergence criterion for the coefficient c is checked. If this coefficient, evaluated at the stability limit, is equal to zero, the calculation ends; otherwise, a new estimate for the volume is generated in an outer loop and the iteration on T is evaluated again. Once T and V have been obtained, the pressure P is evaluated from the equation of state.

After having calculated the critical point, a noncritical equilibrium phase is searched for at constant temperature and pressure conditions. Mushrif and Phoenix (2006) used the stability test implemented by Michelsen (1982a) using the critical composition $\mathbf{z}$ as the reference phase and $\mathbf{Y} = (Y_1, \dots, Y_N)^T$ as the unnormalized trial phase composition with corresponding mole fractions as $y_i = Y_i \big/ \sum_{j=1}^{N} Y_j$.

The tangent plane to the Gibbs energy surface at the trial composition is parallel to the reference-phase tangent plane (critical phase) when

$$\ln f_i - \ln f_i^{\text{crit}} - \theta = \ln(y_i \varphi_i P) - \ln f_i^{\text{crit}} - \theta = 0 \tag{17}$$

where $\theta = -\ln(\sum_{i=1}^{N} Y_i)$ is the dimensionless distance between the two tangent planes and φ_i is the fugacity coefficient evaluated at composition $\mathbf{y}$.

If $\theta = 0$, the trial phase is in equilibrium with the reference phase; if $\theta > 0$, the trial phase is an incipient phase, and if $\theta < 0$, the reference phase is unstable. By combining eq 17 with the definition of θ, the set of N equations to solve for a stationary point can be written as

$$g_i = Y_i - \exp\left[\ln f_i^{\text{crit}} - \ln \varphi_i(\mathbf{y}) - \ln P\right] = 0 \qquad i = 1, \dots, N \tag{18}$$

which can efficiently be solved by Newton iteration or through a minimization method.

A K- or L-point is found if $\theta = 0$. When θ does not meet the convergence criterion (e.g., $\theta^2 < 10^{-12}$), the critical-phase composition of the lightest component (component 1) is updated using the Newton-Raphson method as

$$z_1^{(k+1)} = z_1^{(k)} - \frac{\theta^{(k)}}{\left(d\theta/dz_1\right)^{(k)}} \tag{19}$$

where the derivative $d\theta/dz_1$ is approximated by perturbing the critical composition, recalculating θ from eq 17 and calculating the finite difference analogue of the derivative. Mushrif and Phoenix (2006) have pointed out that failure of the algorithm can occur when a critical composition is updated to a value where no stationary point exist other than the trivial solution.

To calculate a K- or L-point using this algorithm, it is necessary to provide appropriate initial estimates of composition, temperature, and volume. In this case, good initial guesses for critical temperatures and volumes were, depending on the type of calculation, the same as those used by Heidemann and Khalil (1980). The success of the algorithm to locate a K- or L-point strongly depend on (1) the binary interaction parameters used in the equation of state and (2) the initial critical composition $\mathbf{z}^{(0)}$. However, it would seem that the value of the initial critical composition significantly affects the successful convergence of the method to locate a K- or L-point.

5. Thermodynamic models

Modeling of the complex phase behavior of LNG systems requires a suitable thermodynamic model and a robust and efficient computational algorithm for performing phase stability and multiphase flash calculations interwoven in the second element of the TMF. Regarding the thermodynamic models, the SRK EoS and the PC-SAFT EoS have received wide acceptance in the industry because of their ability to predict accurately the phase behavior of oil-gas systems.

5.1 The SRK equation of state

The explicit form of the SRK equation of state (Soave, 1972) can be written as

$$P = \frac{RT}{v-b} - \frac{a(T)}{v(v+b)} \tag{20}$$

where constants a and b for pure-components are related to

$$a = 0.42747\frac{R^2T_c^2}{P_c}\alpha(T_r) \quad ; \quad b = 0.08664\frac{RT_c}{P_c} \tag{21}$$

where $\alpha(T_r)$ is expressed in terms of the acentric factor ω as

$$\alpha(T_r) = \left[1 + \left(0.480 + 1.574\omega - 0.176\omega^2\right)\left(1 - T_r^{1/2}\right)\right]^2 \tag{22}$$

For mixtures, constants a and b are given by

$$a = \sum_{i=1}^{N}\sum_{j=1}^{N} x_i x_j a_{ij} \quad ; \quad b = \sum_{i=1}^{N} x_i b_i \tag{23}$$

and a_{ij} is defined as

$$a_{ij} = \left(1 - k_{ij}\right)\sqrt{a_i a_j} \quad ; \quad k_{ij} = k_{ji}\, ,\, k_{ii} = 0 \tag{24}$$

where k_{ij} is an adjustable interaction parameter characterizing the binary formed by components i and j.

Eq 20 can be written in terms of compressibility factor, $Z = Pv/RT$, as

$$Z^3 - Z^2 + \left(A - B - B^2\right)Z - AB = 0 \tag{25}$$

where $A = aP/(RT)^2$ and $B = bP/(RT)$.

The expression for the fugacity coefficient, $\varphi_i = f_i/y_i P$, is given by

$$\ln\varphi_i = \frac{b_i}{b}(Z-1) - \ln(Z-B) - \frac{A}{B}\left(\frac{2\sum_{j=1}^{N} x_j a_{ij}}{a} - \frac{b_i}{b}\right)\ln\left(1 + \frac{B}{Z}\right) \tag{26}$$

5.2 The PC-SAFT equation of state

In the PC-SAFT EoS (Gross and Sadowski, 2001), the molecules are conceived to be chains composed of spherical segments, in which the pair potential for the segment of a chain is given by a modified square-well potential (Chen and Kreglewski, 1977). Non-associating molecules are characterized by three pure component parameters: the temperature-independent segment diameter σ, the depth of the potential ε, and the number of segments per chain m.

The PC-SAFT EoS written in terms of the Helmholtz energy for an N-component mixture of non-associating chains consists of a hard-chain reference contribution and a perturbation contribution to account for the attractive interactions. In terms of reduced quantities, this equation can be expressed as

$$\tilde{a}^{res} = \tilde{a}^{hc} + \tilde{a}^{disp} \tag{27}$$

The hard-chain reference contribution is given by

$$\tilde{a}^{hc} = \bar{m}\,\tilde{a}^{hs} - \sum_{i=1}^{N} x_i (m_i - 1)\ln g_{ii}^{hs}(\sigma_{ii}) \tag{28}$$

where $\bar{m}=\sum_{i=1}^{N} m_i$ is the mean segment number in the mixture

The Helmholtz energy of the hard-sphere fluid is given on a per-segment basis as

$$\tilde{a}^{hs}=\frac{1}{\zeta_0}\left[\frac{3\zeta_1\zeta_2}{(1-\zeta_3)}+\frac{\zeta_2{}^3}{\zeta_3(1-\zeta_3)^2}+\left(\frac{\zeta_2{}^3}{\zeta_3{}^2}-\zeta_0\right)\ln(1-\zeta_3)\right] \quad (29)$$

and the radial distribution function of the hard-sphere fluid is

$$g_{ij}^{hs}=\frac{1}{(1-\zeta_3)}+\left(\frac{d_i d_j}{d_i+d_j}\right)\frac{3\zeta_2}{(1-\zeta_3)^2}+\left(\frac{d_i d_j}{d_i+d_j}\right)^2\frac{2\zeta_2{}^2}{(1-\zeta_3)^3} \quad (30)$$

with ζ_n defined as

$$\zeta_n=\frac{\pi}{6}\rho\sum_{i=1}^{N}x_i m_i d_i^n \qquad n=0,1,2,3 \quad (31)$$

The temperature-dependent segment diameter d_i of component i is given by

$$d_i=\sigma_i\left[1-0.12\exp\left(-3\frac{\varepsilon_i}{kT}\right)\right] \quad (32)$$

where k is the Boltzmann constant and T is the absolute temperature.

The dispersion contribution to the Helmholtz energy is given by

$$\tilde{a}^{disp}=-2\pi\rho I_1(\eta,\bar{m})\overline{m^2\varepsilon\sigma^3}-\pi\rho\bar{m}\left(1+Z^{hc}+\rho\frac{\partial Z^{hc}}{\partial\rho}\right)^{-1}I_2(\eta,\bar{m})\overline{m^2\varepsilon^2\sigma^3} \quad (33)$$

where Z^{hc} is the compressibility factor of the hard-chain reference contribution, and

$$\overline{m^2\varepsilon\sigma^3}=\sum_{i=1}^{N}\sum_{j=1}^{N}x_i x_j m_i m_j\left(\frac{\varepsilon_{ij}}{k_B T}\right)\sigma_{ij}{}^3 \quad (34)$$

$$\overline{m^2\varepsilon^2\sigma^3}=\sum_{i=1}^{N}\sum_{j=1}^{N}x_i x_j m_i m_j\left(\frac{\varepsilon_{ij}}{k_B T}\right)^2\sigma_{ij}{}^3 \quad (35)$$

The parameters for a pair of unlike segments are obtained by using conventional combining rules

$$\sigma_{ij}=\frac{1}{2}\left(\sigma_i+\sigma_j\right) \quad ; \quad \varepsilon_{ij}=\sqrt{\varepsilon_i\varepsilon_j}\left(1-k_{ij}\right) \quad (36)$$

where k_{ij} is a binary interaction parameter, which is introduced to correct the segment-segment interactions of unlike chains.

The terms $I_1(\eta,\bar{m})$ and $I_2(\eta,\bar{m})$ in eq 33 are calculated by simple power series in density

$$I_1(\eta,\bar{m}) = \sum_{i=0}^{6} a_i(\bar{m})\eta^i \quad ; \quad I_2(\eta,\bar{m}) = \sum_{i=0}^{6} b_i(\bar{m})\eta^i \tag{37}$$

where the coefficients a_i and b_i depend on the chain length as given in Gross and Sadowski (2001).

The density to a given system pressure P^{sys} is determined iteratively by adjusting the reduced density η until $P^{calc} = P^{sys}$. For a converged value of η, the number density of molecules ρ, given in Å^{-3}, is calculated from

$$\rho = \frac{6}{\pi}\eta\left(\sum_{i=1}^{N} x_i m_i d_i^{\,3}\right)^{-1} \tag{38}$$

Using Avogadro's number and appropriate conversion factors, ρ produces the molar density in different units such as $\mathrm{kmol \cdot m^{-3}}$.

The pressure can be calculated in units of $\mathrm{Pa = N \cdot m^{-2}}$ by applying the relation

$$P = ZkT\rho\left(10^{10}\frac{\text{Å}}{m}\right)^{3} \tag{39}$$

from which the compressibility factor Z, can be derived. The expression for the fugacity coefficient is given by

$$\ln\phi_i = \tilde{a}^{res} + \left(\frac{\partial\tilde{a}^{res}}{\partial x_i}\right)_{T,v,x_{j\neq i}} - \sum_{k=1}^{N}\left[x_k\left(\frac{\partial\tilde{a}^{res}}{\partial x_k}\right)_{T,v,x_{j\neq k}}\right] + (Z-1) - \ln Z \tag{40}$$

In eq 40, the partial derivatives with respect to mole fractions are calculated regardless of the summation relation $\sum_{i=1}^{N} x_i = 1$.

6. Results and discussion

Experimental data reported by Llave et al. (1987) for the system nitrogen + methane + ethane, by Hottovy et al. (1981) for the system methane + ethane + *n*-octane, and by Fall and Luks (1988) for the system carbon dioxide + nitrogen + *n*-nonadecane, were used to test and compare the robustness, efficiency and reliability of the two computational techniques and the SRK and PC-SAFT EoS thermodynamic models embedded in the respective elements of the TMF. The prediction and modeling of the phase behavior of these systems demonstrates in a clear-cut way the usual numerical difficulties encountered in the process.

The binary interaction parameters used with the PC-SAFT equation were taken from García-Sánchez et al. (2004) and from Justo-García et al. (2008b), while those used with the SRK EoS were taken from Knapp et al. (1982). Some of the interaction parameters were also obtained from the minimization of the sum of squares of the differences between experimental and calculated bubble-point pressures.

The binary interaction parameters employed are: $k_{C_1-C_2} = -0.0078$, $k_{C_1-C_3} = 0.009$, $k_{C_2-C_3} = 0.0170$, $k_{N_2-C_1} = 0.0278$, $k_{N_2-C_2} = 0.0407$, $k_{N_2-C_{19}} = 0.2714$, $k_{CO_2-N_2} = -0.0205$, and $k_{CO_2-C_{19}} = 0.1152$ for the SRK equation, and $k_{C_1-C_2} = -0.0207$, $k_{C_1-C_3} = 0.0168$, $k_{C_2-C_3} = 0.0195$, $k_{N_2-C_1} = 0.0307$, $k_{N_2-C_2} = 0.0458$, $k_{N_2-C_{19}} = 0.1608$, $k_{CO_2-N_2} = 0.0080$, and $k_{CO_2-C_{19}} = 0.1551$ for the PC-SAFT equation, respectively. The components' physical properties required for the calculations performed with the SRK EoS were taken from DIPPR (Rowley et al., 2006) while the three pure-component parameters (i.e., temperature independent segment diameter σ, depth of the potential ε, and number of segments per chain m) of these compounds for the PC-SAFT equation of state were taken from Gross and Sadowski (2001).

6.1 The nitrogen + methane + ethane system

The three-phase VLL region displayed by this ternary system is bounded from above by a K-point locus, from below by a lower critical solution temperature LCST locus, at low temperatures by a Q-point locus, and, due to the fact that this system contains a binary pair (nitrogen + ethane) which exhibits LLV behavior, its LLV space is truncated. In this case, the partially miscible pair nitrogen + ethane spans the LLV space from a position of the LCST locus to a position on the Q-point space. Because methane is of intermediate volatility compared with nitrogen and ethane, it creates a three-phase LLV space which extends from the binary LLV locus upward in temperature. The topographical nature of the regions of immiscibility for the system nitrogen + methane + ethane is shown in Fig. 1. In this figure it can be seen that the L-L=V and L=L-V critical end-point loci intersect at a tricritical point (L=L=V).

Fig. 2 presents the experimental and calculated L_1-L_2-V phase behavior (in terms of L_1-L_2 nitrogen mole fraction data) for the nitrogen + methane + ethane system at 135 K and different pressures. This figure shows a reasonable agreement between the experimental values of liquid phases L_1 and L_2 and those predicted with both models. Notwithstanding, although the LLV calculations performed with both equations up to a position near the K-point (about 41.25 bar with both models), this point is away from the experimental one (43.05 bar at 135 K).

An attempt to directly calculate either the K- or L-point for this ternary system using the algorithm of Mushrif and Phoenix (2006) was carried out. However, the algorithm was not able to give correct values of these critical end points. This is because the algorithm is strongly initialization dependent and hence gives different values of these points, depending on the initial guess of the critical composition, which meets the convergence criterion. Our preliminary results show that the algorithm advocated by Gregorowicz and de Loos (1996) is more stable than that of Mushrif and Phoenix, even if it also depends on the initial values of the critical composition.

Fig. 2 also shows that at pressures away from the LCST point, the PC-SAFT model gives a better representation of the experimental compositions for liquid phase L_1 while both

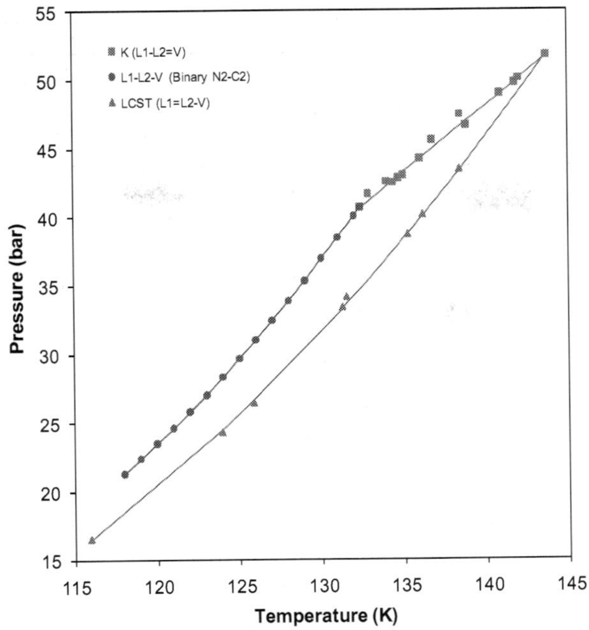

Fig. 1. P-T space of the boundaries of the three-phase L_1-L_2-V regions for the system nitrogen + methane + ethane. Experimental data from Llave et al. (1985, 1987).

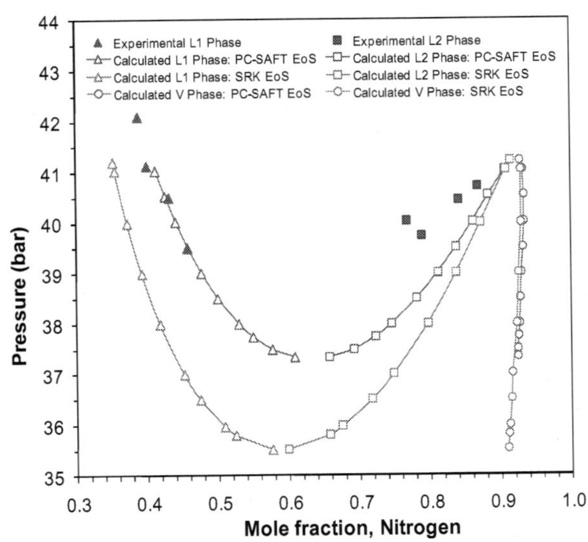

Fig. 2. Comparison of L_1 and L_2 compositional data of nitrogen at 135 K for the system nitrogen + methane + ethane. Experimental data from Llave et al. (1987).

models agree with each other for the liquid L_2 phase but represent the experimental data very closely. Since there are not experimental data below 22.16 bar, comparisons of the models with experiment in the region of the LCST are not possible. However, according to the predictions with both models, the estimated LCST point with the PC-SAFT model (37.20 bar) seems to be closer to the "hypothetical" experimental LCST point (38.37 bar) in comparison with the LCST point estimated from the SRK model (35.45 bar). Of course, these discrepancies can be due to the fact of using binary interaction parameters determined from VL equilibrium data, which, apparently, led to less accurate results.

6.2 The methane + ethane + *n*-octane system

The three-phase LLV region displayed by the methane + ethane + *n*-octane ternary system (a surface in the thermodynamic phase space with two degree of freedom) is bounded from above by a K-point locus (L-L=V), from below by a lower critical solution temperature LCST locus (L=L-V), and at low temperatures by a Q-point locus (S-L-L-V). For the three components in this system, there is no binary immiscibility. The topographical nature of the regions of immiscibility for this system is shown in Fig. 3, where symbols are the experimental data given by Hottovy et al. (1981) identifying the boundaries of the three phase LLV region for this ternary system. This Figure shows also that the L-L=V and L=L-V critical end-point loci intersect at a tricritical point (L=L=V) at the upper temperature limit.

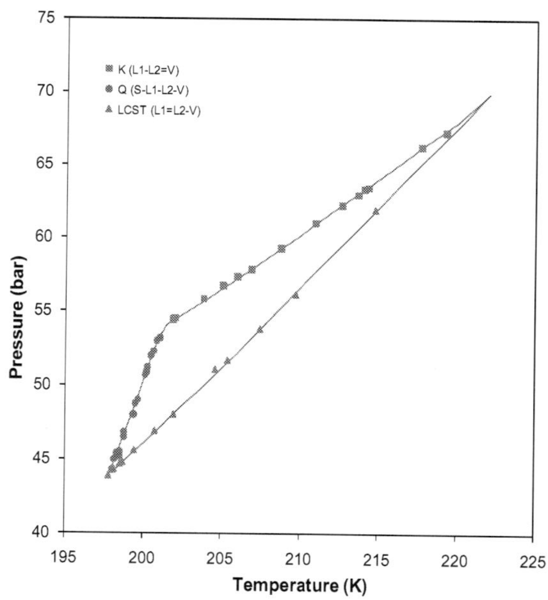

Fig. 3. P-T space of the boundaries of the three-phase L_1-L_2-V regions for the system methane + ethane + *n*-octane. Experimental data from Hottovy et al. (1981).

Following the immiscibility region, a single temperature was chosen to test the capabilities of the PC-SAFT with computational technique 1, and the SRK EoS with computational technique 2, to predict the phase behavior for the system methane + ethane + *n*-octane. Fig. 4 compares the performance of the two methods at 210 K, and at different pressures on the basis of the experimentally measured and calculated nitrogen mole fractions for the liquid L_1 and L_2 phases by the two thermodynamic models employed. In this case, the predictions of the PC-SAFT model are closer to the experimental composition values than those of the SRK model. However, it should be mentioned that it was not possible to continue the calculations with this model to approach either the K- or L-point because the three-phase LLV triangles become so very narrow as pressure decreases or increases that it is extremely difficult to determine an appropriate global composition able to separate this mixture into three-phase LLV equilibria. On the other hand, because the three-phase LLV triangles predicted with the SRK model are wider than those predicted with the PC-SAFT model, it was easier to get a good initial global composition to calculate the three-phase LLV equilibria from the LCST point (52.80 bar) to the K-point (59.09 bar) applying technique 2.

An inspection of this figure shows that the predictions of both EoSs don't follow the behavior of the liquid phase L_1 as well as the variation of liquid phase L_2 as pressure decreases. Also, it is interesting to note that although there is not a true experimental value of the LCST at the temperature considered, Fig. 4 indicates that the estimated LCST point with the PC-SAFT model (56.35 bar) is closer to the "experimental" one (56.64 bar) than that obtained with the SRK model (52.80 bar). Nonetheless, the "experimental" K-point (60.19 bar) is closer to the one calculated with the SRK model.

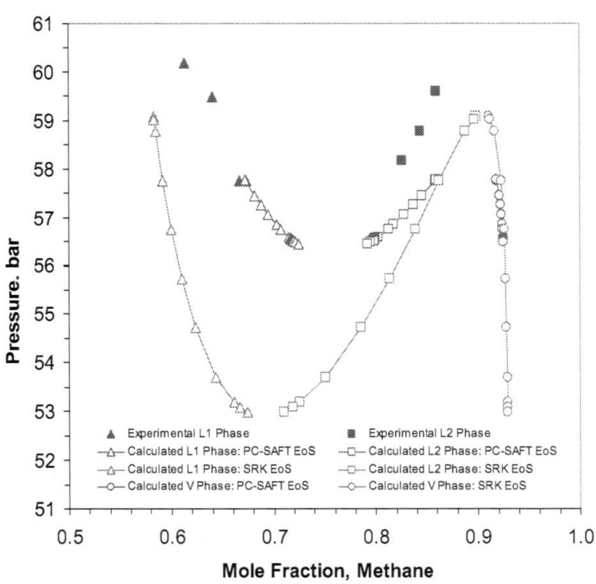

Fig. 4. Comparison of L_1 and L_2 compositional data of nitrogen at 210 K for the system methane + ethane + *n*-octane. Experimental data from Hottovy et al. (1981).

Fig. 4 also shows that at the different pressures the predicted L_1-L_2-V region (in terms of L_1 and L_2 methane mole fraction data) with the SRK model deviates considerably from the experimental one, while the PC-SAFT model predictions are more reasonable. However, since the interaction parameters for the SRK and PC-SAFT models were determined from binary vapor-liquid equilibrium data, the rather poor fit in this region with either model is not unexpected.

6.3 The carbon dioxide + nitrogen + *n*-nonadecane system

As mentioned in Section 2, the type of the LLV region displayed by a ternary system depends on whether it contains an immiscible pair or not. In this context, the system carbon dioxide + nitrogen + *n*-nonadecane exhibits immiscibility in the carbon dioxide + *n*-nonadecane binary pair (Fall et al., 1985), so that its three-phase region is similar to that exhibited by the system nitrogen + methane + *n*-pentane (Merrill et al., 1984b). Therefore, the LLV region is "triangular" and is bounded from above by a K-point locus (L-L=V), at low temperatures by a Q-point locus (S-L-L-V), and, from a position of the Q-point locus to a position on the K-point space, by a binary carbon dioxide + *n*-nonadecane LLV locus.

Fig. 5 presents the experimental pressure-temperature diagram of the LLV space displayed by the system (Fall and Luks, 1986). An examination of the figure shows that this system does not have a LCST (L=L-V) locus and that the Q-point locus terminates at an invariant

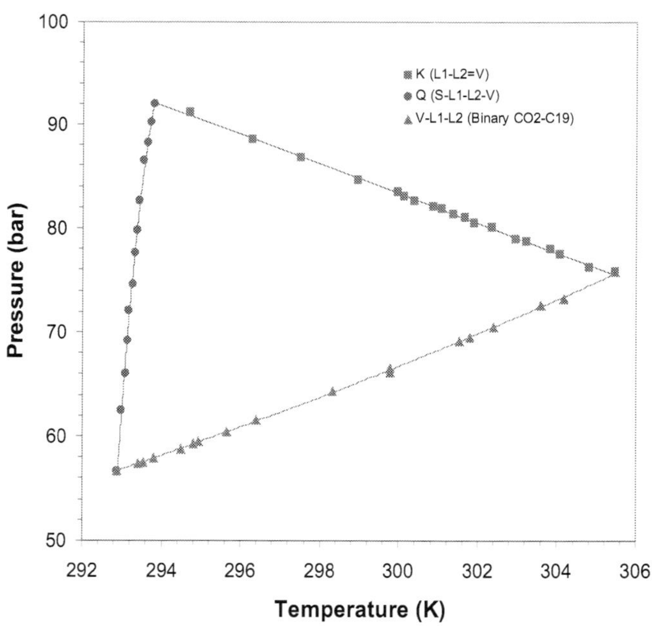

Fig. 5. P-T space of the boundaries of the three-phase L_1-L_2-V regions for the system carbon dioxide + nitrogen + *n*-nonadecane. Experimental data from Fall et al. (1985) and Fall and Luks (1986).

point of the S-L-L=V type. Thus, due to the fact that the carbon dioxide + nitrogen + *n*-nonadecane system contains a binary pair (carbon dioxide + *n*-nonadecane) exhibiting LLV behavior, its "triangular" LLV region is a three-sided space without a tricritical point.

A temperature of 297 K was chosen to study this ternary system and the results obtained are presented in Fig. 6. This figure shows that the PC-SAFT model predicts well the experimental carbon dioxide compositions of the liquid L_2 and vapor phases down to the lowest measured pressure of 62.63 bar (i.e., the binary carbon dioxide + *n*-nonadecane data) for this isotherm. However, the SRK model is superior to the PC-SAFT model in predicting the three phases in equilibrium for this ternary system. In this case, the calculated L_1, L_2, and V phases are close to the experimental ones. Furthermore, though the SRK EoS overpredicts the "experimental" K-point (87.30 bar) by 4.39 bar the performance of the PC-SAFT EoS in this particular case is inferior as it overpredicts by 14.6 bar the "experimental" point.

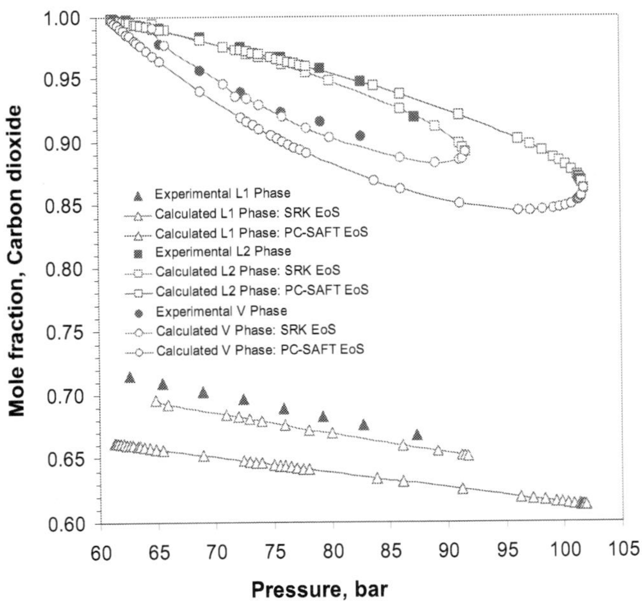

Fig. 6. Comparison of L_1 and L_2 compositional data of carbon dioxide at 297 K for the system carbon dioxide + nitrogen + *n*-nonadecane. Experimental data from Fall and Luks (1986).

Nevertheless, it should be recalled that all calculations were performed using binary interaction parameters obtained from regression of binary experimental VL data, many of them measured at temperatures higher than that studied here. Furthermore, we are confident that the performance of the corresponding thermodynamic models could be improved considerably provided the interaction parameters were obtained from the regression of the experimental data at three phases. Still, if those sets of interaction parameters are used to predict the phase behavior of a given system at conditions different from the original ones then there is the risk that the equilibria predictions and calculations could either give physical meaningless results or fail altogether.

We also tried to directly calculate the K-points at the temperature considered for this ternary system by using the algorithm of Mushrif and Phoenix (2006); however, once again, the algorithm predicted different values of the critical end points, depending on the initial critical composition. In this case, the strategy to find a critical temperature close to the temperature of study was to use a series of initial compositions. Unfortunately, none of the compositions produced a K-point similar to those obtained from the VLL calculations with both models.

Finally, it should be pointed out that for the sake of comparison both equations of state were interwoven into computational procedure 1 and 2, respectively, and that a series of multiphase flash calculations were carried out at different temperatures and pressures for the three ternary systems studied obtaining the same results for the specific EoS, irrespectively of the computational procedure utilized.

The results obtained showed that there are not any essential differences between, or particular advantages of any of two computational procedures, either in their efficiency, effectiveness, robustness or in their convergence behavior. Thus, it can be said that both procedures 1 and 2 can be used to predict the phase behavior of a wide variety of multicomponent nonideal systems over wide ranges of temperature and pressure.

7. Conclusions

Though there has been much progress and advance in two-phase stability and two-phase split calculations with EoSs, there is still not much progress in three-phase split calculations in natural gas systems despite the large number of publications devoted to the subject. The reason behind that is that the difficulties and challenges are dominating over the easy to perform calculations, if any. Thus, it can be accepted that the algorithms and numerical methods advocated are not robust enough for incorporation in a process simulator. In view of this, the further development of a reliable, robust and efficient TMF and its subsequent approbation on model systems, typical representatives of LNG, is of considerable interest both to scientists and engineers.

On the example of three ternary systems (nitrogen-methane-ethane, methane-ethane-*n*-octane, and carbon dioxide-nitrogen-*n*-nonadecane) the capabilities of a TMF, advocated by us, to predict and model complex phase behavior of systems of importance to LNG processing are demonstrated. The TMF employs two numerical techniques which embed different thermodynamic models and stability analysis routines.

The results obtained show that there are many and different challenges and difficulties that are not always possible to overcome completely. For example, the two techniques for multiphase flash calculations cannot always assure steady and non-oscillatory convergence with no tendency towards a strong attraction to the trivial solution, particularly in cases close to the critical lines. Besides, it is known that the phase equilibrium equations are often difficult to converge in the critical region and that the use of inappropriate initial estimates can lead to the trivial solution. In view of this, the availability in a TMF of a robust stability analysis routine that will provide good set of initial estimates for the compositions of possible equilibrium phases and will guarantee steady convergence of the flash routines is of great importance.

Regarding the prediction of K- and L-points with the Gregorowicz and de Loos and Mushrif and Phoenix's methods, it is clear that the both algorithms are strongly initialization dependent. To overcome this problem, for example, Mushrif and Phoenix suggest that the critical phase composition is updated based on the values of parameter θ, calculated from equilibrium phase calculations using the Newton-Raphson iteration. However, during this process of upgrading, the composition may change to a value where there is no phase in equilibrium, particularly when θ differs significantly from iteration to iteration.

The procedure of Gregorowicz and de Loos to calculate K- and L-points seems to be a better fitted method to carry out this task. However, the evaluation of the determinants for solving the conditions of criticality requires the second derivatives of the Helmholtz energy with respect to volume and composition, which makes difficult their evaluation, particularly when these derivatives have to be found analytically applying the PC-SAFT EoS. Still, of course, a possible solution to this problem is to evaluate these derivatives numerically.

Finally, both the PC-SAFT EoS and the RKS cubic EoS are capable of representing with a reasonable accuracy the experimentally observed phase behavior of the ternary systems studied.

8. Acknowledgments

This work was partially supported by the Mexican Petroleum Institute under Research Project D.00406. One of the authors (D. N. J.-G.) gratefully acknowledges the National Polytechnic Institute of Mexico for their financial support through the Project SIP-20110150.

9. References

Baker, L. E. & Luks, K. D. (1980). Critical Point and Saturation Pressure Calculations for Multicomponent Systems. *Society of Petroleum Engineers Journal,* Vol. 20, No. 1, pp. 15–24.

Baker, L. E.; Pierce, A. C. & Luks, K. D. (1982). Gibbs Energy Analysis of Phase Equilibria. *Society of Petroleum Engineers Journal,* Vol. 22, No. 5, pp. 731–742,

Chen, S. & Kreglewski, A. (1977). Applications of the Augmented Van der Waals Theory of Fluids. I. Pure Fluids. *Berichte der Bunsengesellschaft für physikalische Chemie,* Vol. 81, No. 10, pp. 1048–1052.

Fall, J. L. & Luks, K. D. (1986). Effect of Additive Gases on the Liquid-Liquid-Vapor. Immiscibility of the Carbon Dioxide + *n*-Nonadecane Mixture. *Journal of Chemical and Engineering Data,* Vol. 31, No. 3, pp. 332–336.

Fall, D. J.; Fall, J. L. & Luks, K. D. (1985). Liquid-Liquid-Vapor Immiscibility Limits in Carbon Dioxide + *n*-Paraffin Mixtures. *Journal of Chemical and Engineering Data,* Vol. 30, No. 1, pp. 82–88.

Fletcher, R. (1980). *Practical Methods for Optimization. Vol. 1. Unconstrained Optimization,* John Wiley & Sons Ltd, New York.

García-Sánchez, F.; Eliosa-Jiménez, G.; Silva-Oliver, G. & Vázquez-Román, R. (2004). Vapor-Liquid Equilibria of Nitrogen-Hydrocarbon Systems Using the PC-SAFT Equation of State. *Fluid Phase Equilibria,* Vol. 217, No. 2, pp. 241–253.

Gauter, K.; Heidemann, R. A. & Peters, C. J. (1999). Modeling of Fluid Multiphase Equilibria in Ternary Systems of Carbon Dioxide as a Near-Critical Solvent and Two Low Volatile Solutes. *Fluid Phase Equilibria,* Vol. 158-160, pp. 133–141.

Green, K. A.; Tiffin, D. L.; Luks, K. D. & Kohn, J. P. (1976). Solubility of Hydrocarbons in LNG, NGL. *Hydrocarbon Processing,* Vol. 56, No. 5, pp. 251–256.

Gregorowicz, J. & de Loos, Th. W. (1996). Modelling of the Three Phase LLV Region for Ternary Mixtures with the Soave-Redlich-Wong Equation of State. *Fluid Phase Equilibria,* Vol. 118, No. 1, pp. 121–132.

Gross, J. & Sadowski, G. (2001). Perturbed-Chain SAFT: An Equation of State Based on Perturbation Theory for Chain Molecules. *Industrial & Engineering Chemistry Research,* Vol. 40, No. 4, pp. 1244–1260.

Heidemann, R. A. & Khalil, A. M. (1980). The Calculation of Critical Points. *American Institute of Chemical Engineers Journal,* Vol. 26, No. 5, pp. 769–779.

Hottovy, J. D.; Kohn, J. P. & Luks, K. D. (1981). Partial Miscibility of the Methane-Ethane-*n*-Octane. *Journal of Chemical and Engineering Data,* Vol. 26, No. 2, pp. 135–137.

Hottovy, J. D.; Kohn, J. P. & Luks, K. D. (1982). Partial Miscibility Behavior of the Ternary Systems Methane-Propane-*n*-Octane, Methane-*n*-Butane-*n*-Octane, and Methane-Carbon Dioxide-*n*-Octane. *Journal of Chemical and Engineering Data,* Vol. 27, No. 3, pp. 298–302.

Justo-García, D. N.; García-Sánchez, F. & Romero-Martínez, A. (2008a). Isothermal Multiphase Flash Calculations with the PC-SAFT Equation of State. *American Institute of Physics Conference Proceedings,* Vol. 979, pp. 195–214.

Justo-García, D. N.; García-Sánchez, F., Díaz-Ramírez, N. L. & Romero-Martínez, A. (2008b). Calculation of Critical Points for Multicomponent Mixtures Containing Hydrocarbon and Nonhydrocarbon Components with the PC-SAFT Equation of State. *Fluid Phase Equilibria,* Vol. 265, Nos. 1-2, pp. 192–204.

Knapp, H. R.; Döring, R.; Oellrich, L.; Plöcker, U. & Prausnitz, J. M. (1982). *Vapor-Liquid Equilibria for Mixtures of Low Boiling Substances;* DECHEMA Chemistry Data Series, Vol. VI: Frankfurt, Germany.

Kohn, J. P. (1961). Heterogeneous Phase and Volumetric Behavior of the Methane-*n*-Heptane System at Low Temperatures. *American Institute of Chemical Engineers Journal,* Vol. 7, No. 3, pp. 514–518.

Kohn, J. P. & Bradish, W. F. (1964). Multiphase and Volumetric Equilibria of the Methane-*n*-Octane System at Temperatures between -10° and 150°C. *Journal of Chemical and Engineering Data,* Vol. 9, No. 1, pp. 5–8.

Kohn, J. P. & Luks, K. D. (1976). Solubility of Hydrocarbons in Cryogenic LNG and NGL Mixtures. *Research Report RR-22,* Gas Processors Association, Tulsa, Oklahoma.

Kohn, J. P. & Luks, K. D. (1977). Solubility of Hydrocarbons in Cryogenic LNG and NGL Mixtures. *Research Report RR-27,* Gas Processors Association, Tulsa, Oklahoma.

Kohn, J. P. & Luks, K. D. (1978). Solubility of Hydrocarbons in Cryogenic LNG and NGL Mixtures. *Research Report RR-33,* Gas Processors Association, Tulsa, Oklahoma.

Kohn, J. P.; Luks, K. D. & Liu, P. H. (1976). Three-Phase Solid-Liquid-Vapor Equilibria of Binary-*n*-Alkane Systems (Ethane-*n*-Octane, Ethane-*n*-Decane, Ethane-*n*-Dodecane). *Journal of Chemical and Engineering Data,* Vol. 21, No. 3, pp. 360–362.

Llave, F. M.; Luks, K. D. & Kohn, J. P. (1985). Three-Phase Liquid-Liquid-Vapor Equilibria in the Binary Systems Nitrogen+ Ethane and Nitrogen + Propane. *Journal of Chemical and Engineering Data,* Vol. 30, No. 4, pp. 435-438.

Llave, F. M.; Luks, K. D. & Kohn, J. P. (1987). Three-Phase Liquid-Liquid-Vapor Equilibria in the Nitrogen-Methane-Ethane and Nitrogen-Methane-Propane. *Journal of Chemical and Engineering Data,* Vol. 32, No. 1, pp. 14–17.

Luks, K. D. & Kohn, J. P. (1978). The Topography of Multiphase Equilibria Behavior: What Can it Tell the Design Engineer. *Proceedings of the 63rd Annual Convention,* Gas Processors Association, Tulsa, Oklahoma.

Merrill, R. C.; Luks, K. D. & Kohn, J. P. (1983). Three-Phase Liquid-Liquid-Vapor Equilibria in the Methane-*n*-Pentane-*n*-Octane, Methane-*n*-Hexane-*n*-Octane, and Methane-*n*-Hexane-Carbon Dioxide. *Journal of Chemical and Engineering Data,* Vol. 28, No. 2, pp. 210–215.

Merrill, R. C.; Luks, K. D. & Kohn, J. P. (1984a). Three-Phase Liquid-Liquid-Vapor Equilibria in the Methane-*n*-Butane-Nitrogen System. *Advances in Cryogenic Engineering,* Vol. 29, pp. 949–955.

Merrill, R. C.; Luks, K. D. & Kohn, J. P. (1984b). Three-Phase Liquid-Liquid-Vapor Equilibria in the Methane-*n*-Hexane-Nitrogen and Methane-*n*-Pentane-Nitrogen Systems. *Journal of Chemical and Engineering Data,* Vol. 29, No. 3, pp. 272–276.

Michelsen, M. L. (1982a). The Isothermal Flash Problem. Part I. Stability. *Fluid Phase Equilibria,* Vol. 9, No. 1, pp. 1–19.

Michelsen, M. L. (1982b). The Isothermal Flash Problem. Part II. Phase-Split Calculation. *Fluid Phase Equilibria,* Vol. 9, No. 1, pp. 21–40.

Michelsen M. L. (1984). Calculation of Critical Points and Phase Boundaries in the Critical Region. *Fluid Phase Equilibria,* Vol. 16, No. 1, pp. 57–76.

Michelsen, M. L. & Heidemann, R. A. (1988). Calculation of Tri-Critical Points. *Fluid Phase Equilibria,* Vol. 39, No. 1, pp. 53–74.

Mushrif, S. H. & Phoenix, A. V. (2006). An Algorithm to Calculate K- and L-Points. *Industrial & Engineering Chemistry Research,* Vol. 45, No. 26, pp. 9161–9170.

Nghiem, L. X. & Li, Y.-K. (1984). Computation of Multiphase Equilibrium Phenomena with an Equation of State. *Fluid Phase Equilibria,* Vol. 17, No. 1, pp. 77–95.

Orozco, C. E.; Tiffin, D. L.; Luks, K. D. & Kohn, J. P. (1977). Solids Fouling in LNG Systems. *Hydrocarbon Processing,* Vol. 56, No. 11, pp. 325–328.

Peng, D.-Y. & Robinson, D. B. (1976). A New Two-Constants Equation of State. *Industrial & Engineering Chemistry Fundamentals,* Vol. 15, No. 1, pp. 59–64.

Rowley, R. L.; Wilding, W. V.; Oscarson, J. L.; Yang, Y. & Zundel, N. A. (2006). *DIPPR Data Compilation of Pure Chemical Properties Design Institute for Physical Properties;* Brigham Young University, Provo, Utah. http://dippr.byu.edu

Shim, J. & Kohn, J. P. (1962). Multiphase and Volumetric Equilibria of the Methane-*n*-Hexane System at Temperatures between -10° and 150°C. *Journal of Chemical and Engineering Data,* Vol. 7, No. 1, pp. 2–8.

Soave, G. (1972). Equilibrium Constants from a Modified Redlich-Kwong Equation of State. *Chemical Engineering Science,* Vol. 27, No. 6, pp. 1197–1203.

Stateva, R. P. & Tsvetkov S. G. (1994). A Diverse Approach for the Solution of the Isothermal Multiphase Flash Problem. Application to Vapor-Liquid-Liquid Systems. *Canadian Journal of Chemical Engineering,* Vol. 72, No. 4, pp. 722–734.

Wakeham, W. A. & Stateva, R. P. (2004). Numerical Solution of the Isothermal Multiphase Flash Problem. *Review of Chemical Engineering Journal*, Vol. 20, Nos. 1-2, pp. 1–56.

Wilkinson, J. H. (1965). *The Algebraic Eigenvalue Problem*, Clarendon, Oxford.

Zhang, H.; Bonilla-Petriciolet, A. & Rangaiah, G. P. (2011). A Review on Global Optimization Methods for Phase Equilibrium Modeling and Calculations. *The Open Thermodynamics Journal*, Vol. 5, pp. 71-92.

Part 4

Natural Gas Utilization

15

Natural Gas Dual Reforming Catalyst and Process

Hamid Al-Megeren[1] and Tiancun Xiao[2,3,*]
[1]*Petrochemical Research Institute, King Abdulaziz City for Science and Technology, Riyadh,*
[2]*Inorganic Chemistry Laboratory, Oxford University,*
[3]*Guangzhou Boxenergy Technology Ltd, Guangzhou,*
[1]*Saudi Arabia*
[2]*UK*
[3]*PR China*

1. Introduction

Natural gas consists primarily of methane, typically with 0–20% higher hydrocarbons. It is always associated with other hydrocarbon fuel, in coal beds, as methane clathrates, and is an important fuel source and a major feedstock for fertilizers. Natural gas is commercially extracted from oil fields and natural gas fields. Gas extracted from oil wells is called associated gas. The natural gas industry is extracting gas from increasingly more challenging resource types: sour gas, tight gas, shale gas, and coalbed methane. In fact, methane can also be produced from landfill site or anaerobic digestion, which is called biogas, it is in fact a kind of young age natural gas.

Components	(mol %)
CO_2	71
CH_4 + C_2+ hydrocarbons	28
H_2S	0.5
N_2	0.5

Table 1. Typical composition of Natuna gas field (Richardson and Paripatyadar 1990; Suhartanto, York et al. 2001)

In naturel gas, non-hydrocarbon gases (CO_2, N_2, H_2S) can account between 1% to 99% of overall composition. High carbon dioxide (CO_2) concentrations are encountered in diverse areas including South China Sea, Gulf of Thailand, Central European Pannonian basin, Australian Cooper-Eromanga basin, Colombian Putumayo basin, Ibleo platform, Sicily, Taranaki basin, New Zealand and North Sea South Viking Graben. The composition of CO_2 can reach as high as 80% in certain natural gas wells such as wells at the LaBarge reservoir in western Wyoming and the Natuna production field in Indonesia. Besides, purged gas from a

* Corresponding Author

gas-reinjected EOR (Enhanced Oil Recovery) well can contain as much as 90% CO_2 (Ricchiuto and Schoell 1988; Schoell 1995; Ballentine, Schoell et al. 2001; Cathles and Schoell 2007).

So far in industry, there have been several ways to bringing gas to the market. These include gas pipeline, liquid natural gas (LNG), and compressed natural gas (CNG). The costs of different fuel transportation are shown in Fig. 2. Also in gas transportation, any impurities such as CO_2 and H_2S must be fully removed, so as to protect the gas pipeline and container. This again will emit lot s of CO_2, the main green house gas into atmosphere. In terms of cost, as shown in Fig. 2, clearly the smallest gas field has highest transportation cost in both onshore and off shore. In a relatively high gas field, the transportation cost increase not as fast as the small gas field, but still much faster in transportation in cost, while the oil transport cost increases much smaller compared to the natural gas. Normally the gas and oil field are co-existed, and far away from the market, so the transportation cost must be taken into consideration. This may be another reason more and more attentions are being paid to the conversion of the gas into liquid, e.g., gas to liquid, and then the resultant liquid is mixed with the crude oil for transportation.

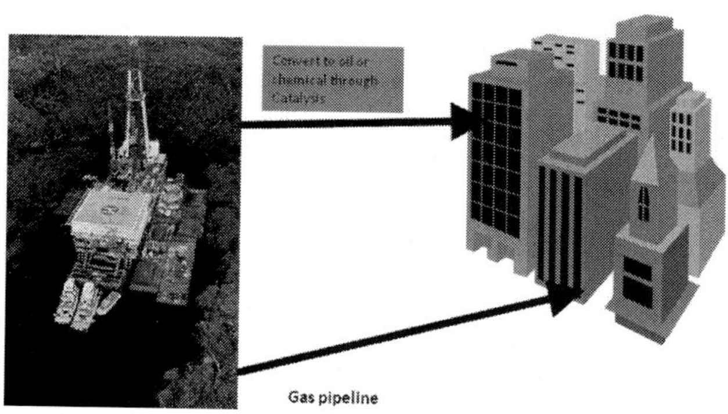

Fig. 1. The ways to bring natural gas into market

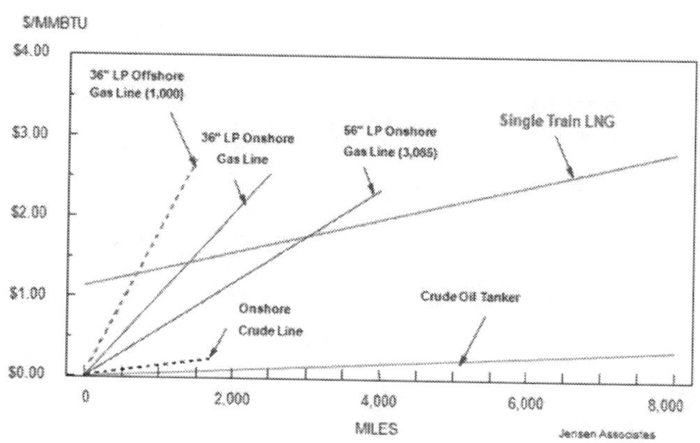

Fig. 2. The dependence of transportation cost on the distance for various fuels.

Indeed, Gas-to-liquids (GTL) has been a technology operated by industry for many years, while continuing improvement. It is a developing technology that converts stranded natural gas into synthetic gasoline, diesel or jet fuel through the Fischer-Tropsch process. Such fuel can be transported to users through conventional pipelines and tankers. Proponents claim GTL burns cleaner than comparable petroleum fuels. Most major international oil companies are in an advanced stage of GTL production, with a world-scale (140,000 barrels (22,000 m^3) a day) GTL plant in Qatar already in production in 2009.

The gas to liquids (GTL) process comprises three main process steps, firstly the reforming of natural gas to syngas, a mixture of gases containing hydrogen, carbon monoxide, carbon dioxide and unreacted methane, secondly the Fischer Tropsch (FT) conversion of carbon monoxide and hydrogen to long chain hydrocarbons and thirdly the upgrading and refining of these hydrocarbons into a specific state of liquid fuels.. The syngas step converts the natural gas to hydrogen and carbon monoxide by partial oxidation, steam reforming or a combination of the two processes. The key variable is the hydrogen to carbon monoxide ratio with a 2:1 ratio recommended for F-T synthesis. Steam reforming is carried out in a fired heater with catalyst-filled tubes that produces a syngas with at least a 5:1 hydrogen to carbon monoxide ratio, as shown in Fig 3. To adjust the ratio, hydrogen can be removed by a membrane or pressure swing adsorption system. Helping economics is the surplus hydrogen is used in a petroleum refinery or for the manufacture of ammonia in an adjoining plant. Because of the steam reforming occurring under very harsh conditions and also giving much higher H_2/CO ratio, it accounts up to 60% of the GTL cost (Shamsi and Johnson 2001). Another way to give ideal H_2/CO mixture syngas is through partial oxidation, which has been well developed in lab scale and going to industrial pilot operation.

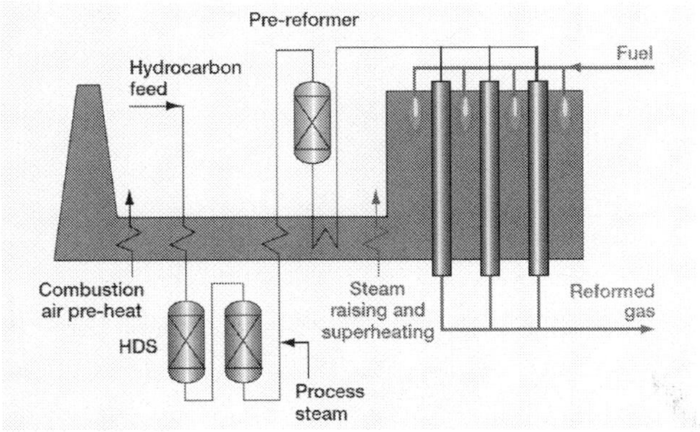

Fig. 3. Diagram of industrial steam reforming process

Another way to give approximate H_2/CO ratio of 2 is to combine dry and steam reforming, e.g., through the following reactions

$$CH_4+H_2O===CO+3H_2 \tag{1}$$

$$CH_4+CO_2===2CO+2H_2 \tag{2}$$

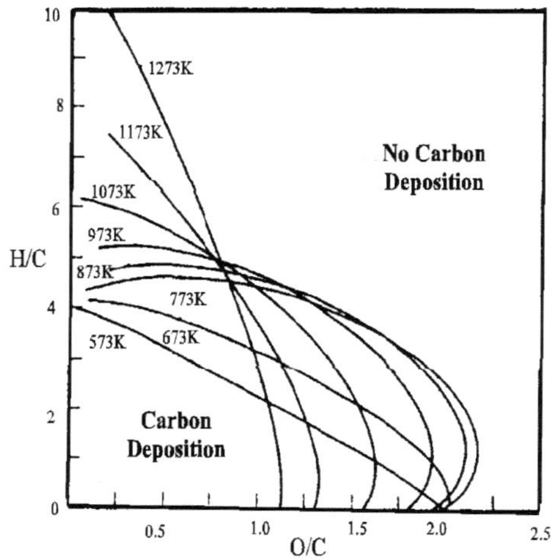

Fig. 4. The dependence of carbon deposition on H/C and O/C in the reforming system (Shimura, Yagi et al. 2002; Zhang, Li et al. 2004)

This combining process here is named as dual reforming process, which can theoretically give H_2/CO in the reaction products to 2:1 ratio, which is suitable for the follow up Fischer Tropsch synthesis. Also this can make use of the CO_2 in the natural gas, especially in the gas field where high CO_2 may be contained. It is well known that transporting the gas through LNG or CNG or even pipeline transportation requires eliminating the CO_2 to avoid the corrosion. (Butts 2006; Do 2007; Bhattacharya, Newell et al. 2009) Converting the CO_2 containing gas into liquid through the dual reforming process not only converts the natural gas into liquid, which can be transported as liquid with low cost, but also can make use of the CO_2 in the gas, thus to reduce green house gas emission. The dual reforming process in principle does not require any modification of the steam reforming reactor, but may have much harsh requirements on the catalyst, because the H_2/C in the products would be much less than the steam reforming alone (Wang, Li et al. 1996; Ding, Yan et al. 2001; Shamsi and Johnson 2003), also the O/C in the products is not very high, carbon thus very easily occurs over the system from the thermodynamic view.

Recently steam/CO_2 methane reforming in the syngas reformer is one of particularities of our new catalysts development. We intend to stick to the currently in-use reforming reactor, but feeding both steam and CO_2, which already exists in the natural gas feedstock ranging from 10% to 60%. Through properly adjusting the CO_2 and H_2O ratio, we hope to utilize CO_2 contained in the natural gas and does not require any CO_2 separation. The Ni based reforming catalyst has been prepared using a special dispersion agent, and an oxygen storage material is introduced into the lattice of spinel support, so as to depress the carbon formation. The catalyst has been successfully applied in steam reforming of natural gas for ammonia industry, and the life time has been as long as 6 years. Here we report the application of the catalyst in dual reforming for syngas production. The optimal operating conditions for dual reforming have been explored.

2. Catalyst design

The preparation of steam reforming catalysts based on nickel as the active component has been extensively and intensively studied for years. Over a given nickel contained catalyst with similar nickel metal dispersion (which can be affected by the preparation, activation and operation conditions), the industrial catalyst activity would depends more in the catalyst shape. In fact, the major geometric factor affecting the activity of catalyst particles is the ratio of the particle's geometric surface area to volume, SA/V. When applying the catalyst to industrial reactor, the optimization of catalyst particles must take the three key factors, e.g., low pressure drop, high radical and high strength into consideration. These factors have direct relations with the shapes and size of the catalyst. For example, the high voidage e.g., large particles gives low pressure drop, however, smaller particles can give higher surface area, leading to more active site exposed to the reactants. Also due to the strong endothermic reaction properties, the reactor requires the catalyst to have good radial mixing so that the heat can be transferred from the reactor wall to the centre for the reaction. Also the strength of the catalyst particles are more important, due to the high pressure and high temperature conditions, the catalyst strength must be strong enough to resist the variations.

In the early stage of reforming catalyst development, ring or rasching shapes have been developed. Normally the smaller the particles the higher the catalytic activity and the better heat transfer properties but the higher the pressure drop as the gases pass through the reformer tube. Since 1990s, more and more attentions are been paid to catalyst geometric shapes to increase the geometrical surface area and hence the catalytic activity. The geometric surface area of a catalyst should be 200m^2/m^3 or above. In recently years, strong shaped reforming catalysts with higher surface areas and voidage has been developed, as shown in Fig 5. Given the same dispersion of the active metal in the overall catalyst, which can give the same intrinsic activity when the mass transfer factors are eliminated, in industrial application, the physical properties of the catalyst in fact exert more influence on the catalyst performance. From Table 2, it is expected that the catalyst with 10 holes would have higher activity. This has also been confirmed by our industrial operation results.

Fig. 5. The novel shapes with high strength of the reforming catalyst and support.

Catalyst Geometry	Surface Area(m^2)	Void Fraction	SA/V(m^{-1})
1-hole	0.00503	0.66	$1.151*10^3$
1-hole -6-grooves	0.0052	0.72	$1.733*10^3$
4-hole	0.00523	0.62	$1.703*10^3$
10-hole	0.00642	0.7	$2.013*10^3$

Table 2. The physical properties of various shaped reforming catalysts.

3. Dual reforming catalyst performance test

3.1 Experiments

The reactions were carried out in a 70cm stainless steel reactor tube with id and od of 1.2cm and 2cm respectively. The furnace, comprising 6 individually heated zones, was specially designed to allow a temperature profile to be set across the reactor mimicking that of an industrial reformer. Four thermocouples were positioned inside a thermowell (inside the catalyst bed) to record the temperature of the catalyst bed at the inlet, exit and two intermediate positions. A diagram of the reactor and test setup is shown in Figure 6.

The rig has a supply of CH_4, CO_2, H_2, N_2 controlled by mass-flow controllers and an HPLC pump delivers H_2O, which is fed and mixed with the gases to flow together through a vaporiser where steam is generated before flowing into the reactor. Temperature programmes can be set on the vaporiser and all reactor furnace zones. Before any reactions were started, it was necessary to set a temperature profile across the reformer tube. The profile was designed to be similar to that of an industrial reformer, ensuring a cooler temperature at the tube inlet and gradually increasing the temperature towards the exit.

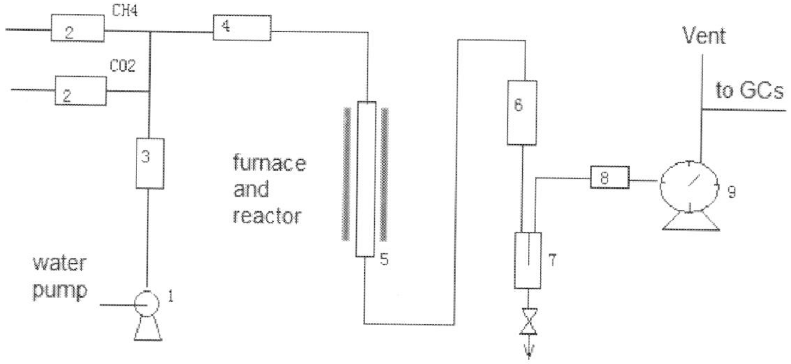

Fig. 6. Pressure Test system setup, 1–water pump, 2–gas MFC, 3–Purifier, 4–Mxier, 5–Tubular reactor, 6–condenser, 7–separator, 8–Back-Pressure regulator, 9–Wet flowing meter

In reforming reaction, carbon will most likely form via the methane cracking reaction which is promoted by high temperature:

$$CH_4 \Leftrightarrow C + 2H_2 \quad \Delta H = 75\ KJmol^{-1}$$

Figure 7 shows equilibrium data for CH_4 cracking with a CH_4 only feed at 3 bar.

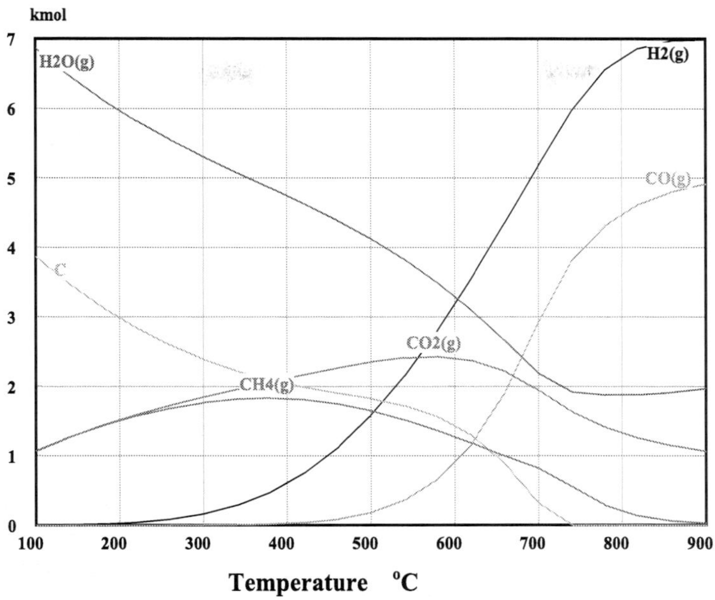

Fig. 7. CH_4 cracking reaction at equilibrium. 3bar and CH_4 only feed at various temperatures

At the catalyst bed inlet there is very little H_2 from reforming available to promote the reverse reaction and so the catalyst inlet is susceptible to coking by CH_4. It is therefore necessary to keep the temperature low enough where the thermodynamics of carbon formation via methane cracking are less favourable (Shamsi and Johnson 2001). However, it must also be sufficiently high to allow some H_2 production to prevent cracking further down the bed. Similar to an industrial reformer, the bed inlet temperature of the current study was 550°C (Valdes-Perez, Fishtik et al. 1999; Shamsi and Johnson 2003; Snoeck, Froment et al. 2003).

Another common carbon forming reaction is the Boudouard reaction (Riensche and Fedders 1991):

$$2CO \Leftrightarrow C + CO_2 \quad \Delta H = -171\ KJmol^{-1}$$

No carbon is predicted from the Boudouard reaction at the bed inlet under the conditions in the present study as there is only CO_2 and no CO.

Towards the exit of the catalyst bed, enough H_2 has been produced from reforming to prevent C from methane cracking reaction and so the temperature can be increased. The reforming reaction is endothermic and favoured by high temperatures so it is desirable to use as high a temperature as possible. Figure 8 shows equilibrium data for a dual reforming reaction at 3 bar with feed CH_4:CO_2:H_2O 1:1:1.

Clearly the high temperature favors the CH_4 conversion and H_2 and CO yields (Treacy and Ross 2004; Maestri, Vlachos et al. 2009), however, it is difficult for CO_2 to be fully converted

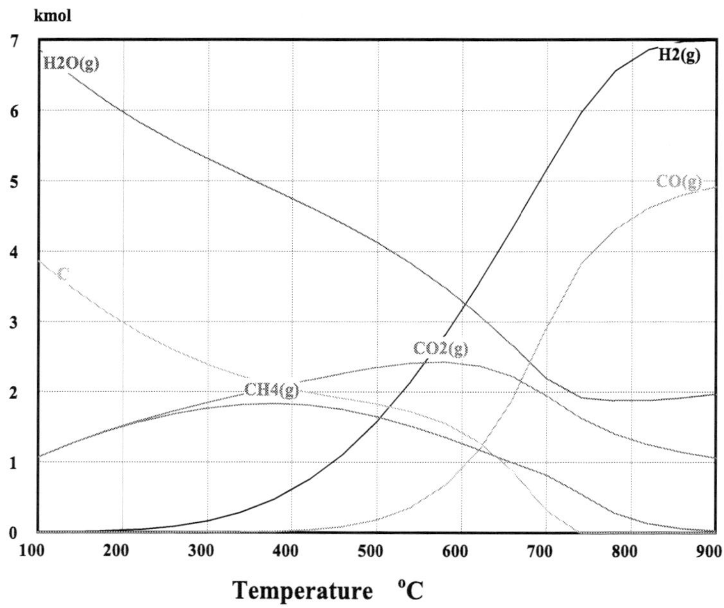

Fig. 8. Dual reforming equilibrium data at increasing temperature. 3bar, CH_4:CO_2:H_2O 1:1:1 feed

into CO even at high temperature. Also at high reaction temperature, such as 750°C or above, no carbon deposition is expected from the thermodynamic views.

3.2 Catalyst charging

The test was carried out in a 100ml pressure reactor system to simulate industrial operation conditions. Natural gas and CO_2 were controlled to feed the reactor using two gas MFCs. De-ionised water was fed using a quantitative pump, first vaporisation then mixing with the gas feedstock. The mixing system was then heated to the required inlet temperature, and further going to pass the catalytic reactor. The reactor converter was a tubular style system, which was heated at different heating zones so as to mimic the industrial reactor temperature distribution. Thermocouples were installed inside the tube to monitor the catalyst bed temperature at different points. The products from the reactor were firstly cooled down to separate liquid, so as to become dry gas. The dry gas was measured and the composition was analysed. The schematic setup can be seen in Fig. 6.

3.3 Catalyst properties

The catalyst sample is from the industrial proved catalyst system for natural gas reforming, whose physical property is shown in table 3. This catalyst combination has been used in both China and abroad in the industries of ammonia, synthetic methanol, industrial hydrogen generation, and the performance is extraordinary.

Indicator	BOXE-1	BOXE-2
Shape and color	10-hole pellet, gray	1-hole grooved pellet gray
Dimensions (od×h) (mm)	Φ10×14	Φ14×14
density (kg/l)	1.10~1.15	1.10~1.15
Radius-strength (N/pellet)	>350	>500
Main composition NiO (m/m) support	>13 Al_2O_3	>13 $CaAl_2O_4$

The catalyst was reduced before use so as to convert nickel oxide into active metal.
$NiO+H_2=Ni+H_2O$

Table 3. The dual reforming BOXE catalyst properties

The physical properties of the steam reforming supports are as follows:

For the 10-hole pellet shaped catalysts, the support consists of pure α- Al_2O_3, surface area: 3.0~3.5m^2/g; pore volume: 0.2~0.25ml/g;

For the 1-hole grooved pellet shaped catalyst, the main phase of the support is $CaAl_2O_4$ Surface area: 5~8m^2/g; Pore volume: 0.2ml/g; average pore size: 1000A;

Pore distribution:

>1500A: 75.70%,
1000~1500A: 10.7%,
500~1000A: 8.58%,
100~500A: 3.85%,
<100A: 1.23%

3.4 Feedstock

Natural gas is civil use grade, whose composition is shown in table 4. CO_2 is industrial grade, and the de-ionised water is normal purified grade. The catalyst reduction hydrogen is industrial grade H_2.

component	Volume percentage %
methane	96.0
ethane	2.35
nitrogen	0.57
CO_2	1.08

Table 4. Typical low CO_2 containing natural gas composition

3.5 Catalyst activation

After catalyst is loaded, pressure test was carried out to ensure no leak occuring. Then the reactor temperature was increased at 10 °C/min to 500°C at inlet, 720°C in the middle and 800°C at the outlet, then hydrogen reduction was started and last for 8 hours

The reduction media is H_2/H_2O at ratio of 1:5, which is the same as industry. Hydrogen GHSV is about =1000h^{-1} at ambient pressure.

4. Results and discussion

4.1 Test under pressure conditions

This series test is to mimic industrial practical process and operation. The conditions are under high pressure at $H_2O/C=2.0$ at low inlet temperature 500°C. These conditions are widely used in industry taking the techo-economical factors.

4.1.1 CO_2 fed amount effect on the product composition at $H_2O/C=2.0$

Pressure 2.6Mpa,
CH_4 GHSV 1200h^{-1},
Inlet T: 500°C
Outlet T: 900°C

The effect of CO_2 feeding (in terms of CO_2 GHSV) on the product composition is shown in Table 5. Each test was carried out for 400 hours to get steady and reproducible results. In industry, the catalyst performance in most case is measured in terms of H_2 and CO yield, and the reactants conversion. The conversion of methane is referred to the slip methane content, meaning the concentration of methane in the products (the dry gas base after steam removal and C elimination). The higher slip methane in the products suggests lower catalyst activity.

No.	NG GHSV H^{-1}	CO_2 h^{-1}	CO_2/CH_4	Inlet T	Outlet T	CH_4 %	CO %	CO_2 %	H_2 %	H_2/CO	CH_4 %
1	1200	2200	1.83	502	901	1.67	27.1	27.6	43.63	1.61	2.31
2	1200	2700	2.25	501	899	1.23	28.2	31.6	38.97	1.38	1.80
3	1200	3200	2.67	500	899	1.00	30.0	34.0	34.10	1.14	1.38
4	1200	3700	3.08	500	896	0.78	32.2	38.0	29.02	0.90	1.28
5	1200	4200	3.50	501	899	0.60	33.3	41.8	24.30	0.73	1.03

Table 5. Dependence of product composition on CO_2 feeding amount

The dependence of H_2/CO ratio on the CO_2/C ratio can be seen in Fig. 9, and the effect of CO_2/C on the slip methane content can be seen in Fig. 10.

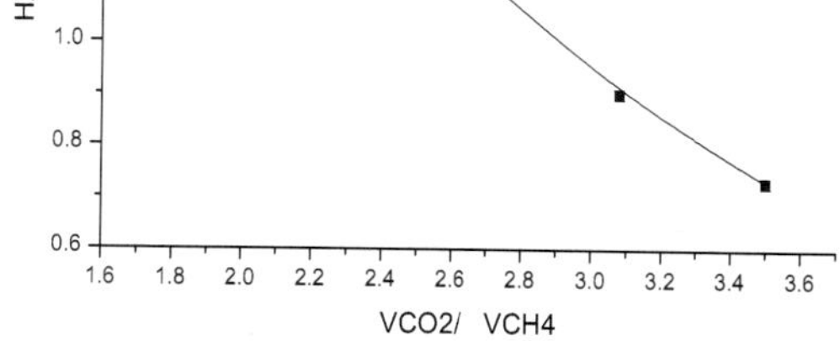

Fig. 9. Dependence of H_2/CO ratio in the product on the CO_2/C ratio in feedstock

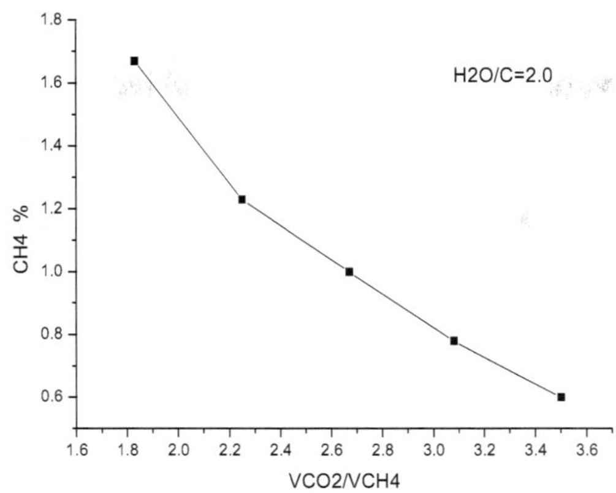

Fig. 10. Effect of CO_2/CH_4 ratio in the feedstock on slip CH_4 content

Clearly Fig 10 shows that the increase of CO_2 in the feedstock leads to the decrease of H_2/CO ratio in the products, and the slip methane content also decrease. This is helpful for the methane conversion and CO production.

4.1.2 Effect of pressure on the composition of products

The reaction conditions were set as follows;

H_2O/C: 2.0,
Methane GHSV: 1200h^{-1},
CO_2 GHSV: 4200h^{-1},
Inlet T: 500°C
Outlet T: 900°C

The effect of pressure on the gas product composition is shown in Table 6. Please note that each test was continued for 100 hours to get the steady state results, during the time on stream test, no catalyst activity drop or selectivity changes were observed, suggesting that the catalyst has high stability under the test conditions

No.	P MPa	Inlet T	Outlet T	CH_4 %	CO %	CO_2 %	H_2 %	H_2/CO	After CO_2 removal CH_4%
1	2.2	501	899	0.50	33.4	40.9	25.2	0.75	0.85
2	2.6	501	899	0.60	33.3	41.8	24.3	0.73	1.03
3	3.0	497	900	0.76	31.9	42.0	25.3	0.79	1.31

Table 6. Dependence of product composition on the pressures

With the increase of pressure, the slip methane content rises in the products, while H_2/CO ratio was little changed. The is in agreement with the thermodynamic equilibrium, because the reforming reaction gives more mole of gases than the reactants, increase the pressure leads to lower methane conversion(Friedmann, Burruss et al. 2003; Shamsi and Johnson

2003; Zhang, Li et al. 2004). However, the overall delta T to the equilibrium is about 5°C, which showed the super performance of the BOXE reforming catalysts.

4.1.3 Effect of outlet temperature on the composition of the product

Conditions:

H_2O/C: 2.0,
P: 2.6Mpa,
Carbon GHSV in terms of CH_4: 1200h^{-1},
CO_2 GHSV: 4200h^{-1},
Inlet T: 500°C

The outlet temperature effect on the product distribution is listed in table 7.

No.	Inlet T	Outlet T	CH_4 %	CO %	CO_2 %	H_2 %	H_2/CO	After CO_2 removal CH_4%
1	500	872	0.89	28.9	42.9	27.3	0.94	1.56
	499	872	0.89	29.0	42.5	27.6	0.95	1.55
2	501	899	0.60	33.3	41.8	24.3	0.73	1.03
	501	899	0.54	33.1	42.1	24.3	0.73	0.93

Table 7. The effect of outlet temperature on the gas product composition

Increasing outlet temperature favors the formation of CO, and helps to decrease the slip CH_4 content in the product. However, the outlet temperature may be restricted by the materials of the tubular reactor and energy consumption. The increase of outlet T may not be always a good solution to increase CH_4 conversion and decrease H_2/CO ratio. So far, most industrial operation for BOXE reforming catalysts uses 870°C as the outlet temperature.

4.1.4 Effect of CO_2 added in the feedstock on the gas product composition

Under of CH_4 GHSV of 1272h^{-1}, the results were shown in Table 8. It is shown that at VCO_2/VCH_4 ratio of 1.51, H_2/CO in the gas product is about 1.5, and slip CH_4 is only 0.1%. Comparison with Table 3 suggests that at the same GHSV of carbon (hydrocarbons), decreasing H_2O/C ratio and reducing P favour the CO generation and reducing CH_4 content.

P Mpa	GHSV h^{-1}		CO_2/CH_4	H_2O/C	T °C		Product dry gas composition %					
	CH_4	CO_2			Inlet	Outlet	CH_4	CO	CO_2	H_2	H_2/CO	After CO_2 removal C_{CH4}
0.3	1272	1283	1.01	1.5	600	900	0.1	27.6	16.1	56.2	2.04	0.12
0.3	1272	1492	1.17	1.5	602	900	0.1	29.5	16.5	53.9	1.83	0.12
0.3	1272	1687	1.33	1.5	601	900	0.1	30.5	17.8	51.6	1.69	0.12
0.3	1272	1916	1.51	1.5	603	900	0.09	31.5	19.8	48.6	1.54	0.11
0.3	1272	2148	1.69	1.5	605	900	0.06	32.5	21.5	45.9	1.41	0.08
0.3	1272	2336	1.84	1.5	605	900	0.05	34.9	22.7	42.4	1.21	0.06

Table 8. Effect of CO_2 added in the feedstock on the product composition at relatively low pressure range

P Mpa	GHSV h^{-1} CH$_4$	GHSV h^{-1} CO$_2$	CO$_2$/C	H$_2$O/C	T ℃ Inlet	T ℃ Outlet	Gas composition % CH$_4$	CO	CO$_2$	H$_2$	H$_2$/CO	After CO$_2$ removal CH$_4$ (%)
0.3	2561	1283	0.5	1.5	602	900	0.25	26.3	8.7	64.8	2.46	0.27
0.3	2561	2596	1.01	1.5	603	900	0.09	31.5	14.8	53.6	1.70	0.11
0.3	2561	2996	1.17	1.5	602	895	0.08	32.5	17.2	50.2	1.54	0.10
0.3	2561	3406	1.33	1.5	602	894	0.07	33.6	18.8	47.5	1.41	0.09
0.3	2561	3867	1.51	1.5	603	900	0.05	34.7	20.8	44.5	1.28	0.06

* Each test lasts for 400 hours to get steady and stable results

Table 9. Under relatively high methane GHSV, the dependence of gas product composition on the CO_2 feeding amount

Figs 10 and table 8 results suggest that low pressure and low H_2O/C ratio are favourable for methane conversion. Even when CH_4 GHSV was increased by one time (2561h^{-1}), slip methane in the gas products does not change significantly. However, the GHSV of methane has significant effect on the H_2/CO ratio. At CH_4 GHSV of 2561h^{-1}, and even when VCO_2/VCH_4 is only 1.17, H_2/CO is already nearly 1.5 in the product, suggesting high GHSV of methane is good for producing syngas with low H_2/CO ratio. This may be more useful for the production of syngas for carbonyl synthesis such as MEG production. It is contrary to the many industrial results.

The effect of GHSV of methane and CO_2/CH_4 ratios in the feedstock on the H_2/CO ratios are summarised in table 10. It is shown that the retention time exerts significant effect on the H_2/CO ratio in the products. High GHSV, e.g., shorter contact time of the reactants would have lower H_2/CO with the increase of VCO_2/VCH_4, which is much lower than the system with lower GHSV. The results are in agreement with the literature (Zhang, Li et al. 2004; Chen, Chiu et al. 2010). This suggests that the reaction is kinetics controlled, It may be worthwhile increasing the pressure so at the narrow the gap between the high and low GHSV for H_2/CO ratios.

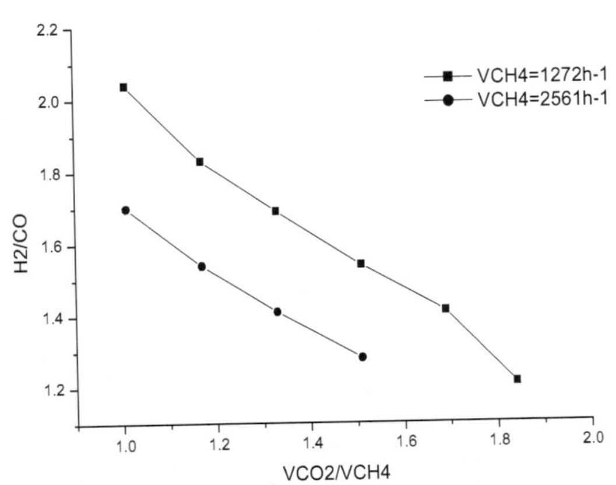

Fig. 11. Effect of GHSV of methane and CO_2/CH_4 ratios in the feedstock on the H_2/CO ratio in the products.

P Mpa	GHSV h^{-1}		CO_2/C	H_2O/C	T °C		Gas composition %					
	CH_4	CO_2			Inlet	Outlet	CH_4	CO	CO_2	H_2	H_2/CO	% CH_4*
0.3	2561	2592	1.01	1.3	610	900	0.13	32.4	13.7	53.8	1.66	0.15
0.3	2561	2592	1.01	1.5	601	900	0.08	31.6	14.8	53.6	1.70	0.11
0.3	2561	2592	1.01	2.0	600	899	0.06	27.6	17.3	55.0	1.99	0.07
0.3	2561	3406	1.33	1.5	602	894	0.07	33.6	18.8	47.5	1.41	0.09
0.3	2561	3406	1.33	2.0	601	900	0.04	30.5	22.9	53.4	1.75	0.05
0.3	2561	2592	1.01	1.5	601	900	0.08	31.6	14.8	53.6	1.70	0.11
0.45	2561	2592	1.01	1.5	603	895	0.35	29.5	15.4	54.8	1.86	0.41
0.6	2561	2592	1.01	1.5	608	890	0.65	29.2	15.7	54.5	1.87	0.77

* The CH_4 concentration refers to the one after excess CO_2 removal.

Table 10. Effect of CO_2 amount in the feedstock on the products compositions, CH_4 GHSV=2561h^{-1}

4.1.5 The effect of CO_2 and H_2O added in the feedstock on the gas product composition under the constant CH_4 GHSV H_2O/C ratio, P and temperatures

The catalyst has been tested at various ratios of H_2O/CH_4 and CO_2/CH_4 at different pressures and outlet temperatures. The highest H_2/CO ratios with a low slip methane results were obtained under the conditions of 0.3Mpa, CO_2/CH_4 1.01 and H_2O/methane 2.0 at inlet temperature of 602°C and outlet temperatures of 899°C. The gas products can be used directly for Fischer Tropsch. However, if a CO rich products are desired, the reaction conditions can be changed to 0.3Mpa, CH_4 GHSV 2561h^{-1}, CO_2 GHSV 3406h^{-1}, inlet temperature of 602°C and outlet temperature of 894°C, under which H_2/CO is 1.41 and slip methane of 0.09.%.

The results in Table 10 also suggests that decreasing H_2O/methane ratio to as low as 1.3 with co-feeding CO_2 can effectively depress the carbon formation, and give a stable catalyst performance. This can significantly reduce the steam feeding in the conventional steam reforming process, which would save energy and simplify the hydrogen plant, also it can make use of CO_2 in the gas stream.

It is seen from Table 10 that when fixing the reaction pressure and CO_2/CH_4 ratio while increasing H_2O/methane ratios, H_2 and H_2/CO increase in the products. Increasing CO_2/CH_4 ratio favors the CO production, giving desirable product compositions for the process where more or pure CO is required.

When the pressure was increased to 0.6Mpa from 0.3Mpa, the slip CH_4 concentration increased more significantly. This can be explained by the thermodynamic model, due to the reforming reaction is volume expansion, while CO WGS is identical volume process (Steinfeld, Kuhn et al. 1993; Tsai and Wang 2008; Maestri, Vlachos et al. 2009).

4.1.6 Effect of lower H_2O/methane ratio on the reactions

At P of 0.3Mpa, and CH_4 GHSV of 2561h^{-1}, CO_2/CH_4 of 1 when decreasing H_2O/methane to as low as 1.0 and 0.8, the catalyst has run for more than 50 hours each , a little carbon deposition were observed in the inlet of the catalyst at H_2O/C at 0.8, but much more carbon deposition in the exit reactor tube has been observed, which caused big pressure drop. This suggests that the steam is very important and essential to depress the carbon formation. A further experiment also proves this. When the pressure drop increased to 2 bars during the reaction with H_2O/CH_4 0.8 for 60hours, the steam feeding was increased to H_2O/CH_4 at 2 with the CH_4 and CO_2 feeding rate unchanging, we can see that the H_2/CO ratio increased to 1.92, and slip methane decreased to 0.05%, also the pressure drop decreased from 2 bar to 0.1bar after 95 hours operation. This suggests that the deposited carbon in the catalyst bed can be removed through adjusting the H_2O/methane ratios.

5. Catalyst deactivation and regeneration study

During the catalyst pilot test, it is found that when the CO_2 is co-fed with the steam at $H_2O/CH_4>1$, the catalyst can effectively convert natural gas and CO_2 into syngas with various H_2/CO ratio, and no pressure drops were observed during the test. However, when the H_2O/CH_4 is decreased to less than 1, pressure drop occurs in the reactor, and increased with the time on stream, but the products distribution and methane conversion remained unchanged. This suggests that even with some carbon deposition, the catalyst active sites can still catalyze the reactions.

Here we collected the carbon deposited catalyst sample and characterized them using XRD, TG and laser Raman. The XRD results showed that nickel is present in highly dispersed metallic metal, while a small amount of crystalline carbon is detected, which show a very small diffraction peak at 28.9°, which is tentatively assigned to the carbon deposition.

The used catalyst has been measured using TG-DTA-MS, the weight change, heat flow and the emitted products were monitored using MS during the thermal analysis. For comparison, a fresh H_2/H_2O reduced Ni catalyst has also been characterised using TG-DTA-MS. The results are shown in Fig. 12. It is clearly shown that with the increase of

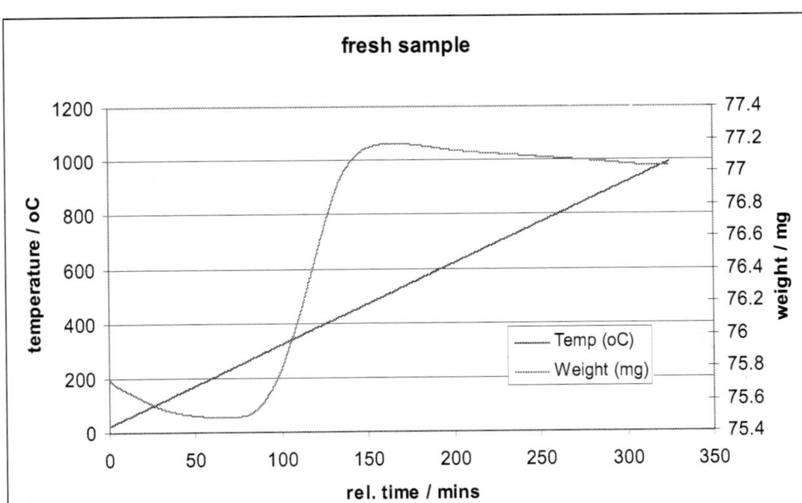

Fig. 12. TG-DTA-MS of the fresh reduced Ni/Al_2O_3 reforming catalyst

temperature, the weight of catalyst sample increase, which can be explained by the oxidation of nickel metal into nickel oxide. (Bhattacharyya and Chang 1994; Gonzalez-Delacruz, Pereniguez et al. 2011).

$$Ni+1/2O_2===NiO$$

Because the nickel metal particles uniformly dispersed over the support, the oxidation occurs in a very narrow range of the temperature from 230°C to 400°C. The weight loss after 500°C may be due to the decomposition of the resultant NiO.

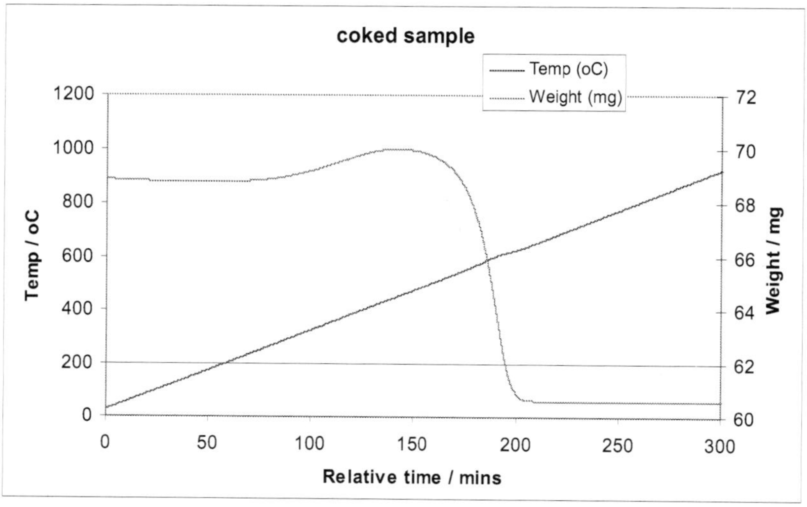

Fig. 13. TG-DTA of the used BOXE reforming catalyst

TG results of the used catalyst (unloaded from the reforming reactor where H_2O/methane=0.5 and operated for 50 hours), It is interesting to see that the catalyst weight started to increase at 210°C until 405°C, this is in agreement with the fresh catalyst sample, which can be explained by the oxidation of nickel metal particles into oxide. When the temperature is increased to 450°C, an abrupt weightloss occurs, with the weight of the sample change from 70mg to 60.5mg, equivalent the weighloss ratio of 13.8wt%. (Riensche and Fedders 1991; Claridge, Green et al. 1994; Verykios 2003; Adachi, Ahmed et al. 2009) Because the TG is carried in static air where oxygen is present, hence the deposited carbon is converted into CO_2 and evolved into the gas products, thus leading to the weightloss. This is further supported by the TG-DTA results as shown in Fig.14.

The TG-DTA results showed that the weighloss of the catalyst corresponds to a strong exothermic reaction, which is due to the oxidation of the carbon, as oxidation reaction gives heat. Also the carbon formation may be present in different forms, so the oxidation reaction is in a broad temperature ranges.

The MS results (Fig. 15) of the burnt gas suggest that only CO_2 is detected, which is in the same temperature for the weighloss of the catalyst sample as shown in Fig 13. No other gas

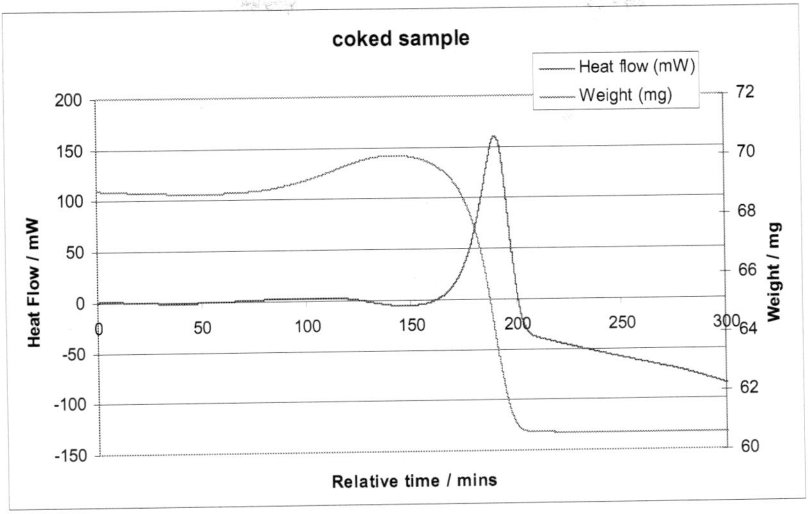

Fig. 14. TG-DTA result of the used BOXE reforming catalysts

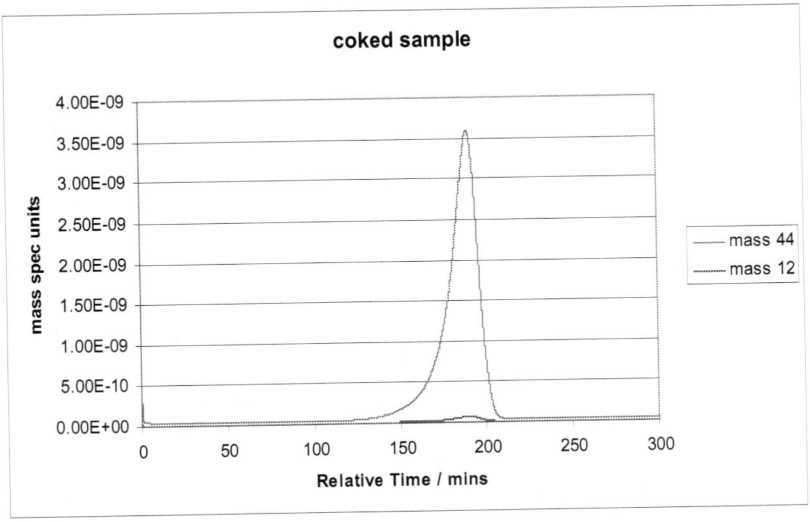

Fig. 15. The evolution of CO_2 from the used reforming catalysts during the thermal analysis

like methane or ethane is detected from the MS, also no aromatics, suggesting that the carbon formation over the BOXE catalyst is though nickel active site, not through the acid site, where aromatic carbon is the coke precursors.

6. Conclusion

6.1 Carbon dioxide is present in most natural gas field, separating CO_2 and transporting the natural gas into market may be more suitable for large gas field, but not economical for small gas field or high CO_2 containing gas field.

6.2 Converting the gas into liquid and transportation together with oil maybe an economical way to utilize the gas, but the conventional steam reforming catalyst requires high H_2O/CH_4 ratio to avoid carbon deposition, however, the resultant syngas has much higher H_2/CO ratio in the gas products.

6.3 BOXE reforming catalysts have been tested in dual reforming reaction to make use of both CO_2 and methane in the gas field, which showed a broad operation window for utilize the CH_4 and CO_2.

6.4 Under proper pressure, H_2O/C 、 CO_2/CH_4 ratios, reforming CH_4 using both H_2O and CO_2 at high temperature to generate syngas with low H_2/CO ratio is feasible. The catalysts showed stable good performance.

6.5 At temperature from inlet of 600°C and outlet of 900°C, given $H_2O/C=1.5$ to get $H_2/CO=1.5$ in the gas products is easy to obtain through adjusting GHSV and pressure and adding the amount of CO_2.

6.6 Reforming CH_4 at H_2O/C as low as 1.0 in the feedstock experience carbon deposition, especially in the reactor wall end. Hence it is not recommended to use the feedstock with low H_2O/C ratio in feedstock.

6.7 To reach long-term stable industrial operation, it is recommended to adopt relatively mild conditions such as lower outlet temperature (<870°C) relative high H_2O/C ratio (ca.1.4-1.5) to do the pilot test.

6.8 Dual reforming with CO_2 and H_2O not only helps to give the desirable H_2/CO ratio for the reformed gas products, but also helps to make use of the CO_2 in the gas fields and reduces GHG emission, The BOXE reforming catalyst showed high activity and selectivity to the syngas production at a very broad operating window.

6.9 The low H_2O/methane feed may result in carbon deposition when CO_2/CH_4 ratio is not high enough. However, the results carbon can be converted into CO through increasing the H_2O feeding during the operation.

6.10 The carbon formation under lean steam/carbon ratio is due to the nickel decomposition of the methane over the catalysts, it can be converted into CO_2 when oxygen is present, this can be another way for the catalyst regeneration outside of the reactor.

7. References

Adachi, H., S. Ahmed, et al. (2009). "A natural-gas fuel processor for a residential fuel cell system." *J. Power Sources* 188(1): 244-255.

Ballentine, C. J., M. Schoell, et al. (2001). "300-Myr-old magmatic CO2 in natural gas reservoirs of the west Texas Permian basin." *Nature* 409(6818): 327-331.

Bhattacharya, S., K. D. Newell, et al. (2009). "Held tests prove microscale NRU to upgrade low-btu gas." *Oil Gas J.* 107(40): 44-48, 50-52.

Bhattacharyya, A. and V. W. Chang (1994). "CO2 reforming of methane to syngas: Deactivation behavior of nickel aluminate spinel catalysts." *Stud. Surf. Sci. Catal.* 88(Catalyst Deactivation 1994): 207-213.

Butts, R. C. (2006). "Processing low BTU gas from the Permian Basin Yates formation." *Annu. Conv. Proc. - Gas Process. Assoc.* 85th: butts clark1/1-butts clark1/10.

Cathles, L. M. and M. Schoell (2007). "Modeling CO2 generation, migration, and titration in sedimentary basins." *Geofluids* 7(4): 441-450.

Chen, W.-H., T.-W. Chiu, et al. (2010). "Enhancement effect of heat recovery on hydrogen production from catalytic partial oxidation of methane." *Int. J. Hydrogen Energy* 35(14): 7427-7440.

Claridge, J. B., M. L. H. Green, et al. (1994). "Methane conversion to synthesis gas by partial oxidation and dry reforming over rhenium catalysts." *Catal. Today* 21(2-3): 455-460.

Ding, R., Z. Yan, et al. (2001). "A review of dry reforming of methane over various catalysts." *J. Nat. Gas Chem.* 10(3): 237-255.

Do, N. T. (2007). "A new approach to utilize associated gas in the upstream business." *Sekiyu Gijutsu Kyokaishi* 72(2): 178-187.

Friedmann, J., R. C. Burruss, et al. (2003). "High CO2 gas fields: Natural analogs to the chemical evolution of CO2 storage in depleted gas fields." *Abstracts of Papers, 226th ACS National Meeting, New York, NY, United States, September 7-11, 2003*: GEOC-140.

Gonzalez-Delacruz, V. M., R. Pereniguez, et al. (2011). "Modifying the Size of Nickel Metallic Particles by H2/CO Treatment in Ni/ZrO2 Methane Dry Reforming Catalysts." *ACS Catal.* 1(2): 82-88.

Maestri, M., D. G. Vlachos, et al. (2009). "A C1 microkinetic model for methane conversion to syngas on Rh/Al2O3." *AIChE J.* 55(4): 993-1008.

Ricchiuto, T. and M. Schoell (1988). "Origin of natural gases in the Apulian Basin in south Italy: a case history of mixing of gases of deep and shallow origin." *Org. Geochem.* 13(1-3): 311-318.

Richardson, J. T. and S. A. Paripatyadar (1990). "Carbon dioxide reforming of methane with supported rhodium." *Appl. Catal.* 61(2): 293-309.

Riensche, E. and H. Fedders (1991). "Conversion of natural gas into carbon monoxide-rich syngases." *Stud. Surf. Sci. Catal.* 61(Nat. Gas Convers.): 541-547.

Schoell, M. (1995). "Geochemical characterization of CO2 containing natural gases in south east asia." *Book of Abstracts, 210th ACS National Meeting, Chicago, IL, August 20-24*(Pt. 1): GEOC-018.

Shamsi, A. and C. D. Johnson (2001). "Carbon deposition on methane dry reforming catalysts at higher pressures." *Abstracts of Papers, 221st ACS National Meeting, San Diego, CA, United States, April 1-5, 2001*: FUEL-049.

Shamsi, A. and C. D. Johnson (2003). "Effect of pressure on the carbon deposition route in CO2 reforming of 13CH4." *Catal. Today* 84(1-2): 17-25.

Shimura, M., F. Yagi, et al. (2002). "Development of a new H2O/CO2 reforming catalyst and process for natural gas utilization." *Prepr. - Am. Chem. Soc., Div. Pet. Chem.* 47(4): 363-365.

Snoeck, J.-W., G. Froment, et al. (2003). "Kinetic evaluation of carbon formation in steam/CO2-natural gas reformers. Influence of the catalyst activity and alkalinity." *Int. J. Chem. React. Eng.* 1: No pp given.

Steinfeld, A., P. Kuhn, et al. (1993). "High-temperature solar thermochemistry: production of iron and synthesis gas by iron oxide (Fe3O4) reduction with methane." *Energy (Oxford)* 18(3): 239-249.

Suhartanto, T., A. P. E. York, et al. (2001). "Potential utilisation of Indonesia's Natuna natural gas field via methane dry reforming to synthesis gas." *Catal. Lett.* 71(1-2): 49-54.

Treacy, D. and J. R. H. Ross (2004). "The potential of the CO2 reforming of CH4 as a method of CO2 mitigation. A thermodynamic study." *Prepr. Symp. - Am. Chem. Soc., Div. Fuel Chem.* 49(1): 126-127.

Tsai, H.-L. and C.-S. Wang (2008). "Thermodynamic equilibrium prediction for natural gas dry reforming in thermal plasma reformer." *J. Chin. Inst. Eng.* 31(5): 891-896.

Valdes-Perez, R. E., I. Fishtik, et al. (1999). "Predictions of activity patterns for methane reforming based on combinatorial pathway generation and energetics." *Prepr. Symp. - Am. Chem. Soc., Div. Fuel Chem.* 44(3): 541-545.

Verykios, X. E. (2003). "Catalytic dry reforming of natural gas for the production of chemicals and hydrogen." *Int. J. Hydrogen Energy* 28(10): 1045-1063.

Wang, S., G. Q. Li, et al. (1996). "Carbon Dioxide Reforming of Methane To Produce Synthesis Gas over Metal-Supported Catalysts: State of the Art." *Energy Fuels* 10(4): 896-904.

Zhang, Q.-H., Y. Li, et al. (2004). "Reforming of methane and coalbed methane over nanocomposite Ni/ZrO2 catalyst." *Catal. Today* 98(4): 601-605.

16

Use of Meso-Scale Catalysts for Bulk-Scale Processing of Natural Gas – A Case Study of Steam Reforming of Methane

Pankaj Mathure[1], Anand V. Patwardhan[2] and Ranajit K. Saha[3]

[1]*Epoxy Division, Aditya Birla Chemicals Ltd.*

[2]*Department of Chemical Engg., Institute of Chemical Technology, Mumbai,*

[3]*Ex-HOD, Department of Chemical Engg., Indian Institute of Technology, Kharagpur,*

[1]*Thailand*

[2,3]*India*

1. Introduction

1.1 Catalytic steam reforming

Catalytic steam reforming of hydrocarbons, alcohols and light oil fractions involves the reaction of steam with methane, ethane, natural gas, LPG, naphtha, gasoline, alcohols like methanol, ethanol, and propanol, over catalysts at elevated temperatures (473 to 1173 K) and pressures (1 to 30 atm) to produce a mixture of hydrogen, carbon monoxide and carbon dioxide (Rostrup–Neilsen, 1984) . The mixture of hydrogen and carbon monoxide is known commonly as syngas.

The technology has matured over the last 55 years. However the endothermic nature of the reactions makes the process energy-intensive. Coke formation which results in catalyst deactivation is also another major concern. Work is still in progress to use milder operating parameters, increase reactor through-put and minimise coke formation increasing catalyst life and thus reduce reformer down time.

It is reported that 90% hydrogen generated globally, is produced by the steam reforming of natural gas and light oil fractions. The major reason for this is the commercial viability of such plants by which hydrogen can be produced at \$2.11/kg of H_2 (Momirlan & Veziroglu 2002, Harayanto et al. 2005). This is by far the most efficient process available in comparison to other processes like, carbon dioxide reforming, coal gasification, pyrolysis, water electrolysis and photobiochemical techniques.

1.2 Uses of hydrogen

Hydrogen has been termed as the "energy carrier of the future" (Das et al. 2001, Momirlan & Veziroglu 2002). It has the highest energy content of 120.7 kJ/gm (Harayanto et al. 2005). It burns producing no polluting emissions like SOx, NOx, CO, VOC etc. However it is available in nature only in the bound form. This makes it necessary to process the primary fuel to obtain H_2 and then use it in energy producing devices such as fuel cells or as fertiliser feedstock. Industry generates some 48 million metric tons of hydrogen globally each year from fossil fuels (Navarro et al., 2007). Nearly 50% of this hydrogen goes into making ammonia which is used in the manufacture of bulk chemicals like urea, phosphates and other fertilisers. Refineries use the second largest amount of hydrogen for chemical processes such as removing sulfur from gasoline (hydro-desulphurisation), converting heavy hydrocarbons into gasoline or diesel fuel, hydro-alkylation, hydro-cracking etc. Apart from these, hydrogen is used in Fischer-Tropsch synthesis, hydrogenation of oils and fats, pharmaceutical manufacture, cryogenic applications, in nuclear reactors, radio isotopes, mineral ore processing, reduction processes, oil processing.

1.3 Production of hydrogen

Bulk manufacturing of hydrogen is done in steam reforming units which use packed bed tubular reactors in which the catalyst (in the form of small pellets, spheres or tablets) is randomly dumped to obtain the desired products (syngas) at elevated temperature and pressure. Steam reforming system for hydrogen production is composed of a steam reformer, a shift converter, and a hydrogen purifier based on the pressure swing adsorption (PSA) or membrane separation unit. A mixture of natural gas and steam is introduced into a catalyst bed in the steam reformer, where the steam reforming reaction proceeds on nickel–based catalyst at 973 to 1073 K The reformed gas is supplied to a shift converter, where carbon monoxide is converted into carbon dioxide to produce more hydrogen by the shift reaction. The reformed gas is passed to PSA unit to separate hydrogen.

A typical steam reformer unit consists of a reformer block containing 40 to 400 tubes of height 6 to 12 m having an inner diameter of 0.7 to 0.16 m and a thickness of 0.01 to 0.02 m (Rostrup–Neilsen, 1984). The tubes are generally made of high alloy nickel chromium steel. As the process is endothermic, the required heat needs to be supplied to the reformer unit. The tubes can be heated externally or internally. In external heating, part of the feed is combusted outside the reformer tube to reach the desired reaction temperature. In internal heating part of the feed is combusted inside the tube of the reformer. This latter process is known as partial oxidation or autothermal reforming depending on the energy supplied to the process.

1.4 Catalysts used

Conventional catalysts are available in the form of cylindrical pellets, tablets and spheres and are simply dumped in the reformer tube. This is called as a fixed bed or packed bed reformer. The catalyst pellets are randomly placed in the reformer tube and do not contribute to a defined structure to the bed. There is no structure to the bed as a whole and it can be imagined as a bed containing a large number of convoluted or tortuous paths through which the feed gas flows. As the gas flows through the path, there is an increased pressure drop (relative to the use of structured catalysts). The quantity of catalyst and its composition is determined on the basis of the feed quantity and quality or composition. The role of the catalyst is to achieve maximum possible conversion of the feed to syngas, a constant pressure drop and to have sufficient mechanical and thermal strength to provide a long process cycle (Rostrup–Neilsen, 1984). Many Indian fertiliser plants use supported nickel catalysts. The support is generally silica, alumina, and/or magnesia. The properties of the catalyst are further modified by addition of components like, calcium oxide, aluminium, molybdenum, special carbides, zirconia, ceria, other alkali earth metals like potassium, etc. to make the catalyst coke resistant, and to increase its mechanical and thermal strength. Effect of promoters such as Cu, La, Mo, Ca, Ce, Y, K, Cr, Mg, Mn, Sn, V, Rh, Pd, and their combinations have been used for improving the stability of Ni catalysts supported on alumina for steam methane reforming. The use of platinum, rhodium, and palladium, along with nickel has also been reported (Harayanto et al. 2005). However, industrial catalysts are usually nickel based catalysts. The most commonly used catalysts are nickel–alumina, nickel–alumina–magnesia, nickel–magnesia, etc. The percentage of nickel varies from as low as 3 % to as high as 52% by weight. Nickel possesses hydrogenation activity but limited water gas shift activity. It is a low-cost but effective catalyst for cleavage of $O-H$, $-CH_2-$, $-C-C-$ and $-CH_3$ bonds. It is these properties that make nickel a favourable choice as a steam reforming catalyst. However, nickel alone cannot be used as a steam reforming catalyst due to its limited water gas shift activity. In order to exploit this secondary reaction, nickel or other metal catalysts are coupled (or supported on) with another metal oxide such as alumina or magnesia.

1.5 Micro and meso–scale reactors

Micro– and meso-scale reactors are a rapidly emerging technology field (Lowe, 2000, Hessel et al., 2005, Holladay et al., 2004). They differ in type / configuration and manufacturing procedures. The most widely used varieties include substrates such as silicon or silicon carbide on which channels are "etched" using well established lithography or deep reactive ion etching (DRIE) techniques. They consist of a stack of plates, onto which channels are introduced. The channels, which have width ranging from 50 to 500 μm, and depths up to 2000 μm, can be coated with active catalytic materials. This offers an advantage over traditional fixed bed reactors as higher transport rates and lower pressure drop can be achieved in a reduced volume. The large surface to volume ratio of these structures leads to effective heat dissipation making these systems suitable for highly exothermic / endothermic reactions with short contact times. There are very few papers that report the use of such micro and meso-scale reformers for the production of hydrogen from steam reforming of methane and ethanol (Pattekar & Kothare, 2004, Lowe, 2000, Hessel et al., 2005, Holladay et al., 2004).

1.6 Concept of the meso–scale, channel reactor

A novel way of developing micro or meso–scaled reactors which can be readily integrated in existing steam reformer plants for bulk manufacture of hydrogen is the use of wash–coated monoliths which can be fitted inside reformer tubes. The size of the monolith and the base material (ceramic or metal) can be selected as per the process requirements. The basic support monolith can be purchased from a number of suppliers like Corning, Degussa, and Emitec in the market and subsequently the wall coating can be modified to suit the feed stock. Monoliths or honeycomb reactors have been used as a support for catalytic converters for treatment of automobile exhaust gases and as filters for processing exhaust gas streams from industries. Monolith reactors act as multifunctional reactor systems having the advantages of low pressure, efficient contact of the reactants with (most frequently) solid catalyst and hence avoiding the problems related to partial wetting of the catalyst with liquid/gas phase; the possible choice of combining reaction and separation processes; the controlled loading of reactants, product removal, etc. Thus in monolith reactors one can combine the advantages of conventional multiphase reactors such as fixed bed, slurry and trickle bed reactors (Vergunst, 1999). This increases process efficiency and its cost effectiveness. One disadvantage of monolith reactors is the poor distribution of liquid and/or gas along the reactor's cross-section. This makes their use difficult, especially under higher flow rates.

Ceramic monoliths are usually made of cordierite which has a composition of $2MgO.5SiO_2.2Al_2O_3$. There are hardly any references (in literature) to the use of meso-scale, cordierite, monolith channels for steam reforming of methane (heat provided external to the reformer tube). It was therefore decided to use ceramic based (cordierite) monoliths having meso–scale rectangular channels as the base material for the monolithic catalyst for steam reforming of methane.

1.7 Objectives of this work

The focus of this work was on the development of novel channel monolithic reactors with meso-scale dimensions which can be coated with suitable catalytic metal oxide and used for steam reforming of methane. The objective of this article is to present the performance characteristics of such a novel reactor and to identify possible strategies for use of novel small dimension reactors for the same process. These meso-scale structures can withstand extreme pressure and temperature conditions. These structures provide the advantages of meso–scale reactors such as high surface area, low pressure drop and can be used in bulk scale manufacture of syngas and hence hydrogen. This work includes a comparison of the performance of these developed reactors with commercial catalysts (used in fixed bed reactors).

The catalyst block (named as PS–CAT) consists of a cordierite meso–scale, channeled, monolithic structure which is wash coated (a specified number of times to achieve the desired catalyst loading) with a nickel–based salt solution (2M concentration), dried, calcined, and reduced with hydrogen. This procedure of wash coating the cordierite substrate and the subsequent processing (drying and calcining of the same) is critical for the durability and quality of the catalyst (Mathure, 2008, Mathure et al., 2007a, Vergunst, 1999). The structured catalyst consists of a number of thin–walled channels which exactly fit the reformer tube inner diameter. A number of such blocks can be stacked one above the other to fill the entire length of the reformer tube. Also, the diameter of such blocks can be as per the process requirements. In this study, separate experiments were conducted with 4 channels (figure 1) as well as for 49 channels, each part of a larger monolith consisting of 100 cpsi (cells per square inch). As the reactor has straight structured channels, the resulting pressure drop is much less than that for a tubular packed bed of catalyst. This reactor concept can be used in the existing reformer infrastructure without any major change in process equipment to significantly increase hydrocarbon feed throughput. With higher throughput, the same steam reformer unit can produce larger quantities of syngas. The main advantages are increased throughput, simplicity of the arrangement, and minimum plant modification. Figure 1a shows a 4–channel PS–CAT. Figure 1b shows a schematic of the a 49–channel PS–CAT

Fig. 1a. PS–CAT (4–channel): A meso–scale, channel, structured catalyst.

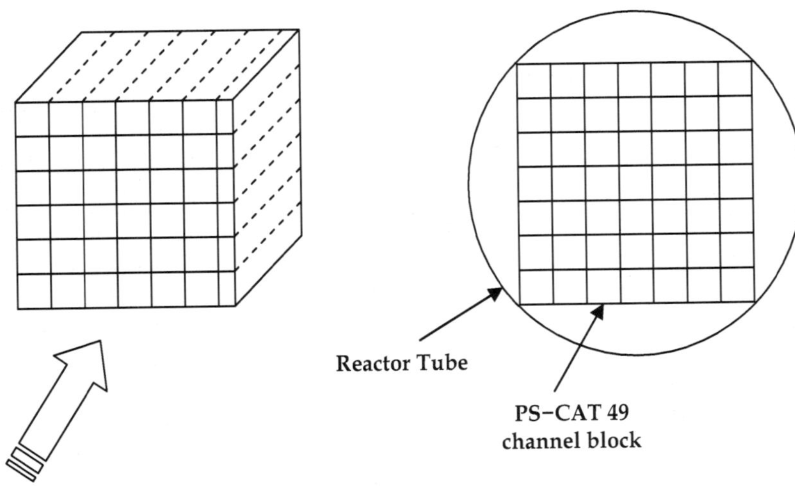

Fig. 1b. Typical schematic of a meso–scale, channeled monolith. The figure on right shows the block placed in a reactor tube.

2. Literature review

The main focus of this section is review work on the use of monolithic, channel reactors for steam reforming of methane for production of hydrogen. Tomasic and Jovic, 2006 have published a state of the art review on the use of monolithic catalysts. According to the authors development of monolithic catalysts and/or reactors has been one of major achievements in the field of heterogeneous catalysis and chemical reaction engineering of the recent years. The work highlights the advantages of monolithic catalysts and/or reactors with respect to the conventional ones, with particular focus on the integral approach to the catalyst and reactor design. The paper is detailed and gives basic definitions and classification of monolithic catalysts, including basic features of the monoliths and factors that have proceeded to the development and application of the monolith structures. The paper also discusses the preparation of monolithic catalysts and their commercial application with particular emphasis on the less known applications, and those which are still under development. These applications include the use of monolithic catalysts for Fischer–Tropsch synthesis, selective hydrogenation, oxidation, catalytic fuel combustion units, and gas cleaning devices for industries and automobiles. The authors have concluded that futher research should be directed towards the development of new constructional designs by changing geometries to resolve the distribution problems at the entry of monolithic structures.

Hessel et al., 2005 have reviewed work on micro–structured devices (both etched channels and monoliths) developed for reforming of methane and other hydrocarbons. They have reported a number of devices which can have been used mainly for partial oxidation or autothermal reforming of methane, propane, isooctane and gasoline.

Meille, 2006 has discussed the methods used to deposit a catalyst on structured surfaces. The review outlines both physical methods such as physical vapour deposition and chemical methods (sol–gel, chemical vapour deposition, direct synthesis, etc.) The coating of catalysts based on oxide, zeolite or carbon support is detailed on various surfaces such as silicon or steel microstructured reactors, cordierite monoliths or foams, fibres, tubes, etc.

Zhou et al., 2006 have investigated alumina supported Pt group metal monolithic catalysts for selective oxidation of CO in hydrogen–rich methanol reforming gas for proton exchange membrane fuel cell (PEMFC) applications. The results show that Pt/Al_2O_3 was the most promising candidate to selectively oxidize CO from an amount of about 1 vol% to less than 100 ppm. They have investigated the effect of the O_2 to CO feed ratio, the feed concentration of CO, the presence of H_2O and/or CO_2, and the space velocity on the activity, selectivity and stability of Pt/Al_2O_3 monolithic catalysts. The Pt/Al_2O_3 catalyst was scaled up and applied in 5 kW hydrogen producing systems based on methanol steam reforming and autothermal reforming.

Giroux et al., 2005 have published a review paper for the technical and market potential of hydrogen for use in varied applications namely, stationary fuel cell power generation units, on-site generators of hydrogen for industrial uses, and hydrogen fueling stations for fuel powered automobiles. They have proposed the use of monolithic structures as alternatives to particulate catalysts for the reforming of hydrocarbons for hydrogen generation. They have further proposed that new process designs for fuel processing can take advantage of the successful experience of environmental catalysis in automotive and stationary applications using monolithic catalysts and other substrates such as heat exchangers.

Bobrova et al., 2005 have studied syngas formation by selective catalytic oxidation of liquid hydrocarbons in a short contact time adiabatic reactor. Their research involved pilot plant scale exploration of syngas formation from liquid fuels like isooctane and gasoline by selective catalytic oxidation at short contact times at nearly adiabatic reactor conditions. Monolithic catalysts with different supports (micro–channel ceramics and metallic honeycomb structure) have been used in the experiments. Their results demonstrated that over the range of the operational parameter O_2/C = 0.50–0.53 required for syngas generation, equilibrium syngas was produced over the catalysts employed, thus proving the evidence of their high activity and selectivity.

Qia et al., 2005 have studied the performance of La-Ce-Ni-O monolithic perovskite catalysts for gasoline autothermal reforming system. Autothermal reforming of gasoline or its surrogates, n−octane with or without thiophene additive, was carried out on either bulk perovskite pellet or monolithic perovskite catalyst. During the 220−h long-term test, the pellet catalyst exhibited high thermal stability and activity with hydrogen yield approaching to the theoretical maximum value and only minor amount of CH_4 slipping through. It possessed fairly good sulfur tolerance (almost immune to 5 ppmw sulfur) although it could still be seriously poisoned when subjected to high concentration of sulfur. Their raw cordierite monolith could be an effective catalyst for autothermal reforming of gasoline, exhibiting superior performance to the catalyst of 0.3 weight % Rh/CeO_2-ZrO_2/cordierite at a temperature range of 923 to 1073 K.

Pasel et al., 2004 have studied different water−gas−shift catalysts for combined autothermal reforming with water−gas−shift (WGS) reaction. The autothermal reforming of liquid hydrocarbons using different conditions have been studied. In the first step, the dry reformate from autothermal reforming and a separate stream of steam was used to conduct the WGS reaction. Strong differences concerning the catalytic activity between the three investigated commercial monolithic catalysts could be observed. The most active one showed a very promising catalytic behaviour. At a gas hourly space velocity (GHSV) of 12,250 h^{-1}, CO conversion amounted to 85% at 553 K. CO concentration in the reformate reduced from 6.1 to about 0.9 vol. %. In another step, unconverted water from autothermal reforming was fed to the reactor for the WGS reaction together with additional components of the reformate. No catalyst deactivation was observed within almost 90 h under autothermal reforming conditions generating only traces of carbon dissolved in the water.

Lindstrom et al., 2003 have carried out an experimental investigation on hydrogen generation from methanol using monolithic catalysts. They have evaluated the activity and carbon dioxide selectivity for the reforming of methanol over various binary copper−based materials, Cu/Cr, Cu/Zn and Cu/Zr. The methanol reforming was performed using steam reforming and combined reforming (a combination of steam reforming and partial oxidation). This combined reforming process was carried out at two modes of operation: near autothermal and at slightly exothermal conditions. Their results show that the choice of catalytic material has a great influence on the methanol conversion and carbon dioxide selectivity of the reforming reaction. The zinc−containing catalyst showed the highest activity for the steam reforming process, whereas the copper/chromium catalyst had the highest activity for the combined reforming process. The copper/zirconium catalyst had the highest CO_2 selectivity for all the investigated process alternatives.

Apart from these there are a few patents which refer to the use of ceria-coated zirconia monoliths for catalytic partial oxidation process of hydrocarbons for syngas production and the use of platinum and palladium monoliths for autothermal reforming of hydrocarbons (Jiang et al, 2006, Hwang et al., 1985).

Vergunst, 1999 has published a detailed thesis on the preparation aspects of carbon−coated monolithic catalysts and three phase hydrogenation of cinnamaldehyde using the same [3]. The author has covered preparation of monolithic catalysts, testing of monolithic catalysts in the (selective) hydrogenation of benzaldehyde, cinnamaldehyde, methylstyrene, benzene, and γ−butyrolactone, testing of monolithic catalysts in solid acid catalyzed reactions (acylation and esterification) and in the selective oxidation of cyclohexanol into adipic acid, development of internally finned monolithic structures, hydrodynamics, and scale up. The work is extremely important for any work related to monolithic catalysts. However the manufacture of these catalysts and their subsequent application is totally different from the current application (steam reforming) being discussed.

As can be observed from the above work, wash–coated monoliths with meso or micro–scaled channel sizes are a promising substitute for conventional large and small scale fixed bed reactors. Once the coating procedure and coating material is finalized, these monoliths provide cost-effective reactor geometries to retrofit existing reformer units. To the best of our knowledge, there are hardly any references (in literature) to the use of meso–scale, ceramic, monolith channels for steam reforming of methane (heat provided external to the reformer tube). It was therefore decided to use ceramic based (cordierite) monoliths having meso–scale rectangular channels as the base material for the monolithic catalyst for steam reforming of methane. As nickel possesses hydrogenation activity and is a low-cost but effective catalyst for cleavage of $O-H$, $-CH_2-$, $-C-C-$ and $-CH_3$ bonds, it was decided to exploit these properties and use nickel nitrate solution for the coating of the monoliths.

3. Development of the monolith structures

3.1 Coating of the monolith structures

The catalyst block (named as PS-CAT) developed in the course of our work consisted of a cordierite substrate with a square-channel, monolithic structure which was wash-coated with a nickel-based salt solution (2M concentration), dried at 393 K, calcined at 673 K, and reduced with hydrogen (Mathure, 2008). The catalyst block is shown in figure 1. The structured catalyst consisted of a number of thin-walled channels which exactly fitted the experimental reformer tube inner diameter (Mathure et al. 2007b). The average wall thickness was found to be about 270 μm. The channel dimensions were determined to be 1500 μm × 1500 μm (hence the term meso-scale). In this study, experiments were conducted with a block having 4 channels cut out from a larger monolith consisting of 100 cpsi (channels per square inch). Figure 1 shows a 4-channel PS-CAT. The performance of the developed catalyst was tested extensively using a multi-feed, multi-scale, fully instrumented experimental rig fabricated by M/s. Texol Engineering Private Limited, Pune, Maharashtra, India & M/s Dampf Kolben , Pune, India. The details of experimental rig and procedure have been reported elsewhere (Mathure et al., 2007b, Mathure, 2008). Figure 2 from shows the actual photograph of the experimental rig and its components such as the mass flow meter, HPLC pumps, reactors and Reactor furnace assembly. Figure 3 shows the schematic arrangement of the experimental rig. PS-CAT contained a nickel oxide loading of around 8.5% (weight% of the coated block). The elemental nickel loading was around 6.7% (weight% of the coated block) as determined by EDS -X-ray analysis.

Each block was weighed before being dipped into the coating solution, and after the calcination, to find out the amount of catalyst (nickel oxide) loading on the block. This weight was taken as W for calculation of W/F_{CH4} during the steam reforming experiment. The blocks were activated using a hydrogen stream of flow rate 40 N ml/minute in the steam reformer unit just before use in the experiment.

Fig. 2. Experimental rig and its components used for steam reforming of methane.

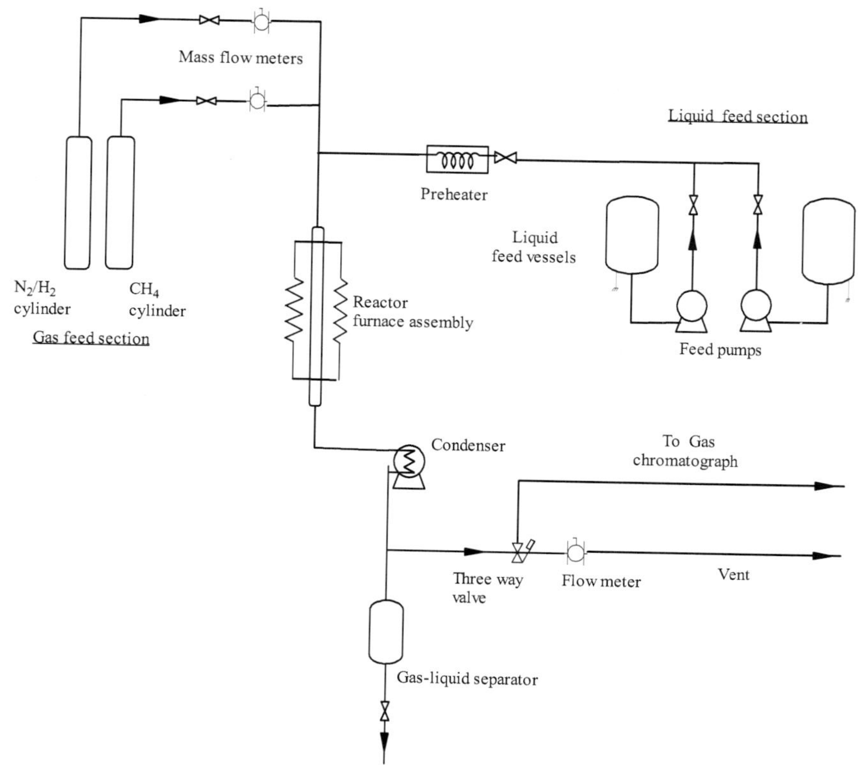

Fig. 3. Schematic of the experimental rig used for steam reforming of methane and ethanol.

3.2 Characterisation of the structures

The characterisation was done by EDS–X-ray analysis, BET surface area measurement and XRD analysis. A detailed discussion of the characterization can be found the work of Mathure, 2008.

3.2.1 X-ray diffraction study

The crystal phases in the uncoated cordierite substrates and the coated PS-CAT were identified by a X-ray powder diffraction study. XRD spectra were obtained with a Rigaku–Miniflex diffractometer using monochromatic Cu-Kα radiation. The 2θ scanning was performed from 10 to 80° at a rate of 2°/min. Figure 4 shows the XRD patterns for the uncoated cordierite substrate and Figure 5 shows the XRD pattern for the coated PS-CAT. The only active ingredient in the coated catalyst was the nickel oxide species. An observation of both XRD patterns shows that the extra peaks present in Figure 5 are of nickel oxide as verified using JCPDS database. These peaks are not seen in the uncoated raw substrate.

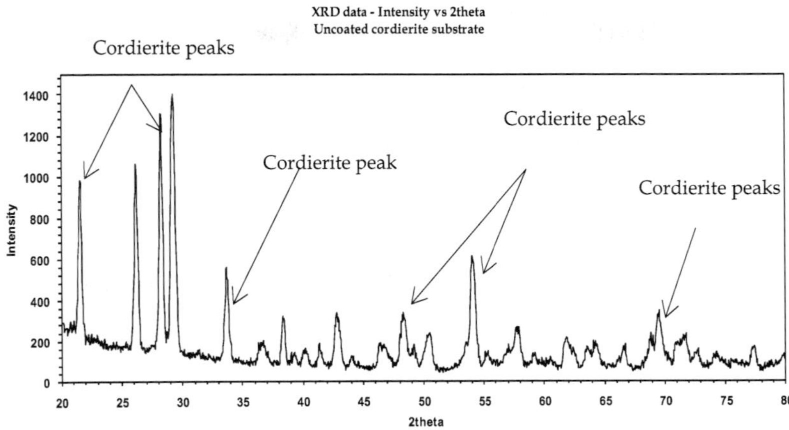

Fig. 4. X-ray diffraction spectra of the raw uncoated cordierite substrate.

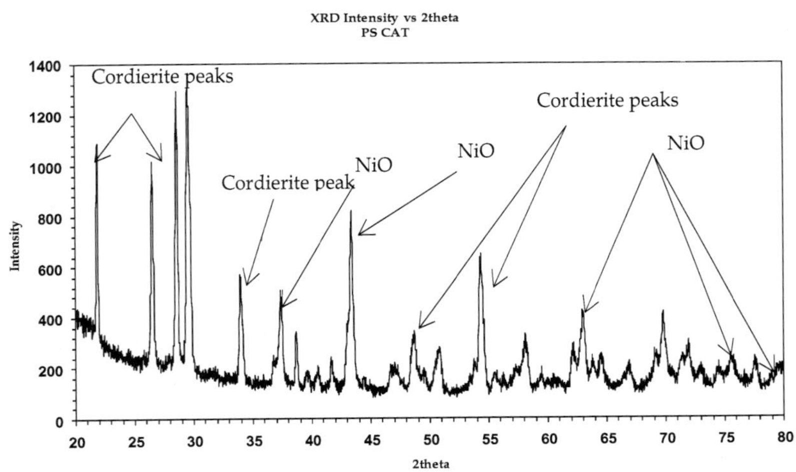

Fig. 5. X-ray diffraction spectra of the coated catalyst block PS–CAT.

3.2.2 Energy dispersive spectroscopic X-ray analysis (EDS–X-ray)

The metal loading was estimated by EDS– X-Ray unit (Link ISIS 300, Japan). Nickel oxide which acts as the active ingredient content was found to be 8.5% (weight % of the coated block). The elemental nickel loading was found to be around 6.7% (weight % of the coated block).

3.2.3 BET surface area measurement

The Brunauer–Emmett–Teller (BET) specific surface area was measured using nitrogen as an adsorbent in a Quantachrome make AUTOSORB 1C machine. The samples were degassed for 12 hours at 573 K before analysis. The values obtained were as follows;

Uncoated (raw) cordierite substrate	: 3.14 m^2/gm
Coated substrate, PS–CAT	: 1.59 m^2/gm
Used (post-experimentation) PS–CAT	: 1.18 m^2/gm

The extremely low values have been observed due to the use of cordierite ceramic blocks which have very low specific surface area in the first place. Subsequent and prolonged dip coating further decreases this value. It must be noted here that the blocks were directly coated with the metal salt solution. No binder or support material such as γ–Al_2O_3 sol was used to enhance the surface area.

4. Experimental study

The performance of the developed catalyst was tested extensively using the experimental set up described detail in Mathure, 2008 & Mathure et al., 2007a. The testing was carried out by varying the process parameters such as time–on–stream analysis (100 hours on stream), temperature (673 to 1048 K), space time (W/F_{CH4} = 10 to 55.0 gm–cat min/mol), and the number of channels (4 channels and 49 channels in parallel). However as the number of monoliths available to us were limited, the bulk of the experiments were performed using the 4 channel PS–CAT. Time–on–stream studies were carried out for both the blocks. The length of each block (both 4 channels and 49 channels) was 0.025 m. In case of experiments with the 4 channel PS–CAT, the tubular reactor housing with inner diameter 6 mm was used. Five blocks each having 4 channels were placed end to end within the tubular reactor housing such that the centre of the catalyst bed corresponded to the central heating zone of the furnace. Rest of the tubular reactor housing (both above and below the catalyst blocks) was filled up with inert particles. As 5 blocks were used in series, the total length of the meso–scale channel reactor was 0.125 m. In case of experiments with the 49 channel PS–CAT, the tubular reactor housing with inner diameter 22 mm was used. Two blocks each having 49 channels were placed end to end within the tubular reactor housing such the centre of the catalyst bed corresponded to the central heating zone of the furnace. Rest of the tubular reactor housing (both above and below the catalyst blocks) was filled up with inert particles. As 2 blocks were used in series, the total length of the meso scale channel reactor was 0.05 m. The reactor was mounted in the furnace assembly. Fresh catalyst was used for each experimental run.

The catalyst was activated in a hydrogen flow of 40 N ml/min with a steady increase of temperature. The catalyst was maintained at the desired reaction temperature for two hours. Methane gas (make: BOC, XL grade, purity: 99.9 %, specific gravity: 0.553, calorific value: 55530 kJ/kg) was fed to the reactor at the desired flow rate using a Bronkhorst make mass flow meter. The preheater temperature was maintained at 443 K to ensure complete vaporisation of the feed water even at high liquid flow rate.

A HPLC (high pressure liquid chromatography Make: SSI, USA) pump was used to feed water to the reactor at the desired flow rate (between 0 to 5 ml/min). The product stream was condensed using the double pipe condenser at the outlet of the reactor. The liquid product (mostly water) was collected in a separator while the outlet gas stream flow rate was measured using a wet gas flow meter.

The gas was periodically sent to the GC (gas chromatograph) for analysis using a three-way valve. Gas analysis was done using a thermal conductivity detector (TCD) with Spherocarb column (1/8″ diameter and length 8′) to find the composition of the outlet gas stream. The product gas was found to contain methane, carbon monoxide, carbon dioxide and hydrogen. No other hydrocarbon species was detected. The results of the gas analysis of the product gases indicated whether steady state conditions had been achieved. The measurements for the study were done once similar gas compositions were obtained after the start of the experiment. As this is a fast reaction in general steady state conditions were observed within 15 to 20 minutes from the start of the experiment. The pressure for all experimental runs was maintained at 1 atmosphere.

The liquid product was weighed and analysed periodically in the GC using the TCD in a Porapak Q column (1/8″ diameter and length 6′). The liquid product was found to be only water. The reaction was stopped after 3 to 6 hours of steady state operation (except for time–on–stream studies which were conducted for 100 hours of operation).

Methane conversion was calculated using equation 1.

$$\frac{mols_{CH4}in - mols_{CH4}out}{mols_{CH4}in} \times 100 = X_{CH_4} \qquad (1)$$

Reproducibility of the experiments was verified by conducting repeat runs for a few random reaction conditions. The results showed a maximum error of ±4.0%.

5. Results and discussion

5.1 Effect of temperature

Steam reforming of methane was carried using PS–CAT in a temperature range of 673 K to 1048 K keeping the molar ratio of feed to steam (MR) and the W/F_{CH4} constant. Figure 6 shows the effect of temperature on the conversion of methane. As temperature increases the conversion of methane gradually increases from 30% to nearly 100%. The catalyst exhibited a methane conversion of more than 90% for a temperature above 973 K. The catalyst showed nearly complete conversion (above 98 %) of methane at a temperature above 1048 K at a W/F_{CH4} close to 55.0 g–cat min/mol and a molar ratio (CH_4:H_2O) close to 3.0. With increase in temperature the carbon monoxide content in the product gas stream increase from as low as 0.21% to 22.43% (v/v). The carbon dioxide content decreases with increase in temperature. Thus the catalyst promotes the endothermic steam reforming reaction but does not promote the exothermic water–gas–shift reaction which is favoured at low temperature. This result is in accordance with the presence of nickel as the active ingredient of the catalyst which promotes bond cleavage in the steam reforming reaction but has limited water–gas–shift activity (Rostrup–Neilsen, 1984, Vaidya & Rodrigues, 2006).

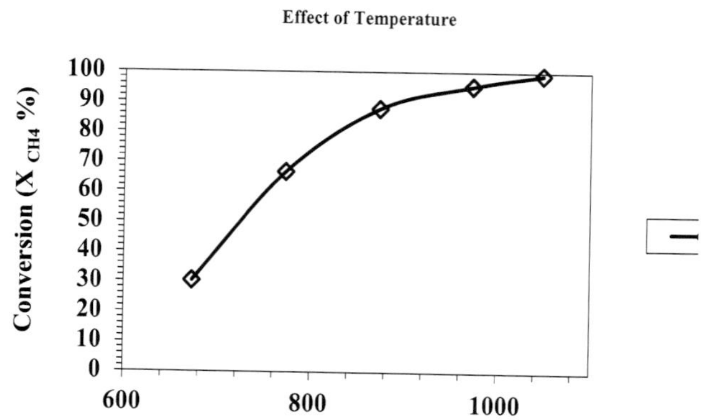

Fig. 6. Effect of temperature on methane conversion. Other conditions: PS–CAT 4 channel block, 5 blocks in series, W/F_{CH4} = 56 gm-cat min/mol, steam to methane molar ratio in feed = 3:1.

5.2 Time–on–stream study

The most important feature of industrial catalysts is their ability to perform steam reforming of methane for hundreds of hours of continuous operation without significant loss of activity (Rostrup–Neilsen, 1984). The study to test the performance of the catalyst for long duration of continuous operation is called as time–on–stream study.

PS–CAT was tested for steam reforming of methane at a steam to methane molar ratio of 3:1 at atmospheric pressure and temperature of 1048 K for a over 100 hours on stream. Both the 4 channel PS–CAT as well as the 49 channel PS–CAT were tested. The resulting exit gas was analysed and the results of the same for four channels is shown in figure 7 and figure 8. As can be seen a methane conversion of more than 98 % is obtained over the entire period (more than 100 hours) of operation. The product gas composition also shows steady formation of all species. In case of the 49 channel PS–CAT the temperature of operation was 1113 K. The higher temperature was necessary as due to increase in the number of channels, the time required to acquire steady state conditions increased by 60 minutes. The reason for this is the low thermal conductivity of the cordierite substrate (2.76 W/m K). Thus the 49 channel block required more time to "soak– up" heat and attain a uniform temperature. Once a uniform temperature was attained, the performance of this block was also very good and exhibited a steady methane conversion of more than 97%. Therefore the catalyst block can readily be used in industrial reformers simply by using a block of higher diameter to fit the inside of a reformer tube with minimum modification to the existing infrastructure of the plant. In both cases the amount of CO produced was around 20 % (v/v of the dry product gas). This shows that the catalyst exhibits limited water gas shift capability. This aspect has been included in the scope of future work wherein the catalyst composition can be varied to enhance water gas shift reaction also. The amount of hydrogen produced was nearly 70 % (v/v of the dry product gas) in both cases. Thus PS–CAT is a very promising substitute for packed bed reactors for production of syngas and hydrogen which can be used for manufacture of fertilisers.

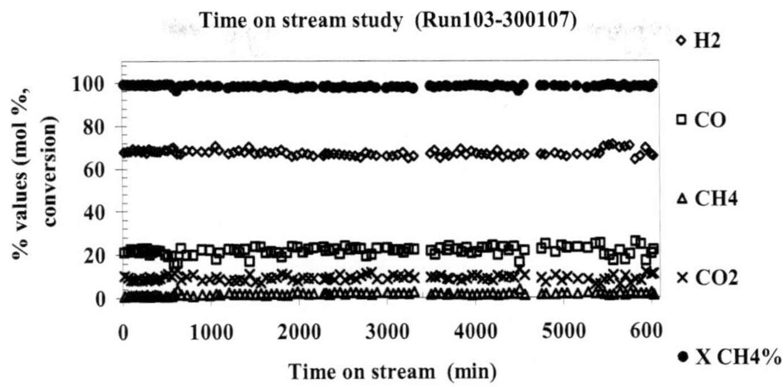

Fig. 7. Product gas composition of the exit gases. Other conditions PS−CAT 4 channels, 5 blocks in series, feed to steam molar ratio of 1:3, W/F = 56 gm−cat min/mol, pressure = 1 atm and temperature 1048 K, X_{CH4} = 98.27%, Yield 3.4 mol of H_2/mol of CH_4 reacted.

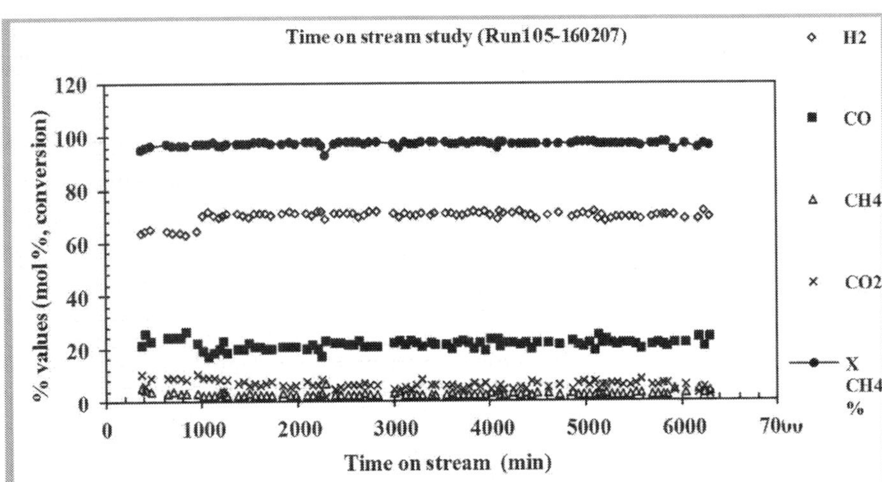

Fig. 8. Product gas composition of the exit gases. Other conditions PS−CAT 49 channels (7×7) channels, 2 blocks in series, feed to steam molar ratio of 1:3, W/F = 56 gm−cat min/mol, pressure = 1 atm, T =1113 K, X_{CH4} = 98.27%, Yield 3.14 mol of H_2/mol of CH_4 reacted.

5.3 Effect of space time

The effect of space time was studied by varying the W/F_{CH4} ratio from 10 to 55.0 gm-cat min/mol for three temperatures namely 873, 973 and 1048 K at a molar ratio of steam to methane of 3.0. Figure 9 shows the effect of space time on methane conversion. The

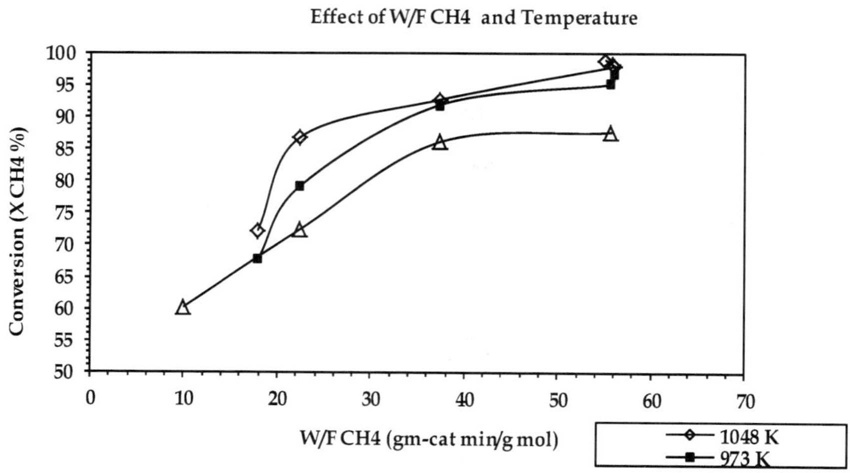

Fig. 9. Effect of space time on methane conversion. Other conditions: PS–CAT 4 channel block, 5 blocks in series, steam to hydrocarbon molar ratio in feed = 3.0.

conversion of methane increases with increase in W/F_{CH4} ratio. The catalyst exhibits a conversion of more than 90% for a W/F_{CH4} ratio above 37.3 gm-cat min/mol for temperature above 923 K.

5.4 Comparison of performance with a commercial catalyst

As mentioned in the literature review section, to the best of our knowledge there are hardly any references (in literature) to the use of meso-scale, ceramic, monolith channels specifically for steam reforming of methane (heat provided external to the reformer tube). Thus direct comparison with other data was not possible.

In order to compare the performance of the developed catalyst, experimental runs were conducted using a commercial nickel containing catalyst in a tubular fixed bed reactor of inner diameter 6 mm and 3 mm. In case of the 3 mm and 6 mm reactors the experiment was conducted maintaining the ratio of the length of the catalyst bed to the diameter of the reactor to enable a common basis of comparison between the two cases (Hoang and Chan, 2006). This ratio was maintained at 10. Table 1 gives the details of the process parameters used to compare the performance of PS–CAT with a commercial catalyst. The table also gives the composition of the catalyst determined using EDS-X-ray analysis. Swamy, 2007 has reported a detailed kinetic study of steam reforming of methane using the same catalyst. The commercial catalyst was crushed, sieved to obtain the required size of particles (approx 1.0 mm). The required weight of catalyst (0.1 gm) was filled in the tubular packed bed reactor of diameter 3 mm or 6 mm such that the centre of the catalyst bed corresponded to the central heating zone of the furnace. The remainder of the reactor was filled with inert ceramic material. The reactor was mounted in the furnace assembly. The catalyst was activated in a hydrogen flow of 40 N ml/min with a steady increase of temperature. The

Composition of active components		Structure	Weight	Residence Time	Molar Ratio	Temp.	Conversion of Methane
			gm	gm cat min / mol	Steam to Methane	K	(%)
Commercial Catalyst							
compound	wt– %	Particles of average size of 1 mm. Packed bed reactor used	0.1	56.00	3	1048	**79.73**
Ni	25						
Al_2O_3	15						
MgO	55						
PS-CAT							
compound	wt– %	Meso Scale Channel dimension 1.5 x 1.5 mm 100 cpsi 4 channels used, 5 blocks of 25 mm length used in series	0.1 (coating only)	56.00	3	1048	**98.27**
NiO	8.5						

Table 1. Process parameters used to compare the performance of PS–CAT with a commercial catalyst.

catalyst was maintained at the desired reaction temperature for two hours or till steady state conditions were achieved which can be determined from the temperature of the bed and the reactor skin temperature (outer wall of the reactor). At identical conditions (feed to steam molar ratio of 1:3 at a W/F = 56 gm–cat min/mol , pressure = 1 atm and temperature 1048 K), when a commercial nickel–based steam reforming catalyst was tested for steam reforming of methane in a 6 mm packed bed reactor the corresponding conversion obtained was 79.7 %. The conversion of methane using PS–CAT at identical conditions was found to be 98.27 % exhibiting a massive increase in the conversion of methane. Figure 10 shows the effect of temperature on methane conversion while comparing the performance of PS–CAT with commercial catalyst at identical process conditions. Figure 11 compares the performance of PS–CAT with commercial catalyst exhibiting the effect of W/F_{CH4} on methane conversion. In both cases, PS–CAT outperforms the commercial catalyst.

The reactor exhibited a minimum increase of 10% and a maximum increase of 22% in the conversion of methane that directly translates into increased production of syngas and hydrogen with existing plant infrastructure. Thus PS-CAT is a very promising substitute for packed bed reactors for production of syngas and hydrogen which can be used for manufacture of fertilisers. The better contact between the reactants and the active catalytic ingredient in case of the use of the structured catalyst shows a substantial increase in the performance in comparison to a fixed bed catalyst consisting of randomly packed catalyst. In the case of the 3 mm and 6 mm diameter fixed bed reactors, the conversion of methane in the 3 mm reactor was higher than the 6 mm reactor. This is in agreement with principles of use micro or meso–scale devices (Lowe, 2000). Decreasing the linear dimensions, for a given

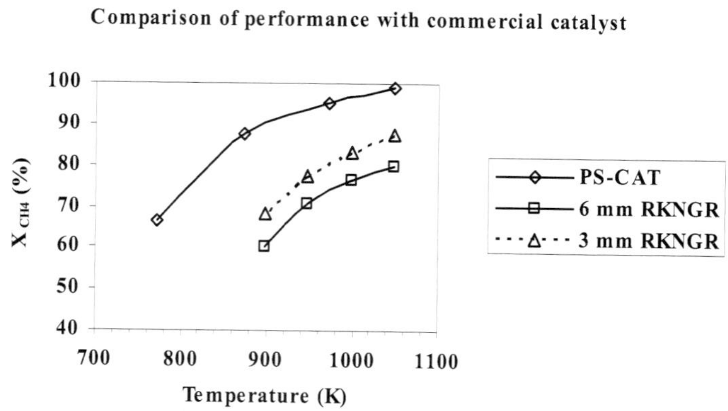

Fig. 10. Comparison of performance of PS−CAT with commercial catalyst, effect of temperature on methane conversion. Other conditions: PS−CAT 4 channel block, 5 blocks in series, W/F_{CH4} = 56 gm−cat min/mol, steam to hydrocarbon molar ratio in feed = 3.0.

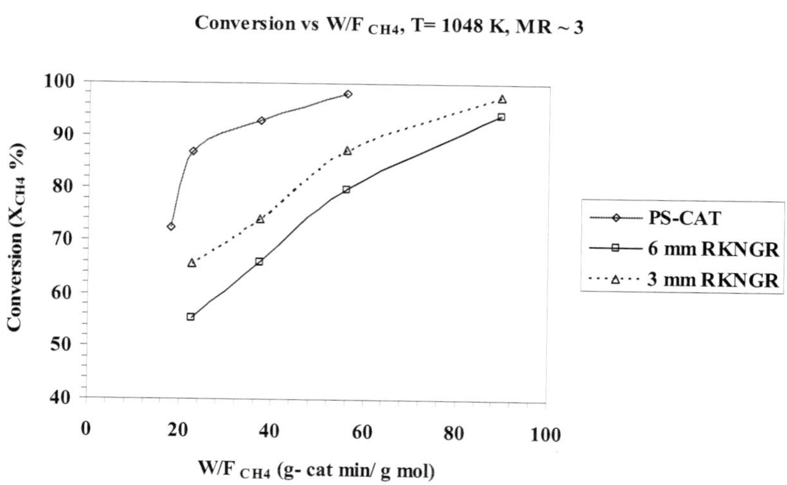

Fig. 11. Comparison of performance of PS−CAT with commercial catalyst, effect of W/F_{CH4} on methane conversion. Other conditions: PS−CAT 4 channel block, 5 blocks in series, T= 1048 K, steam to hydrocarbon molar ratio in feed = 3.0.

difference in a physical property (like temperature or concentration) increases the respective gradient. Consequently the driving forces for heat transfer, mass transport or diffusion flux per unit volume or unit area increase.

Literature reports the effect of changing diameter of a membrane reactor system on methane conversion for the partial oxidation process (Hoang and Chan, 2006). The authors have reported an increase in methane conversion with decrease in reactor diameter (at the same length to diameter ratio). Even though the system reported is different from this study the reasoning for the increase in methane conversion for smaller reactor diameters has a similar basis and can explain the increase in conversion for the 3 mm reactor.

6. Conclusions

A novel, meso-scale, structured, channel reactor was developed and successfully tested for steam reforming of methane. The structured catalyst called PS-CAT is a cordierite based monolithic structure coated with nickel oxide. The coating procedure was finalised on the basis of a set trials. The characterisation was done by EDS-X-ray analysis, BET surface area measurement and XRD analysis. The testing was carried out by varying the process parameters such as time-on-stream analysis (100 hours on stream), temperature (673 to 1048 K, space time (W/F_{CH4} = 10 to 55.0 gm-cat min/mol), and the number of channels (4 channels and 49 channels *in parallel*). PS-CAT was tested for steam reforming of methane at a steam to methane molar ratio of 3:1 at atmospheric pressure and temperature of 1048 K for over 100 hours of operation.

The conversion of methane observed with the said conditions was above 97 %. The catalyst exhibited good conversion (above 90%) of methane for a molar ratio of steam to methane from 2.5 to 7. The catalyst showed a conversion of more than 90% for a W/F_{CH4} ratio above 37.3 gm-cat min/mol for temperature above 923 K. Maximum hydrogen yield of 3.8 mol of H_2/mol of CH_4 reacted was observed at a W/F_{CH4} of 55.1 gm-cat min/mol at a temperature of 1048 K and molar ratio of steam to methane of 3.0.

Finally the performance of this developed catalyst was compared with the performance of commercial nickel-based catalyst in small diameter (3 mm and 6 mm) fixed bed reactors. The catalyst exhibited a minimum increase of 10% and a maximum increase of 22% in the conversion of methane that directly translates into increased production of syngas and hydrogen with existing plant infrastructure.

The main advantages of use of PS-CAT are increased throughput, simplicity of the arrangement, and minimum plant modification. Thus PS-CAT is a very promising substitute for packed bed reactors for production of syngas and hydrogen which can be used for manufacture of fertilisers. This case study also emphatically demonstrates the application of meso-scale catalytic structures for bulk scale production of commodity gases such as syngas

7. Nomenclature

F_{CH4}	=	molar flow rate of methane, mol/min
F_{H2O}	=	molar flow rate of water, mol/min
Q_{CH4}	=	volumetric flow rate of methane, N ml/min
Q_{H2O}	=	volumetric flow rate of water, ml/min
T	=	temperature, K
v/v	=	composition of gas species expressed as volume percent of the total product gas %
W	=	weight of catalyst, gm
W/F	=	ratio of weight of catalyst to molar flow rate of feed (space time), gm−cat min/mol
X_{CH4}	=	fractional conversion of methane

8. Abbreviations

BET–SSA	=	Brunauer–Emmett–Teller specific surface area method
cpsi	=	cells per square inch
EDS X-ray	=	energy dispersive spectroscopic X–ray analysis
GC–TCD	=	thermal conductivity detector of gas chromatograph
MR	=	molar ratio of steam to feed hydrocarbon
XRD	=	X–ray powder diffraction
WGS	=	water–gas–shift
DRIE	=	deep reactive ion etching
LTS	=	low temperature shift
PSA	=	pressure swing adsorption

9. Acknowledgments

I wish to acknowledge the Department of Fertilizers, Ministry of Chemicals and Fertilizers, Government of India, for providing the financial assistance for the execution of the project titled, *"Studies on reforming of methane to synthesis gas using micro–reactors for production of hydrogen (MPH)"*, sanctioned *vide* letter - 15011/1/2004–FP, dated 05/03/2004.

I wish to thank the Council of Scientific and Industrial Research (CSIR) for the award of Senior Research Fellowship (Individual) sanctioned *vide* letter 9/81(624)/07 EMRI dated 09/03/2007 from April 2007 onwards.

I would like to thank my co-authors & my ex-colleague Prof. Anand V. Patwardhan, Prof. Ranajit K. Saha & Shouvik Ganguly for their invaluable insights and unwavering support.

I would like thank my sister & my brother-in-law Tejas & Sameer Oundhakar for providing financial assistance in the publication of this book chapter.

10. References

Bobrova, L., Zolotarskii, I., Sadykov, V., Pavlova, S., Snegurenko, O., Tikhov, S., Korotkich, V., Kuznetsova, T., Sobyanin, V. and Parmon, V., Syngas formation by selective catalytic oxidation of liquid hydrocarbons in a short contact time adiabatic reactor, *Chemical Engineering Journal*, (2005), *107*, 171–179.

Das, D., Veziroglu, T. N., Hydrogen production by biological processes: A survey of literature, *International Journal of Hydrogen Energy* (2001), *26*, 13–28.

Giroux, T., Hwang, S., Liu, Y., Ruettinger, W. and Shore, L., Monolithic structures as alternatives to particulate catalysts for the reforming of hydrocarbons for hydrogen generation, *Applied Catalysis B: Environmental*, (2005), *56*, 95–110.

Harayanto, A., Fernando, S., Murali, N., Adhikari, S., Current status of hydrogen production techniques by steam reforming of ethanol: A review, *Energy and Fuels* (2005), *19*, 2098–2106.

Hessel, V., Lowe, H., Muller, A. and Kolb, G., *Chemical Micro Process Engineering – Processing and Plants*, Wiley–VCH, Germany (2005).

Hoang D.L., Chan S.H., Effect of reactor dimensions on the performance of an O_2 pump integrated partial oxidation reformer–a modeling approach, *International Journal of Hydrogen Energy*, (2006), *31*, 1–12.

Holladay, J.D., Wang, Y., Jones, E., Review of developments in portable hydrogen using microreactor technology *Chemical Reviews* (2004), *104* , 4767–4790.

Hwang, H., Heck, R.M., Yarrington, R. M., Fuel cell electric power production *United States Patent 4,522,894*, (1985)

Jiang, W., Park, S., Tomczak, M. S., Acharya, D. R., Tamhankar, S. S., Ramachandran R., Monolith based catalytic partial oxidation process for syngas production *United States Patent 7,090,826*, (2006)

Lindström, B., Agrell, J., Pettersson, L.J., Combined methanol reforming for hydrogen generation over monolithic catalysts, *Chemical Engineering Journal*, (2003), *93* ,91–101

Lowe H. V., *Microreactors New technology for modern chemistry*, Wiley–VCH, Germany (2000).

Lowe, H. V., *Microreactors New technology for modern chemistry*, Wiley–VCH, Verlag GmbH, Weinheim, (2000).

Mathure P.V., Ganguly S., Patwardhan A.V. and Saha R.K., Steam reforming of ethanol using a commercial Nickel-based catalyst, *Industrial and Engineering Chemistry Research* (2007b), *46*, 8471-8479.

Mathure P.V., Swamy B., Patwardhan A.V., Saha R.K., Ganguly S., A structured catalyst for steam reforming of methane for production of syngas *Application filed in the name of I.I.T., Kharagpur –Indian Patent Application No. 1515/KOL/2007* dated 02/11/2007a.

Mathure, P.V., Studies on steam reforming for the production of hydrogen, Ph.D. Thesis, I.I.T., Kharagpur, India (2008).

Meille, V., Review on methods to deposit catalysts on structured surfaces, *Applied Catalysis A: General*, (2006), *315*, 1–17.

Momirlan, M., Veziroglu, T.N., Current status of hydrogen energy, *Renewable and Sustainable Energy Reviews* (2002), *6*, 141–179.

Navarro, R. M., Pena, M. A., Fierro, J. L. G., Hydrogen production reactions from carbon feedstocks: fossil fuels and biomass, *Chemical Reviews* (2007), 107, 3952–3991.

Pasel, J., Cremer, P., Wegner, B., Peters, R. and Stolten, D., Combination of autothermal reforming with water–gas–shift reaction - small scale testing of different water–gas–shift catalysts, *Journal of Power Sources*, (2004), *126*, 112–118.

Pattekar, A.V., Kothare, M.V., A microreactor for hydrogen production in micro fuels cell applications, *Journal of Microelectromechanical Systems* (2004), *13*, 7–18.

Qia, A., Wang, S., Fub, G., Nib C. and Wub, D., La-Ce-Ni-O monolithic perovskite catalysts potential for gasoline autothermal reforming system", *Applied Catalysis A: General*, (2005), *281*,233–246.

Rostrup–Neilsen, J.R., Catalytic Steam Reforming in *Catalysis, Science and Technology, vol.5*, ed. Anderson, J. R. and Boudart, M., Springer–Verlag, Berlin (1984).

Swamy, B., *Studies on methane steam reforming for production of hydrogen*, Master's Thesis, I.I.T., Kharagpur, India (2007).

Tomasic, V., Jovic, F., State-of-the-art in the monolithic catalysts/reactors, *Applied Catalysis A: General* (2006), *311*, 112–121.

Vaidya, P.D., Rodrigues, A.E., Insight into steam reforming of ethanol to produce hydrogen for fuel cells, *Chemical Engineering Journal* (2006), *117*, 39–49.

Vergunst, T., *Carbon coated monolithic catalysts preparation aspects and three phase hydrogenation of cinnamaldehyde*, PhD Thesis, Technical University, Delft, The Nethelands (1999).

Zhou, S., Yuan, Z. and Wang, S., Selective CO oxidation with real methanol reformate over monolithic Pt group catalysts: PEMFC applications, *International Journal of Hydrogen Energy,* (2006), *31,* 924 - 933.

17

Innovative Technologies for Natural Gas Utilization in Power Generation

Victor M. Maslennikov, Vyacheslav M. Batenin, Yury A. Vyskubenko and Victor Ja. Shterenberg
Joint Institute for High Temperatures of the Russian Academy of Sciences, Russian Federation

1. Introduction

Numerous publications are dedicated to the development of traditional power engineering and to the possibility for essentially new power generation methods to appear. But if we shell a try to make detailed analysis of this issue for near perspective, the imagination has to be limited substantially. First of all, these limitations are connected with the great scale of the problem.

Appearance of a new technology in field of power generation, apparently may be considered as real, only when the first demonstration plant will comes into service or when the way of the transition from ideas and experiments to creation of the new plant is clear , together with needed time and resources. According to the progress in the last 40 years, a few of the principled new power production technologies have been realized [1,3]. Here, gas-turbine and combined cycle steam-gas technologies are the best examples. For the most parts progress consists of step-by-step enhancement and parameter increase, that had resulted in the qualitative growth of technical and economic indicators. In general, power engineering remains a very conservative field.

The second boundary condition of the origination of a new power production technology is the presence of the necessary volume of primary resources. However, the excess some of these resources can slow down the development of new technologies, based on the other primary resources.

Finally, with the start of XXI century, power generation development is more and more coordinated by the world community, which connects the increase of fossil based power production with the possible global climate change.

In respect to Russia, these boundary conditions are added by internal problems of new economic structure formation, which take rather long period of solution, according to the experience of the last two decades.

The estimations of organic fuel resources give their inevitable depletion in the foreseeable future. However, the exact figures are under permanent correction. Crude oil resources, apparently, are minimal even with taking into account the volumes which are difficult to extract and the development of new methods of extractions. Concerning natural gas, the estimations are more optimistic. The resources of natural gas assured and perspective for recovery grow (methane of coal beds and shale gas, for example). With respect to possibility of gas hydrates utilization, the reserves of gas seem to be practically unlimited for power production in the oncoming century.

The above mentioned is also true for the reserves of coal. Thereby, natural gas and coal are the most important primary energy resources.

With this, analyzing the perspectives of all power production technologies based on fossil fuels, it is worth taking into account the following:

- The problem of transportation is very actual because the gas and coal fields in most cases are distant from the areas of electric energy consumption;
- Transformation efficiency increase connected not only with fuel saving, but also with the reduction of greenhouse gases emission;
- Mono-production of electric energy does not correspond with long-term energy strategy;
- Even at keeping the present level of electric power production in Russia up to 2030, to compensate the retirement of power of thermal power plants only, it is necessary to put new 7 - 8 million of Kw/year in operation, where 4 - 5 million of Kw/year using natural gas, mainly at high capacity steam-gas plants.

Taking into account that not less than 10 years are to pass from draft technical proposal of new power production technology to the creation of big-scale demonstration plant, the problem of advance thermal efficiency at power production, natural gas first of all, should be solved in two directions:

1. By the upgrade at existing power plants, as a first stage;
2. The creation of principally new high efficient ecologically clean power units to replace the worked-out power units and for further development of fossil fuel power production.

Below is the description of several new technologies for natural gas utilization developed in Joint Institute for High Temperatures of Russian Academy of Sciences (JIHT RAS) in both areas.

2. Efficiency increase of existing power production facilities

2.1 Repowering of the existing water-heating boiler

Repowering of the existing gas fueled water-heating boiler by adding of gas-turbine units with its outlet to the boiler combustor for the first time was proposed by group of JIHT RAS specialists headed by Academician Mikhail Styrikovich in the 80th. At present this proposal is recognized and accepted for implementation, however big discrepancy exists in the reading of its technical essence, up to absurd (creation of gas-turbine units with exhaust-heat boilers at boiler facilities). The essence of the proposal is in the following: products of combustion from gas-turbine unit, containing 17% of free oxygen are fed to boiler combustor (Fig.1).

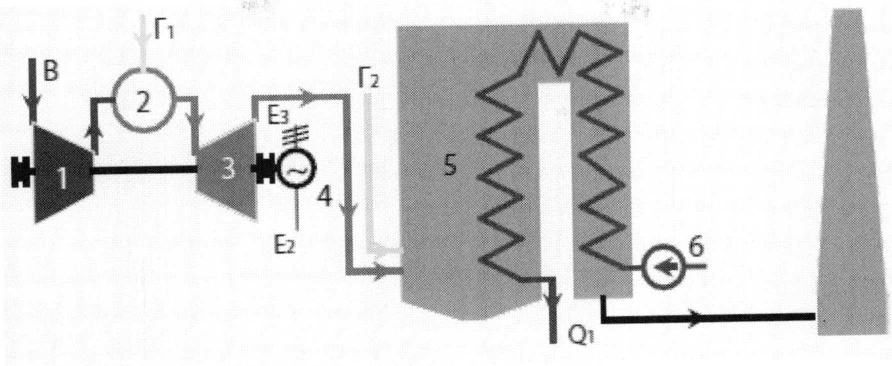

B – air, Γ1 – gas to GT combustor, Γ2 – gas to boiler, 1 – compressor, 2 – combustion chamber, 3 – gas turbine, 4 – generator, 5 – boiler, 6 – feed pump.

Fig. 1. Water boiler with gas-turbine buildup.

With this, at minimal thermal load regime (20%) the heat is generated due to the cooling of the gas turbine unit products of combustion. At the increase of thermal load (up to nominal), additional fuel is after-burned in the boiler in the flow of products of combustion due to free oxygen combustion. With this, the efficiency of electric power production changes from 60% (summer regime) to 90% (winter regime). The efficiency is defined as:

$$\eta_e = \frac{N_e}{H_T - \frac{Q_m}{\eta_b}} \tag{1}$$

Water heating boilers of 100 Gcal/h capacity could be built-up by gas-turbine units of 16-20 MW (defined by the products of combustion flowrate). Only in Moscow at district heating units there are about 100 of such boilers.

The described technology has one peculiarity, namely, the gas flow rate to boiler combustor changes in a wide range at near constant flow rate of oxidant, represented by the gas-turbine products of combustion. This can be implemented by step-by-step some burners switch off or by the installation of special burners. The technology can be applied without additional research at test units and the cost of electricity generated is lower than at perspective steam-gas plants.

2.2 Repowering of the existing gas-fired steam turbine units

There are some variants to retrofit the steam turbine units with using gas turbines. A few obstacles need to be overcome for these possibilities to be namely:

1. the extent, to which the equipment of the power plant to be retrofitted is worn out, needs to be taken into consideration, so that its remaining service life would not differ too much from the service life of the newly installed equipment;
2. free space must be available for installing new equipment at the existing power plant;
3. it is necessary that the newly installed equipment should not bring about a substantial reduction in the capacity and efficiency of the existing basic power-generating equipment; and
4. the new process scheme should not result in the loss of reliability of the object being retrofitted; in the worst case, it should not increase the environmental impact on the region, while in the best case it should considerably relieve this impact.

The technical offer made by OIVTAN was aimed at solving this actual problem and involved the retrofitting of existing, relatively new steam turbine units by way of integrating gas turbines into these units and employing an original technology referred to as "partial oxidation" [2], [4], [5]. This technology essentially consists in that the natural gas, utilized by the steam-turbine unit, is preconverted in the combustors of the gas-turbine unit to carbon monoxide and hydrogen, and the resultant fuel gas expands in the gas turbine and then burned in the boiler furnace.

2.2.1 List of the units under consideration

The investigation, presented in this chapter, deals with a comparative feasibility study of the most representative alternative options. The main items under comparison are:

- thermal efficiency additional power production,
- relative cost of electricity in base mode operation.

Five alternative options have been studied:

1. A conventional 315 MW condensing steam-turbine unit fired with gas and fuel oil, for the steam parameters of 24 MPa and 540/540 oC. (It was used as a standard for comparison).
2. A 150 MW binary-cycle steam-gas unit (SGU) incorporating two W251B12 gas-turbine units by Westinghouse and a 51 MW steam turbine.
3. Gas-turbine topping with partial oxidation process.
   a. 7 variants with Aircraft Engines.
   b. 4 variants with Heavy-duty GTU.
4. Topping with dumping of GTU gas into the boiler furnace (Hot-Windbox).
   a. 4 variants with Aircraft Engines.
   b. 2 variants with Heavy-duty GTU.
5. Topping with GTU combined with STU in a single steam-generating circuit.

Standard steam-turbine unit

Units of this type were commissioned widely since the 1970s. Following the period of debugging, they have exhibited high reliability and efficiency and are at present employed most extensively in the energy systems of Russia and republics of the former USSR. The present-day characteristics of this unit are listed in the following table 1.

Generator power, MW	317.9
Steam parameters:	
-At turbine inlet	
Flow rate, kg/s	276.7
Pressure, kg/cm2	240.0
Temperature, oC	540.0
-At reheater outlet	
Flow rate, kg/s	221.3
Pressure, kg/cm2	38.6
Temperature, oC	540.0
-At condenser inlet	
Flow rate, kg/s	169.7
Pressure, kg/cm2	0.035
3. Feedwater temperature, oC	280.8
4. Pressure in deaerator, kg/cm2	7.0
5. Number of regenerative feedwater heaters	
- low pressure	4
- high pressure	3
6. Feedpump turbodrive power, MW	11.9
7. Specific heat consumption, KJ/KWh	8055
8. Efficiency of boiler , %	94.5
9. Unit net efficiency, %	40.6

Table 1. The main characteristics of standard steam-turbine unit.

2.2.2 An existing aircraft engine used as a topping unit in partial oxidation scheme

The schematic diagram is given in Fig. 2.

The existing engine in this particular case is used as a "gas generator". The combustion products containing up to 17% free oxygen are passed from the gas turbine exhaust to a special converter (11), where natural gas is fed in excess.

In the converter (11), natural gas at a temperature of about 1100 oC (possibly in the presence of a catalyst), is subjected to partial oxidation to hydrogen and carbon monoxide. The products of partial oxidation are expanded in the power gas turbine (3) and discharged as fuel to the upper tiers of the steam boiler. (The lower tier of the boiler may be operated on the former fuel).

Steam extracted from the STU exhaust to the GTU can be used in the topping for the following purposes:

- Cooling the power gas turbine blades;
- Increasing power output of the gas turbine (being fed into the combustion chamber of the aircraft engine (2);
- Decreasing the soot formation (being fed into the converter(11).

In order to enhance the efficiency some part of the combustion products may by-pass and fed to the additionally installed boiler feedwater preheaters. In doing so, the steam extracted from the steam turbine exhaust for feedwater preheating is changed to optimize thermal circuit.

The main characteristics of two variants of this unit are listed in the table 2.

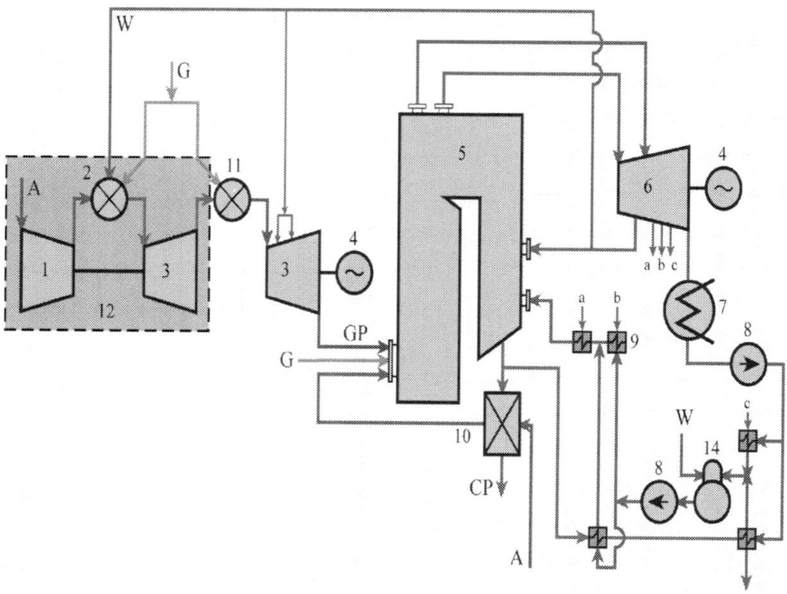

1 – compressor, 2 - combustion chamber, 3 - gas turbine, 4 - electric generator, 5 - steam boiler, 6 -steam turbine, 7 – steam condenser, 8 - feed pump, 9 - steam regeneration, 10 - air heater, 11 – partial oxidation chamber, 12 - gas generator (AGTE block), 13 - deaerator.
A -air, G -natural gas, CP - combustion products, GP - natural gas conversion products, W - water, steam

Fig. 2. Repowering by addition of topping SGU based on aircraft gas turbine with using partial oxidation.

Steam injection into GTU, kg/s	**0.0**		**19.0**	
MODES	1	2	1	2
Main characteristics				
1. GTU total power, MW	38.2	38.2	74.5	74.5
2. Additional gas flow to the boiler, kg/s	7.78	7.35	8.95	8.37
as % of the total flow	48.1	46.7	53.2	51.5
3. Efficiency of boiler , %				
- boiler per se	91.9	91.7	89.9	89.7
- system comprising boiler+gas regenerator	94.2	94.1	93.4	93.3
4. SGU total power, MW	320.4	311.1	300.4	288.8
5. Parasitics, MW	9.6	9.6	9.6	9.6
6. Additional useful power of the unit, MW	41.1	31.8	57.4	45.9
7. Total fuel consumption by the unit, kg/s	16.18	15.76	16.83	16.25
8. Efficiency of additional el. power production, %	69.0	82.6	62.5	72.8

Mode 1 - With gas regeneration. **Ne add = Nmax.**
Mode 2 - With gas regeneration. **Efficienci = Max.**

Table 2. The main characteristics of existing Aircraft Engine used as a topping unit in partial oxidation scheme.

2.2.3 A gas turbine with discharge of combustion products into the boiler furnace for repowering

This scheme (Fig. 3) is well established and it is mentioned here as alternative, the comparison being carried out for the same initial assumptions. In distinction from fig 1 addition topping gas turbine to steam generation block.

The combustion products containing up to 17% free oxygen are discharged to the boiler burners where they are used as an oxidizer of the additionally fed fuel. To control the boiler steam production rate, additional amounts of air and fuel are added to some burners.

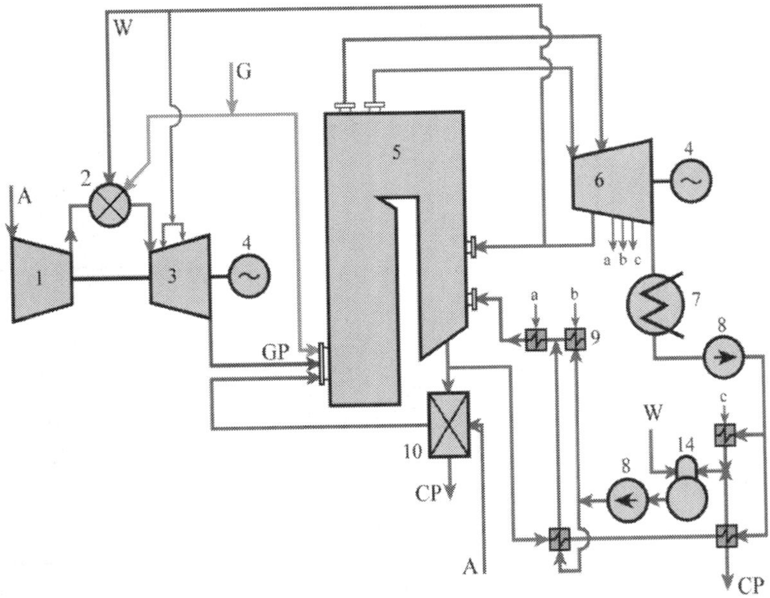

1 – compressor, 2 - combustion chamber, 3 - gas turbine, 4 - electric generator, 5 - steam boiler, 6 -steam turbine, 7 – steam condenser, 8 - feed pump, 9 - steam regeneration, 10 - air heater, 11 – deaerator. A -air, G -natural gas, CP - combustion products, GP - natural gas conversion products, W - water, steam

Fig. 3. Repowering by addition of topping SGU, based on Westinghouse GTU.

Since the air flow through the air preheater (10) decreases significantly, part of the combustion products is removed via a by-pass for heating feedwater in parallel with the steam regeneration system. The flow rate of the by-passed combustion products in this unit is much higher, then in units based on partial oxidation and, by consequence, a larger fraction of the steam regeneration is forced out, which leads either to a loss of efficiency, or to a decrease in power output of the steam turbine. In the comparative thermodynamic analysis given hereafter we shall analyze the effect of amount of the combustion products discharged into the boiler downstream of the GTU as well as changes in the steam regeneration system on the efficiency of additional production of electric power and change in the steam turbine power output.

2.2.4 GTU integrated with Steam Turbine Unit in a single steam generating circuit

The schematic diagram of this option is given in Fig. 4. In this case the steam and the GTUs are operated to a large extent independently. However, the heat of combustion products downstream of the GTU is utilized for heating feedwater of the STU with partial forcing out the steam regeneration. Naturally, the efficiency of this scheme is lower than that of the above options, nevertheless it exhibits a number of important advantages:

- The problem of GTU location is simplified greatly.
- There is no need in revamping the major boiler components.

The main characteristics of both units are listed in the table 3.

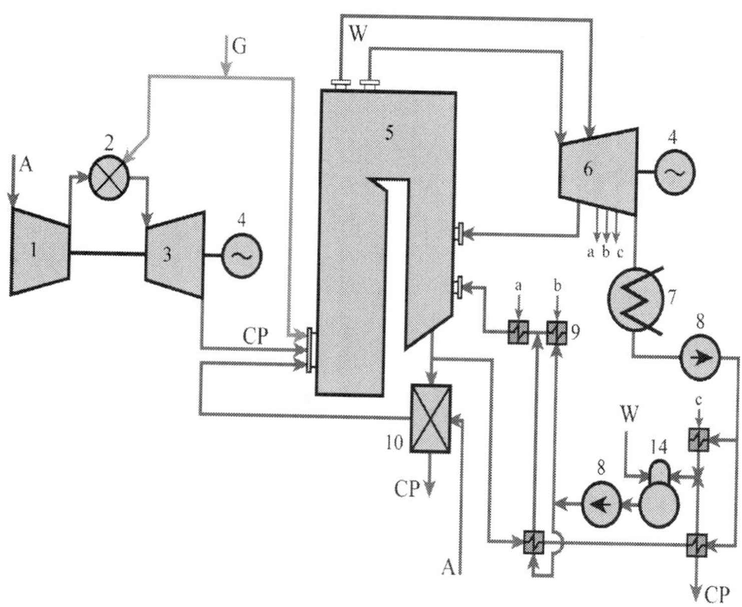

1 - compressor, 2 - combustion chamber, 3 - gas turbine, 4 - electric generator, 5 - steam boiler, 6 -steam turbine, 7 – steam condenser, 8 - feed pump, 9 - steam regeneration, 10 - air heater, 11 – deaerator. A -air, G -natural gas, CP - combustion products, GP - natural gas conversion products, W - water, steam

Fig. 4. Repowering by addition of GTU with combustion products exhaust to boiler.

Type of scheme	GTU with discharge of comb. products into the boiler furnace		GTU integrated with STU in a single steam generating circuit	
MODES	1	2	3	4
Main characteristics				
1. GTU total power, MW	48.5	48.5	48.5	48.5
2. Additional gas flow to the boiler, kg/s	14.0	12.6		
As % of the total flow	82.3	80.7		
3. Efficiency of boiler , %				
- boiler per se	88.9	87.4	94.5	94.4
- system comprising boiler+gas regenerator	88.9	93.6		
4. STU total power, MW	317.9	294.1	317.9	292.2
5. Parasitics, MW	10.1	10.1	10.4	10.4
6. Additional useful power of the unit, MW	48.5	24.7	48.2	22.5
7. Total fuel consumption by the unit, kg/s	17.0	15.59	17.99	15.86
8. Efficiency of additional electric power production, %	48.3	82.9	32.1	51.4

Mode 1 - Without gas regeneration.
Mode 2 - With gas regeneration. Maximum heat use in gas regeneration system.
Mode 3 - Separate location of GTU and STU.
Mode 4 - GTU and STU are integrated in a single steam generating circuit.
Maximum heat use in gas regeneration system.

Table 3. The main characteristics of gas turbine with discharge of combustion products into boiler and gas turbine integrated with steam turbine in a single steam-generating circuit.

2.2.5 Results of the thermodynamic analysis

The following characteristics of the various schemes were used for comparison:

1. The additional useful power output of the steam-gas unit, SGU, (Ne add, MW),
2. Efficiency of Generation of the Additional Electric Power (*Effadd*).

$$Eff_{ad} = (N_{SGU} - N_{STU}) / (\Sigma g_i \cdot Q_{iSGU} - \Sigma g_i \cdot Q_{iSTU}) \quad (2)$$

When a topping gas turbine unit is added to the STU quite natural desire is to introduce minimum changes into the flowsheet of the STU, and to retain the possibility of independent operation of the STU in case of shut-down of the GTU.

The efficiency of such repowering depends to a large extent on how successfully the following problems will be solved:

- Compatibility of the two units in flow rates of the working fluid, i.e. the possibility of passing a new volume of gases through the STU boiler so that no changes occur in the boiler working surfaces temperatures and the basic parameters of the boiler. In connection with the fact that the suggested Partial Oxidation Technology is of notably specific nature, in the calculation that will follow the "Xf" parameter will be introduced, equal to the ratio of fuel flow to the topping unit to the nominal fuel flow to the STU.

- Any one of the topping GTU causes reduction of air flow through the regenerative air heater (RAH), and by consequence additional heat losses with the flue gases. To make up for these losses some part of the flue gases upstream of the RAH may be used for heating feedwater in a special heat exchanger with corresponding forcing out the steam regeneration.
- In the topping GTU using partial oxidation the gas turbines cannot use air as coolant due to both explosion danger and cooling efficiency. Therefore, in all calculations steam extracted from the steam turbine was used for this purpose. The cooling steam flow was determined to be such as to maintain the same temperature level of the components being cooled.

When optimizing the technological scheme in certain cases it is advisable to feed additional extracted steam to the GTU; in doing so the GTU power is appreciably increased. Although the efficiency of additional electric power production in this case may be lowered the total fuel savings in the repowered unit may be larger.

In this work the amount of steam injected into the gas turbine varies as an independent variable. The diagram in Fig. 5 illustrate the effect of using various schemes of gas regenerative heating of feedwater and also different amounts of steam injected into the GTU on power and economic efficiency of different repowering topping schemes with partial oxidation based on use of the aircraft engine AL-31 GTU.

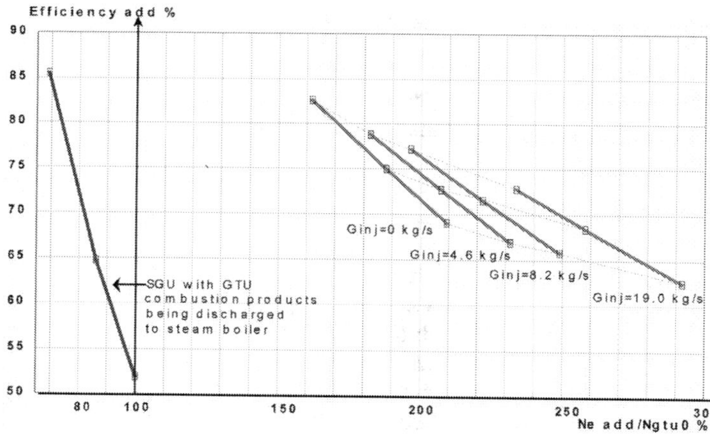

Fig. 5. Repowering by addition of AL-31 aircraft GT.

The additional power in the diagram is normalized in respect to NGTU0 -the power of the AL-31 GTU, operating in a conventional mode with one combustion chamber.

For comparison purposes the above diagrams show the indices of the scheme with the combustion products of the same aircraft GTU being discharged to the STU boiler.

2.2.6 Results of feasibility study analysis

Analysis made by OIVTRAN jointly with Mosenergoproyekt Power-Plant Design Institute reveals that the existing building of power plants may at best accommodate only one gas-turbine unit per steam-turbine plant. No acceptable technical solutions could be found involving two, to say nothing of three, gas-turbine units; in so doing, the capacity of gas turbine proper does not appear very critical (naturally, within reasonable limits).

The cost of power generation and the efficiency of capital investment are calculated with the following preconditions:

Δ_1 - the share of depreciation charges (from specific capital costs)
- standard STU = 0.07
- gas-turbine topping = 0.08

Δ_2- the share of maintenance repair deduction (from depreciation charges) = 0.2
Δ_3- the share of wage deduction (from specific capital cost) = 0.01
c- bank interest = 0.06
z- deductions from capital costs = 0.06
n- number of hours of operation in year = 6000
j- operational factor = 0.95

The calculations were performed for three values of the relative cost of fuel: low, medium and high. The results are given graphically in Fig. 6 (relative cost of electricity).

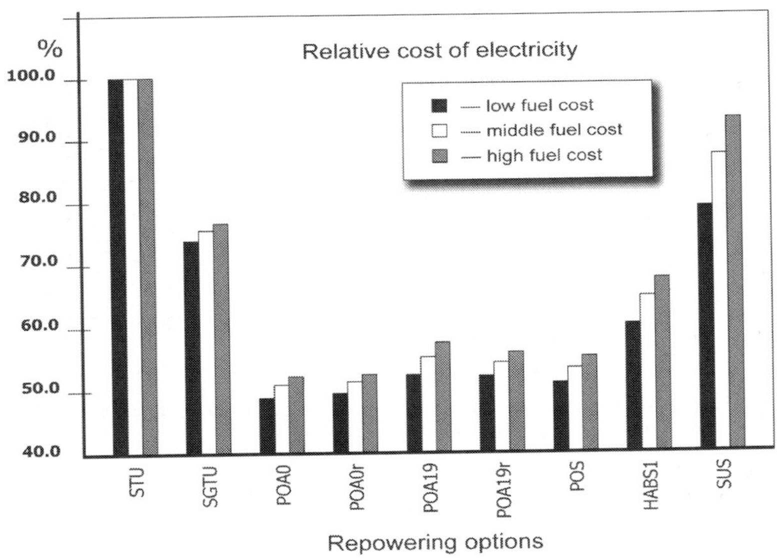

STU - conventional steam power unit, SGTU – steam-gas power unit
P0A0 - topping unit, based on aircraft engine, steam injection=0; without gas regeneration
P0A0r - topping unit, based on aircraft engine, steam injection=0; with gas regeneration
P0A19 - topping unit, based on aircraft engine, steam injection=19 kg/c; without gas regeneration
P0A19r - topping unit, based on aircraft engine, steam injection=19 kg/c; with gas regeneration
P0A19 - topping unit, based on stationary gas turbine engine
HABS1 - topping unit with GTU combustion products being discharged to steam boiler
SUS - topping unit with GTU combined with STU in a single steam-generating circuit.

Fig. 6. Relative cost of electricity.

2.2.7 Conclusions

1. Repowering the existing steam-turbine plants fired with gas and fuel oil with the aid of gas-turbine toppings is much more feasible technically and economically than the closest rival option, that of construction of advanced steam-gas plants.
2. Of all of the alternative technologies for repowering the best performance is offered by facilities with partial oxidation of fuel [5]. This technology permits of:
- raising the capacity of the gas-turbine unit employed by 30-50% for heavy-duty GTU and by a factor of two or two-and-a-half for aeroderivative GTU;
- reaching the efficiency of additional generation of electricity of 60 to 80%.
- reducing to a minimum the NOx emission with the stack gas of steam power plants [6,7];
- reducing the cost of additional generated electricity two times as compared with a steam-turbine unit.

3. The creation of new highly efficient power production technologies

3.1 Environmentally friendly combined-cycle unit for cogeneration

Steam-gas plants for electric power generation have undoubted advances compared to steam-turbine plants. When efficiency of modern condenser steam-turbine plants accounts for 38-39%, the same for perspective steam-gas plants reaches 55-60%. However, when speaking about cogeneration of heat and electricity, the advantages of steam-gas plant are not that evident, because the efficiency of electric power production at heat consumption and in the nominal regime become practically equal and heat production by steam-turbine unit is much more higher than in steam-gas one [9].

For the needs of characterization of the efficiency of heat producing units, the following terms are used:

- Fuel heat utilization factor (FHUF):

$$\text{FHUF} = \frac{N_e + Q_t}{H_t} \tag{3}$$

here FHUF - fuel heat utilization factor, Ne - useful electric power (MW), Qt - district heating power (MW), Ht - total energy of the consumed fuel.

- Efficiency of electric power production at heat consumption regime η_e:

$$\eta_e = \frac{N_e}{H_t - \dfrac{Q_t}{\eta_k}} \tag{4}$$

here η_e - efficiency of electric power production at heat consumption regime, η_k - efficiency of boiler of district heating facility.

The technology scheme of heat production steam-gas unit on the base of aviation AL-31 engine, produced by "Salut" Moscow Engineering Production Enterprise, was developed by JIHT RAS [10]. This unit has some advantages compared to traditional heating steam-turbines and steam-gas units. Short description and schematic diagram are given in Fig. 7.

Ambient air is compressed in the compressor НД (1) up to 6.3 kg/cm2 (abs) and fed in the mixing type air cooler (3), where the processes of saturation by steam and cooling down by evaporation take place. The use of air cooler gives the opportunity to keep the air temperature at the compressor high pressure (2) outlet at the rated level of initial gas-turbine engine (446 0C) and keep the normal mode of blade row operation.

Compressed air, natural gas and steam with temperature 285oC are fed into combustion chamber of steam-gas unit, providing the following parameters of the working fluid at the compressor common drive turbine inlet: pressure in 64 ata, temperature 1310 oC, at excess air factor α=1,12. Steam is used for the cooling of turbine group elements, thus the temperature of cooling agent can be increased over the design value for the original gas-turbine unit at keeping the temperature of blade row metal temperature below the design value.

At the outlet from the group of driving turbines, the steam-gas mixture is fed into the regenerative heat exchanger (6) where the generation and superheat of the injected steam take place together with the heating of network water for cogeneration purposes.

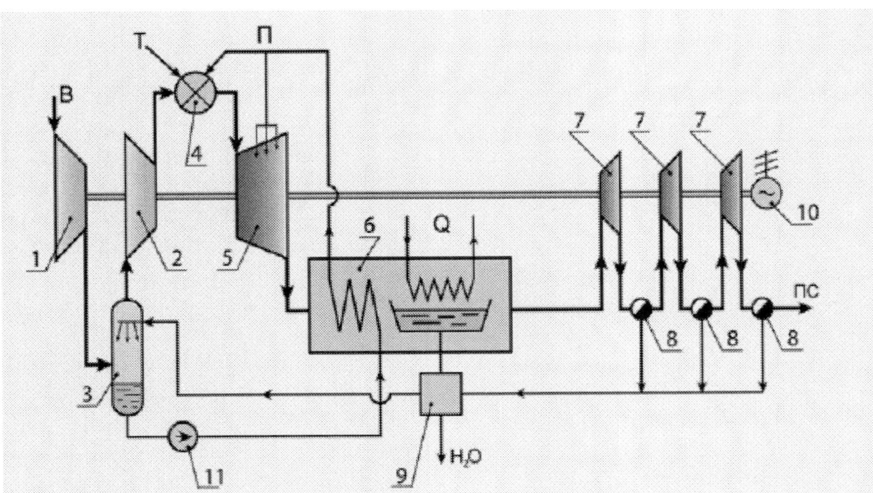

1 - low-pressure compressor, 2 - high-pressure compressor, 3 - air cooler, 4 - combustion chamber, 5 - gas turbine, 6 - heat exchanger, 7 - expander stage, 8 - drift eliminator, 9 - purifier, 10 - electric generator, 11 - feed pump, В - air, Т - natural gas, П - steam, ПС - products of combustion, Q - district heating system water

Fig. 7. The scheme of steam-gas unit for co-production of heat and electricity.

The main peculiarity of the proposed scheme is higher pressure in the heat exchanger (3.05 kg/cm2 (abs)) which increases partial pressure of water vapor and gives the opportunity to condensate it at the temperature level that fits the requirements of standard district heating system for network water.

After the heat exchanger (6) steam-gas mixture is expanded up to ambient pressure in the gas expansion stage (7) with the generation of useful work and additional water due to condensation. The additional water is captured by drift eliminator (8).

Together with the main flow of condensate from the heat exchanger (6) this water is fed to the purifier (9) and after that is fed to the air cooler (3) and after that to the steam generation and steam superheat in the heat exchanger (6). The excess water condensed from the combustion products can be used for any purpose.

The steam-gas unit based on the described scheme and using the equipment of АЛ-31 gas-turbine unit will have the following features (at nominal regime):

1. Useful electric power - 62.4 MW,
2. Useful thermal power - 77.8 MW,
3. Efficiency at heat consumption - 115.9%.
4. FHUF - 103.3%,
5. Air flow - 52.4 kg/s,
6. Condensate excess - 8.6 t/h.

The values of efficiency and FHUF exceed 100%, which is not a paradox and resulting from the term "the fuel lower calorific value", where the heat of condensation of steam generated at hydrogen fuel oxidation is not taken into account.

Since the products of combustion in the expansion machine are cooled below the dew point, the heat of condensation is transformed in work. The main parameters comparison for alternative heating units is given in Table 4.

Characteristics	Type of the cogeneration system			
	Steam-turbine	Combimed steam-gas plant		
	T-250/300-240	SGU 450T	SGU 230	Plant on fig.7
Net electric power, MW	231,9	433,8	213,6	62,4
Net heating power, MW	384,0	410,5	160,5	77,8
Fuel consumption, MJ/s	726,8	1033,1	460,2	135,7
FHUF	84,7	72,2	73,3	103,3
Efficiency at heat consumption	71,9	72,2	73,3	115,9

Table 4. The main parameters comparison for alternative heating units.

3.1.1 The main backgrounds of equipment selection and of the calculation of principle scheme of the unit

The main principle of the prototype selection for separate parts of the steam-gas unit and of the further calculation of thermal flow scheme is the priority of domestic aggregates and assemblies both, existing or under implementation by power engineering, i.e. the equipment without substantial difficulties concerning design and manufacture. As the main aggregate of the steam-gas unit of turbo -compressor group, the gas generator of АЛ-31 gas-turbine unit, produced by NPO "Saturn" is proposed.

The preliminary analysis of the proposed scheme shows the necessity to make the following changes in the design of the original gas-turbine unit:

1. the nominal air flow should be decreased from 63.5 kg/s down to 52.4 kg/s according to the requirement of keeping the maximal mechanical stress in the blade group of compressors within the design value;

2. the low pressure compressor must be added by a group of stages to increase the total compression rate;
3. the cooling air in the cooling system of gas turbine must be replaced by steam without any change of the design. In this case the value of steam flow rate in the cooling system is defined by initial steam pressure and by hydraulic resistance of the cooling system. Numerous literature data prove that at the transition to steam cooling, the temperature of the working fluid can be increased on 90-110 °C at the same temperature of blade group metal. In the proposed scheme, the initial temperature of the working fluid is 1310 °C, being in 53 °C higher than the design value for АЛ-31 unit. In this case, the temperature of metal in the most loaded parts of steam-gas units is lower than the design value, thus increasing the unit operation reliability.

The preliminary estimations done in cooperation with the group of designers of NPO "Saturn" showed that the air-gas channel is able to let through the design flow of working fluid; the power turbine must be redesigned due to the increase of last stage blades length. The use of power turbine from another NPO "Saturn" product could be the alternative.

The designs of mixing air cooler and of heat exchangers are based on the commercial elements and have been elaborated by the specialists of NPO "Raduga".

At the calculations within the design of the expansion machine, the average aerodynamics of low-pressure cylinder stage of condensing steam turbines was taken into account with respect to the decrease of air-gas channel efficiency due to drop precipitation.

The general principle used at calculations and design is the creation of the unit consisting of restrained number of commercial parts with the opportunity of transportation by standard railroad platform. The necessity to use intensified heat exchanges, because of this, positively influenced the mass and the cost of the equipment but decreased thermal efficiency.

4. Power and Chemical Complex for co-production of electricity and synthetic liquid fuel

Power and Chemical Complex for co-production of electricity and synthetic liquid fuels (methanol, dimethyl ether, gasoline) – a promising way for saving of fuel resources.

Up to now, the progress in any branch is connected with the upgrade of one product type production technology. The thematic example for this is the progress in power production connected with the modernization of steam-gas binary cycle utilizing natural gas as a fuel. Modern steam-gas units have 55 % efficiency with the perspective of its growth up to 60%. The upgrade is taking place both due to technological scheme optimization and by the increase of actuating medium parameters at the inlet of gas turbine.

However, this way practically has reached the limit and though the further increase of working fluid parameters can increase the efficiency, the cost of produced energy most likely cannot be decreased due to investment cost growth. To our opinion the better perspective in the utilization of fossil fuels and the natural gas first of all can be obtained by the creation of power and chemical complexes with technological scheme optimized for the mass of several end products [11,12].

Certainly, it is not possible to say that in all technologies only one end product have been manufactured up to now. The production of several types of products exists but in the way of by-product production with the main mono-product being the target for the technological scheme. The basic principle for the creation of power and chemical complex technological scheme is solving several target problems by one technological process producing synergy effect by the combination of several technologies.

With this, the positive feasibility effect is obtained by the following:

- high economy efficiency due to the synergy effect induced by integration of production of several type of mass products,
- the use of the most up-to-date technologies both newly developed or taken from various branches of industry,
- multiple decrease of ecological damage to the regions of the unit sites due to decrease or in some cases total elimination of harmful emission,
- involvement, together with traditional energy sources, such primary energy resources as low- natural gas, oil gases and coal-mine methane,
- the opportunity to create feasible energy-dependent medium and small capacity plants in case of need.

The essence of the considered energy and technology complexes is the partial oxidation reaction of the initial hydrocarbons with production of energy and synthesis gas, which includes of CO and H_2. The released heat can be used for power production. The CO/H_2 ratio in the synthesis gas and the presence of H_2O, CO_2, N_2 depends on the initial hydrocarbon composition and on the technology of partial oxidation reaction.

Thermodynamic analysis and the study of kinetics of the process resulting in synthesis gas origination show that the process occurs efficiently and quickly at relatively high temperatures without catalyst. The mathematical model developed in IVTAN helps to select the optimal parameters of the process, initial component composition and determine the time process duration needed for the reaction of partial oxidation .

The first synthesis gas generator has been created in IVTAN on the base of upgraded diesel engine, where the cylinder is a unique device combining the features of compressor, high temperature chemical reactor and thermal energy to mechanical work transformer [11,13].

Fig. 8 shows the appearance of Д-245 of 80 kW diesel-based synthesis gas generator.

Fig. 8. Д-245 diesel-based synthesis gas generator.

The generator utilizes natural gas or natural gas with butane-propane thus imitating the possible compositions of oil gases. The air is used as oxidant. Excess air factor, necessary for synthesis gas production accounts for 0,4-0,6 which is absolutely inappropriate for the regular diesel operation. The transformation of diesel to synthesis gas generator demanded additional efforts, which were successful.

The synthesis gas generator in combination with catalytic reactor of methanol synthesis (Fig. 9) received the name "Sintop-300" plant. The plant capacity is ~ 800 l of methanol per day. The plant was involved in the wide range of experimental investigations.

Fig. 9. "Sintop-300" plant.

The following are the main outcomes of these investigations:

- the opportunity to organize non-catalytic conversion of natural hydrocarbon gases by partial oxidation by air at the parameters that are technically achievable,
- the possibility for efficient catalytic methanol synthesis from synthesis gas ballasted by air nitrogen,
- the possibility to create feasible energy-dependent commercial low and middle capacity units for the conversion of gaseous hydrocarbons to liquid chemical products.

It is worth mentioning that the modified diesel based synthetic gas generator is able to operate with low-pressure gases, such as off-balance or worked-out gas fields, coal bed methane, oil gases. With the use of the obtained results in 2008, the pilot plant was designed with capacity of 12000 tons of methanol per year. The main equipment was selected or partly manufactured, namely synthesis gas generator (Fig. 10), compressor, heat exchangers etc. The generator was based on 16-piston H26/26 diesel engine with net capacity 2000 KW, produced by Kolomna mechanical engineering holding.

The further development of energy and technology complexes is connected with the development of continuous-flow generator variant, allowing the use of gas-turbine and thus to increase substantially the unit capacity and start the production of the units with capacities compared with modern level of power plants. The process flow sheet of such a perspective energy and technology complex is presented in Fig.11.

Fig. 10. Synthesis gas generator.

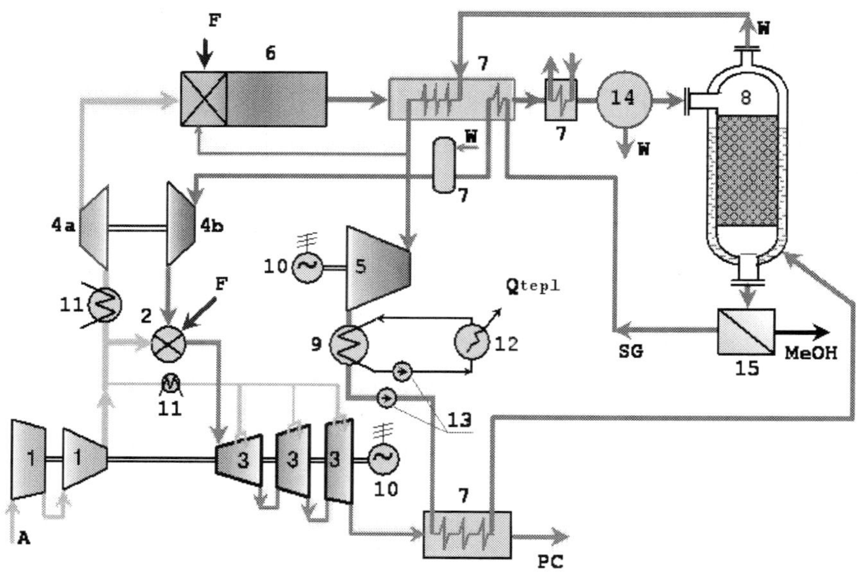

1 - compressor, 2 - the gas turbine combustor, 3 - gas turbine, 4a - air buster compressor, 4b - turbo-detander, 5 - steam turbine, 6 - partial oxidation chamber, 7- syngas cooler and waste-heat boiler, 8-synthesis CH_3OH reactor, 9 - steam condenser, 10 - electric generator, 11 - air cooler, 12 - water cooling tower, 14 - separator, 15 - product separator.

Fig. 11. Combined utilization of natural gas with the use of steam-gas turbine with the production of methanol, thermal and electric energy.

Conversion АЛ-31 aircraft engine is used as power unit. Compressed air after the compressor of the gas-turbine unit is additional compressed by compressor of the turbo-expander to 5.0 MPa and is fed to synthesis gas generator to where all the natural gas consumed by the plant is fed. The partial oxidation of natural gas is conducted in synthesis gas generator at temperatures 1200-1300 °C.

The obtained synthesis gas is cooled with the heat regeneration in power production cycle and fed in single-pass catalytic methanol synthesis reactor, where about 50-60% is converted to methanol.

The remaining gas, ballasted by nitrogen is expended in turbine of the turbo-expander and fed to the combustion chamber of gas-turbine unit.

Let us name all the advantages of the proposed scheme:

- compressed air, being the working fluid of the gas-turbine unit, is fed to the synthesis gas generator without substantial energy consumption,
- the single-pass methanol synthesis reactor with low degree of synthesis gas conversion is used, because the remaining gas is the fuel for gas-turbine unit,
- there are no energy losses at the recuperation of compressed gas energy after the methanol synthesis reactor,
- the power producing unit becomes automatically ecologically clean because the reburning of low calorific gas takes place in the gas-turbine unit combustion chamber practically without origination of toxic nitrogen oxides, both "thermal" (Zeldovich mechanism) or «prom + NOx).

The cogeneration of two products gives the opportunity to decrease considerably the cost of electricity even compared to the existing steam-gas plants. Fig. 12 shows the comparison results of relative feasibility indicators of various power plants, where the reference is the cost of electricity generated by 300 MW steam-turbine power plant.

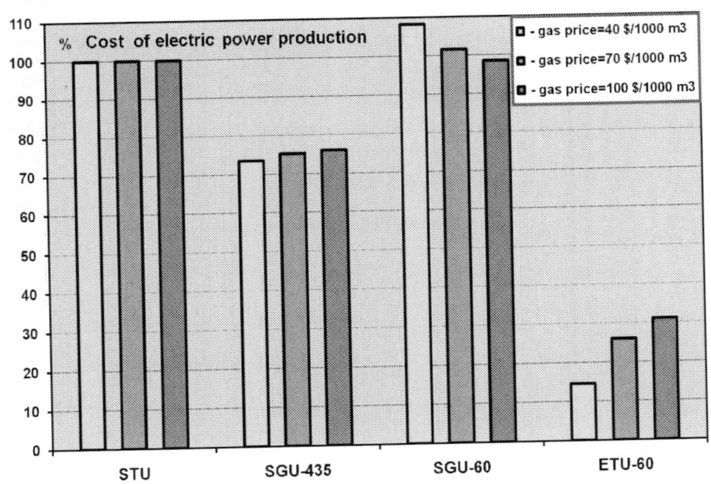

Fig. 12. The comparison results of relative feasibility indicators of various power plants.

The comparison of the cost of electricity after deduction of methanol cost of production at modern big enterprise is conducted for four alternatives at the variation of natural gas cost:

- steam-turbine power unit for supercritical steam parameters with 300 MW capacity STU,
- binary cycle steam-gas unit with capacity 435 MW (SGU 435),

- 60 MW binary cycle steam-gas unit (SGU 60),
- 62 MW(e) power and chemical complex (ETU-60).

The creation of such energy and technology complexes demands the salvation of a range of problems, where in the first place, are fine-tuning of natural gas conversion in the continuous flow reactor and the turbine operation with the use of synthesis gas ballasted by nitrogen. In order to solve these problems together with the range of other tasks, experimental-industrial unit with 1.2 MW gas turbine was created in accordance with state contract financed by the Ministry of Education and Science.

The unit includes practically all important elements of the scheme of energy and technology complex. The appearance of the unit under creation is presented in Fig.13.

1 - The synthesis gas generator by partial gas oxidation with syngas cooler, 2 - gas turbine, 3- soot filter, 4 - water heating boiler, 5 - water cooling tower, 6 - block of methanol synthesis.

Fig. 13. Experimental-industrial unit with 1.2 MW gas turbine.

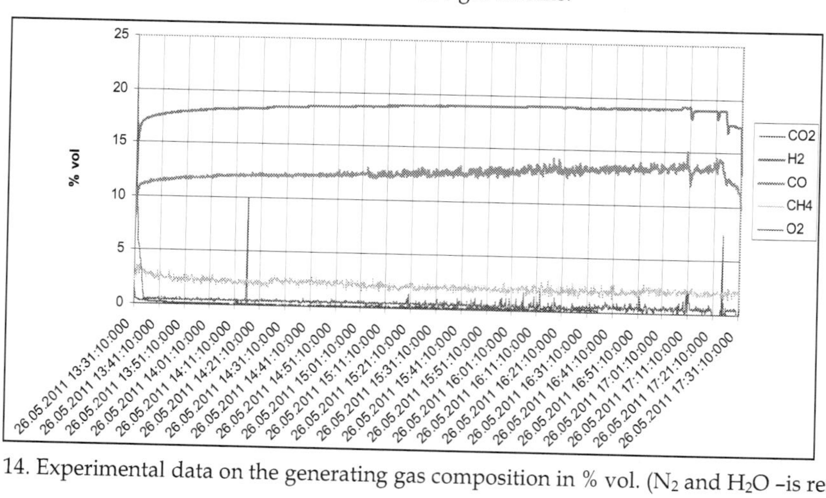

Fig. 14. Experimental data on the generating gas composition in % vol. (N_2 and H_2O –is rest).

The industrial energy and technology complex could be created, for example, on the base of 20 MW gas-turbine unit produced by "Salut" Moscow Engineering Production Enterprise. Using two of such gas-turbine units, the energy and technology complex can have electric capacity of 50 – 60 MW and produce up to 150 000 tons of methanol per year. The cost of electricity generated could be 2.5 – 3 times lower than at perspective high-capacity steam-gas unit and the exhaust gases practically free from toxic NO.

5. Total conclusions

1. The new approaches described above do not limit all the possibilities for the upgrade of the existing and for the creation of new power production technologies that increase the efficiency of natural gas utilization and decrease the harmful emission to the environment.
2. The implementation of relatively simple technical solutions in upgrade of existing power production technologies often give bigger saving rate per the investment unit, than at the creation of new traditional power production technologies.
3. The creation of energy-technology complexes of co-generation of energy, synthetic liquid fuel and other valuable accompanying products of mass demand is the perspective way for rational use of the fossil fuels (natural gas, coal, shale).

6. Acronyms and abbreviations

GTU - Gas Turbine Unit;
SGU – Combined-cycle Steam-Gas Turbine Units;
STU – Steam Turbine Unit;
Ne – electric power, MW;
H_T – energy of the fuel combusted, MW;
Q_T – thermal capacity of the boiler, MW;
η_b -boiler efficiency;
η_e – electric power production efficiency at heat combustion;
Ne add, MW- additional useful power output of the steam-gas unit after repowering;
Eff_{ad} – thermal efficiency of Generation Additional Electric Power after repowering;
Ne – useful electric power (MW);
Qt – district heating power (MW);
Ht – total energy of the consumed fuel.
FHUF - fuel heat utilization factor;
N_{SGU} and N_{STU} are the useful power output of the repowered and original unit, respectively, operated at nominal rate;
$\Sigma g_i{}^*Q_{iSGU}$ and $\Sigma g_i{}^*Q_{iSTU}$ are the energy of all kinds of fuel burnt in the repowered and original units, respectively, at nominal rate;

7. References

[1] V. Bushuev, A. Troitskii "The Energy Strategy of Russia until 2020 and Real Life. What Is Next?" Thermal Engineering Vol.54, Number 1, January 2007, pp 1-7.

[2] V.M. Maslennikov, V.Ja. Shterenberg, "Method of the heat transforming into useful mechanical work in the cycle with multistage combustion", Patent of Russia 1809141 Apr., 1991/

[3] V.M. Maslennikov (USSR), R.G. Smithson (USA) et al. "Steam-and-Gas Units with Fuel Gasification within the Cycle and Ecological Problems of Power Engineering" Battele Memorial Institute, USA, 1992.

[4] V.M. Maslennikov, V.Ja. Shterenberg, "Environmentally Clean Efficient Steam - Gas Units for Repowering of Existing Power Plants", "Energeticheskoe Stroitel'stvo", Moscow, 8, 1992/

[5] V.M. Maslennikov, V.Ja. Shterenberg "Advanced gas turbine system utilizing partial oxidation technology for ecologically clean power generation", International Journal of Low-Carbon Technologies, Jan., 2011.

[6] S. Daw, K. Charkavathy, J. Pihl, J. Conkin (Oak Ridge National Laboratory) "The Theoretical Potential of Thermochemical Exhaust Heat Recuperational for Improving the Fuel Efficiency of Internal Combustion Engines", Presentation to the Chicago Section of AICHE, may 17.2011/

[7] P. Childs "Chemical looping combustion for high efficiency and carbon capture", Gas Turbine World, May-June 2011, pp 24-27.

[8] V. de Biasi "1.8:1 steam-to-fuel ratio reduces NOx output of LM 2500 to 5 ppm", Gas Turbine World, May-June 2011, pp 12-17.

[9] V. Dobrokhotov, Yu. Zeigarnik "Cogeneration: Problems and Possibilities of Realization under Present Conditions" Thermal Engineering Vol.54, Number 1, January 2007, pp 8-9.

[10] V.M. Maslennikov, Yu.A. Vyskubenko, E.A. Tsalko, V.Ya. Shterenberg. E.G. Shadek. "Heat generation in steam-gas cycle and steam-gas unit for its implementation", RF Patent №217942 F01K 26/06 of 25.04.2001.

[11] V.M. Batenin, V.M. Maslennikov, L.S. Tolchinsky "Technology of combined production of electricity and liquid synthetic fuel with the use of gas-turbine and steam-gas units", RF Patent №2250872, 27.04.2005.

[12] V.M. Batenin, V.M. Maslennikov, Yu.A. Vyskubenko "Technology for the complex use of solid fuel in power production units with cogeneration of energy and sideline products", RF Patent №2364737, 20.08.2009 г.

[13] I. I. Lishchiner, O. V. Malova, A. L. Tarasov, V. M. Maslennikov, Yu. A. Vyskubenko, L. S. Tolchinskii, Yu. L. Dolinskii "Synthesizing Methanol from Nitrogen-Ballasted Syngas" . ISSN 2070-0504 Catalysis in Industry, 2010, vol.2, No.4, pp. 368-373.

Part 5

Natural Gas Combustion

18

Application of Natural Gas for Internal Combustion Engines

Rosli Abu Bakar[1], K. Kadirgama[1], M.M. Rahman[1], K.V. Sharma[1] and Semin[2]

[1]Faculty of Mechanical Engineering, University Malaysia Pahang,
[2]Department of Marine Engineering,
Institut Teknologi Sepuluh Nopember, Surabaya,
[1]Malaysia
[2]Indonesia

1. Introduction

It is well known that the fossil-fuel reserves in the world are diminishing at an alarming rate and a lack of crude oil is expected at the early decades of this century (Aslam et al., 2006). Gasoline and diesel fuel becomes scarce and most expensive (Catania et al., 2004). Alternative fuel becomes more conventional fuel in the coming decades for internal-combustion engines. Nowadays, the alternative fuel has been growing due to concerns that the reserves of fossil fuel all over the area are limited. Furthermore, the world energy crisis made the fossil-fuel price increases.

Natural Gas (NG) has been found in various locations in oil and gas bearing sands strata located at different depths below the earth surface (Catania et al., 2004). NG is a gaseous form of NG was compressed. It has been recognized as one of the promising alternative fuels due to its significant benefits compared to gasoline fuel and diesel fuel. These include reduced fuel cost, cleaner exhaust gas emissions and higher octane number. Therefore, the numbers of engine vehicles powered by NG were growing rapidly (Poulton, 1994; Pischinger, 2003). NG is safer than gasoline in many respects (Cho and He, 2007; Ganesan, 1999; Kowalewicz, 1984). The ignition temperature of NG is higher than gasoline fuel and diesel fuel. Additionally, NG lighter than air and dissipate upward rapidly. Gasoline fuel and diesel fuel will pool on the ground, increasing the risk of fire. NG is nontoxic and will not contaminate groundwater if failed. Advanced NG engines undertake significant advantages over the conventional gasoline engine and diesel engine (Kato et al., 1999). NG is a commonly available type of fossil energy. However, the investigation of applying NG as an alternative fuel in engines will be a beneficial activity, because the liquid fossil fuels will be finished and will become scarce and expensive (Catania, 2004; Sera, 2003). NG has some advantages compared to gasoline and diesel from the environmental perspective. It is a cleaner fuel than either gasoline or diesel as far as emissions are concerned. NG is considered to be an environmentally clean alternative to those fuels (Cho and He, 2007; Kato et al., 1999; Shashikantha and Parikh, 1999; Wayne, 1998). Advantages of NG as a fuel its octane numbers are extraordinarily suitable for spark ignition (SI) engines. NG engine can be operated in high compression ratio (Ganesan, 1999).

2. Natural Gas engine

2.1 Natural Gas engine development trend

There are four NG engine types, the traditional premixed charge spark ignition engine, the port injection lean burn engine, the dual-fuel/pilot injection engine, and the direct injection engine (Ouellette, 2000; Shashikantha and Parikh, 1999). Significant research has been done on these engines, the most promising of these, the injection engine requires further development in order to investigate the injection full potential. Shashikantha and Parikh (1999), studied a 17 kW, stationary, direct injection diesel engine converted to operate as a gas engine using producer-gas and NG as the fuels on two different operational modes called SIPGE (Spark Ignition Producer Gas Engine) and DNGE (Compressed NG Engine). Shashikantha and Parikh (1999) results of conversion to SIPGE (or DNGE) can be regarded as a success since comparable power and efficiency could be developed. NG operation of SIPGE yielded almost comparable power and higher efficiency, which establishes the fuel flexibility of the machine under spark ignition performance. The spark advance needed for producer-gas operation is much higher at 35° BTDC as compared to NG operation which is 22° BTDC, with compression ratio being same, i.e., 11.5:1.

Kato et al. (1999) has developed a new engine Toyota Camry that uses NG as fuel by modifying the base 2.2-liter gasoline engine in the unmodified engine, torque and power for NG decrease compared to gasoline. The new engine has adopted a high compression ratio, intake valves with early closed timing, intake and advanced exhaust valves with increased lift and a small back pressure muffler, which thereby restores the loss of engine power. Fig. 1 shows a multi port injection or multi point injection system was chosen by Czerwinski et al. (2003), and the injectors and pressure regulator have been recently developed in order to significantly reduce exhaust emissions. At the same time, precise air-fuel (A/F) ratio control and special catalysts for NG exhaust gas have been utilized. The resulting NG engines output power has been restored to approach that of the gasoline base engine. Wang and Watson (2000) have developed of a NG engine with ultra-lean-burn low emissions potential, hydrogen-assisted jet ignition (HAJI) is used to achieve reliable combustion and low NOx emissions, whilst direct injection is used to improve thermal efficiency and reduce hydrocarbon (HC) emissions. It is found that port-inducted propane, port-inducted NG and directly injected NG all produce negligible levels of CO and NOx.

The vast majority of NG engines in use today is premixed charge spark ignition engines (Chiu, 2004). Spark ignited (SI) engines have significant advantages over diesel engines in terms of particulate and NOx emissions, there are some drawbacks with respect to performance. Premixed SI engines allow 30% lower power output than equivalent size diesel engines due to knock limitations (Kato et al., 1999). In addition, SI engines receive high pumping losses, due to the need to throttle the intake air at part load conditions. These factors result in a 15 to 30% reduction in volumetric efficiency compared diesel engines (Brombacher, 1997). In diesel engine, Ouellette (2000) developed high pressure direct injection (HPDI) of NG in diesel engines, the result shown, that NG or methane are reduced by about 40% over diesel operation NOx. Peak torque loss 9% when running on NG compared to gasoline (Durell et al., 2000). Although peak power was not obtained on gas (due to the limitations of the injectors) there is also a predicted loss 9% on peak.

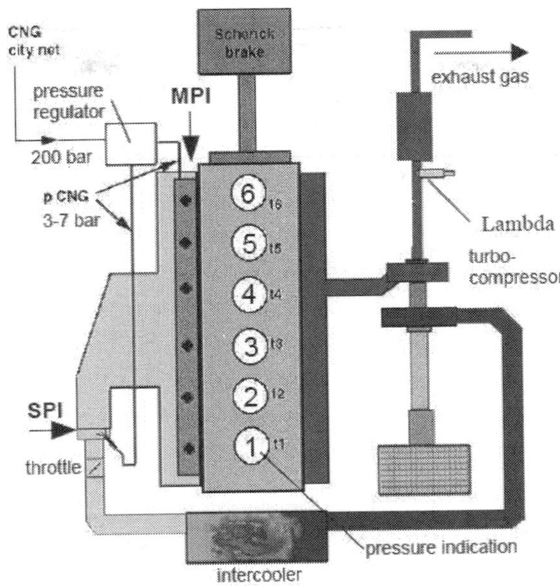

Fig. 1. Gas injection of NG engine.

The NG engine is best operated if such conditions as listed by Bakar et al. (2002) in Fig. 2. The principal operations are operated in high volumetric efficiency, turbulent flame speed, high compression ratio and proper air-fuel ratio. In operation in high volumetric efficiency and suitable of the air-fuel ratio is based on turbulent effect, injector type and lean burn operation. According to Bakar et al. (2002), injector is the important component in the best operation of NG engine. Another that, many researchers and institutions have contributed in improving the NG engine performance. In the area of increasing volumetric efficiency, Kubesh et al. (1995) developed an electronically controlled NG fueled engine with a turbocharged-aftercooled engine controlled by an electronic control system. Tilagone et al. (1996) found an increase up to 16% of thermal efficiency on a turbocharge spark ignition NG fuelled engine with multi point injection and optimized ignition timing with spark advance 200 higher running on stoichiometric A/F ratio.

In designing a turbulent effect in order to increase the flame speed combustion, Johansson and Olsson (1995) developed ten different geometries of the combustion chamber (CC). Their results showed a strong correlation between in cylinder turbulence and rate of heat release in the combustion process. However, the results also showed that geometries that gave the fastest combustion would gave the highest NOx values. In their further analysis, Johansson and Olsson (1995) developed six different CC to observe its effect on the combustion performance. The results showed different geometrical CC, with the same compression ratio (12:1), have extremely different combustion performance. The Quartette type of CC gave the highest peak turbulence. A squish-generated charge motion combustion chamber had its effect to the burning rates. High levels of turbulent generated from the

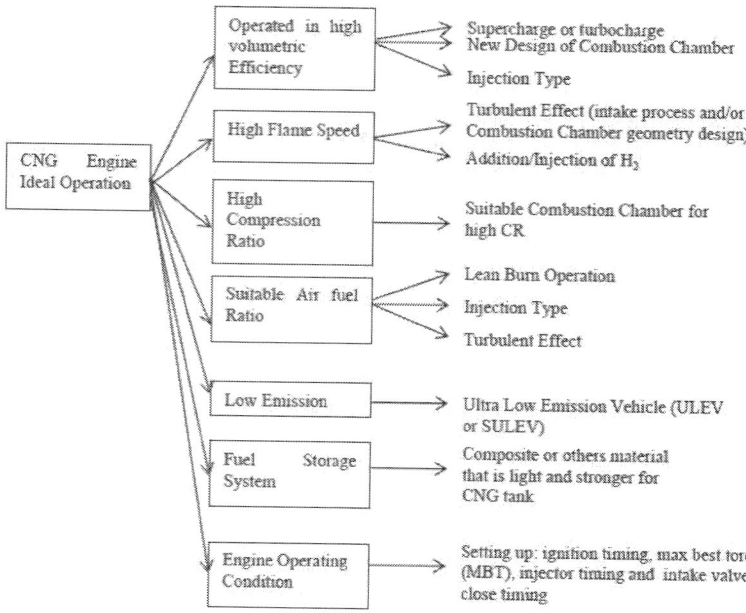

Fig. 2. The ideal NG engine operating conditions.

squish pretend to faster burning rates, which resulted in improvement of thermal efficiency. Evan et al. (1996) proved that the faster burning rates led to an average of 1.5% reduction in brake specific fuel consumption (BSFC) or 1.5% increase in power output under wide open throttle condition, as compare to the slowest burning cases. However the highest turbulence intensity combustion chamber also showed the highest emission.

In optimizing to the NG engine performance, Duan (1996) proposes the modification of setting up MBT, higher compression ratio and the use of gaseous fuel injection systems. Meanwhile, Ford introduces the NG Vehicle (NGV) truck by modifying fuel storage, fuel metering and emission control system. The injector timing, fuel control, spark advance, and exhaust gas recirculation (EGR) were also changed (Vermiglio, 1997). The simulations areas also conducted to increase the performance of NG engine. Oullette (1998) had simulated the combustion process and provides a better understanding of the injection and combustion process of the pilot-ignited directly-injected NG. The numerical simulation was expected to optimize the injection process by looking in especially at the geometry and the injection delay between two fuels. The model includes modifications for under expanded NG jets and includes a turbulent combustion model.

2.2 Injection methods of Natural Gas engine

There are four methods to inject the NG into the engine cylinder (Zastavniouk, 1997). First type is gas mixer / carburetor injection, second type is the single point injection, third type is multi point injection and fourth type is direct injection. The illustration of the four methods of NG injection is shown in Fig. 3.

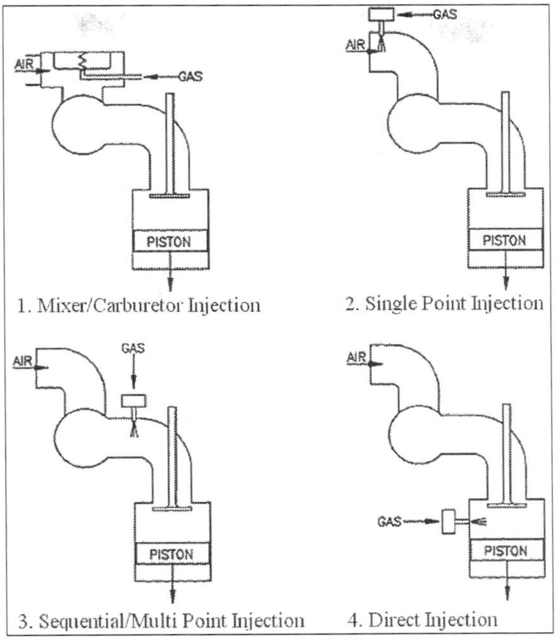

Fig. 3. Injection methods of NG engine.

The existing metering and mixing of the fuel may be accomplished using either a mechanical gaseous fuel mixer or carburetor, or an electronically controlled gaseous fuel metering system. This approach strives to achieve a homogeneous mixture of air and fuel before the air flow splits in the intake manifold. As discussed by Klimstra (1989), failure to obtain a homogenous mixture at this point can cause significant cylinder-to-cylinder variations in the air-fuel ratio. According to Zastavniouk (1997), Klimstra (1989) and Lino et al. (2008), this injection option can be increases emissions and the possibility of knock phenomena. Single point injection is use gaseous fuel injector to mix the gaseous fuel with the intake air in the manifold at one location for all cylinders of the engine. In this case, fuel is injected in a single location much like a gas mixer or carburetor. Single point electronic injection offers the advantage of more precise control of the amount of gaseous fuel entering the intake charge of the engine as well as the economy of using a minimum number of injectors (Zastavniouk, 1997). Multi point injection (MPI) is to inject the fuel into the each cylinder via intake port before intake valve (Czerwinski, 1999; 2003; Zastavniouk, 1997). This system uses one or more fuel injectors for each cylinder intake port of an engine and allows the designer to remove the fuel supply from the air supply area of the intake manifold. Direct injection is to inject the gaseous fuel directly into each combustion chamber of the engine.

In the MPI methods of NG, it is necessary to develop considerable turbulence during the compression stroke to obtain adequate air-fuel mixing. A high-turbulence, high swirl combustion chamber and high air-fuel mixing are useful for this type of injection to increase the engine performance.

2.3 Multi point injection system of Natural Gas

According to Lino et al. (2008), the main elements of the NG injection system are a fuel tank storing high pressure gas, a pressure reducer, a common rail and electro-injectors. The MPI system of NG is shown in Fig. 4.

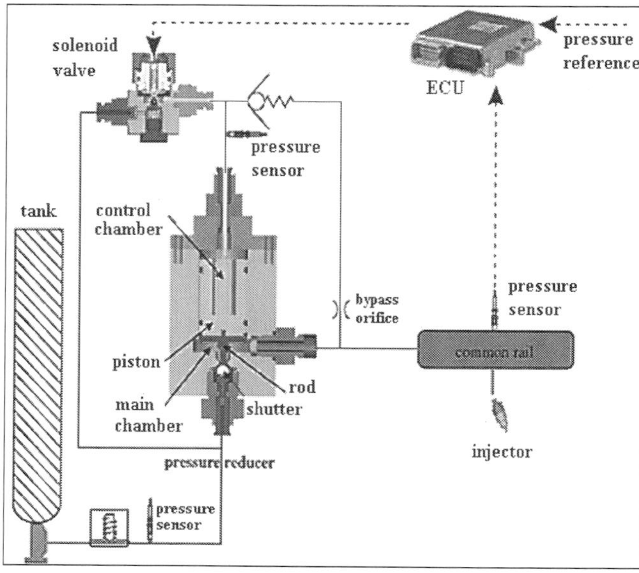

Fig. 4. Multi point injection system of NG engine.

The fuel coming from the tank supplies the pressure reducer before reaching the common rail and feeding the electronically controlled injectors. By supplying gas to the intake manifolds, injectors lead to the proper air/fuel mixture (Lino et al., 2008; Czerwinski et al., 1999; 2993). The large volume of the common rail helps in damping the oscillations due to the operation of both pressure controller and injectors. Namely, combining the electronic control of rail pressure with optimum design of the rail volume reduces the pressure oscillations inside the rail and leads to a more accurate fuel metering. The flow rate depends only on the rail pressure. Hence, the injected fuel quantity can be metered acting on rail pressure and injection timings are driven by the electronic control unit (ECU).

2.4 Diesel engine convert to multi point injection Natural Gas engine

In the diesel engines converted to run on NG, there are two main options discussed. The first is dual-fuel engine and the second is NG engine. Dual-fuel engine is referred to diesel engines operating on a mixture of NG and diesel fuel. NG has a low cetane rating and is not so suited to compression ignition, but if a pilot injection of diesel occurs within the gas/air mixture, standard ignition can be initiated. Between 50% and 75% of conventional diesel consumption can be replaced by gas when operating in this mode. The engine can also revert to 100% diesel operation. NG engines are optimized for the NG fuel. They can be derived from gasoline engines or may be designed for the purpose. According to Poulton (1994), until manufacturer original equipment (OE) engines are more readily available, however, the practice of converting diesel engines to spark ignition will continue, which involves the replacement of diesel fuelling equipment by a gas carburetor and the introduction of an ignition system and spark plugs. For compression ignition engines conversions to spark ignition, the pistons modified to reduce the actual compression ratio and a high-energy ignition system fitted (Czerwinski et al., 1999; 2003). The system is suitable for NG and is ideally suited to MPI system but can also be used for single point and low pressure in-cylinder injection. Gas production provides greater precision to the timing and quantity of fuel provided, and to be further developed and become increasingly used to provide better fuel emissions (Poulton, 1994).

The port injection NG produces negligible levels of CO, CO2 and NOx (Suga et al., 2000). In order to significantly reduce exhaust gas emissions, a port injection system was chosen by Czerwinski et al. (1999, 2003), Hollnagel et al. (1999, 2001) and Kawabata and Mori (2004), and the injectors and pressure regulator have been recently developed. In the same time, precise air-fuel (A/F) ratio control and specific catalysts NG exhaust gas has been utilized. By using it, NG engines output power near the gasoline base engine.

With the multi point injection, a high-speed gas stream is pulsed from the intake port through the open intake valve into the combustion chamber, where it causes effects of turbulence and charge stratification particularly at engine part load operations. The system is able to reduce the cyclic variations and to develop the border of lean operation of the engine. The flexibility of gas pulse timing offers the potential advantage of reduced emissions and fuel consumption. With three types of port injectors available on the market, Czerwinski et al. (2003) compared for stationary and transient engine operation. There are several advantages of port injection, e.g., better possibility to balance the air-fuel ratio of the cylinders, optimization of the gas injection timing and of the gas pressure for different operating conditions. The port injection has an injector for each cylinder, so the injectors can be placed in proximity to the cylinder's intake port. It also enables fuel to be delivered exactly as required for each individual cylinder (multi point injection) and enables more sophisticated technologies such as skip-firing to be used. Skip-firing is when only some of the cylinders are operating (the other cylinders are being skipped). This enables even more efficient use of the fuel at low loads, further lowering fuel consumption and unburned hydrocarbon (Czerwinski et al., 2003; Zastavniouk, 1997).

2.5 Multi point injection gas injector

In principle, the utilization of an optimal fuel-air mixture should provide the required power output with the lowest fuel consumption that is consistent with smooth and reliable operation (Zhao et al., 1995). Over past decades, MPI system has evolved into an electronic, pulse-width-modulated system that utilized multi point injectionly-timed separate injections into each intake port. According to Zhao et al. (1995), these transient sprays of 2.5 to 18.0 ms duration have a constant phase relative to the intake valve event, either for start on injection or end of injection, and provide significant advantages in engine transient response and hydrocarbon (HC) emissions. It is necessary to note, however, that the meteoric expansion in the use of such systems has generally out-paced the basic knowledge and understanding of the complex, transient fuel sprays that they produce.

According to Shiga et al. (2002), improvement of NG injector nozzle holes geometries and understand of the processes in the engine combustion is a challenge because the compression-ignition combustion process is unsteady, heterogeneous, turbulent and three dimensional and exceedingly complex. In MPI NG engines, NG is injected by fuel nozzle injector via intake port into the combustion chamber and mixing with air must occur before ignition of the gas fuel. To improve the perfect of the mixing process of NG fuel and air in the combustion chamber is arranging of nozzle holes geometry, nozzle spray pressure, modified of the piston head, arranging of piston top clearance, letting the air intake in the formation of turbulent and changing the NG fuel angle of spray (Mbarawa et al., 2001). The NG fuel spraying nozzle is the amount of earning variation so that can be done by research experimentation and computational of engine power, cylinder pressure, specific fuel consumption and missions which also the variation of them. Czerwinski et al. (1999, 2003) has researched the multi point injection injection of NG offers several advantages to increase the NG engine performance. The injector multi holes geometries development is to provide optimum fuel air mixing of the engine that will promote a similar engine performance (Ren and Sayar, 2001). According to Czerwinski et al. (2003) NG MPI has advantages for the more efficiency. The power, fuel consumption and thermal efficiency of the engine are higher than carburetor and single point. In the port injection NG engine, every cylinder has least one injector and the fuel are injected from the intake manifold into the engine cylinder when the intake valve is opened.

3. Development of multi point injection Natural Gas engine

The development of MPI NG engine is using diesel engine as a baseline engine. The fuel in the diesel engine is changed to NG. The ignition system is compression ignition changed to spark ignition. The fuel injection system is from direct injection mechanical system changed to multi point injection system and managed by electronic control unit. The NG engine is using throttle to control the intake air. The development of NG engine is reducing the compression ratio by modified the piston surface. The fuel is injected from the intake manifold into the engine cylinder when the intake valve is opened. The engine performance investigation is based on experimental and computational.

4. Multi point injection Natural Gas engine performance

4.1 Introduction

This chapter is exploring the engine performance based on experimental and computational. The engine computational model is used in the preliminary design to simulate the compression ratio effect of multi point port injection NG engine converted from diesel engine. The compression ratio has given the significant impact on engine power performance. In the engine computational model, if the compression ratio of the diesel engine convert to multi point port injection NG engine is designed in 12.5:1, 13.5:1, 14.5:1, 15.5:1, 16.5:1, 17.5:1, 18.5:1, 19.5:1 and 20.28:1, the brake power of the engine has been reduced 42.2%, 41.71%, 41.37%, 41.51%, 41.43%, 41.48%, 41.78%, 42.0% and 42.23%. Based on this brake power performance reduction effect from the compression ratio, the compression ratio with lower reduce brake power will be used in the engine conversion. The compression ratio 14.5:1 will be used in the development of multi point port injection NG engine. The engine conversion data are shown in Table 1.

Engine Parameter	Diesel Engine	NG Engine
Bore (mm)	86.0	86.0
Stroke (mm)	70.0	70.0
Displacement (cc)	407.0	407.0
Compression ratio	20.28:1	14.5:1
Ignition system	Compression Ignition	Spark Ignition
Engine Management	Mechanical Control	Electronic Control
Fuel system	Direct Injection	Multi point Port Injection
Fuel	Diesel	Natural Gas

Table 1. Specification the engine conversion.

4.2 Cylinder pressure of multi point injection Natural Gas engine

The results of cylinder pressure performance of direct injection diesel engine, compression ratio modified direct injection diesel engine and multi point port injection NG engine are shown in Fig. 5.

The results investigation of engine cylinder pressure is based on crank angle degree. The negative 180 to 0 crank angle degree is the engine compression stroke and the 0 to 180 crank angle degree is the engine power stroke for original diesel engine (ODE), compression ratio modified diesel engine (14.5CR DE) and MPI NG engine (NGE).

The engine cylinder pressure profile investigation results are shown in Fig. 5 are shown that the cylinder pressure is increasing in compression stroke to combustion ignition in crank angle negative180 degree bottom dead center (BDC) until around in crank angle 0 degree top dead center force (TDCF). In the compression stroke, the air-fuel volume is compressed from BDC to TDC. The simulation and experiment results are not similar. The simulation results are higher than the experimental results. The deviation is in average 2% for NG engine (NGE) and original diesel engine (ODE). The compression ratio of original direct injection diesel engine is 20.28:1, the compression ratio of modified direct injection diesel engine is 14.5:1 and the compression ratio of port injection NG engine is 14.5:1. From 1500 to 4000 rpm engine speed are shown that the original direct injection diesel engine cylinder pressure is higher than the modified direct injection diesel engine and multi point port injection NG engine. The highest of cylinder pressure is around in crank angle 0 degree (TDCF). From the cylinder pressure performance can be predicted that the product of engine power from the air-fuel combustion of original direct injection diesel engine is higher than modified direct injection diesel engine and the multi point port injection NG engine. The original direct injection diesel engine cylinder pressure is higher than modified direct injection diesel engine and multi point port injection NG engine because the compression ratio of original diesel engine is higher.

The highest of maximum cylinder pressure in the combustion process both of diesel engines and for multi point port injection NG engine are shown in Fig. 5a. In the original diesel engine, the maximum cylinder pressure is 84.0 bar declared in 1500 rpm engine speed. In the modified diesel engine, the maximum cylinder pressure is 61.1 bar declared in 1500 rpm engine speed. In the multi point port injection NG engine, the maximum cylinder pressure is 76.23 bar and declared in 1500 rpm engine speed. In this operating condition, both of diesel engines and NG engine combustion process are most excellent than the other condition. In the diesel engine, the 1500 rpm engine speed condition is not higher and not lower for the combustion of diesel fuel. Burned diesel fuel rate in 1500 rpm is most excellent to product the higher pressure and power. In the multi point port injection NG engine, the 1500 rpm engine speed condition is not higher and not lower for the combustion of NG engine. Burned NG fuel rate in 1500 rpm is most excellent and product the higher pressure and torque of the engine. The trend of the maximum cylinder pressure for original direct injection diesel engine, modified direct injection diesel engine and port injection NG engine are decrease if the engine speed is increased.

Fig. 5f shows the lowest of maximum cylinder pressure of direct injection diesel engine, modified direct injection diesel engine and MPI NG engine. The lowest maximum cylinder pressure in combustion process of original direct injection diesel engine, modified direct injection diesel engine and port injection NG engine are shown in 4000 rpm engine speed and the nominal is 72.82 bar for original diesel engine, 52.29 bar for modified diesel engine and 25.00 bar for port injection NG engine. In this case the combustion of diesel engines and NG engine are in lately so the combustion process is not excellent and unburned fuel is highest, this phenomenon can be decreasing the engine cylinder pressure performance. The port injection NG engine maximum cylinder pressure is lowest because the natural gas fuel is lower in density, hydrocarbon and energy than the diesel fuel. So, the cylinder pressure in the same compression ratio, the NG engine is lower than modified diesel engine if the engines are operated in high speed. The lowest cylinder pressure of original diesel engine is higher than modified diesel engine because the compression ratio of original diesel engine is higher than modified diesel engine.

The maximum cylinder pressure effect of the diesel engine converted to multi point port injection NG engine in the similar or higher compression ratio and in variation engine speed is shown in Fig. 5. In the 1500 rpm engine speed, the conversion of diesel engine to NG engine is increase the maximum cylinder pressure 8.97 %. In the 2000 rpm engine speed, the conversion of diesel engine to NG engine is decrease the maximum cylinder pressure 1.70 %.

In the 2500 rpm engine speed, the conversion of diesel engine to NG engine is decrease the maximum cylinder pressure 13.53 %. In the 3000 rpm engine speed, the conversion of diesel engine to NG engine is decrease the maximum cylinder pressure 39.12 %. In the 3500 rpm engine speed, the conversion of diesel engine to NG engine is decrease the maximum cylinder pressure 51.40 %. At the 4000 rpm, the conversion of diesel engine to NG engine is decrease maximum cylinder pressure 58.56 %.

The maximum cylinder pressure for NG engine is lower than the original diesel engine. It caused the compression ratio of NG engine is lower than the original diesel engine and the combustion energy output of diesel fuel is produces highest power than the natural gas fuel. Another that, the density of natural gas fuel is lower than the diesel fuel. So, in the same volume, the diesel fuel is has higher pressure than the gas fuel. In this engine conversion, the NG engine better to operate at low speed. In the low speed the maximum cylinder pressure increasing is higher dramatically than at the medium and high speed. For all of engine speed, the conversion of modified diesel engine to NG engine is increase the cylinder pressure in low speed, but in the high speed the engine conversion can be decreasing the cylinder pressure. In the high speed NG engine, the fuel energy is reduced and the combustion is not completely, but in the low speed the combustion of NG engine is completely because the combustion ignition is assisted by spark plug system and the

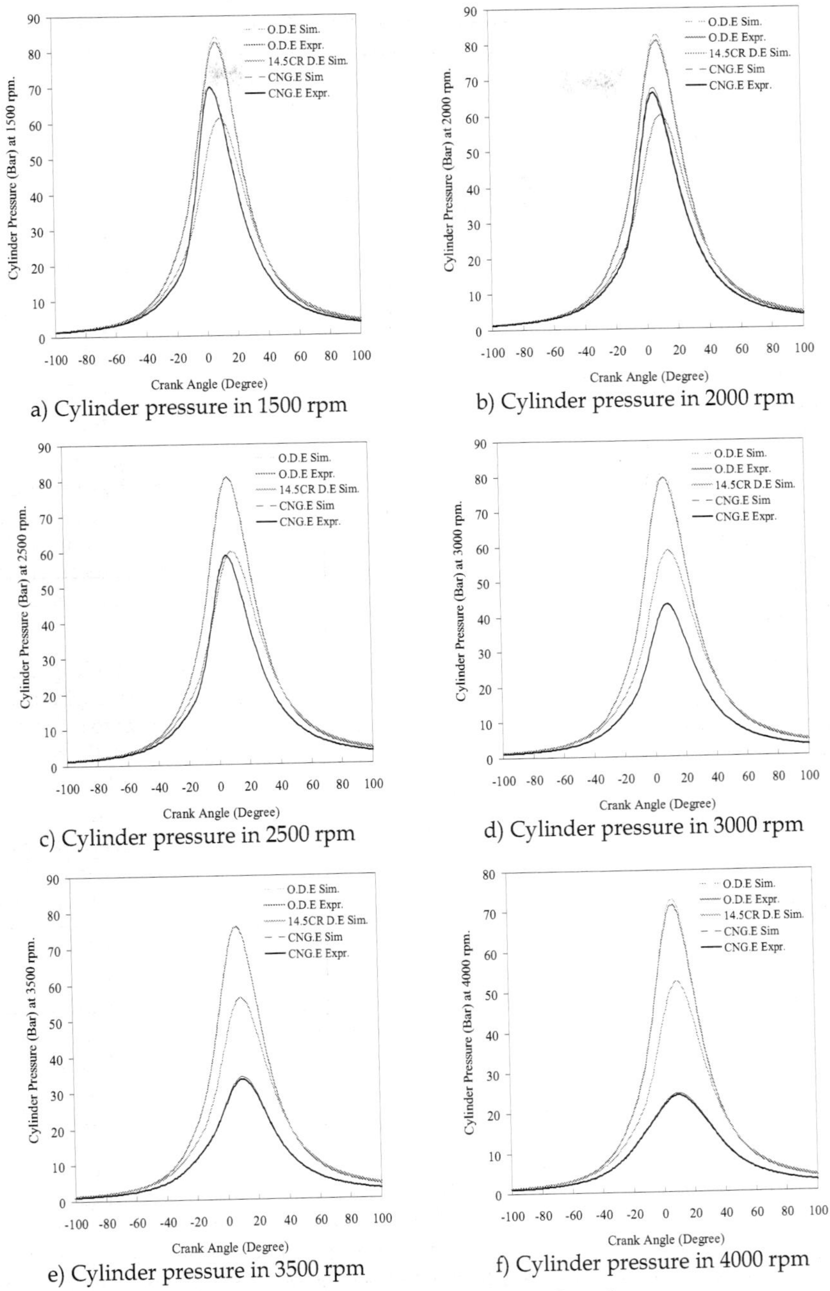

a) Cylinder pressure in 1500 rpm

b) Cylinder pressure in 2000 rpm

c) Cylinder pressure in 2500 rpm

d) Cylinder pressure in 3000 rpm

e) Cylinder pressure in 3500 rpm

f) Cylinder pressure in 4000 rpm

Fig. 5. Cylinder pressure of diesel engine convert to MPI NG engine.

ignition point of natural gas fuel is higher than the diesel fuel, so it can be producing the higher engine cylinder pressure.

4.3 Cylinder temperature of multi point injection Natural Gas engine

The investigation results of the engine cylinder temperature characteristics of original direct injection diesel engine (O.D.E), modified direct injection diesel engine (14.5CR D.E) and multi point port injection NG engine (NG.E) are shown in Fig. 6. In these figures, negative 180 to 0 degree is compression stroke and the 0 to 180 degree is power stroke for diesel engines and NG engine. The average deviation result of simulation and experiment is 2% for ODE and NGE.

In the low speed, the engine cylinder temperature of NG engine is higher than original diesel engine and modified diesel engine as shown in Fig. 6. In the high speed, the engine cylinder temperature fro both of diesel engines are higher than NG engine as shown from Fig. 6a to Fig. 6f.

The results are shown that increasing engine speed of diesel engine can be increase the maximum temperature in-cylinder engine. Unfortunately, the increasing engine speed of NG engine will be decrease maximum temperature in-cylinder engine. The decreasing engine speed of diesel engines will be decrease maximum temperature in-cylinder engine. Decreasing engine speed of NG engine will be increase maximum temperature in-cylinder engine. In this investigation results are shown that the highest maximum in-cylinder temperature in combustion process is not declared in the highest engine speed. In the both of diesel engines, the highest maximum temperature in-cylinder is declared in 3500 rpm engine speed, because in this case the combustion is most excellent than the other condition and unburned fuel is lowest, so the temperature product from the combustion is the highest. In the both of diesel engines, the lowest maximum temperature in combustion process is in 1500 rpm engine speed.

In this engine speed, the combustion process is not excellent and unburned fuel is highest than the other condition for compression stroke of compression ignition diesel engines. In the NG engine, the highest maximum temperature in-cylinder is declared in 1500 rpm engine speed, because in this case the combustion is most excellent than the other condition and unburned fuel is lowest, so the temperature product from the combustion is the highest. In the NG engine, the lowest maximum temperature in combustion process is in 4000 rpm engine speed. After 1500 rpm, the increasing engine speed, the combustion process is not excellent, the gas fuel density is lower, the air-fuel volume is lower and unburned fuel is highest than the other condition for compression stroke of natural gas spark assisted combustion engine. In the low speed, the engine cylinder temperature of NG engine is higher than original diesel engine and modified diesel engine as shown in Fig. 6a. In the high speed, the engine cylinder temperature fro both of diesel engines are higher than NG engine as shown from Fig. 6a to Fig. 6f.

The results are shown that increasing engine speed of diesel engine can be increase the maximum temperature in-cylinder engine. Unfortunately, the increasing engine speed of NG engine will be decrease maximum temperature in-cylinder engine. The decreasing engine speed of diesel engines will be decrease maximum temperature in-cylinder engine. Decreasing engine speed of NG engine will be increase maximum temperature in-cylinder

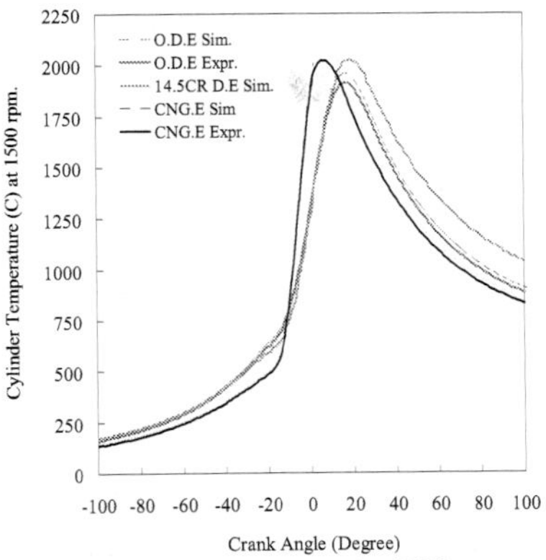

a) Cylinder temperature in 1500 rpm

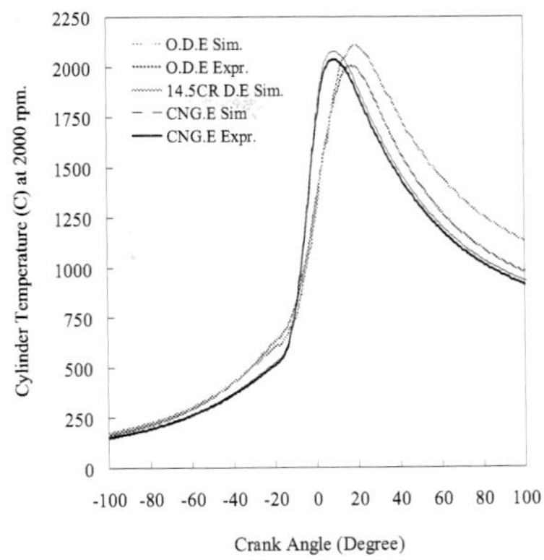

b) Cylinder temperature in 2000 rpm

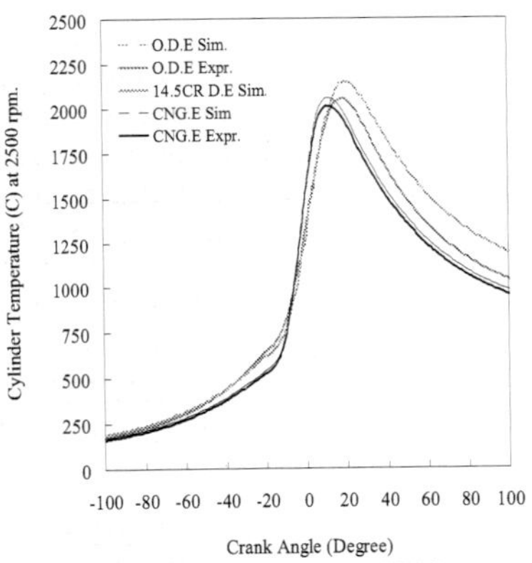

c) Cylinder temperature in 2500 rpm

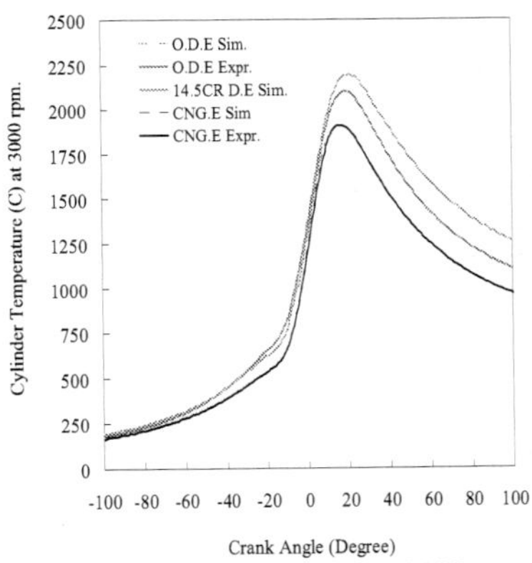

d) Cylinder temperature in 3000 rpm

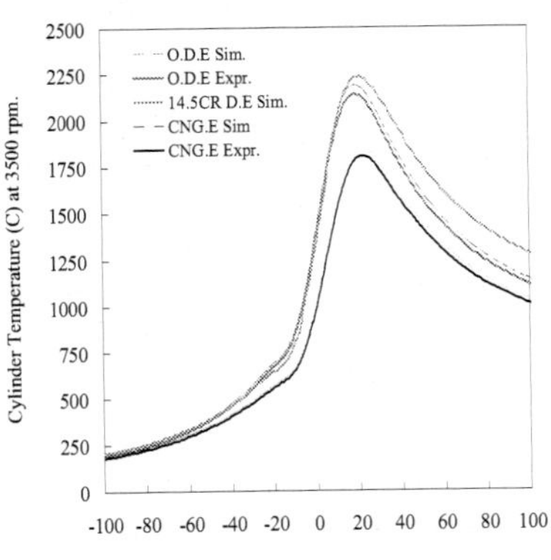

e) Cylinder temperature in 3500 rpm

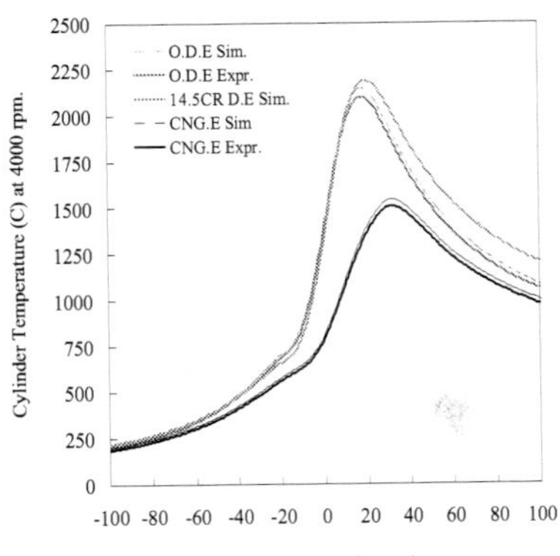

f) Cylinder temperature in 4000 rpm

Fig. 6. Cylinder temperature of diesel engine convert to MPI NG engine.

engine. In this investigation results are shown that the highest maximum in-cylinder temperature in combustion process is not declared in the highest engine speed. In the both of diesel engines, the highest maximum temperature in-cylinder is declared in 3500 rpm engine speed, because in this case the combustion is most excellent than the other condition and unburned fuel is lowest, so the temperature product from the combustion is the highest. In the both of diesel engines, the lowest maximum temperature in combustion process is in 1500 rpm engine speed. In this engine speed, the combustion process is not excellent and unburned fuel is highest than the other condition for compression stroke of compression ignition diesel engines. In the NG engine, the highest maximum temperature in-cylinder is declared in 1500 rpm engine speed, because in this case the combustion is most excellent than the other condition and unburned fuel is lowest, so the temperature product from the combustion is the highest. In the NG engine, the lowest maximum temperature in combustion process is in 4000 rpm engine speed. After 1500 rpm, the increasing engine speed, the combustion process is not excellent, the gas fuel density is lower, the air-fuel volume is lower and unburned fuel is highest than the other condition for compression stroke of natural gas spark assisted combustion engine.

The effect of diesel engine converted to multi point port injection NG engine on the maximum engine cylinder temperature is shown in Fig. 6a - Fig. 6f. In the 1500 rpm, conversion of modified diesel engine to NG engine has been increase the maximum engine cylinder temperature 5.29% and 1.94 %. In the 2000 to 4000 rpm engine speed, conversion of engine has been decrease the maximum engine cylinder temperature 2.18%, 5.88%, 15.34%, 23.36% and 28.15%.

4.4 Torque performance of multi point injection Natural Gas engine

The engine torque performance investigation results of diesel engine convert to multi point port injection NG engine are shown in Fig. 7a - Fig. 7c.

The indicated torque results are shown in Fig. 7a. The simulation and experimental investigation results are not similar in 0.1 to 1.0 %. The simulation results are higher than the experimental results, it caused by the not excellently setting and reading data experiment and the assumption in simulation with not have losses. The indicated torque represents the thermodynamic work transferred from the gas to the piston converted via geometry to a torque applied to the crankshaft. In the original diesel engine, the highest indicated torque is 24.3453 Nm and declared at 3000 rpm engine speed. In the modified diesel engine, the highest indicated torque is 23.76 Nm and declared at 3000 rpm engine speed. The diesel engines indicated torque performance profile shows that from the minimum engine speed at 1500 rpm to 3000 rpm as the point of the highest indicated torque, the indicated torque performance is increase if the engine speed is increased until 3000 rpm engine speed. After 3000 rpm engine speed, the increasing engine speed can be decrease the indicated torque. In the port injection NG engine, the highest indicated torque is 20.4798 Nm and declared at 2000 rpm engine speed. After 2000 rpm engine speed, the increasing engine speed can be decrease the indicated torque. Based on Fig. 7a, the conversion diesel engine to NG engine can reduce the engine torque performance. The maximum indicated torque of diesel engine convert to NG engine is reduced 15.88%. The increasing engine speed can be increase the deviation percentage of the indicated torque of diesel engines compared to NG engine. On 1500 to 4000 rpm, the NG engine has reduce indicated torque of 11.08%, 13.43%, 23.51%,

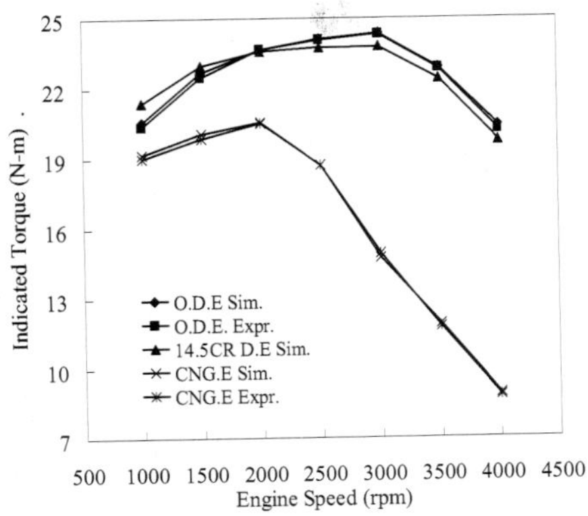

a) Indicated torque of NG engine compared to diesel engines

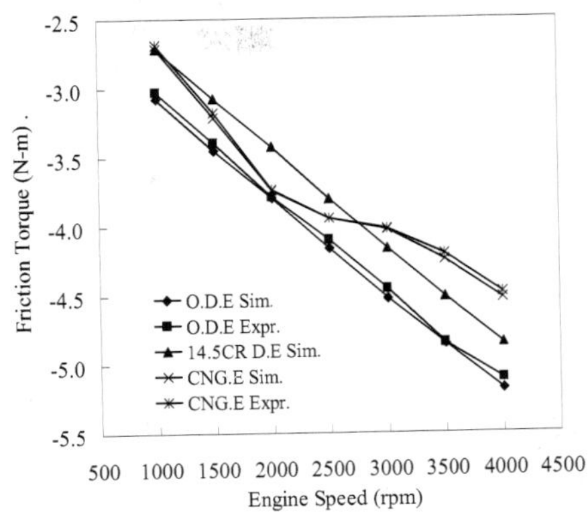

b) Friction torque of NG engine compared to diesel engines

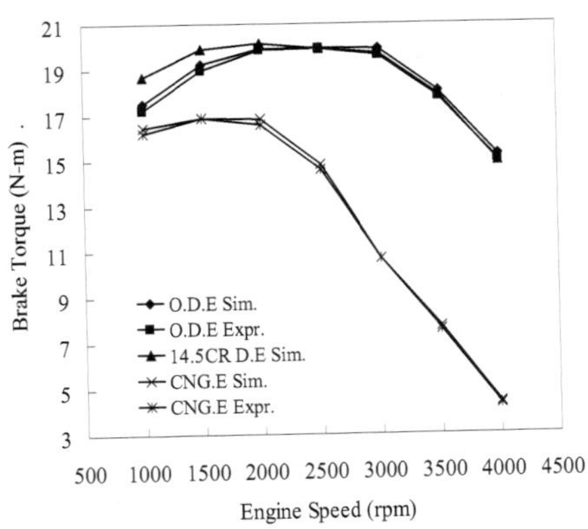

c) Brake torque of NG engine compared

Fig. 7. Torque performance of diesel engine convert to MPI NG engine.

41.29%, 50.34% and 56.54%. It meant that the thermodynamics energy were resulted from the diesel fuel combustion is higher than the NG fuel. It caused by the hydrocarbon chain, density and energy of diesel fuel of diesel engine is higher than the gas fuel of NG engine were ignited using spark assistant.

The friction torque result is sown in Fig. 7b. The simulation results are higher than the experimental results, it caused by the not excellently setting and reading data experiment and the assumption in simulation with not have losses. The highest friction torque in the original diesel engine is negative 5.18 Nm and modified diesel engine is negative 4.85 Nm declared at 4000 rpm engine speed. In the diesel engines, the friction torque is increase if the engine speed is increased. In the NG engine, the highest friction torque is negative 4.44 Nm and declared at 4000 rpm engine speed. In the NG engine and diesel engine, the friction torque is increase if the engine speed is increased. The conversion diesel engine to NG engine can increase the friction torque. The increasing engine speed can be increase the friction torque. Based on engine speed increasing, the percentage friction torque of NG engine is higher than the diesel engine in every engine speed. In 1500 rpm, the friction torque of NG engine is 15.02% and diesel engine is 15.21% from the indicated torque, where in this engine speed the reducing torque of diesel engine is higher than NG engine. In 2000 rpm, the friction torque of NG engine is 17.45% and diesel engine is 16.03% from the indicated torque, where in this engine speed the reducing torque of diesel engine is lower than NG engine. In 2500 rpm, the friction torque of NG engine is 20.57% and diesel engine is 17.29% from the indicated torque, where in this engine speed the reducing torque of NG engine is higher than diesel engine. In 3000 rpm, the friction torque of NG engine is 27.22% and diesel engine is 18.6% from the indicated torque, where in this engine speed the reducing torque of NG engine is exactly continue higher than diesel engine and the percentage deviation is increase. In 3500 rpm, the friction torque of NG engine is 36.4% and diesel engine is 21.23% from the indicated torque, where in this engine speed the reducing torque of NG engine is exactly continue higher than diesel engine and the percentage deviation is increase. In 4000 rpm, the friction torque of NG engine is 50.17% and diesel engine is 25.44% from the indicated torque, where in this engine speed the reducing torque of NG engine is exactly continue higher and the percentage deviation is increase. If the engine is running in higher than 1500 rpm engine speed, the NG engine friction torque is higher than diesel engine. In these cases, increasing engine speed will be increase the friction torque both of the engines, but the NG engine give the higher friction torque. It meant that the conversion of diesel engine to NG engine can be increase the friction torque of engine. It is caused by the natural gas as a fuel is less lubrication compared to diesel fuel as a liquid fuel and has the lubrication.

Brake torque of original diesel engine, modified diesel engine and NG engine are shown in Fig. 7c. The simulation results are higher than the experimental results, it caused by the not excellently setting and reading data experiment and the assumption in simulation with not have losses. Brake torque represents the torque available at the flywheel, after accounting for all friction and attachment losses as well as the acceleration of the crank train inertia. In the original diesel engine, the highest brake torque is 19.89 Nm declared at 2500 rpm engine speed. In the modified diesel engine, the highest brake torque is 20.12 Nm declared at 2000 rpm engine speed. In the diesel engine, the brake torque performance profile shows that from the low engine speed to medium engine speed, the brake torque performance is increase if the engine speed is increased. After 2500 rpm engine speed, the increasing engine speed can be decrease the brake torque. In the NG engine, the highest brake torque is 17.14 Nm and declared at 1500 rpm engine speed. After 1500 rpm engine speed, the increasing engine speed can be decrease the brake torque. The conversion of diesel engine to NG engine can reduce the engine brake torque. The maximum brake torque of modified diesel engine convert to NG engine is reduced 16.18%. The increasing engine speed can be increase the deviation point or percentage of the brake torque of diesel engine compared to NG engine. If the engines are run on 1500 to 4000 rpm, the conversion diesel engine to NG engine reduced brake torque 15.14%, 16.47%, 25.67%, 45.68%, 57.04% and 70.68%. Lower brake torque NG engine is caused by lower energy of and higher friction NG fuel.

4.5 Power performance of multi point injection Natural Gas engine

Indicated power performance of original diesel engine, modified diesel engine and port injection NG engine are shown in Fig. 8a. Indicated power represents the thermodynamic power transferred from the gas to engine. In the NG engine, the highest indicated power is 4.9 kW declared on 2500 rpm engine speed. From the minimum engine speed to 2500 rpm engine speed, the increasing engine speed can be increase the indicated power performance. After 2500 rpm to maximum engine speed, the increasing engine speed can be decrease the indicated power performance. The simulation and experimental investigation results are not similar in 0.05 to 0.5 %, it caused by the not excellently setting and reading data experiment and the assumed in simulation with not have losses. In the original diesel engine and modified diesel engine, the highest indicated power is 8.54 kW and 8.27 kW declared on 4000 rpm engine speed. The maximum indicated power of modified direct injection diesel engine convert to port injection NG engine is reduced 40.7%. The increasing engine speed can be increase the deviation point or percentage deviation of the indicated power of NG engine compared to diesel engine. From the 1500 to 4000 rpm, the conversion of modified diesel engine to NG engine can reduce indicated power 12.58%, 12.77%, 21.1%, 38.14%, 46.84% and 54.92%. It meant that the NG engine not applicable to run on high speed and very good power performance in medium speed. The combustion in high speed is produce lower power than the medium speed because the NG engine is developed in low compression ratio and thermodynamic energy. The effect of natural gas combustion of NG engine is lower in thermodynamic energy. The effect of lower thermodynamic energy is lower in indicated power.

Friction power of original diesel engine, modified diesel engine and port injection NG engine are shown in Fig. 8b. The simulation and experimental investigation results are not similar in 0.5 to 1.5 %. The simulation results are higher than the experimental results, it caused by the not excellently setting and reading data experiment and the assumption in simulation with not have losses. In the original and modified diesel engine, the highest friction power is negative 2.17 kW and 2.03 kW declared at 4000 rpm engine speed. The friction power profile shows that from the minimum engine speed at 1500 rpm to maximum engine speed on 4000 rpm, the friction power is increase if the engine speed is increased. In the multi point injection NG engine, the highest friction torque is negative 1.9 kW and declared on 4000 rpm engine speed. The friction power profile shows that increasing engine speed from the minimum engine speed at 1500 rpm to maximum engine speed at 4000 rpm is increase the friction power. In both of the NG engine and diesel engine, the friction power is increase if the engine speed is increased.

The conversion diesel engine to NG engine can increase the friction power performance percentage. The increasing engine speed can be increase the friction power of NG engine compared to diesel engine. The percentage friction power of NG engine is higher than the diesel engine in every engine speed. In 1500 rpm, the friction power of NG engine is 15.02% and diesel engine is 15.21% from the indicated power. In 2000 rpm, the friction power of NG engine is 17.45% and diesel engine is 16.03% from the indicated power. In 2500 rpm, the friction power of NG engine is 20.57% and diesel engine is 17.29% from the indicated power. In 3000 rpm, the friction power of NG engine is 27.22% and diesel engine is 18.6% from the indicated power. In 3500 rpm, the friction power of NG engine is 36.4% and diesel engine is

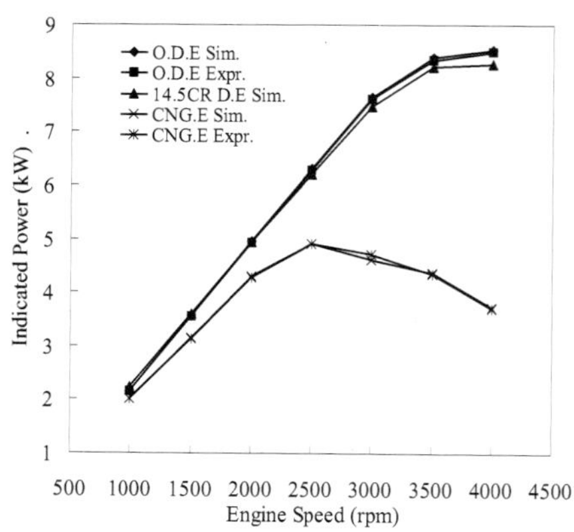

a) Indicated power of NG engine compared to diesel engines

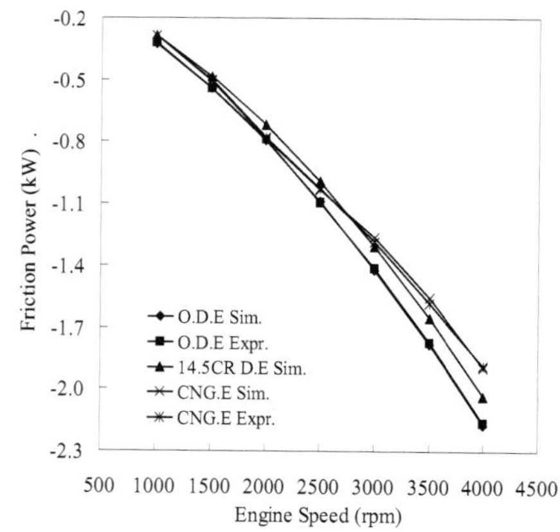

b) Friction power of NG engine compared to diesel engines

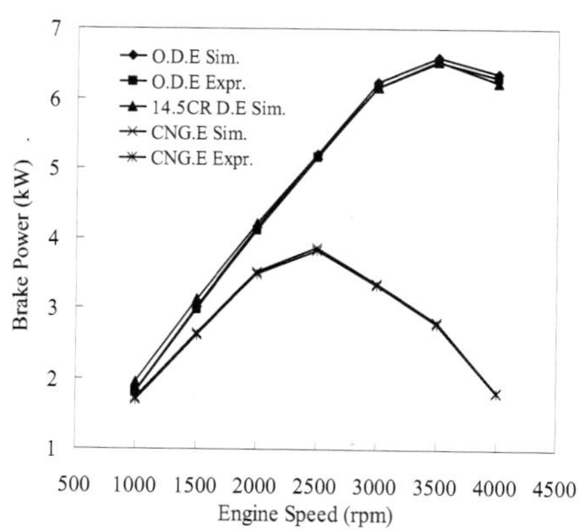

c) Brake power of NG engine compared

Fig. 8. Power performance of diesel engine convert to MPI NG engine.

21.23% from the indicated power. In 4000 rpm, the friction torque of NG engine is 50.17% and diesel engine is 25.44% from the indicated power. The percentage deviation of friction power compared to the indicated power of NG engine is lower than diesel engine in low engine speed until 1500 rpm. If the engine is running in higher than 1500 rpm engine speed, the NG engine friction power is higher than diesel engine. In these cases the increasing engine speed will be increase the friction power both of the engines and the NG engine give more friction power. It meant that the conversion of diesel engine to NG engine can be increase the friction power of engine. It is caused by the natural gas properties aspects, where in the natural gas as a fuel, the engine is less of the lubrication liquid compared to diesel fuel. Diesel fuel as a liquid fuel is have lubrication to reduce the friction.

Brake power performance of original diesel engine, modified diesel engine and port injection NG engine are shown in Fig. 8c. The brake power represents the power available at the flywheel, after accounting for all friction and attachment losses as well as the acceleration of the crank train inertia for brake torque. In the port injection NG engine, the highest brake power is 3.87 kW declared at 2500 rpm engine speed. In the NG engine, from 1500 to 2500 rpm, the increasing engine speed is increase brake power. After 2500 rpm engine speed, the increasing engine speed can be decrease the brake power. The simulation and experimental investigation results are not similar average in 1.0 %. The simulation results are higher than the experimental results, it caused by the not excellently setting and reading data experiment and the assumption in simulation with not have losses. In the original and modified diesel engine, the highest brake power is 6.6 kW and 6.54 kW declared at 3500 rpm engine speed. In the diesel engines, the brake power performance profile shows that from the minimum engine speed at 1500 rpm to 3500 rpm as the point of the highest brake power, brake power performance is increase if the engine speed is increased until 3500 rpm engine speed. After 3500 rpm engine speed, the increasing engine speed can be decrease the brake power. The conversion of four stroke direct injection diesel engine to multi point injection NG engine can reduce the engine brake power performance. The maximum brake power of direct injection diesel engine convert to multi point injection NG engine is reduced 41.374%. The increasing engine speed can be increase the deviation point or deviation percentage of the brake power of diesel engine compared to NG engine. So, the increasing engine speed can be increase the percentage deviation of brake power of the NG engine and diesel engines. If both of the engines are run on 1500 to 4000 rpm, the conversion of modified direct injection diesel engine to port injection NG engine has been reduced brake power 15.14%, 16.47%, 25.67%, 45.68%, 57.04% and 70.68%. The increasing engine speed can be increase the percentage deviation of power performance of NG engine compared to diesel engine. The reduction of the brake power in the NG engine is caused by low of brake torque. The low brake torque is caused by the low density of natural gas, low energy of natural gas, low volumetric efficiency, low flame speed, low compression ratio and higher friction of NG as an alternative fuel.

The investigation results on indicated power, friction power and brake power of the diesel engine converted to NG engine are shown that the engine conversion development can be decrease engine power performance. Increasing of engine speed over medium speed can be decreasing engine power performance of NG engine.

4.6 Fuel consumption of multi point injection Natural Gas engine

The simulation and experimental investigation results of the fuel consumption investigation of the diesel engines and NG engine are focuses on the indicated specific fuel consumption and brake specific fuel consumption. The investigation results are shown in Fig. 9.

Indicated specific fuel consumption (ISFC) of original diesel engine, modified diesel engine and port injection NG engine are shown in Fig. 9a. The ISFC is the nominal total fuel were used the engine to product their indicated power output. The simulation results are higher

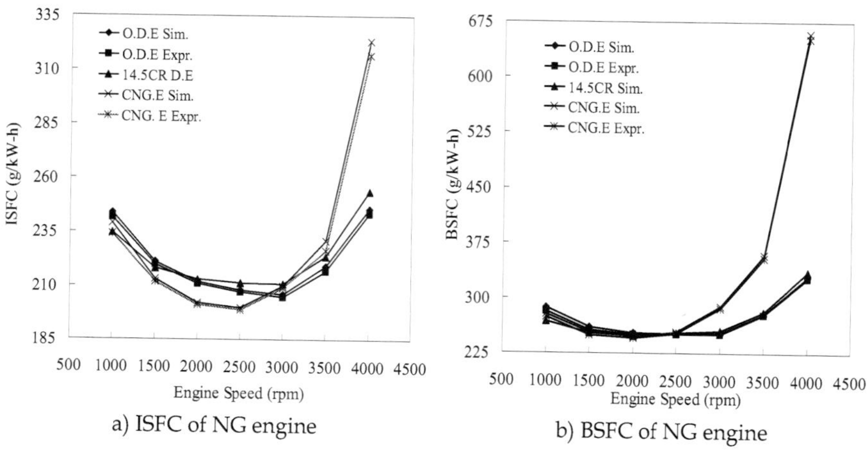

Fig. 9. Specific fuel consumption of diesel engine convert to MPI NG engine.

than the experimental results, it caused by the not excellently setting and reading data experiment and the assumption in simulation with not have losses. In the port injection NG engine, the fuel is entering to engine cylinder via intake port, the lowest ISFC is 199.593 g/kW-h declared on 2500 rpm engine speed and the highest ISFC is 323.532 g/kW-h

declared on 4000 rpm engine speed. From the minimum engine speed to 2500 rpm engine speed, the increasing engine speed can be decrease the ISFC. After 2500 to 4000 rpm, the increasing engine speed can be increase the ISFC. In the original and modified direct injection diesel engine, the fuel is injected directly to engine cylinder, the lowest ISFC are 205.93 g/kW-h and 210.98 g/kW-h declared on 3000 rpm engine speed. Then, the highest ISFC are 245.981 g/kW-h and 254 g/kW-h declared on 4000 rpm engine speed. The ISFC profile shows that from the minimum engine speed at 1500 rpm to medium speed, the increasing engine speed is decrease the ISFC. From medium speed to maximum engine speed at 4000 rpm, the increasing engine speed is increase the ISFC. The conversion of diesel engine to NG engine can be reduce the ISFC in the low to medium engine speed, but in the medium engine speed to high speed can be increase the ISFC. The minimum ISFC of multi point injection NG engine is reduced 2.9% compared to direct injection diesel engine. The maximum ISFC of port injection NG engine has increase 31.53% compared to direct injection diesel engine. The increasing engine speed can be increase the ISFC of NG engine compared to both of diesel engines. It meant that the NG engine not applicable to run on high speed, but it is very good ISFC in medium speed. The combustion in high speed is not excellent, not completely and high in unburned fuel, so the engine is produce lower power than the medium speed because the NG engine is developed in low energy and low density. The effect of the lower energy, density and power is increase the ISFC to the higher.

Brake specific fuel consumption (BSFC) of original diesel engine, modified diesel engine and port injection NG engine are shown in Fig. 9b. The BSFC is the nominal total fuel were used in the engine to product their brake power output. The simulation results are higher than the experimental results, it caused by the not excellently setting and reading data experiment and the assumption in simulation with not have losses. In the original and modified diesel engine, the lowest BSFC is 252 g/kW-h and 249.18 g/kW-h declared in 2500 and 2000 rpm engine speed. Then, the highest BSFC both of the diesel engines are 329.678 g/kW-h and 336.62 g/kW-h declared in 4000 rpm engine speed. In the original diesel engine, the BSFC profile shows that from 1500 rpm to 2500 rpm, the increasing engine speed can be decrease the BSFC. After 2500 rpm engine speed, the increasing engine speed can be increase the BSFC. In the modified diesel engine, the BSFC profile shows that from 1500 rpm to 2000 rpm, the increasing engine speed can be decrease the BSFC. But, after 2000 rpm engine speed, the increasing engine speed can be increase the BSFC. In the port injection NG engine, the lowest BSFC is 247.23 g/kW-h declared in 2000 rpm engine speed and the highest BSFC is 659 g/kW-h declared in 4000 rpm engine speed. From 1500 to 2000 rpm engine speed, the increasing engine speed of NG engine has been reduced the BSFC. After 2000 to 4000 rpm engine speed, the increasing engine speed can be increase the BSFC extremely. Its means, that the conversion of direct injection diesel engine to multi point injection NG engine can reduce the BSFC in lowest to medium engine speed and increase the BSFC in medium to highest engine speed. The minimum BSFC of modified direct injection diesel engine convert to port injection NG engine is reduced 0.78%. The maximum BSFC of modified direct injection diesel engine convert to port injection NG engine is increase 95.79%. The increasing engine speed can be increase the deviation point or deviation percentage of the BSFC of diesel engine compared to NG engine. In the low to medium engine speed, the unburned fuel in NG engine combustion is decrease so the out put can be product the higher torque and power. From the medium to the highest engine speed, the combustion of NG engine is not excellently, so the fuel consumption is increase because the unburned fuel is increase in the medium to highest speed. The effect of the increasing unburned fuel is can be decrease the engine brake torque and brake power. The effect of lower brake torque and brake power is increase the BSFC.

4.7 Mean effective pressure of multi point injection Natural Gas engine

The IMEP result is shown in Fig. 10a. The simulation and experimental investigation results are not similar average in 1.5 % for both of the engines. The simulation results are higher than the experimental results, it caused by the not excellently setting and reading data experiment and the assumption in simulation with not have losses. In the NG engine, the highest IMEP is 6.34952 bar declared on 2000 rpm engine speed. From the minimum engine speed to 2000 rpm engine speed, the increasing engine speed can be increase the IMEP. After 2000 rpm to maximum engine speed, the increasing engine speed can be decrease the IMEP. In the original and modified diesel engine, the highest IMEP is 7.5242 bar and 7.34 bar declared on 3000 rpm engine speed. The IMEP performance profile of diesel engines shows that from the minimum engine speed at 1500 to 3000 rpm engine speed, the increasing engine speed can be increase the IMEP, but from 3000 rpm to maximum engine speed at 4000 rpm, the increasing engine speed can be decrease the IMEP. Based on Fig.10a, the conversion diesel engine to NG engine can reduce the IMEP. The maximum IMEP of modified direct injection diesel engine convert to port injection NG engine has been reduced

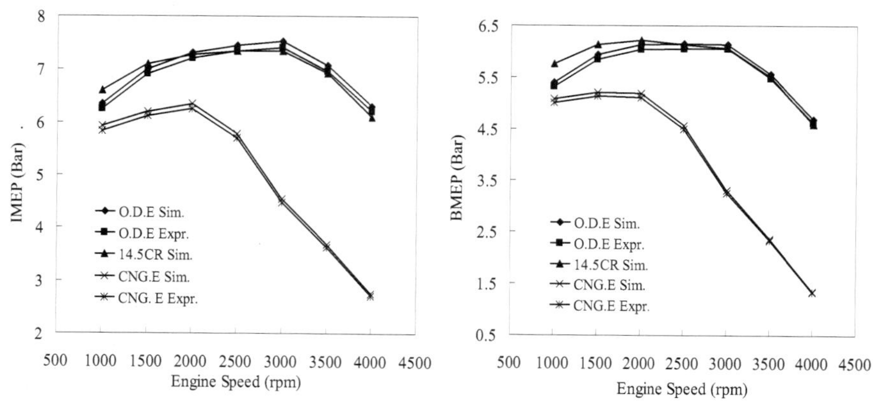

a) IMEP of NG engine b) BMEP of NG engine

Fig. 10. Mean effective pressure of diesel engine convert to MPI NG engine.

13.54%. The increasing engine speed can be increase the deviation point or percentage deviation of the IMEP of NG engine compared to diesel engine. If the engine is run on 1500 to 4000 rpm, the conversion of modified diesel engine to port injection NG engine It meant that the NG engine not applicable to run on high speed and very good IMEP performance in low to medium speed. The combustion in high speed produces lower IMEP than the medium speed because the NG engine is developed in low energy. Effect of lower energy is causing the lower indicated torque. The lower indicated torque is can be reducing the lower IMEP.

Brake mean effective pressure (BMEP) performance of original diesel engine, modified diesel engine and port injection NG engine are shown in Fig. 10b. The BMEP is the external shaft work done per unit displacement. The simulation results are higher than the experimental results, it caused by the not excellently setting and reading data experiment and the assumption in simulation with not have losses. In the port injection NG engine, the highest BMEP is 5.21 bar declared on 1500 rpm engine speed. From 1500 to 1500 rpm, the increasing engine speed is increase BMEP. After 1500 rpm engine speed, the increasing engine speed can be decrease the BMEP. In the original and modified diesel engine, the highest BMEP are 6.148 bar and 6.22 bar declared at 2500 and 2000 rpm engine speed. The BMEP performance profile of original and modified diesel engine shows that the BMEP performance is increase if the engine speed is increased until 2500 and 2000 rpm engine speed. After 2500 and 2000 rpm engine speed, the increasing engine speed can be decrease the BMEP. The conversion of four stroke direct injection diesel engine to multi point injection NG engine can be reducing the engine BMEP performance. The maximum BMEP of modified direct injection diesel engine convert to port injection NG engine is reduced 16.18%. If the engines are run on 1500 to 4000 rpm, the conversion of diesel engine to port injection NG engine reduced BMEP 15.14%, 16.48%, 25.67%, 45.68%, 57% and 70.68%.

The investigation results on mean effective pressure performance such as indicated and break mean effective pressure of direct injection diesel engine converted to port injection NG engine are shown that the engine conversion development can be decrease the engine power performance. The reduction of the mean effective pressure in the NG engine is caused by lower energy, density and higher friction of compressed natural gas as an alternative fuel for engines. The increasing of engine speed over the medium speed can be decrease the engine mean effective pressure performance of NG engine. The highest of engine mean effective pressure of NG engine is declared in medium engine speed and the lowest mean effective pressure is declared in highest engine speed. The increasing engine speed will be increase the percentage deviation of mean effective pressure performance of NG engine compared to diesel engine.

5. Conclusion

The original diesel engine cylinder pressure is higher than the modified diesel engine and CNG engine. It caused the compression ratio of NG engine is lower than the original diesel engine and the combustion energy output of diesel fuel has to produce the highest power than natural-gas fuel. Another that, the density of natural-gas fuel is lower than the diesel fuel. The increasing engine speed of the diesel engine has increase the maximum temperature in engine. Unfortunately, the increasing engine speed of NG engine has been decrease maximum temperature inin- engine. The decreasing engine speed of diesel engines has been decrease maximum temperature inin- engine. Decreasing engine speed of NG engine has increase maximum temperature in engine. Engine torque, power, mean effective pressure and efficiency performance of original direct injection diesel engine cylinder pressure is higher than the modified diesel engine and sequential port injection dedicated NG engine. It meant that the thermodynamics energies were resulted from the diesel fuel combustion is higher than the NG fuel. It caused by the hydrocarbon chain, density and energy of diesel fuel of the diesel engine is higher than the gas fuel of NG engine were ignited using to spark assistant. The fuel consumption of NG engine is higher than the diesel engine. Fuel consumption is increased because the unburned fuel is increased in the medium to the highest speed of NG engine. The effect of the increasing unburned fuel is can decrease the engine brake torque and brake power. The effect of lower brake torque and brake power increase the brake specific fuel consumption..

6. References

Aslam, M.U.; Masjuki, H.H.; Kalam, M.A.; Abdesselam, H.; Mahlia, T.M.I.; Amalina, M.A. (2006). An experimental investigation of CNG as an alternative fuel for a retrofitted gasoline vehicle. *Fuel* 85: 717–724.

Bakar, R. A.; Sera, M.A.; Mun, W.H. (2002). Towards The Implementation of CNG Engine: A Literature Review Approach To Problems And Solutions. *Proc. of BSME-ASME International Conference on Thermal Engineering*. December 31, 2001 – January 2, 2002. Dhaka. BSME-ASME. 2002.

Brombacher, E.J. (1997). Flow Visualisation of Natural Gas Fuel Injection, *MSc Thesis*, University of Toronto.

Catania, A.E.; Misul, D.; Spessa, E.; Vassallo, A. (2004). Analysis of combustion parameters and their relation to operating variables and exhaust emissions in an upgraded multivalve bi-fuel CNG SI engine. *SAE Paper*. 2004-01-0983.

Chiu, J.P. (2004). Low Emissions Class 8 Heavy-Duty, On-Highway Natural Gas and Gasoline Engine. *SAE Paper*. 2004-01-2982.

Cho, H. M.; He, Bang-Quan (2007). Spark ignition natural gas engines - A review. *Energy Conversion and Management* 48: 608–618.

Czerwinski J.; Comte P.; Janach.W.; Zuber P. (1999). Sequential Multipoint Trans-Valve-Injection for Natural Gas Engines. *SAE Paper*. 1999-01-0565.

Czerwinski, J.; Comte, P.; Zimmerli, Y. (2003). Investigations of the Gas Injection System on a HD-CNG-Engine. *SAE Paper*. 2003-01-0625.

Duan, S.Y. (1996). Using Natural Gas in Engines: Laboratory experience with the use of natural gas fuel in IC engines. *IMechE Seminar Publication*, 1996: 39-46.

Durrel, Elizabeth.; Allan, Jeff.; Law, Donald.; Heath, John. (2000). Installation and Development of a Direct Injection System for a Bi-Fuel Gasoline and Compressed Natural Gas. *Proceeding of ANGV Conference 2000*. October 17 – 19, 2000. Yokohama: ANGV.

Evan, R.L.; Blaszczyk, J.; and Matys, P. (1996). An Experimental and Numerical Study of Combustion Chamber Design for Lean-Burn Natural Gas Engines. *SAE Paper*. 961672.

Ganesan, V. (1999). *Internal Combustion Engines*. N.Delhi: Tata McGraw-Hill.

Hollnagel, Carlos.; Borges, Luiz.H.; Muraro, Wilson. (1999). Combustion Development of the Mercedes-Benz MY1999 CNG-Engine M366LAG. *SAE Technical Paper* 1999-01-3519.

Hollnagel, C.; Neto, J. A. M.; Di Nardi, M. E.; Wunderlich, C.; Muraro, W. (2001). Application of the natural gas engines Mercedes-Benz in moving stage for the carnival 2001 in Salvador City. *SAE Paper*. 2001-01-3824.

Johansson, B. and Olsson, K. (1995). Combustion Chambers for Natural Gas SI Engines Part I: Fluid Flow and Combustion. *SAE Paper*. 950469.

Kato, K.; Igarashi, K.; Masuda, M.; Otsubo, K.; Yasuda, A.; Takeda, K.; Sato, T. (1999). Development of engine for natural gas vehicle. *SAE Paper*. 1999-01-0574. 1999.

Kawabata, Y and Mori, D. (2004). Combustion diagnostics and improvement of a prechamber lean-burn natural gas engine. *SAE Paper*. 2004-01-0979.

Klimstra, J. (1989). Carburetors for Gaseous Fuels - On Air-to-Fuel Ratio, Homogeneity and Flow Restriction. *SAE Paper* 892141.

Kowalewicz, Andrzej. (1984). *Combustion System of High-Speed Piston I.C. Engines*. Warszawa: Wydawnictwa Komunikacji i Lacznosci..

Kubesh, J.T.; Podnar, D.J.; Gugliemo, K.H. and McCaw, D. (1995). Development of an Electronically-Controlled Natural Gas-Fueled John Deere Power Tech 8.1L Engine. *SAE Paper* .951940.

Kubesh, John. T.; Podnar, Daniel. J. (1998). Ultra Low Emissions and High Efficiency from On-Highway Natural Gas Engine. *SAE Paper* 981394.

Lino, Paolo.; Maione, Bruno.; Amorese, Claudio. (2008). Modelling and predictive control of a new injection system for compressed natural gas engines. *Control Engineering Practice*, 2008. 16 (10): 1216-1230.

Mbarawa, M.; Milton, B.E.; Casey, R.T. (2001). Experiments and modelling of natural gas combustion ignited by a pilot diesel fuel spray. *Int. J. Therm. Sci.* 40: 927–936.

Oullette, P.; Mtui, P.L.; Hill, P.G. (1998). Numerical Simulation of Directly Injected Natural Gas and Pilot Diesel Fuel in a Two Stroke Compression Ignition Engine. State of Alternative Fuel Technologies. *SAE Paper* 981400.

Ouellette, Patric. (2000). High Pressure Direct Injection (HPDI) of Natural Gas in Diesel Engines. Proceeding ANGVA 2000 Conference. Yokohama.

Pischinger, S.; Umierski, M.; Hüchtebrock, B (2003). New CNG concepts for passenger cars: High torque engines with superior fuel consumption. *SAE Paper*. 2003-01-2264.

Poulton, M.L. (1994). *Alternative Fuels for Road Vehicles*, Computational Mechanics Publication, London.

Ren, W and Sayar, H. (2001). Influence of nozzle geometry on spray atomization and shape for port fuel injector. *SAE Paper*. 2001-01-0608.

Sera, M.A.; Bakar, R.A.; Leong, S.K. (2003) CNG engine performance improvement strategy through advanced intake system. *SAE Paper*. 2003-01-1937.

Shashikantha.; Parikh, P.P. (1999). Spark ignition producer gas engine and dedicated compressed natural gas engine-Technology development and experimental performance optimization. *SAE Paper*.1999-01-3515.

Shiga, S.; Ozone, S.; Machacon, H. T. C.; Karasawa, T. (2002). A Study of the Combustion and Emission Characteristics of Compressed-Natural-Gas Direct-Injection Stratified Combustion Using a Rapid-Compression-Machine, *Combustion and Flame* 129:1–10.

Suga, T.; Muraishi, T.; Brachmann, T.; Yatabe, F. (2000). Potential of a natural gas vehicle as EEV. *SAE Paper*. 2000-01-1863.

Tilagone, R.; Monnier, G.; Chaouche, A.; Baguelin, Y. and Chauveron D.S. (1996). Development of a High Efficiency, low Emission SI-CNG Bus Engine. *SAE Paper*. 961080.

Vermiglio. E; Jenskins, T.; Kleliszewski, M.; Lapetz, J.; Povinger, B.; Willey, R.; Herber, J.; Sahutske, K.; Blue, M, and Clark, R. (1997). Ford's SULEV Dedicated Natural Gas Trucks. *SAE Paper*. 971662.

Wang, D. E and Watson, H. C. (2000). Direct injection compressed natural gas combustion and visualization. *SAE Paper* 2000-01-1838.

Wayne, W. S.; Clark, N. N.; Atkinson, C. M. (1998). A parametric study of knock control strategies for a bi-fuel engine. *SAE Paper*. 980895.

Zastavniouk, Oleg. (1997). Study of Mixing Phenomena in a Dual Fuel Diesel Engine Air Intake Manifold. *MSc Thesis*. University of Alberta.

Zhao, F.; Lai, M.; Harrington, D.L. (1995). The Spray Characteristics of Automotive Port Fuel Injection - A Critical Review. *SAE Paper*. 950506.

19

The Influence of Modified Atmosphere on Natural Gas Combustion

Małgorzata Wilk and Aneta Magdziarz
AGH University of Science and Technology, Krakow, Poland

1. Introduction

Coal technology, dominant in Poland, ensures efficient production of electricity and heat. However, a large exploitation of existing reserves and the growing demand for electricity causes interest in other available fuels, primarily natural gas. Environmentally friendly modern technologies are constantly looking for the possibility of using energy sources other than coal. Research is carried out on various innovative technologies such as CO_2 capture and storage or unconventional combustion. Combustion in the oxygen-enriched gas mixtures at the first was conducted in the steel industry and metallurgy, which require very high temperatures to heat the metal and pig iron. Such a process can be classified as the future technology known as the clean combustion technologies.

The use of natural gas in the metallurgical processes requires the precise technological regime, therefore, it is difficult to make changes in the process (Amann, Kanniche & Bouallou, 2009; Choi & Katsuki, 2001; Flamme, 2001). However, appropriately organized gas combustion process can lead to a reduction in CO_2 emissions, which is required by applicable laws imposed by the European Union. The European Union environmental priorities promote the development of the new technologies of energy production from the conventional sources (coal and natural gas), e.g. oxy-combustion (Andersson & Johnsson, 2007; Czakiert et al., 2006; Davidson, 2007; H.K. Kim, Kim, Lee & Ahn, 2007; Kotowicz, 2007; Lampert & Ziębik, 2007; Li, Yan & Yan, 2009; Muskał et al., 2008; Seepana & Jayanti, 2009; Simpson & Simon, 2007; Szlęk et al., 2009; Tan et al., 2002). The oxy-fuel combustion process is conducted in an atmosphere enriched in oxygen which means that the reactor is supplied by the combustion gas mixture, in which the oxygen concentration is higher than the concentration of oxygen in the air. The study of the modified air combustion appears in scientific literature. There are various modified oxidizing atmospheres like O_2/N_2, $O_2/N_2/CO_2$, O_2/CO_2. It should be emphasized that the majority of results of the oxy-combustion process refer to coal combustion, because of large amount of deposited coal in Poland (Buhre et al., 2005; Chen, Liu & Huang, 2007; Croiset, Thambimuthu & Palmer, 2000; Czakiert, Nowak & Bis, 2008; Kim et al., 2007; Normann et al., 2008). The research effort is focused on the conventional coal, fluidised bed combustion and co-generation solutions with the coal gasification (Lampert & Ziębik, 2007; Tan et al., 2002). Oxy-combustion is the subject of research in many international research institutions (Chalmers University of Technology, Sweden, University of Leeds UK, CANMET Energy Technology Centre, Canada, Chicago USA Research Centre, Tokyo Institute of Technology, Japan, University of Newcastle Australia). In Poland, the research of oxy-combustion concerns mainly coal (Czestochowa University of Technology), but the natural gas research is also carried out in the field of high-temperature combustion gas HTAC technique (Normann, 2008; Seepana & Jayanti, 2009). Fuel combustion processes are the main source of environmental pollution. In many branches of industry, the main fuel is natural gas, because of the possibility of obtaining a high temperature process e.g. in the glass industry and in the manufacture of cement, the process of oxy-combustion can be applied, or air combustion enriched with oxygen. This of course raises the temperature of the combustion process, which is associated with increased concentrations of NO_x. Therefore, oxy-combustion is used simultaneously with the process of eliminating nitrogen from the air combustion (largely responsible for the formation of NO) and replacing it into exhaust gas (RFG - Recycled Flue Gas.). Despite the very high temperatures in the chamber (e.g. 1600 °C in the melting process), the concentration of nitrogen oxides may be lower, due to the elimination of nitrogen from combustion air. The primary obstacle to the propagation of oxy-combustion has so far been the high cost of obtaining pure oxygen. Since the pure oxygen production technologies have been improved and costs have been reduced, oxy-combustion can be applied in many industrial processes. Industry may be interested in this technology, where the conditions in the very high temperature contribute to the formation of large amounts of thermal NO. An additional advantage of the oxy-combustion is high combustion efficiency, lower volume of exhaust gases, less fuel consumption and therefore lower CO_2 and NO_x. Combustion of natural gas in O_2/CO_2 atmosphere allows optimising the combustion process. The results of experimental studies indicate that the combustion of natural gas in O_2/CO_2 with recycling exhaust gases has positive effects on reducing CO_2 emissions, a noticeable reduction or even elimination of NO_x and improve the efficiency of the furnace (Lampert & Ziębik, 2007; M.Wilk, Magdziarz & Kuźnia, 2010). It was noted that the major advantage of this technology is the ability to apply it in an existing energy plants (H.K. Kim, Kim, Lee & Ahn, 2007; Simpson & Simon, 2007).

Therefore, the problem seems to be both interesting and promising. The complex nature of the combustion process of natural gas causes the obtained experimental results which are not always repeatable so this issue requires further study.

2. Mechanisms of CO and NO_x formation in natural gas combustion processes

In many industries, because of the possibility of obtaining high temperatures, the primary fuel is natural gas, whose main component (ca. 98%) is methane. Combustion of natural gas is the source of the formation of many pollutants. During the combustion of natural gas in metallurgical furnaces the air pollutants are formed. These are nitrogen oxides, carbon oxides, and possibly trace amounts of hydrocarbons. The composition of natural gas varies slightly. The number and types of pollutants emitted from combustion are related to the composition of the fuel, the type of oxidation atmosphere used and the temperature prevailing in the combustion chamber.

2.1 The mechanism of CO formation

The relatively high CO emissions in combustion processes of natural gas occur in the following cases:

- Staging combustion for reduction of NO_x emissions,
- Inadequate mixing of air and fuel,
- Very rapid cooling of combustion products in the cold boundary layer of the combustion chamber.

The formation of CO in the flame is one of the main paths of reaction in the mechanism of combustion of hydrocarbons. Fuel hydrocarbons during the chemical degradation can be partially converted to CO. The formation of CO is done quickly, right at the beginning of the flame. The proposed overall reaction of formation of CO is as follows (for gaseous and liquid hydrocarbons) and volatiles of solid fuels. (Wilk, 2002; Bartok & Sarofim, 1991)

$$C_nH_m + \frac{n}{2}O_2 \rightarrow nCO + \frac{m}{2}H_2 \tag{1}$$

Oxidation of CO is strongly catalysed by even small amounts of hydrogen or its compounds with oxygen. There are two paths of CO oxidation. Main path of oxidation takes place at $T > 1500$ K (at $p \approx 0{,}1$ MPa) and is as follows

$$CO + OH \rightarrow CO_2 + H \tag{2}$$

The second path CO oxidation takes place at T = 1000 - 1500 K and at p > 1 MPa is as follows

$$CO + HO_2 \rightarrow CO_2 + OH \tag{3}$$

HO_2 radical is produced in the recombination reaction:

$$H + O_2 + M \rightarrow HO_2 + M \tag{4}$$

HO_2 radical concentration is comparable to the concentration of OH radical reactions and rapid reactions (Bartok & Sarofim, 1991)

$$H + O_2 \leftrightarrow OH + O \tag{5}$$

$$O + H_2 \leftrightarrow OH + H \tag{6}$$

$$O + H_2O \leftrightarrow OH + OH \tag{7}$$

$$H + H_2O \leftrightarrow OH + H_2 \tag{8}$$

Competitive reactions to the above and CO oxidation reactions are the following recombination reaction

$$H + H \rightarrow H_2 \tag{9}$$

$$H + OH \rightarrow H_2O \tag{10}$$

$$H_2 + OH \rightarrow H_2O + H \tag{11}$$

H and OH radicals can meet together on the walls where it comes to their exhaustion, and it causes stopping of the CO oxidation reaction. In practice it is the result of too rapid cooling of exhaust gases below 1000 K. This is the main reason of no oxidation of CO to CO_2. The direct oxidation of CO by reaction:

$$CO + O_2 \rightarrow CO_2 + O \tag{12}$$

is very unlikely, because this reaction is very slow (the activation energy of this reaction is very high).

2.2 The mechanisms of NO formation

The primary adverse products of high temperature combustion are the nitrogen oxides NO_x. Knowledge of the mechanism of the NO_x formation can identify thermal and chemical conditions of furnaces and control of combustion processes, which affects prevention or reduction of the harmful substances emissions. The source of nitrogen oxides is nitrogen in the fuel and molecular nitrogen from the air. In combustion processes, there are two main types of nitrogen oxides: nitrogen monoxide, NO, and dioxide, NO_2. The main component of NO_x produced during natural gas combustion is NO, whose share in total NO_x emissions is typically at least 95%, and the rest is NO_2. The concentration of other oxides N_2O, N_2O_3 and N_2O_5 is low. The amount of NO_x in the exhaust gases depends mainly on the combustion temperature, excess air ratio and residence time in the reaction zone. There are four different mechanisms of the formation of NO:

- the thermal mechanism,
- the prompt mechanism,
- by means of N_2O
- the fuel NO_x.

During the combustion of natural gas, containing mostly methane, and not containing chemically bound nitrogen, the main way of NO_x formation mechanism in natural gas combustion is the thermal mechanism.

2.2.1 The thermal mechanism of NO formation

The thermal mechanism is based on the oxidation reactions of nitrogen from the air supplied for combustion, the rate becomes significant above 1400 °C. These reactions were first described by Zeldovich (Bartok & Sarofim, 1991; Warnatz et al., 2006; Tomeczek & Gradoń, 1997; Muzio & Quartucy, 1997;. Flamme, 1998):

$$O + N_2 \xrightarrow{k_1} NO + N \quad k_1 = 1{,}8 \cdot 10^{14} \exp(-318\,kJ \cdot mol^{-1} / (RT))\ cm^3 / (mol \cdot s) \tag{13}$$

$$N + O_2 \xrightarrow{k_2} NO + O \quad k_2 = 6{,}4 \cdot 10^{9}\, T \exp(-26\,kJ \cdot mol^{-1} / (RT))\ cm^3 / (mol \cdot s) \tag{14}$$

$$N + OH \xrightarrow{k_3} NO + H \quad k_3 = 3{,}8 \cdot 10^{13} \tag{15}$$

The term "the thermal NO" is connected with a very high activation energy due to a strong, triple-atomic bond in the molecule N_2. It is a highly endothermic reaction that runs with considerable speed only at temperatures higher than 1400 °C. The reaction of a hydrocarbon radical OH plays an important role in the combustion of humidified hydrocarbon fuels.

The rate of formation of NO is expressed by the equation (Warnatz et al., 2006; Gardiner, 2000)

$$\frac{d[NO]}{dt} = k_1[O][N_2] + k_2[N][O_2] + k_3[N][OH] \tag{16}$$

Atomic nitrogen is formed by the reaction (13) and is consumed in the reaction (14) and (15), hence the rate of formation is:

$$\frac{d[N]}{dt} = k_1[O][N_2] - k_2[N][O_2] - k_3[N][OH] \tag{17}$$

Taking into account the fact that reactions (14) and (15) are so fast that their products reach the equilibrium state, the preliminary assumption can be given:

$$\frac{d[N]}{dt} = 0 \tag{18}$$

and then equation (16) takes the form

$$\frac{d[NO]}{dt} = 2k_1[O][N_2] \tag{19}$$

k_1, k_2, and k_3 are the rate constants of the reaction.

The rate of formation of NO is controlled by the first, slow reaction of Zeldovich (13). If one molecule of one atom of NO and N is produced in this reaction, it immediately becomes the second particle produced by rapid reaction of NO (14). The formation of thermal NO takes place just behind the flame front in a zone of high temperatures (t > 1400 °C). The basic ways of reducing emissions of thermal NO in combustion processes are the reduction of the temperature, shortening the stay of the reagents in the zone of high temperatures and reducing the local concentration of N_2 and O_2.

Malte and Pratt proposed the mechanism of taking into account the role of N_2O in NO formation by the following reaction at temperatures lower than 1800 K (Kordylewski, 2008 (in Polish), Steele et al, 1995)

$$O + N_2 + M \rightarrow N_2O + M \tag{20}$$

$$N_2O + O \rightarrow NO + NO \tag{21}$$

$$N_2O + O \rightarrow N_2 + O_2 \tag{22}$$

These reactions, together with the reactions (13) and (14) in the literature are called "the extended thermal mechanism". The formation of NO by a mechanism of N_2O formation is particularly important at lower temperatures (T < 1200 °C) in the flames rather poor ($\lambda > 1$). Important role in the formation of N_2O plays the kind and characteristics of the third body M. It can be assumed that H_2O or its dissociation products (O, H, OH) affect the course of the reaction (20) (Wilk, 2002).

2.2.2 The prompt mechanism of NO formation

Fenimore conducting the experimental research on the combustion of rich mixtures ($\lambda < 1$) with various hydrocarbons (methane, ethane, propane) found that relatively high concentrations of NO occur in combustion zone just before the flame where the temperature does not exceed 750 °C. Fenimore was confident that there should be another mechanism for the generation of thermal NO and called him a "prompt" which is immediate. In the hydrocarbon flames there are not only O, H, OH radicals but also hydrocarbon radicals, where its highest concentration was found in the reaction zone of the flame. The hydrocarbon radicals are capable of activating N_2 reaction with the formation of nitrogen oxides in the flame. Fenimore assumed that CH_i hydrocarbon radicals react with nitrogen air according to the following reaction:

$$CH_2 + N_2 \rightarrow HCN + NH \tag{23}$$

$$CH + N_2 \rightarrow HCN + N \tag{24}$$

$$C + N_2 \rightarrow CN + N \tag{25}$$

Generally, these reactions can be written:

$$CH_x + N_2 \rightarrow HCN \text{ and other radicals } (CN, NH, N...) \tag{26}$$

The forming of amino and cyano compounds, among which the most important are HCN, NH, and CN, is oxidized to NO in the flame with the participation of radicals H, O, OH (Glarborg, Alzueta & Dam-Johansen, 1998). The prompt NO is formed very quickly during combustion. Velocity of its formation is like combustion velocity. The amount of formed NO depends weakly on temperature, but strongly depends on the local concentration of N_2. The NO prompt participates in further reactions running along the flame and lose their individuality.

2.2.3 The fuel NO mechanism

The amount of nitrogen in the fuel composition is very diverse. Nitrogen, in the gaseous fuel, is not chemically bonded with the combustible gas. However, it can occur as free molecular nitrogen N_2, which is the source of thermal or prompt nitrogen oxides. It is assumed therefore, that during the gas combustion the fuel nitrogen oxides do not occur.

2.2.4 The formation of NO_2

Miller and Bowman gave the most plausible explanation of the mechanism of formation of NO_2. They assumed that as a result of diffusion of H radicals from the flame in the area of low temperature ($T < 750$ °C) and high concentration of O_2 the reaction occurs (Miller & Bowman, 1989; Hori, 1986).

$$H + O_2 + M \rightarrow HO_2 + M \tag{27}$$

At the same time from the flame to low temperature zone NO diffuses, which comes in a rapid reaction with a peroxide radical reaction of HO_2 by:

$$NO + HO_2 \leftrightarrow NO_2 + OH \tag{28}$$

In parallel, at higher temperatures, the NO_2 decomposition reactions may occur:

$$NO_2 + H \rightarrow NO + OH \tag{29}$$

$$NO_2 + O \rightarrow NO + O_2 \tag{30}$$

Therefore, under normal conditions of combustion (T = 1000 - 1700 °C, $\lambda \leq 1{,}3$) the final of NO_2 emission is low and does not exceed 5% of the total NO_x emissions. Significant impact on the formation of NO_2 next to a low temperature is connected with a high combustion pressure and the presence of hydrocarbons. The increase in the pressure favours the growth of the concentration of NO_2.

3. Experimental apparatus

The investigation of oxy-combustion process was conducted on a laboratory reactor containing a specially designed combustion burner, oxidizer preparation system, and temperature system, flow rate of combustion substrates control system, exhaust gas analysis system and exhaust gas system (Fig. 1).

The study included the characterization of the basic parameters of combustion, and above all took into account the effect of oxygen and carbon dioxide concentrations in the oxidizer on the exhaust gas composition and temperature profile along the combustion chamber.

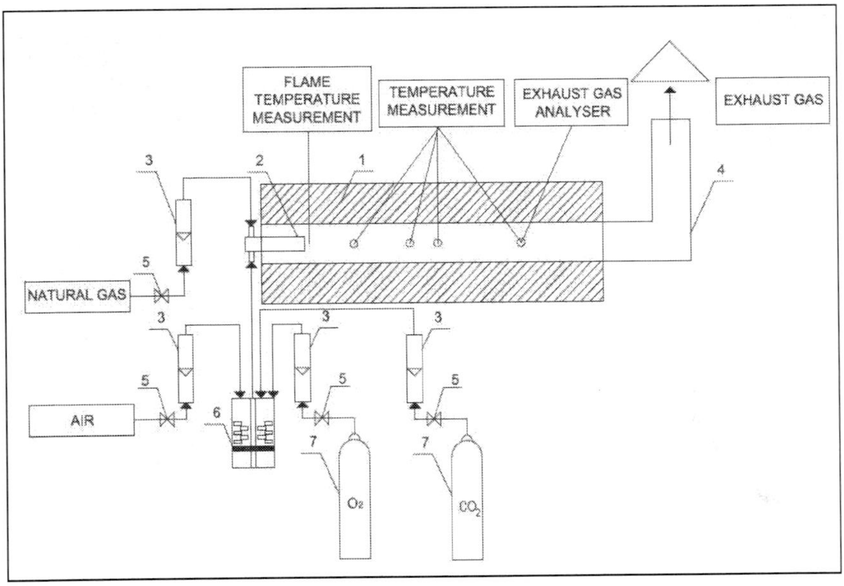

Fig. 1. Scheme of the experimental apparatus: 1 - combustion chamber, 2 - burner, 3 - rotameter, 4 - exhaust gas system, 5 - control valves, 6 - mixer, 7 - cylinder with oxygen, 8 - cylinder with carbon dioxide.

The specially designed kinetic burner, so called "pipe in pipe", was used. The inside diameter of burner was 34 mm and the outer diameter was 47 mm (Fig. 2). The combustion chamber was made from the heat-resistant steel with a length of 1310 mm and a diameter of 160 mm. Thermal isolation chamber was made of ceramic fibre with a thickness of 150 mm. Along the combustion chamber were holes, which allowed the measurement or analysis of the exhaust gas temperature inside the furnace. Oxidizer preparation system consisted of oxidizer pressure cylinders containing oxygen, air supply system with a fan and a mixer filled with the ceramic fittings, which enable more efficient mixing of the streams brought air and oxygen. The flow rates control system of combustion substrates included the rotameters and control valves. Mixing of fuel with an oxidizer took place in space between two pipes inside the burner, and was enforced by the system of holes in the outer pipe of the burner. The homogeneous mixture was introduced into the combustion chamber.

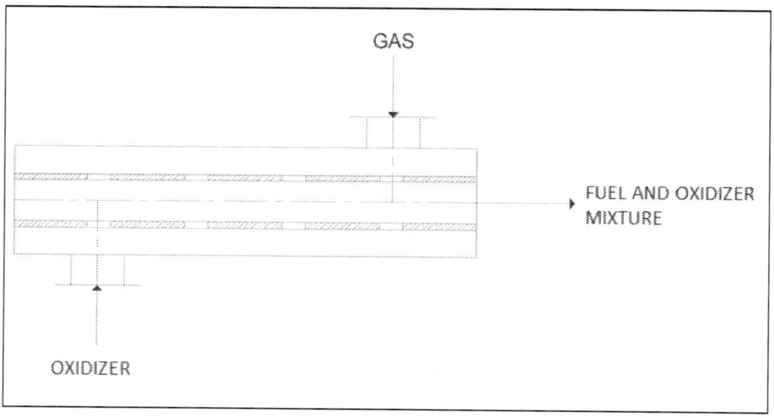

Fig. 2. Scheme of a specially designed burner.

To measure the flame temperature the thermocouple (PtRh-Pt) is installed in the wall of the chamber combined with a digital millimetre. The temperature profile along the length of the furnace was measured by NiCr-NiAl thermocouples at four points connected to a multichannel temperature recorder Czaki WRT-9 consisting of a microprocessor thermometer EMT 200, the switch places the PMP test. Concentrations of the combustion products (O_2, CO, CO_2, NO) were measured by the means of a gas analyser Testo 350 XL.

4. Results and discussion

The investigation of the oxy-combustion of natural gas is takes into account three types of oxidizing mixtures with an increased oxygen contents: 25% O_2, 27% O_2, 29% O_2. The parameters of the combustion process of the natural gas with the addition of oxygen to the combustion air are shown in Table 1. The study concerned the natural methane rich gas from the city with the following average composition: CH_4 - 98%, C_2-C_4 - 0,9%, N_2 - 1%, CO_2 - 0,1%.

To study the combustion of natural gas in modified atmosphere three options were carried out: the first - for the selected excess air ratio, the second - assumed a steady stream of gas

$\dot{V}_g$, m³/h	$\dot{V}_{oxidizer}$, m³/h	O_2, %	$\dot{V}_{CO_2}$, m³/h	T_{flame}, °C	$T_{exhaust\ gas}$, °C (in measured point)
0,8	6,3 - 9,5	21 - 29	0 - 2	1231 - 1316	885 - 1277

Table 1. The parameters of the combustion process of natural gas.

and air mixture for each oxygen-enriched air, and the obtained values λ resulted from the mixture of air and oxygen, and third option concerned the study of CO_2 addition to oxidizer whereas the air combustion was oxygen-enriched up to 27% and excess air ratio was: $\lambda_1 = 1{,}25$. The excess air ratio was calculated by taking into account the increased oxygen content in the mixture.

The first investigations were conducted for three values of the air excess ratio: $\lambda_1 = 1{,}15$, $\lambda_2 = 1{,}20$, $\lambda_2 = 1{,}25$. The flow of gas was constant $\dot{V}_g = 0{,}8$ m³/h. For each value of λ, by choosing an appropriate air and oxygen flows ratio, the air was oxygen-enriched in the range 21 - 29%. Table 2 and 3 shows the average values of measured experimental data.

The effect of oxygen addition to the combustion air on CO concentration for three different excess air ratios λ_1 - λ_3 was presented in Figure 3. For all the cases an increase in CO concentration with increasing oxygen concentration in the oxidising mixture was observed. The higher the excess air ratio, the lower was the concentration of carbon monoxide. It was observed that a small addition of oxygen around 4% slightly increases the concentration of CO, and the addition of 6% - 8% strongly increases the CO concentration. There was no

$\lambda_1 = 1{,}15$				
$\dot{V}_{air}$, m³/h	$\dot{V}_{O_2}$, m³/h	O_2, %	CO, ppm	NO, ppm
8,67	0	21	175	125
7,014	0,38	25	192	231
6,317	0,534	27	230	347
5,715	0,663	29	301	533
$\lambda_2 = 1{,}20$				
$\dot{V}_{air}$, m³/h	$\dot{V}_{O_2}$, m³/h	O_2, %	CO, ppm	NO, ppm
9,04	0	21	161	114
7,32	0,40	25	170	228
6,59	0,56	27	224	352
5,96	0,69	29	293	521
$\lambda_3 = 1{,}25$				
$\dot{V}_{air}$, m³/h	$\dot{V}_{O_2}$, m³/h	O_2, %	CO, ppm	NO, ppm
9,421	0	21	112	78
7,620	0,418	25	118	206
6,866	0,580	27	179	305
6,212	0,720	29	215	458

Table 2. The results of experimental studies of natural gas combustion in oxygen enriched atmosphere for $\dot{V}_g = 0{,}8$ m³/h, $\dot{V}_{CO_2} = 0$.

λ_3 = 1,25; O_2 = 27 %		
$\dot{V}_{CO_2}$, m^3/h	CO, ppm	NO, ppm
0	179	305
0,5	182	144
1	210	73
1,5	242	43
2	355	24

Table 3. The results of experimental studies of natural gas combustion in CO_2 and oxygen enriched atmosphere for $\dot{V}_g$ = 0,8 m^3/h, λ_3 = 1,25; O_2 = 27 %.

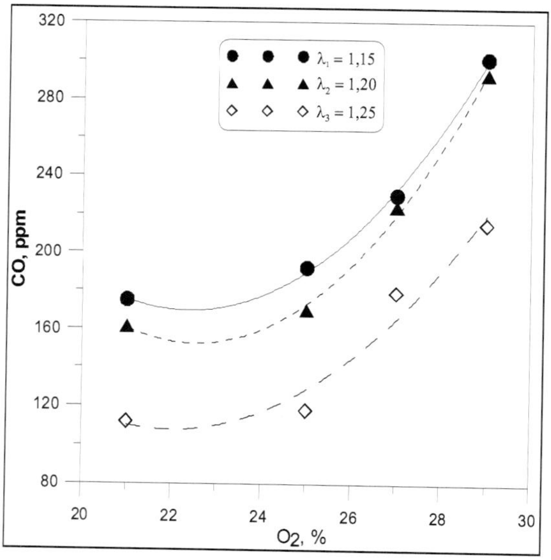

Fig. 3. The effect of oxygen addition to the combustion air on CO concentration in the natural gas combustion process with the oxygen enriched air depending on the excess air ratio.

difference in the concentration of CO depending on the excess air ratio used if air has been enriched to 29% O_2. It should be noted that the increase in O_2 concentration decreased the flow of oxidising mixture, so the observed increases of the concentrations of CO and NO are quite large. Figure 4 shows the effect of the oxygen addition to the combustion air for three different excess air ratios λ_1 - λ_3 of on the concentration of nitrogen oxide NO. Addition of oxygen to the oxidizer also causes an increase in NO concentration in the exhaust gas. NO concentration for all cases is at the same level.

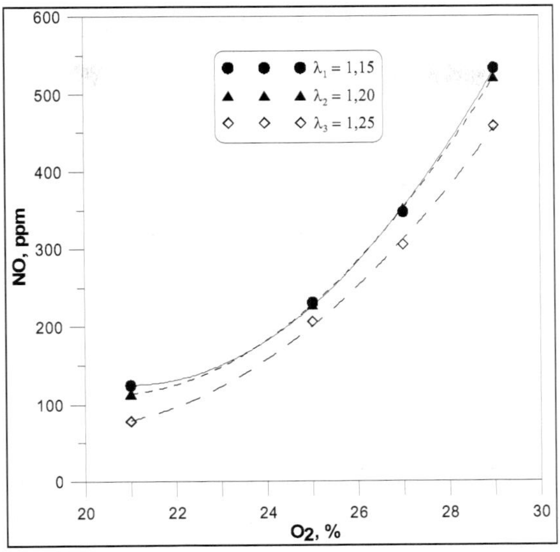

Fig. 4. The effect of oxygen addition to the combustion air on NO concentration in the natural gas combustion process with the oxygen enriched air depending on excess air ratio.

In the second series, the investigations were carried out for a fixed flow of gas and air in natural gas combustion in oxygen-enriched atmosphere. The combustion air was enriched from 21 to 29% oxygen, thereby generating the excess air ratios λ from 1,18 to 1,63. The results are presented in Table 3 and the graphs (Fig. 5 - 7).

λ	O_2, %	CO, ppm	NO, ppm	$T_{exhaust}$, °C
1,18	21	23	85	991
1,40	25	42	167	987
1,52	27	50	203	985
1,63	29	71	290	983

Table 4. The results of natural gas combustion in oxygen-enriched atmosphere for $\dot{V}_g$ = 0,8 m³/h and $\dot{V}_{air}$ = 9 m³/h, T_{flame} = 1140 - 1160 °C.

Figure 5 shows the effect of oxygen addition to the CO concentration in the combustion of natural gas in oxygen enriched air for constant flows of gas and air ($\dot{V}_g$ = 0,8 m³/h and $\dot{V}_{air}$ = 9 m³/h).

The increase of oxygen concentration in the oxidizer from 21 to 29% O_2, an increase in CO concentrations up to 300% was observed. Addition of oxygen in the oxidizer does not improve the complete combustion of the gas. CO molecule is more stable than CO_2 and its oxidation by oxygen is very slow, it is unknown whether pure carbon monoxide CO could be burnt (Kotowicz & Janusz, 2007). However, the oxidation of CO is possible by means of even a small concentration of hydrogen and its compounds, e.g. addition of a small concentration of water vapour would cause the CO combustion.

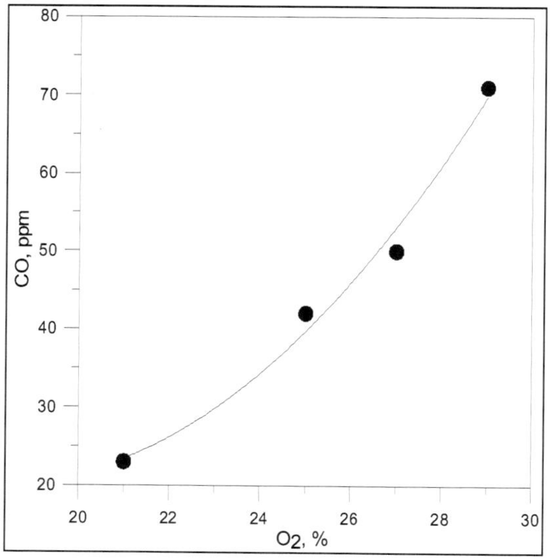

Fig. 5. The effect of oxygen addition on the CO concentration in the natural gas combustion in oxygen-enriched oxidizer for $\dot{V}_g$ = 0,8 m³/h and $\dot{V}_{air}$ = 9 m³/h.

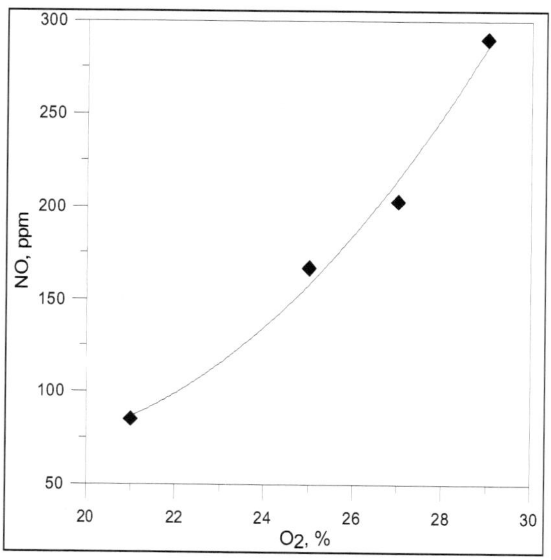

Fig. 6. The effect of oxygen addition on the NO concentration in the natural gas combustion in oxygen-enriched oxidizer for $\dot{V}_g$ = 0,8 m³/h and $\dot{V}_{air}$ = 9 m³/h.

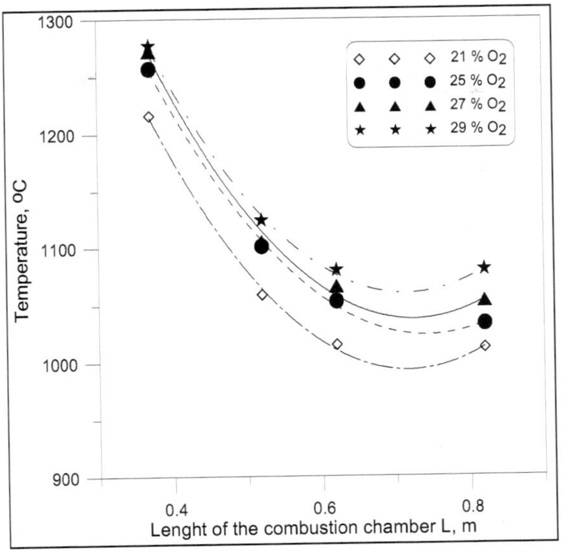

Fig. 7. The exhaust gas temperature profile along the length of the combustion chamber during combustion of natural gas in oxygen-enriched air.

The consequence of fuel combustion in oxygen-enriched atmospheres in O_2 + CO_2 system (replacing nitrogen by CO_2), is the reduction of pollutants emissions, mainly NO_x (Amann et. al, 2009; Li, Yan & Yan, 2009; Muskał et al., 2008). In the studied system (modified atmosphere: oxygen-enriched air combustion) lower concentrations of nitrogen oxide NO are not obtained, but on the contrary, more than threefold increase in NO concentration was observed (Fig. 6). Addition of oxygen increases the flame temperature, and therefore it also increases the NO concentration. Increased concentration of NO may also result from the larger concentration of oxygen in the reacting system, making easier the connection between the air nitrogen and oxygen at high temperature.

The temperature profile along the length of the furnace was also performed (Fig. 7). It was found that the exhaust gas temperature decreased along the furnace chamber in the measured points of the furnace, as well as an increased concentration of oxygen in the oxidizer. Oxygen enrichment of combustion air is done in order to raise the temperature of combustion in the furnace and to raise the growth rate of fuel combustion, which causes shortening of the flame. The reduction of the flame length explains the decrease of the exhaust gas temperature in the measured fixed points along the length of the furnace. The larger was the addition of oxygen, the flame was shorter and the temperature was lower in the test point. The flame temperature with the addition of oxygen increased from 1140 to 1160 °C.

In the course of the experiment it was observed that addition of oxygen resulted in visible changes in shape and colour of the flame, the bright crown of the burner nozzle and a very bright flame colour. Opportunity to observe these changes undoubtedly comes from reduction of the flame length. Adding oxygen to the combustion air also caused a change in the sound of the furnace operation to the louder, more intense one, associated with changes in the fuel and oxidizer mixture flow within the combustion chamber.

The study of natural gas combustion was also conducted in another modified atmosphere: $O_2/CO_2/N_2$. The process was studied under excess air ratio λ_3 = 1,25 and the oxidizer was oxygen-enriched up to 27%. CO_2 was added to the oxidizer in the range of 0 to 2 m^3/h.

The effect of carbon dioxide addition on the CO and NO concentrations of the studied process were investigated (Figure 8 and Figure 9). The CO concentration increased with CO_2 addition. The addition of 2 m^3/h of CO_2 has generated two times larger CO concentration comparing to the process operated in conventional atmosphere O_2/N_2. NO concentration, in contrary, decreased with increasing addition of CO_2. The maximum of CO_2 addition (2 m^3/h) decreased NO concentration ca. 12 times. The decrease of NO was connected with lower temperature obtained in the combustion chamber, because of large CO_2 thermal capacity. That fact confirms the flame temperature measured close to the burner nozzle in the axes of the flame presented in Figure 11. The decrease of flame temperature was observed with increasing CO_2 addition to the oxidizer. The thermal NO formation, main way of NO formation during natural gas combustion, takes place just behind the flame front in a zone of high temperature (t > 1400 °C). Therefore, the efficient method of the NO reduction is the lower range of temperature used, shortening the stay of the reagents in the zone of high temperatures and reducing the local concentration of N_2 and O_2.

The exhaust gas temperature profile was conducted along the combustion chamber including the CO_2 addition (Figure 10). The exhaust gas temperature decreases with CO_2 addition along the combustion chamber. The more CO_2 is added the lower temperature is obtained.

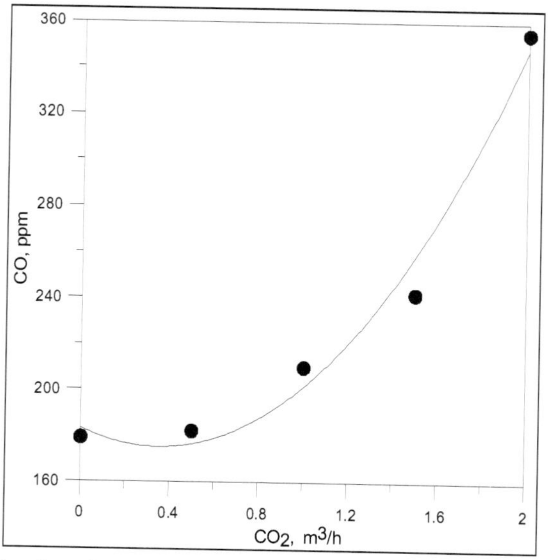

Fig. 8. The effect of carbon dioxide on the concentration of CO in the combustion of natural gas with oxygen-enriched oxidizer up to 27% O_2 and λ_3 = 1,25.

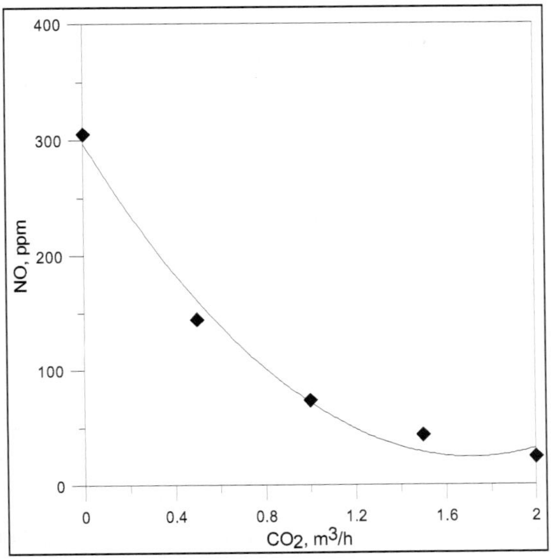

Fig. 9. The effect of carbon dioxide on the concentration of NO in the combustion of natural gas with oxygen-enriched oxidizer up to 27% O_2 and $\lambda_3 = 1{,}25$.

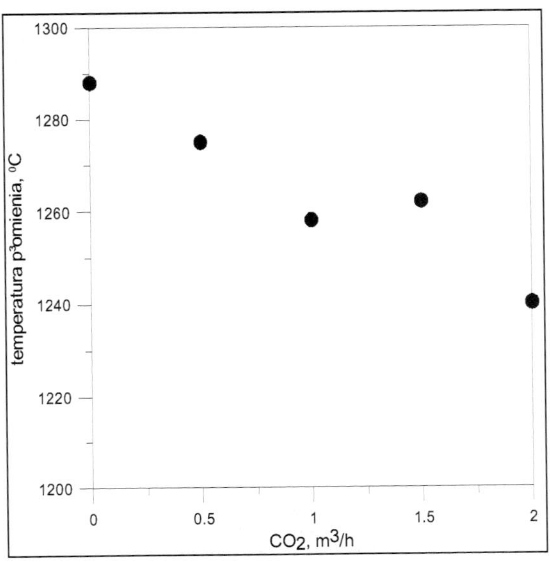

Fig. 10. The effect of carbon dioxide on the flame temperature in the combustion of natural gas with oxygen-enriched oxidizer up to 27% O_2 and $\lambda_3 = 1{,}25$.

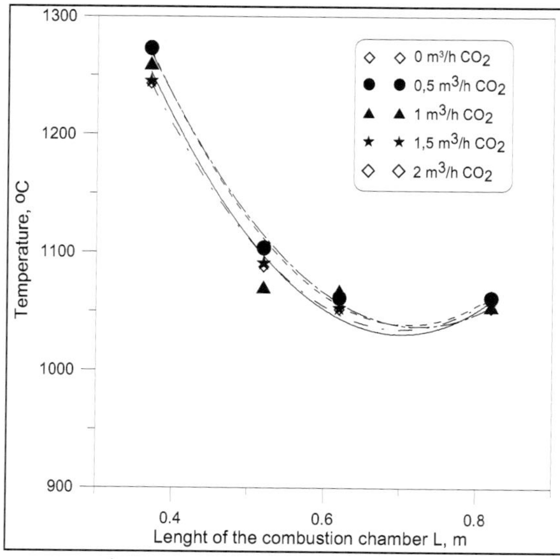

Fig. 11. The effect of carbon dioxide on the exhaust gas temperature measured along the combustion chamber in the combustion of natural gas with oxygen-enriched oxidizer up to 27% O_2 and λ_3 = 1,25.

5. Conclusion

In industrial processes, which are required to maintain very high temperatures such as in glass or in the production of cement, the oxy-combustion process can be used, or oxygen-enriched combustion air. This of course raises the temperature of the combustion process, which is associated with an increase in the concentration of NO_x. Therefore, the oxy-combustion should be used simultaneously with the process of elimination of nitrogen from combustion air (largely responsible for the formation of NO). Otherwise, the addition of oxygen increases the NO concentration, and therefore undesirable effect is achieved. A mixture of oxygen and carbon dioxide by replacing the air can lead to lower concentrations of nitrogen oxides by eliminating nitrogen from combustion air.

6. References

Amann J.-M., Kanniche M., Bouallou C. (2009). Natural gas combined cycle power plant modified into an O_2/CO_2 cycle for CO_2 capture, *Energy Conversion and Management*, Vol. 50, pp. 510-521,

Andersson K., Johnsson F. (2007). Flame and radiation characteristics of gas-fired O_2/CO_2 combustion, *Fuel*, Vol. 86, pp. 636-668,

Bartok W., Sarofim A. F. (1991). Fossil fuel combustion. A source book. *John Wiley and Sosns, INC*, New York, Chichester, Brisbane, Toronto, Singapore,

Buhre B.J.P., Elliott L.K., Sheng C.D., Gupta R.P., Wall T.F. (2005). Oxy-fuel combustion technology for coal-fired power generation, *Progress in Energy and Combustion Science*, Vol. 31, pp. 283-307,

Chen J.C., Liu Z.S., Huang J.S.: Emission characteristics of coal combustion in different O_2/N_2, O_2/CO_2 and O_2/RFG atmosphere, Journal of Hazardous Materials 2007, 142, s. 266-271,

Choi G., Katsuki M. (2001). Advanced low NO_x combustion using highly preheated air, *Energy Conversion and Management*, Vol. 42, pp. 639-652,

Croiset E., Thambimuthu K., Palmer A. (2000). Coal Combustion in O_2/CO_2 mixtures compared with air, *The Canadian Journal of Chemical Engineering*, Vol. 78, pp. 402-407,

Czakiert T., Bis Z., W. Muskała, Nowak W. (2006). Fuel conversion from oxy-fuel combustion in a circulating fluidized bed, *Fuel Processing Technology*, Vol. 87, s. 531-538,

Czakiert T., Nowak W., Bis Z (2008). Spalanie w atmosferach modyfikowanych tlenem, kierunki rozwoju dla kotłów CWF, *Energia i Ekologia*, Vol. 10, pp. 713-718 (in Polish),

Davidson J. (2007). Performance and cost of power plants with capture storage of CO_2, *Energy*, Vol. 32, pp. 1163-1176,

Flamme M. (1998): Low NO_x Combustion Technologies for High Temperature Application. *Proceedings of 2nd International Symposium on Advanced Energy Conversion System and Related Technologies*, (December 1-3), Nagoya, Japan, pp. 152-159,

Flamme M. (2001). Low NO_x combustion technologies for high temperature applications, *Energy Conversion and Management*, Vol. 42, pp. 1919-1935,

Gardiner W.C. (2000). Gas-Phase Combustion Chemistry. *Springer-Verlag*, New York,

Glarborg P., Alzueta M., Dam-Johansen K. (1998). Kinetic Modeling of Hydrocarbon/Nitric Oxide Interactions in a Flow Reactor. *Combustion and Flame*, Vol. 115, pp. 1-27,

Hori M. (1986). Experimental Study of Nitrogen Dioxide Formation in Combustion Systems. *Twenty-first Symposium (International) on Combustion/The Combustion Institute*/,pp. 1181-1188,

Kim H.K., Kim Y., Lee S.M., Ahn K. Y. (2007). NO reduction in 0.03-0.2 MW oxy-fuel combustor using flue gas recirculation technology, *Proceedings of the Combustion Institute*, Vol. 31, pp. 3377-3384,

Kordylewski W. (2008). Combustion and fuels, *Oficyna wydawnicza Politechniki Wrocławskiej*, Wrocław, Poland (in Polish),

Kotowicz J., Janusz K. (2007). Sposoby redukcji emisji CO_2 z procesów energetycznych, *Rynek Energii*, Vol. 1 (in Polish),

Lampert K., Ziębik A. (2007). Comparative analysis of energy requirements of CO_2 removal from metallurgical fuel gases, *Energy*, Vol. 32, pp. 521-527,

Li H., Yan J., Yan J., Anhenden M. (2009) Impurity impacts on the purification process in oxy-fuel combustion based CO_2 capture and storage system, *Applied Energy*, Vol. 86, pp. 202-213,

Miller J. A., Bowman C. T. (1989). Mechanism and Modeling of Nitrogen Chemistry in Combustion. *Progress in Energy and Combustion Science*, Vol. 15, pp. 287-338,

Muskał W., Krzywański J., Czakiert T., Sekret R., Nowak W. (2008). Spalanie w atmosferach modyfikowanych O_2 i CO_2, *Energetyka*, Vol. 10, pp. 669-671,

Muzio L.J., Quartucy G.C. (1997). Implementing NO_x Control: Research to Application. *Progress in Energy and Combustion Science*, Vol. 23, pp. 233-266,

Normann F., Andersson K., Leckner B., Johnsson F. (2008). High-temperature reduction of nitrogen oxides in oxy-fuel combustion, *Fuel*, Vol. 87, pp. 3579-3585,

Seepana S., Jayanti S. (2009). Flame structure and NO generation in oxy-fuel combustion at high pressures, *Energy Conversion and Management*, Vol. 50, pp. 1116-1123,

Simpson A., Simon A.J. (2007). Second law comparison of oxy-fuel combustion and post combustion carbon dioxide separation, *Energy Conversion and Management*, Vol. 48, pp. 3034-3045,

Steele R.C., Malte P.C., Nicol D.G., Kramlich J.C. (1995). NO_x and N_2O in Lean-Premixed Jet-Stirred Flames. *Combustion and Flame*, Vol. 100, pp. 440-449,

Szlęk A., Wilk R.K, Werle S., Schaffel N. Clean technologies generating energy from coal and the prospect of flameless combustion, *Rynek Energii*, Vol. 4, pp. 39-45 (in Polish),

Tan Y., Douglas M.A., Thambimuth K.V. (2002). CO_2 capture using oxygen enhanced combustion strategies for natural gas power plants, *Fuel*, Vol. 81, pp. 1007-1016,

Tomeczek J., Gradoń B. (1997). The Role of Nitrous Oxide in the Mechanism of Thermal Nitric Oxide Formation within Flame Temperature Range. Combustion Science and Technology, Vol. 125, pp. 159-180,

Warnatz J., Maas U., Diable R.W. (2006). Combustion. Physical and Chemical Fundamentals, Modeling and Simulation, Experimentals, Pollutant Formation. *4th Edition Springer-Verlag*, Berlin Heidelberg,

Wilk M., Magdziarz A., Kuźnia M. (2010). The influence of oxygen addition into air combustion on natural gas combustion process, *Rynek Energii*, Vol. 5, pp. 32-36 (in Polish),

Wilk R. (2002). Low-emission combustion. *Wydawnictwo Politechniki Śląskiej*, Gliwice, Poland,

Zhang N., Lior N. (2008). Two novel oxy-fuel power cycles integrated with natural gas reforming and CO_2 capture, *Energy*, Vol. 33, pp. 340-351.

20

Modelling a SOFC Power Unit Using Natural Gas Fed Directly

Nguyen Duc Tuyen and Goro Fujita
Shibaura Institute of Technology
Japan

1. Introduction

1.1 Literature review and objective of this chapter

Completed models covering dynamic characteristics of those types of DGs are not openly available. The necessary task is to study these dynamic models based on the literature and any available operational data on DGs. Simulation of various types of DGs in a suitable software environment is the key step in analyzing the dynamic characteristics of DGs and designing the control strategies. In fact, computer simulation plays a vital role in the design and analysis of power system. Designing power systems without computer simulation is extremely laborious, time consuming, error-prone and expensive. Especially, in the new research field as DGs, computer simulation in an industrial environment with regard to the time in shortening the overall design process as compared to assembling and testing the components in the laboratory and deciding on the optimum values for components and controller parameters.

Among many types of FC, high-temperature fuel cells such as the solid oxide fuel cells (SOFC) have the potential for centralized power generation as well as combined heat and power. This chapter employs the SOFC model method. Especially, in this simulation, the rate of temperature change and load following ability will be included. In short, it consists of 3 main mathematical models, namely, the electrochemical model, the heat balance model and the power conditioning unit model.

The electrochemical model is to calculate output voltage, to regulate the fuel and air streams and to represent the ability to follow the load chance of SOFC.

The heat balance model is to calculate thermal energy inside SOFC stack as well as operating temperature. The heat exchangers are included in this simulation to represent the practical application when using to increase the temperature of input air and methane. Because of the high SOFC operating temperature, if there have no preheat, manufacture still have to set up a preheat system using electricity from other source or a small part of SOFC output power to elevate input species temperature to prevent thermal shock which can damage materials. Therefore, using heat generated inside SOFC stack for preheat can increase overall efficiency.

Finally, there are two important results that this chapter points out. The first is load following ability of SOFC power unit. The fuel cell control is achieved by adjusting the input volume of gas and air and controlling real power output. The two control loops which are in SOFC itself and in the DC-DC inverter make the SOFC power unit strongly following flexible change of load. And the second is heat balance inside fuel cell system with HX included.

1.2 Different modelling approaches

Depending upon the application, different models are available in the open literature and there are large differences in the level of details in the models presented. This section presents a review of the work of selected authors relevant to the model developed in the present work. Research work of SOFCs modelling has been begun since 1980's and there have been a lot of models developed so far. Initial models were lumped mass models and there were a lot of uncertainties in the results due to lack of experimental data as well as mature approaches. Increasing experimental research during early 1990's focused many such issues and established many empirical relationships to accurately predict the performance of SOFC. Also, due to the increase in calculation capabilities, it was possible to create more detailed models. During the late 1990's, several projects were initiated for detailed single SOFC modelling. Today modelling research is pursued in both detailed single SOFC modelling and system level stack modelling. Nevertheless, lumped models still continue to attract the attention of researchers due to their simplicity and small calculation time. Lumped models are considered over detailed models when it comes to predict accurately the FCs overall thermodynamic and electrical performance. A large amount of experimental data and mathematical relations exists for components such as air and fuel compressors, heat exchangers, thus these components can be modeled fairly accurately upon the lumped approach. Accordingly, lumped models are also easier to adjust to experimental data. The disadvantage of lumped SOFC models is that they can only account for mean values of the parameters and more detailed investigation of the cell is needed to check for undesirable effects such as thermal cracking, coking or exceeding temperature limits locally. This problem may be partly solved by using a detailed model to test the validity of the results after using a lumped model for system calculations. Obviously, implementing a detailed SOFC model in the system model gives the most accurate results.

References (Achenbach & Elmar, 1995), (Wang & Nehrir, 2007) provide a basic approach for fuel cell modelling suitable for distributed generation, however not discussed in details about SOFC. A SOFC model has been developed by various researchers in (Li & Chyu, 2003), (Ali Volkan Akkaya & Erdem, 2009), (N. Lu, 2006), (David L. Damm, 2005), (Mitsunori Iwata, 2000), (Tomoyuki Ota, 2003), (Xiongwen Zhang, 2007), (Takanobu Shimada, 2009), (S. Campanari, 2004) and (A.C. Burt, 2004) taking its thermodynamic effect into consideration which concentrated on the effects of temperature changes on the output voltage response. Heat balance is considered in specific model configuration more than in general and the detail calculation seems to be complex. Some did not consider the dynamics of the chemical species. Reference (Tadashi Gengo, 2007) points out empirical responses of real model considering temperature inside SOFC combine with output voltage, current and power. (Takanobu Shimada, 2009), (D. Sanchez, 2008) considers internal CH_4 reformer, fuel and air input temperature are increased. But there is not any research caring about using SOFC exhaust to take full advantages of high operating temperature. (M. Uzunoglu, 2006), (Caisheng Wang, 2007), (Caisheng Wang, 2005) take the double layer charging effect into account but not SOFC. Recently research (Takanobu Shimada, 2009), (P. Piroonlerkgul, 2009),

(Graham M. Goldin, 2009) approach this chapter ideal when taking HX into account but calculate heat balance in another way and for specific SOFC configuration. David investigated the transient behavior of a stand-alone SOFC caused by a load change in (Achenbach & Elmar, 1995), (Kourosh Sedghisigarchi, 2004), (J. Padulles, 2000). However, the built model is simple for evaluate the real response but these simplified models consider constant cell temperature. (M.Y. El-Sharkh, 2004) considers only the dynamic characteristic of Power Conditioning Unit system. (P.R. Pathapati, 2005) represents PEM dynamic model which does not consider concentration loss. (S.H.Chan, 2002) deals with HX model. A physically based model for tubular SOFC was developed in (Caisheng Wang, 2007).

A transient dynamic model of SOFC will be proposed in this chapter. Electrochemical and thermal simulations of a SOFC reported in all reference will be used to identify the key parameters of this SOFC system from a single cell to a N_0 single fuel cell connected in series. The cell's terminal voltage during a load change was discussed. Overall heat balance inside SOFC power unit effecting on operating temperature afterward on output voltage will be pointed out.

1.3 Fuel cell

Fuel cells will be important components of distribution system due to their high efficiency and low environmental pollution. Generally, efficiency of the fuel cells ranges from 40-60% can be improved to 80-90% in co-generation applications. The waste heat produced by the lower temperature cells is undesirable since it cannot be used for any application and thus limits the efficiency of the system. The higher temperature fuel cells have higher efficiency since the heat produced can be used for heating purposes. Due to an electric-chemical process of power generation, there is no noise develop usually in mechanical members of conventional generator. All of these features will without any doubt lead to their wide application in the power industry in the near future. Several types of fuel cells have been reported in the literatures: phosphor acid fuel cell (PAFC), solid oxide fuel cell (SOFC), molten carbonate fuel cell (MCFC) and proton exchange membrane fuel cell (PEMFC). The PAFC has been commercially used in hospitals, nursing homes, utility power plants, etc. The SOFC can be used in large-power applications such as central electricity generation station. SOFC has the highest potential in large power application.

1.4 Fuel for fuel cells

Each of these FC types differs in the electrolyte and fuel used, operating temperature and pressure, construction materials, power density and efficiency.

The most important component of a FC is the fuel processor and the reformer since hydrogen is not readily available. Fossil fuels such as gasoline, natural gas and coal gases need to be processed and reformed to obtain enriched hydrogen. Natural gas is the most easily available fuel source. Bio-fuels can also be used as a source to obtain hydrogen. Biological methods such as photosynthesis and fermentation can be used to produce hydrogen. Though there are different methods to produce hydrogen, a proper and feasible method which can be commercialized is not yet available.

Storage of hydrogen is an important aspect of the FC systems because the fuel has to be readily available for continuous supply of electric power. Sometimes, electrical energy is used to divide water into hydrogen and oxygen with the help of electrolysers during times of high supply and low demand. FCs have to be compact and portable for mobile applications; hence storage of hydrogen is essential for such applications. Hydrogen needs to be handled with great care because it is a highly volatile and flammable gas. It has a high leak rate due to which the gas tends to escape through small orifices, faster than the other gases. Hence storage of hydrogen plays a key role in the FC systems.

A fuel processor converts the primary fuel source (hydrocarbons) into the fuel gas (hydrogen) required by the FC stack. The processor uses a catalytic reaction to break the fuel into hydrogen and separate it from the carbon based gases. Each of the FC types has specific fuel requirements. Natural gas and petroleum liquids contain sulphur compounds and have to be desulphurized before they can be used as a fuel. The anode catalysts are intolerant of sulphur and it must be removed before it degrades catalyst performance. There is a risk of carbon formation in fuel cell systems which can be reduced by carrying out pre-reforming of the fuel gas before it is fed to the reformer reactor. Carbon monoxide can be used as a fuel for SOFC and MCFC because it can be internally converted to hydrogen whereas the PEMFC should be completely free from it. CO has high affinity for anode catalyst (especially platinum) and it prevents the flow of fuel in the PEMFC. Ammonia is a poison for all the FC types due to its adverse effects on the cell life except for SOFC, where it can be internally reformed.

Lower-temperature FCs require an external reformer to obtain the hydrogen rich fuel, thus increasing the cost and thereby reducing the efficiency. Higher temperature FCs do not require an external reformer; its high temperature allows direct conversion of natural gas to hydrogen. High temperature requires stringent materials which increases the cost of the fuel cells. Hence, researchers are working to combine the benefits of the PEMFC and the PAFC to obtain intermediate temperature cells, often referred to as high temperature PEM.

2. Solid oxide fuel cell

The SOFC technology, dates from Walther Nernst, who around 1890 discovered that stabilized zirconia is an isolator at room temperature, but turns into an ionic conductor between 600 – $1000^{o}C$ and an electronic and ionic conductor around $1500^{o}C$. The first SOFC based on zirconia was introduced by Baur and Preis in 1937. Since then, research on SOFC has been steadily increasing until today.

2.1 Fundamentals

There are different types of FCs that have been mentioned above and are currently in use and development. Among them, SOFCs that works in the simplest structure have grown in recognition as a viable high temperature FC technology. SOFC can be improved to create a hydrogen fuel with heat inevitably occur in the cell, the power structure is a device that requires no external reformer, and without the need for transformer and reactor. The SOFC has a few typical advantages compared to other FCs.

1. Higher efficiency compared to other FCs
2. Easy to handle with simple structure which is composed of all solid
3. The reforming system is simple
4. Carbon monoxide (CO) can be used as a fuel

5. No precious metals as catalysts (platinum: Pt, etc.)
6. Structure for carbon dioxide recovery

In addition, while maintaining high efficiency, to lower the temperature of the SOFC is driving development issues. Because 1000^oC high operating temperature of the material will degrade cells, reduce the choice of construction materials. Operating temperature of $700-1000^oC$ from the conventional $500-800^oC$, these issues are resolved smaller, lower cost, can improve endurance. High-temperature operation removes the need for a precious-metal catalyst, thereby reducing the cost. It also allows SOFCs to reform fuels internally, which enables the use of a variety of fuels and reduces the cost associated with adding a reformer to the system. Although a SOFC produces electricity, it only produces DC power and utilizes only processed fuel. Therefore, a SOFC based power generation system requires the integration of many other components beyond the SOFC stack itself. Moreover, to recover the high quality waste heat from the SOFC stack, an efficient integration of co-generation or bottoming system with the FC section is crucial for a SOFC based power generation plant. Since the balance of plant will directly impact the overall system efficiency and may cost more than the SOFC stack itself, it is obvious that the design of a SOFC power generation system involves more than the optimization of the SOFC unit with respect to efficiency or economics. It also involves balance of plant studies. With SOFC materials and stacks approaching a commercialization stage, there is a need to explore various process designs to obtain optimal efficiency and economics based on specific applications and fuel availability.

2.2 SOFC based power generation systems

As other types of FCs, a SOFC produces only DC power and requires processed fuel. It also produces high quality heat due to its high operating temperature. Beyond the SOFC stack itself, a typical SOFC power system basically includes: a reformer to start the hydrogen production process, a fuel conditioner to clean up the pollutants that could otherwise poison the fuel cell elements, a power conditioner to convert direct current from the fuel cell to the appropriate voltage range and current type depending on the application, and a cogeneration or bottoming cycle to utilize the rejected heat to achieve high system efficiency. The system also requires the most common balance of plant equipments such as heat exchangers, air blower and fuel compressors, controls systems, and safety systems. Fig.1 illustrates fundamental parts in a FC power unit.

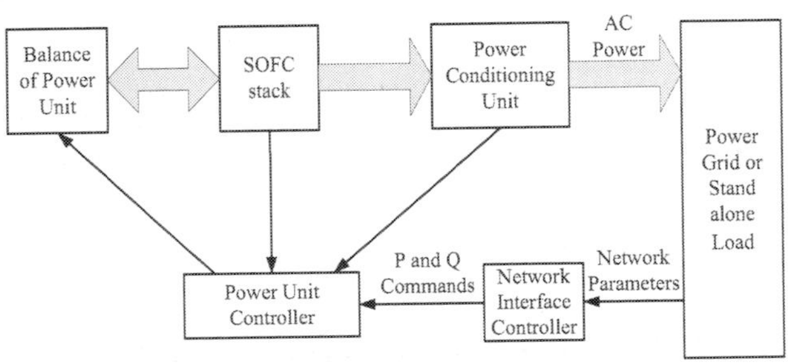

Fig. 1. Structure of SOFC power generation unit

2.3 Fuel processing

Fuel processing converts a commercially available fuel to a fuel gas suitable for the SOFC anode reaction. Typical fuel processing steps include:

- Desulphurization, where a catalyst is used to remove sulphur contaminants in the fuel. Sulphur compounds are noxious, and they can also bind catalysts used in later stages of fuel reformation poisoning the catalyst.
- Reformation, where the fuel is mixed with steam and then passed over a catalyst to break it down into hydrogen, as well as carbon dioxide and carbon monoxide.
- Shift conversion, where the carbon monoxide reacts with steam over a catalyst to produce more hydrogen and carbon dioxide.

However, high operating temperature SOFCs can accommodate internal reforming by means of a CO-tolerant nickel catalyst, so they can operate on natural gas with minimum pre-processing of the fuel. This will not only reduce the capital cost of the SOFC system, but also can be beneficial to system efficiency because there is an effective transfer of heat from the exothermic cell reaction to satisfy the endothermic reforming reaction.

Hydrogen sulfide, hydrogen chloride and ammonia are impurities typically found in coal gas. Some of these substances maybe are harmful to the performance of SOFCs. Therefore, a SOFC system will require fuel cleanup equipment such as desulfurizer depending on the raw fuel components.

2.4 Rejected heat utilization

At 1000*C* operating temperature, SOFCs produce a tremendous amount of waste heat while generating electricity. In order to obtain the highest possible system efficiency, the heat must be recovered by producing hot water, steam, or additional electricity. In a large SOFC power system (>100MW), production of electricity via a steam turbine bottoming cycle is maybe advantageous.

2.5 Power conditioning unit

While used as a power generator, FCs usually are connected to the load or distribution system via Power Conditioning Unit basically including DC-DC converter and DC-AC inverter. Therefore, low cost and high efficiency inverters are required together with acting controllers for fast tracking of real and reactive power demands. The inverter serves as the interface between the SOFC and the power distribution system. It is controlled in order to provide real and reactive power set point tracking and to adjust the power factor as well as frequency. Transient response control equipment may also be included. The efficiency of the power conversion is typically on the order of 94 to 98%.

2.6 Electrochemistry of SOFC

Fig.2 shows the processes taking place in a SOFC with hydrogen.

The SOFC fundamentally consists of two porous electrodes (anode and cathode) separated by a ceramic electrolyte in the middle, and flow channels for fuel and air delivery and collection.

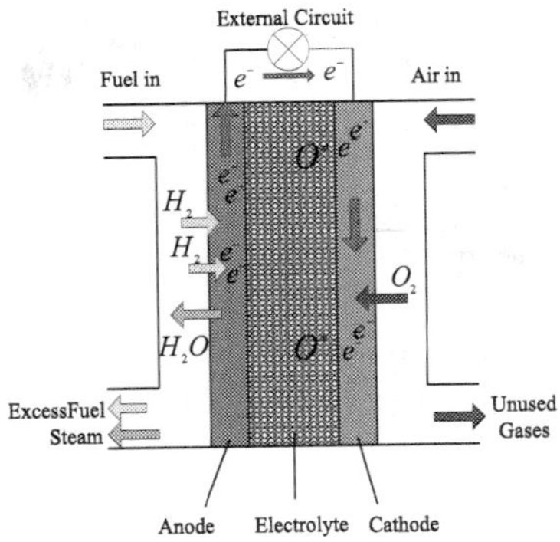

Fig. 2. Reaction process inside SOFC

Air flows along the cathode. When an oxygen molecule inside air flow contacts the cathode/electrolyte interface, it acquires 4 electrons from the cathode and splits into 2 oxygen ions. The reaction occurring at the cathode is:

$$O_2 + 4e^- \rightarrow 2O^{2-} \tag{1}$$

The oxygen ions diffuse into the electrolyte material and migrate to the other side of the cell where they encounter the anode. These oxygen ions travel through the porous electrolyte and react with H2 (fuel) to produce water and most importantly, electrons at the anode and the following reaction may occur at the anode:

$$2H_2 + 2O^{2-} \rightarrow H_2O + 4e^- \tag{2}$$

The electrons transport through the anode to the external circuit and back to the cathode, thus an electro motive force (EMF) is generated between two electrodes. The two electrodes can be connected via an external circuit and an electrical current can be generated. At the same time, we get heat when the reaction occurs. The overall reactions are:

$$H_2 + \frac{1}{2}O_2 \rightarrow H_2O \tag{3}$$

In cases using methane as a fuel, the CH4 reacts with O2 via internal reformer which will be discuss later that produces H2O and CO2. The final reaction equation is:

$$CH_4 + 2O_2 \rightarrow 2H_2O + CO_2 \tag{4}$$

2.7 Fuel cell voltage and nernst equation

Before we begin to look at how the electromotive force (EMF) and thus work is produced in a FC, it is necessary to understand some basic thermodynamic concepts. The Gibbs free energy is the energy required for a system at a constant temperature with a negligible volume, minus any energy transferred to the environment due to heat flux. Gibbs free energy is the energy available to do external work which involves moving electrons around an external circuit. In FCs, change in Gibbs free energy of formation (ΔG) is considered, as this change is responsible for the energy released. This change is the difference between the free energy of the products and the reactants, as shown in equation.

$$\Delta G = \Delta G_{products} - \Delta G_{reactants} \quad (5)$$

Consider the following thermodynamic identity for a reversible process when there is no shaft work extracted and the system is restricted to do only expansion work: $dG = VdP - SdT$, and if the process is isothermal, the above equation reduces to: $dG = VdP$. Using the ideal gas equation, $V = nRT$, we have $dG = nRTdP/P$. Integrating this equation from state 1 to state 2, we get $G_1 - G_2 = nRTln\frac{P_2}{P_1}$. If the state 1 is replaced with some standard reference state, with Gibbs free energy G_0 and standard pressure P_0, the Gibbs free energy per unit mole at any state 'i' is given by,

$$g_i = g_0 + RTln\frac{P_i}{P_0} \quad (6)$$

Consider that the following chemical reaction takes place at constant pressure and temperature, $aA + bB \leftrightarrow mM + nN$ Where a, b, m and n are the stoichiometric coefficients of the reactants A and B and the products M and N, respectively. Now, Equation 6 takes the following form,

$$\Delta G = \Delta G_0 + RTln(\frac{P_M^m P_N^n}{P_A^a P_B^b}) \quad (7)$$

ΔG_0 is the standard Gibbs free energy change for the reaction ($\Delta G_0 = mg_M^0 + ng_N^0 - ag_A^0 - bg_B^0$ and g_i^0 are the standard Gibbs free energies of the constituents).

Equation 7 gives the Gibbs free energy change for the reaction. We are interested on how is that energy change is related to the work of the SOFC system performed. To find that relation, consider the following thermodynamic identity for a reversible process, ($dQ = TdS$)

$$dG = -\delta W + PdV + VdP - SdT \quad (8)$$

At constant temperature and pressure, the above equation can be written as,

$$dG = -\delta W + PdV \quad (9)$$

Since it is a non-expansion work, Equation (2.9) takes the form,

$$dG = -\delta W_e \quad (10)$$

Equation 10 means the change in Gibbs free energy of the reaction is equal to the maximum electrochemical work, W_e, that can be extracted when reactants A and B react to give products M and N under constant temperature and pressure conditions through a reversible reaction.

Now, we can focus on how the maximum electrochemical work relates to the EMF of the cell.

For the SOFC, n_e (n_e=8 with Equation 4(of Energy, 2004)) electrons pass through the external circuit for each CH_4 molecule used. In a lossless system, electrical work done is equal to the change in Gibbs free energy which has been proved previously. Further, electrical work done to move a charge of n_e F (to move n_e electrons) for a voltage of E is given by below equation.

$$\text{Electrical work done} = -\text{n}_\text{e}\text{FE}_\text{cell}\text{joules} \quad (11)$$

(1e charge 1.602×10^{-19}, therefore 1 mole CH_4 which is equivalent to n_e mole e (1mol has $N = 6.022 \times 10^{23}$ e) will charge $-n_e \times 6.022 \times 10^{23}$,$1.602 \times 10^{-19} = -n_e \times 96485 = n_e \times F$)

The EMF produced due to half-cell reactions drives the electrons to move from the anode to the cathode. If ne mole of electrons move from anode to cathode per unit time and the EMF of the cell is E, the power extracted is simply EMF multiplied by the current,

$$W_e = n_e F E_{cell} \tag{12}$$

where F is the total charge of 1 mole of electrons, known as Faraday's constant. Now if we look at the integral form of Equation 7), 10 combined with 12, we get,

$$\Delta G = n_e F E_{cell} \tag{13}$$

Applying Equation 7 to Equation 13 we get what is known as Nernst equation,

$$E_{cell} = \frac{\Delta G_0}{n_e F} + \frac{RT}{n_e F} ln(\frac{P_M^m P_N^n}{P_A^a P_B^b}) \tag{14}$$

For the reaction occurring in an SOFC with methane in Equation 4:

$$E_{cell} = \frac{\Delta G_0}{n_e F} + \frac{RT}{n_e F} ln(\frac{P_{CH_4} P_{O_2}^2}{P_{H_2O}^2 P_{CO_2}}) = E_0 + \frac{RT}{8F} ln(\frac{P_{CH_4} P_{O_2}^2}{P_{H_2O}^2 P_{CO_2}}) \tag{15}$$

With E_0: Ideal Voltage at Standard Pressure and: $\Delta g_0 = 2g^0_{H_2O} + g^0_{CO_2} - 2g^0_{O_2} - g^0_{CH_4} = -980(kJ.mol^{-1})$(at standard pressure 1 atm and temperature 298 K). (Δg_0 changes with reaction like Equation 4), it is fairly constant with temperature)

Therefore:

$$E_0 = \frac{980.000}{8 \times 96485} = 1.27[V] \tag{16}$$

Actually E_0 depends on temperature:

$$E_0^T = E_0 - k_E(T - 298) \tag{17}$$

This maximum theoretical voltage, E, is also known as "Open Circuit Voltage" (OCV) and can be measured when there is no current in the circuit. Also, it can be observed, that to get the maximum OCV, a high concentration of reactants is required.

2.8 Voltage loss

When the FC is under load (a current is flowing), the voltage supplied at the electrodes will be different from the E_{cell} calculated from Equation 15. The dependency of these losses on temperature, current density and species concentrations mainly determine the characteristics of a FC. The output voltage is therefore lower than the circuit voltage when the FC is operated. Three main mechanisms of voltage losses exist: activation/polarization loss (η_{act}), Ohmic loss (η_{ohmic}), and concentration/diffusion loss (η_{con}).

A typical curve of the cell electrical voltage against current density is shown in Figure 3. It can be seen that there exists a linear region where the voltage drop is linearly related with the current density due to the Ohmic contact. Beyond this region the change in output voltage varies rapidly. At very high current density, the voltage drops significantly because of the gas exchange efficiency (η_{con}). At low current level, the Ohmic loss becomes less significant, the increase in output voltage is mainly due to the activity of the chemicals (η_{act}).

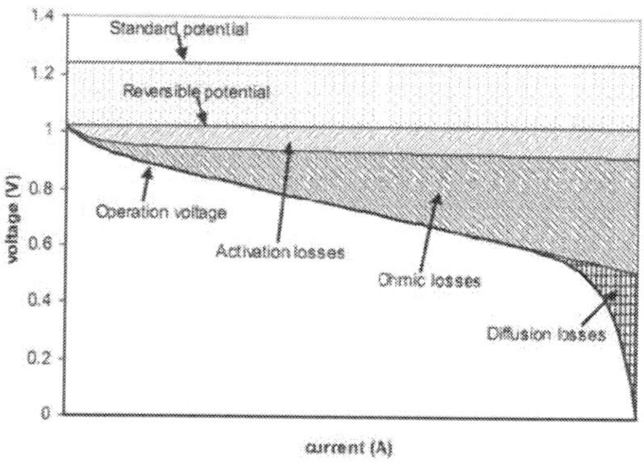

Fig. 3. FuelCell Output Voltage (of Energy, 2004)

The output voltage of a cell, V_{cell} can, therefore, be written as:

$$V_{cell} = E_{cell} - \eta_{act} - \eta_{ohmic} - \eta_{con} \tag{18}$$

The output voltage of the FC stack including N_0 individual single FC connected in series can be obtained as:

$$V = N_0[E_{cell} - \eta_{act} - \eta_{ohmic} - \eta_{con}] \tag{19}$$

To calculate the FC output voltage, the above three voltage drops should be calculated.

2.8.1 Activation loss

Action voltage loss is caused by an activation energy barrier that must be overcome before the chemical reaction occurs. At open circuit, no outer current is flowing. However, reactions are still taking place, but at equal rates in both directions. Just regarding the current which flows into one of the directions, we find the "exchange current density". In order to achieve an outer current higher than this, an extra potential is required to achieve the desired reaction rate, called activation voltage. The voltage drop is increasing fast at low reaction rates and is from a certain level almost constant. Activation is the dominant source of loss for low-temperature FCs, while their influence is smaller for SOFCs.

Butler-Volmer equation is normally used to calculate the activation voltage drop. To avoid the ambiguity of simplified model, such as the Tafel equation or a linear potential-current relation, used under different operating conditions, the following general Butler-Volmer equation is used to calculate the respective overpotential of anode and cathode:

$$i = i_{0,k}\left\{e^{\beta\frac{n_eF\eta_{act,k}}{RT}} - e^{-(1-\beta)\frac{n_eF\eta_{act,k}}{RT}}\right\} \tag{20}$$

- β: transfer coefficient. (is considered to be the fraction of the change in polarization that leads to the change in reaction rate constant; its value is usually 0.5 in the context of a FC).
- i_0: the 'apparent' current exchange density
- k: is anode (a) or cathode (c)

Under high activation polarization, the first exponential term in Equation 19 will be much less than unity and the exponential terms can be excluded from the equation. Rearranging the simplified equation yields $\eta_{act} = \frac{RT}{\beta n_e F} ln \frac{i}{i_0}$ which is the well-known Tafel equation. This equation will yield an unreasonable value for $\eta_a ct$ when $i = 0$ (in the case of load is be disconnected to SOFC because of faults). Therefore, we will use the Butler-Volmer equation. Hence,

$$i = 2i_{0,k} sinh(\frac{n_e F \eta_{act,k}}{RT}) \rightarrow \eta_{act,k} = \frac{2RT}{n_e F} sinh^{-1}(\frac{i}{2i_{0,k}}) \tag{21}$$

From (S.H. Chan, 2001),

$$i_{0,a} = 5300[A/m^2] = 5300.1000/10000 = 530[mA/cm^2] \tag{22}$$

$$i_{0,c} = 2000[A/m^2] = 2000.1000/10000 = 200[mA/cm^2] \tag{23}$$

Actually, the temperature effects on exchange current density. However in this simulation, this influence is so small that can be neglected. Hence,

$$\eta_{act} = \eta_{act,a} + \eta_{act,c} = \frac{2RT}{n_e F}[sinh^{-1}(\frac{i}{2i_{0,a}}) + sinh^{-1}(\frac{i}{2i_{0,c}})] \tag{24}$$

The equivalent activation resistance can then be defined as:

$$R_{act} = \frac{\eta_{act}}{i} = \frac{2RT}{i n_e F}[sinh^{-1}(\frac{i}{2i_{0,a}}) + sinh^{-1}(\frac{i}{2i_{0,c}})] \tag{25}$$

Compared with Anode, the Cathode exhibits higher activation overpotential, which is due to the poor "apparent" exchange current density at the electrode/electrolyte (LSM-YSZ/YSZ) interface. Since the Cathode exchange current density directly affects the electrochemical reaction rate at the Cathode, it can be understood that the low electrochemical reaction rate in the Cathode lead to high cathode activation polarization in the SOFC.

According to Equation 24, the activation voltage drop will be zero when load current is zero. The Ohmic and concentration voltage drops (will be discussed) are also zero when the fuel cell is not loaded ($i = 0$). However, even the open-circuit voltage of an SOFC is known to be less than the theoretical value given by Equation 25). Therefore, a constant and a temperature-dependent term can also be added to Equation (2.24) for activation voltage drop computation of SOFC as follows (P.R. Pathapati, 2005):

$$\eta_{act}^T = \xi_1 + \xi_2 T + iR_{act} = \eta_{act,1} + \eta_{act,2} \tag{26}$$

where $\eta_{act,1} = \xi_1 + \xi_2 T$ is the part of activation drop affected only by the FC internal temperature, while $\eta_{act,2} = iR_{act}$ is both current and temperature dependent.

2.8.2 Ohmic overpotential

Ohmic overpotential, which contributes by the electrolyte, electrodes and interconnector of the FC, occurs because of the resistance to the flow of ions in the ionic conductors and the resistance to electrons through the electronic conductors. At a given temperature and geometry, the voltage loss is proportional to the current. Since these resistances obey Ohm's law, the overall Ohmic overpotential can be written as:

$$\eta_{Ohmic} = iR_{ohm} \tag{27}$$

The resistances of these FC components are determined by the resistivity of the materials used and their respective thickness. The results show that the resistances of the cathode, electrolyte and interconnector decrease with increase in temperature. By contrast, the anode resistance displays the opposite trend. The resistance of each material used in the SOFC components can be calculated from its respective resistivity, which is a function of temperature. The electrical resistance R_{ohm} is calculated simply as the sum of the anode R_a, electrolyte R_e, cathode R_c and interconnector resistance R_i:

$$R_{ohm} = R_a + R_e + R_c + R_i \tag{28}$$

However, the main contribution to the Ohmic polarization is from the transport resistance of O^{2-} in the electrolyte. The resistance of electrolyte strongly depends on the temperature, and its effect cannot be ignored. The dependence of electrolyte resistance on temperature is given by the following equation:

$$R_e = R_{e0}e^{10100\left(\frac{1}{T}-\frac{1}{1273}\right)} \tag{29}$$

The resistances of other parts are assumed to be constant because of the weak dependence on temperature and their contributions to the total voltage drop are small (Susumu Nagata, 2001). Hence,

$$R_{ohm} = R_{e0}e^{10100\left(\frac{1}{T}-\frac{1}{1273}\right)} + R_a + R_c + R_i \tag{30}$$

According to (Takanobu Shimada, 2009),

$$\begin{aligned} R_{ohm} &= R_{e0}e^{10100\left(\frac{1}{T}-\frac{1}{1273}\right)} + R_a + R_c + R_i \\ &= \frac{0.00745}{0.3} \times e^{10100\left(\frac{1}{T}-\frac{1}{1273}\right)} + (0.01003 + 0.0184) \times 0.3 + \frac{0.00022}{0.3} \\ &= 0.0248 \times e^{10100\left(\frac{1}{T}-\frac{1}{1273}\right)} + 0.0093[\Omega/cm^2] \end{aligned} \tag{31}$$

2.8.3 Concentration overpotential

Reactants must flow through the porous electrodes to the TPB, and products must flow into the other direction, driven by diffusion. During the reaction process, concentration gradients can be formed due to mass diffusion from the flow channels to the reaction sites (catalyst surfaces). The effective partial pressures of hydrogen and oxygen at the reaction site are less than those in the electrode channels, while the effective partial pressure of water at the reaction site is higher than that in the anode channel. Thus, the calculated potential will be lower and the difference is called diffusion or concentration losses. At high-current densities, slow transportation of reactants (products) to (from) the reaction site is the main reason for the concentration voltage drop. Any water film covering the catalyst surfaces at the anode and cathode can be another contributor to this voltage drop. The voltage drop increases with increasing current against an asymptotic maximum current. At this point, the concentration of one of the reactants at the TPB is zero and no further current increase is possible. This equation is entirely empirical which has become more favored lately after many research finding out accurate equations, and yields an equation that fits the results very well. It provided the constants $m = 3.10^{-5}[V]$ and $n = 8.10^{-3}[mA^{-1}.cm^2]$ are chosen properly.

$$\eta_{con} = me^{ni_{fc}} \tag{32}$$

The equivalent resistance for the concentration voltage drop can be calculated as:

$$R_{con} = \frac{\eta_{con}}{i_{fc}} \tag{33}$$

2.9 Double-layer charging effect

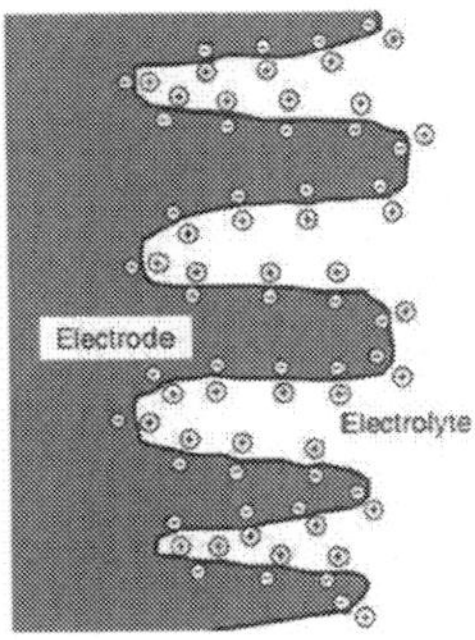

Fig. 4. The charge double layer at the surface of a fuel cell electrode (James Larminie, 2003)

In an SOFC, the two electrodes are separated by the electrolyte (Figure 4), and two boundary layers are formed, e.g., anode-electrolyte layer and electrolyte-cathode layer. These layers can be charged by polarization effect, known as electrochemical double-layer charging effect, during normal fuel cell operation. The layers can store electrical energy and behave like a super-capacitor. The model for SOFC considering this effect can be described by the equivalent circuit shown in Figure 5 (Caisheng Wang, 2005).

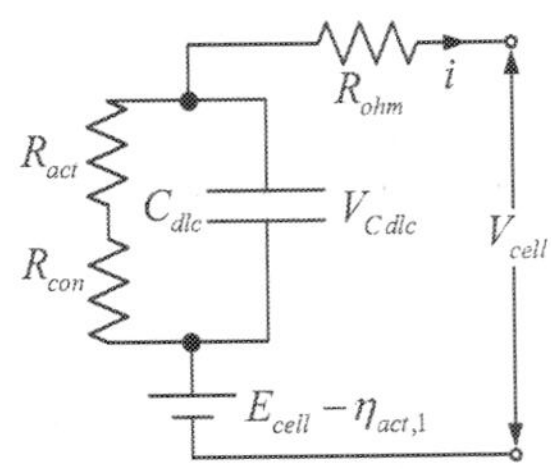

Fig. 5. Equivalent electrical circuit of the double-layer charging effect inside an SOFC

In the above circuit, R_{ohm}, R_{act} and R_{con} are the equivalent resistances of Ohmic voltage drop, activation, and concentration voltage drops, which can be calculated according to Equation

31, 25, and 33, respectively. C_{dlc} is the equivalent capacitor due to the double-layer charging effect. The capacitance of a capacitor is given by this formula:

$$C_{dlc} = \varepsilon \frac{A}{d} \tag{34}$$

where ε is the electrical permittivity, A is the surface area, and d is the separation of the plates. In this case, A is the real surface area of the electrode, which is several thousand times greater than its length×width since the electrodes of a SOFC fuel cell are porous. Also d, the separation, is very small, typically only a few nanometers. The result is that, in some fuel cells, the capacitance will be very large (can be in the order of several Farads) which is high in terms of capacitance values. (In electrical circuits, a 1 μF capacitor is on the large size of average). The voltage across C_{dlc} is :

$$V_C = (i - C_{dlc}\frac{dV_C}{dt})(R_{act} + R_{con}) \tag{35}$$

The double-layer charging effect is integrated into the modelling, by using V_C instead of $\eta_{act,2}$ and η_{con} to calculate V_{cell} . The fuel cell output voltage now turns out to be:

$$V_{cell} = N_0[E_{cell} - V_C - \eta_{ohmic} - \eta_{act,1}] \tag{36}$$

Recent approaches show that cathode activation and Ohmic overpotentials are responsible for the major losses in the SOFC over normal operating range.

2.10 Reforming

A great advantage of the SOFC is its possibility for internal reforming of hydrocarbon fuel. Sulphur-free natural gas (mainly a mixture of the alkanes methane, ethane and propane), which is technically available today, may be used as fuel. Due to the high temperature and the existence of nickel as a catalyst at the anode, the fuel cell reforms the alkanes to hydrogen and carbon monoxide internally through the steam reforming reaction:

$$C_nH_{2n+2} + nH_2O \Leftrightarrow (2n+1)H_2 + nCO$$

The equilibrium of this reaction is at the right hand side for elevated temperatures. As the reforming reaction is strongly endothermic, it severely decreases the temperature where it takes place in the fuel cell and therewith the local current density.

The model is considering reforming the internal reformer, but rather a reaction when hydrogen and oxygen that is generated from methane. Equation 37 shows the internal reforming reaction formula

$$CH_4 + H_2O \rightarrow 3H_2 + CO \tag{37}$$

1 [mol] CH_4 generates 3 [mol] H_2 and 1 [mol]CO. These products react with O_2. The following equations:

$$3H_2 + 3/2O_2 \rightarrow 3H_2O \tag{38}$$

$$CO + 1/2O_2 \rightarrow CO_2 \tag{39}$$

These three equations match the combustion reaction of methane shown in this equation 40. $CH_4 + H_2O + 2O_2 \rightarrow 3H_2O + CO_2$ Or

$$CH_4 + 2O_2 \rightarrow 2H_2O + CO_2 \tag{40}$$

The Equation 40 is equivalent to Equation 4. In short, internal reforming model is used to model the reaction of methane combustion.

3. Heat exchangers

Heat exchangers are used extensively in the energy and process industry. In power cycles they are called recuperators and their use is to recover heat from exhaust streams for preheating the process streams and therewith saving part of the fuel. SOFC systems in particular involve recuperation of heat due to the high gas inlet temperatures required and the high amount of heat in the exhaust. In fact the inlet temperatures of air and fuel have to be increased somehow to get out of heat shock with the reactants. The reason here is that it is cannot to fed the inlet species with low temperature such as ambient temperature while the stack temperature is quite high. By any kind of means, SOFC power unit has to increase the inlet species temperature, for example, using several percentage of output SOFC power. In this simple SOFC power unit model, waste heat recovery used for preheating the fuel and air directly, the system efficiency therefore can be improved.

3.1 Configuration

There are different flow configurations, depending on the application. Analogous to the fuel cell, flow configurations may be co-flow, counter-flow or cross-flow. Exegetically, counter-flow is most efficient, because the cold fluid outlet may closely approach the hot fluid inlet temperature if the flow rates and HX surface are suitably chosen. A co-flow configuration may be more effective for HXs with a huge temperature difference between hot and cold fluid and only small temperature changes. In this model, both counter flow and parallel flow types of HX are selected as pre-heaters for comparison.

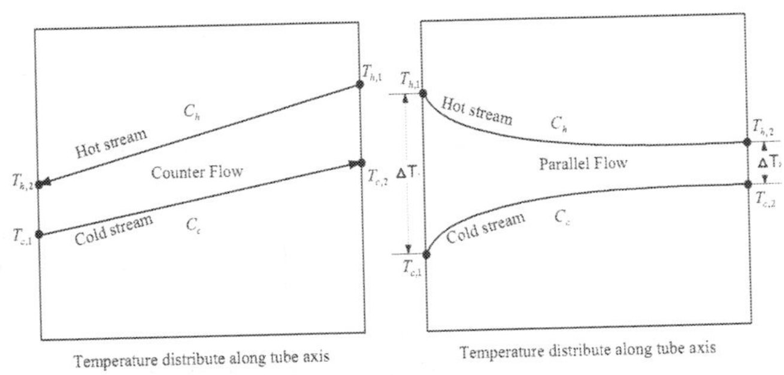

Fig. 6. Temperature distribution of flows in HX

The Simulink Model has been designed to assume that the inlet temperatures of air and fuel are equal to ambient temperature, i.e. $T_1 = 298[\mathrm{K}]$.

To determine the outlet temperatures of the pre-heats, which vary with the inlet conditions, the heat capacity rates of the cold and hot gas streams are calculated:

$$\begin{cases} C_c = \sum q_c C_{p,c} \\ C_c = \sum q_c C_{p,c} \end{cases} \left[\frac{mole}{s} \cdot \frac{J}{mole.K} = W/K \right] \tag{41}$$

From Equation 4: $CH_4 + 2O_2 = 2H_2O + CO_2$

Species	Heat Capacity C ($J/mole.K$)
CH_4	75.264
CO_2	57.112
H_2O	75.312
O_2	35.84
N_2	33.964
Air ($21\%O_2 + 78\%N_2$)	34.018

Table 1. Heat capacity of SOFC species

- HX1

$$C_c[\text{Air}] = q^{\text{in}}_{O_2} C_{O_2} + (\frac{78}{21}) q^{\text{in}}_{O_2} C_{N_2} \tag{42}$$

$$C_h[\text{CH}_4 + \text{N}_2 + \text{CO}_2 + \text{H}_2\text{O}] = \\ = \left(\frac{1-U_{opt}}{U_{opt}}\right) q^{in}_{CH_4} + (\tfrac{78}{21}) q^{in}_{O_2} C_{N_2} + \left(\frac{1-U_{opt}}{U_{opt}}\right) q^{in}_{O_2} C_{O_2} + q^{out}_{H_2O} C_{H_2O} + q^{out}_{CO_2} C_{CO_2} \tag{43}$$

- HX2

$$C_c[\text{CH}_4] = q^{in}_{CH_4} C_{CH_4} \tag{44}$$

$$C_h[\text{HX}_2] = C_h[\text{HX}_1] \tag{45}$$

The effectiveness & number of transfer units method (ε-NTU methodology: ε=f[NTU,C_r])) is used to model the pre-heaters in CH_4 SOFC system, which makes use of two non-dimensional groups: the number of heat transfer units - NTU, and the effectiveness, ČÃ defined below.

By comparing two heat capacities of hot and cold streams, the lower and higher values are assigned as C_{min} and C_{max}, respectively. The ratio of heat capacity rates is then available. Thus,

$$C_r = C_{min}/C_{max} \tag{46}$$

Number of heat transfer unit,

$$NTU = UA/C_{min} \tag{47}$$

where, U is overall heat transfer coefficient $[W/cm^2K]$, which is defined largely by the system and in many cases it proves to be insensitive to the operating conditions of the system. With our simulation, we take U to be a constant value and $U = 0.5[W/cm^2K]$ with high pressure gas. This is fairly reasonable in compact single-phase heat exchangers; and A-total heat transfer area.

The heat exchanger effectiveness is defined as the ratio of actual heat transfer rate, q, and the maximum possible heat transfer rate between the 2 streams, q_{max},

$$\varepsilon = q/q_{max} \tag{48}$$

Hence, the heat exchange rate between the hot and the cold gas stream is:

$$q = \varepsilon q_{max} \tag{49}$$

Where the theoretical maximum heat transfer is:

$$q_{max} = C_{min}(T_{h,1} - T_{c,1}) \tag{50}$$

With the value $(T_{h,1} - T_{c,1})$ is simply the temperature difference between the 2 inlet ports, i.e. the largest temperature difference between 2 streams and hence defines the ceiling value for the heat transfer rates between the 2 streams.

$$\varepsilon = \frac{C_h(T_{h,1} - T_{h,2})}{C_{min}(T_{h,1} - T_{c,1})} = \frac{C_h(T_{c,2} - T_{c,1})}{C_{min}(T_{h,1} - T_{c,1})} \tag{51}$$

And we have the ε-NTU relation by these equations:

- In case of counter flow:

$$\varepsilon = \frac{1 - e^{[-NTU(1-C_r)]}}{1 - C_r e^{[-NTU(1-C_r)]}} \tag{52}$$

- In case of parallel flow:

$$\varepsilon = \frac{1 - e^{[-NTU(1+C_r)]}}{1 + C_r} \tag{53}$$

Therefore, once effectiveness is calculated, based on the energy balance, the exit temperature of the hot and cold gas streams from the heat exchanger are:

$$T_{h,2} = T_{h,1} - q/C_h \tag{54}$$

$$T_{c,2} = T_{c,1} + q/C_c \tag{55}$$

3.2 Gas temperature of SOFC exhaust

Because relatively high amount of input gas flow not be used and they pass though SOFC without reaction. With the reactants, it can be considered their output temperate is operating temperature. However, non-reactants do not have the operating temperature when they come out from SOFC stack. Assuming that the SOFC configuration is as a Heat Exchanger for these gases, it is similarly to calculate their output temperature as in case of HX in the above section. Following figure represents the concept of this idea.

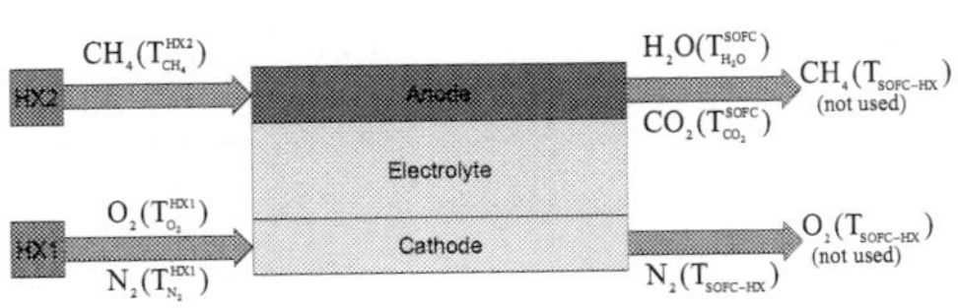

Fig. 7. Exhaust temperature calculation concept

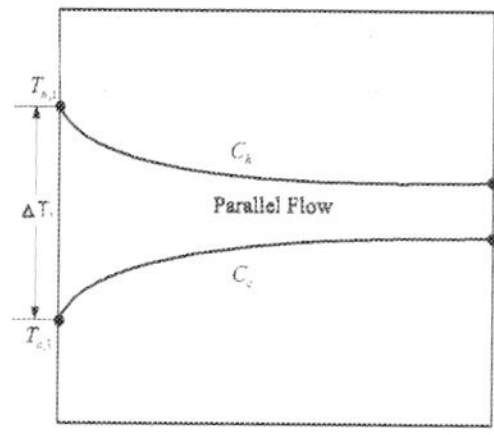

Fig. 8. Considering temperature change of SOFC exhaust as in HX

And SOFC is considered as HX to calculate non-reactants output temperature. Heat capacity rates of the cold and hot gas streams are calculated:

$$C_h[CO_2 + H_2O] = q^{out}_{CO_2} C_{CO_2} + q^{out}_{H_2O} C_{H_2O} \tag{56}$$

$$C_c[CH_4 + N_2 + O_2] = q^{out}_{CH_4} C_{CH_4} + q^{in}_{N_2} C_{N_2} + q^{out}_{O_2} C_{O_2} \tag{57}$$

Following the previous section, these values can be calculated:

$C_r = C_{min}/C_{max}, q_{max} = C_{min}(T_{h,1} - T_{c,1})$

Assuming the SOFC configuration as parallel, the efficiency can be calculated:

$$\varepsilon = \frac{1 - e^{[-NTU(1+C_r)]}}{1 + C_r} \tag{58}$$

and $q = \varepsilon q_{max}$

Hence,

$$T_{h,2} = T_{h,1} - q/C_h \tag{59}$$

$$T_{c,2} = T_{c,1} + q/C_c \tag{60}$$

where,

- $T_{h,1}$=T:operating temperature
- $$T_{c,1} = \frac{T^{HX_2}_{CH_4} q^{out}_{CH_4} + T^{HX_1}_{O_2}\left(q^{out}_{O_2} + q^{in}_{N_2}\right)}{q^{out}_{CH_4} + q^{out}_{O_2} + q^{in}_{N_2}} \tag{61}$$
- $T_{h,2} = T^{SOFC}_{CO_2} = T^{SOFC}_{H_2O}$ (reactants output temperature)
- $T_{c,2} = T_{SOFC-HX}$ (non-reactants output temperature)

The isothermal temperature of SOFC exhaust which will be fed into HX therefore can be calculated as:

$$T_{SOFC-exhaust} = \frac{T^{SOFC}_{H_2O}\left(q^{out}_{H_2O} + q^{out}_{CO_2}\right) + T_{SOFC-HX}\left(q^{out}_{CH_4} + q^{out}_{O_2} + q^{in}_{N_2}\right)}{q^{out}_{H_2O} + q^{out}_{CO_2} + q^{out}_{CH_4} + q^{out}_{O_2} + q^{in}_{N_2}} \tag{62}$$

3.3 Overall system energy efficiency comparison

As statement before, the aim to simulate HX included in SOFC simulation is to increase overall system energy efficiency. This section is going to point out the way to calculate the efficiency for SOFC model with HX versus one without HX. The efficiency of a chemical process must be evaluated differently than the conventional heat engine.

$$\eta_{withoutHX} = Electrical\,Energy_{per\,second} \times 100/\Delta H[\%] \tag{63}$$

$$\eta_{withHX} = \frac{(\Delta H - Heat_{HXExhaust})_{per\,second} \times 100}{\Delta H}[\%] \tag{64}$$

$Heat_{HXExhaust}$ = Gas sensitive heat of HX exhaust.

4. Thermal dynamic model

4.1 Thermal balance model

Thermodynamics is the study of energy changing from one state to another. The predictions that can be made using thermodynamic equations are essential for understanding and modelling SOFC performance since SOFCs transform chemical energy into electrical energy. Basic thermodynamic concepts allow one to predict states of the SOFC system, such as potential, temperature, pressure, volume, and moles in a fuel cell.

The first few concepts relate to reacting systems in SOFC thermal balance analysis: absolute enthalpy, specific heat, entropy, and Gibbs free energy. The absolute enthalpy includes both chemical and sensible thermal energy. Chemical energy or the enthalpy of formation (h_f) is associated with the energy of the chemical bonds, and sensible thermal energy (Δh) is the enthalpy difference between the initial reactants and products of reaction. The next important property is specific heat, which is a measure of the amount of heat energy required to increase the temperature of a substance by 1^0C (or another temperature interval). Entropy (S) is another important concept, which is a measure of the quantity of heat that shows the possibility of conversion into work. Gibbs free energy (G) is the amount of useful work that can be obtained from an isothermal, isobaric system when the system changes from one set of steady-state conditions to another.

The temperature-voltage and voltage drop affect the life of material, the model should reflect this. The first step in determining the heat distribution in a fuel cell stack is to perform energy balances on the system. The total energy balance around the fuel cell is based upon the power produced, the fuel cell reactions, and the heat loss that occurs in a fuel cell. Heat losses include the convective heat transfer occurs between the solid surface and the gas streams, and the conductive heat transfer occurs in the solid and/or porous structures. The reactants, products, and electricity generated are the basic components to consider in modelling basic heat transfer in a fuel cell, as shown in Figure 9. Notice that in this model we have inserted the gas sensitive of the heat exchangers which will be explained in details in next section.

The general energy balance states that the enthalpy of the reactants and gas sensitive after HXs entering the fuel cell equals the enthalpy of the products leaving the cell plus the sum of the heat generated by the power output, and the rate of heat loss to the surroundings. The basic heat transfer calculations will aid in predicting the temperatures and heat in overall fuel cell stack and stack components.

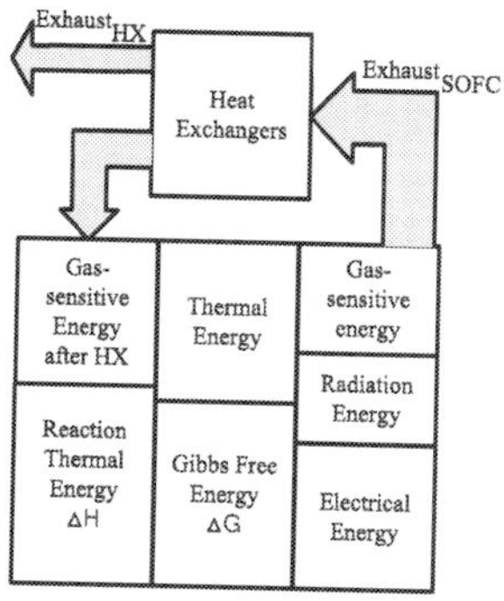

Fig. 9. Heat balance inside SOFC power unit with HX included

The parameters appear in thermal balance model are defined here.

- ΔH: Heat of reaction (Hess Energy) [J/mol] (Total energy produced by the reaction)
- ΔG: Thermal generation (Gibbs energy) [J/mol] (Electrical energy can retrieve a theoretical maximum)
- V: Cell terminal voltage [V]
- I_{fc}: Current [A]
- q_i: Species emissions [mol/s]
- R_i: Specific heat of water [J/mol.K]
- R_{FC}: Specific heat of the fuel cell system [J/K] (depends on manufacture)
- $\Delta H - IV$: Thermal Energy [J/mol]
- K_h: Coefficient of heat [W/K]
- T_0: Ambient temperature [K]
- T_{1i}, T_{2i}: The temperate at first and second state of species [K]
- T: Operating temperature [K]
- T_{ini}: The initial value Temperature [K]
- GSH_{HX}: Gas sensitive heat after HX [J/mol]
- GSH_{SOFC}: Gas sensitive heat after SOFC stack [J/mol]

Based on the diagram in Figure 9, the temperature equation and the flow of energy per unit time are as this representable expression.

$$GSH_{HX} + \Delta H - I_{fc}V = GSH_{SOFC} + R_{FC}\frac{dT}{dt} + K_h(T - T_0) \tag{65}$$

IV is the output power, $K_h(T - T_0)$ is considered to be the energy dissipation. In the steady-state:

$$GSH_{HX} + \Delta H - I_{fc}V = GSH_{SOFC} + K_h(T - T_0) \tag{66}$$

The dynamic operating temperature can be derived from Equation (2.65) shown in the following Equation.

$$T = T_{ini} + 1/R_{FC}\int_{0}^{t}\left\{(GSH_{HX} + \Delta H - I_{fc}V) - K_h(T - T_0) - GSH_{SOFC}\right\}dt \qquad (67)$$

Thermal characteristics model is the model that uses the expression 67. A summary of this equation in Figure 10.

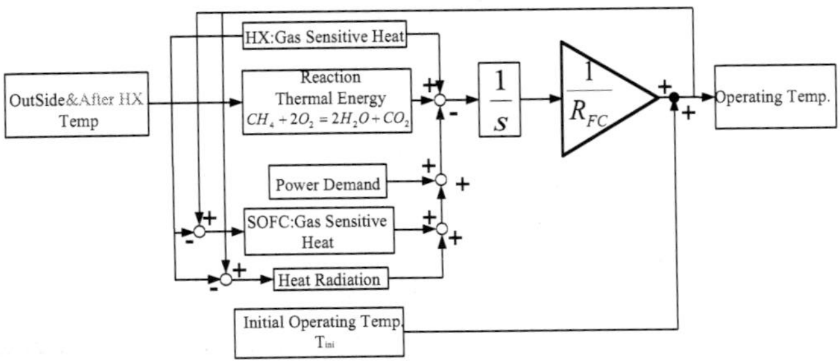

Fig. 10. Concept of thermal dynamic model

4.2 Heat factor

From the model shown in Figure 10, it is capable to calculate K_h by a independent model because this coefficient depend on the actual model and is independent with the change of operating temperature. It can be consider a constant with the chance of operating temperature. While Equation 66 is used, T can be considered as T_{ini}. Energy generated by fuel cells shown in Figure 10, and to assume that to change all the remaining heat energy to electrical energy extracted from the reaction of energy to representable fuel. Gibbs free energy is theoretically converted into electrical energy that can be used. Energy used in fuel cells is actually considered to be divided into three parts. And extract energy from the reaction energy as electricity output, the remaining energy is divided into heat radiation energy and heat energy is used to increase the temperature of the gas. From the Equation 65, the stack heat loss coefficient is obtained by dividing the amount of temperature change in the heat of the stack.

$$K_h = \frac{GSH_{HX} + \Delta H - I_{fc}V - GSH_{SOFC}}{T_{ini} - T_0} \qquad (68)$$

The conceptual model of above Equation is shown in Figure 11 which is derived by calculating the coefficient of heat energy from the stack number of moles of each gas equivalent reaction.

5. Simulation results of SOFC model implemented in Matlab/Simulink

The presented work is an attempt to model a SOFC system for DG applications. The aim of authors is to develop an efficient tool in Matlab-Simulink, which could simulate a SOFC system with sufficient accuracy. The reasons to use SIMULINK, are that the Matlab package

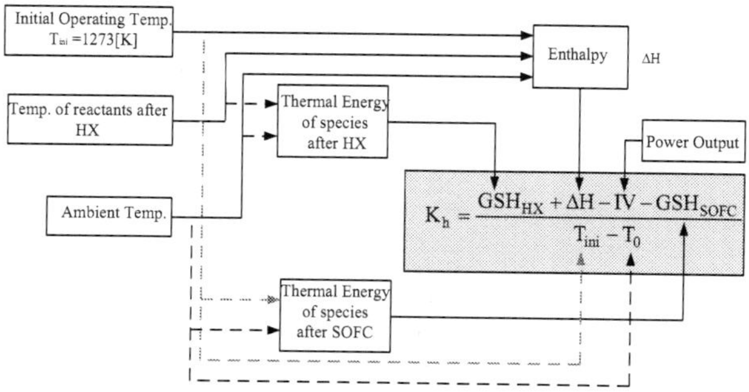

Fig. 11. Concept of heat loss coefficient calculation model

is commonly used among academic institutions and a graphical user interface with the high level of capabilities. Figure 12 shows the structure of the transient model.

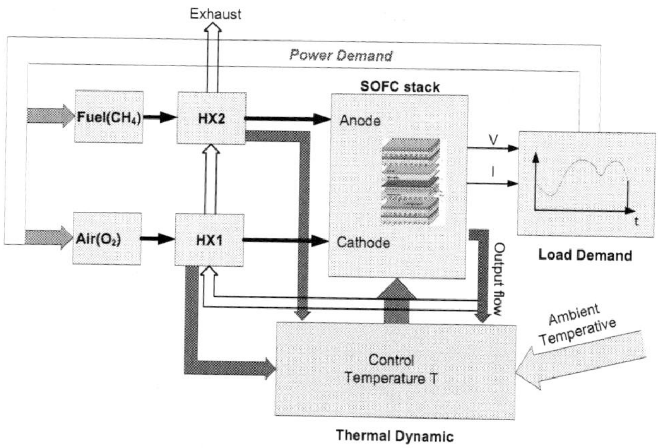

Fig. 12. Outline of single SOFC model

This SOFC model has simulated the internal reforming cell type, which generates hydrogen from methane by the high temperature of the cell. And its target is expected to follow changes in output load (50 [s] output at 70 [kW] to 100 [kW]). The transient analysis of the output is a directive issued. And the model also focuses on operating temperature control over long time (10.000s). The switch which has been set up in the center can simulate both if not considering the heat change for comparison. Each of the major SOFC components is be built as a complex sub-model (the load following control, air supply, fuel supply, the partial pressures, the voltage generator, the voltage drop, the temperature property, the heat exchangers, the efficiency calculation).

In order to carry out the SOFC model, it has to calculate partial pressures of 4 species and the chemical reaction for the SOFC terminal output voltage. The value of orders issued to meet the demand by the inverter power supply at the demand side by repeating this calculation, and numerical analysis. A feedback system has to be inserted for control the load following ability. The ac real power injection into the utility grid is considered to be the reference power for the fuel cell. The stack voltage and the reference power are used to determine the reference current which in turn is used to determine the fuel cell stack current (fuel and air supply). The fuel/air flow is proportional to the stack current. Throughout calculation system, the heat change affection is considered.

5.1 Load following ability

In this study, SOFC 70 [kW] power unit operating at 1273 [K] temperature works at normally stable load, 70 [kW] and the output command 100 [kW] is suddenly required for study of the response of SOFC. Observing SOFC load following ability, we need to consider the change of operating temperature.

Figure 13(a) is diagram of the output of current, voltage, power and operating temperature in short period test time, 200 [s]. This case considers robust following load change in small time scale. The output power takes 30 seconds to follow the increasing of load. That is equivalent to 1 [kW/s] velocity. The slow response of the fuel cell is due to the slow and gradual change in the fuel flow and the chemical reaction which is proportional to the stack current.

Figure 13(b) is diagram of the output of current, voltage, power and operating temperature in long period test time, 10000 [s]. This case considers the operating temperature. The temperature respond velocity with increase of load is 10 [K]/100 [s]. The final stable operating temperature is about 1050 [K] obtained at 2000s instance while the voltage and current output is about 437 [V], 230 [A].

Volt-amp characteristics of SOFC: The number of cells is taken to be 384 and the output voltage is 430 [V] which decreases as the load current increases. The drop is fairly linear in the middle region, known as region of Ohmic polarization. This is the operating region for the fuel cell.

5.2 Heat exchangers

5.2.1 Comparison between SOFC with and without heat exchangers

The simulation results are showed in Figure 14(a). Once we used HXs, according to the compared simulation results, the operating temperature be reduced by about 50[K].

5.2.2 Comparison between counter and parallel heat exchangers

The clear results are pointed out by Figure 14(b). The exhaust temperature into atmosphere of Counter-HX is lower than that of Parallel HX mean the higher energy efficiency. However, the operating temperature of SOFC in case using Counter-HX is higher than using Parallel-HX. The higher operating temperature makes stack materials work in severer condition. Based on this conclusion, the manufactures will decide which configuration is suitable for their real model.

5.2.3 Efficiency comparison

Figure 15(a) below shows the result that the efficiencies will change versus the times. The figure points out that the efficiencies change with the SOFC operation status within small range and when using HXs, the energy efficiencies are much higher.

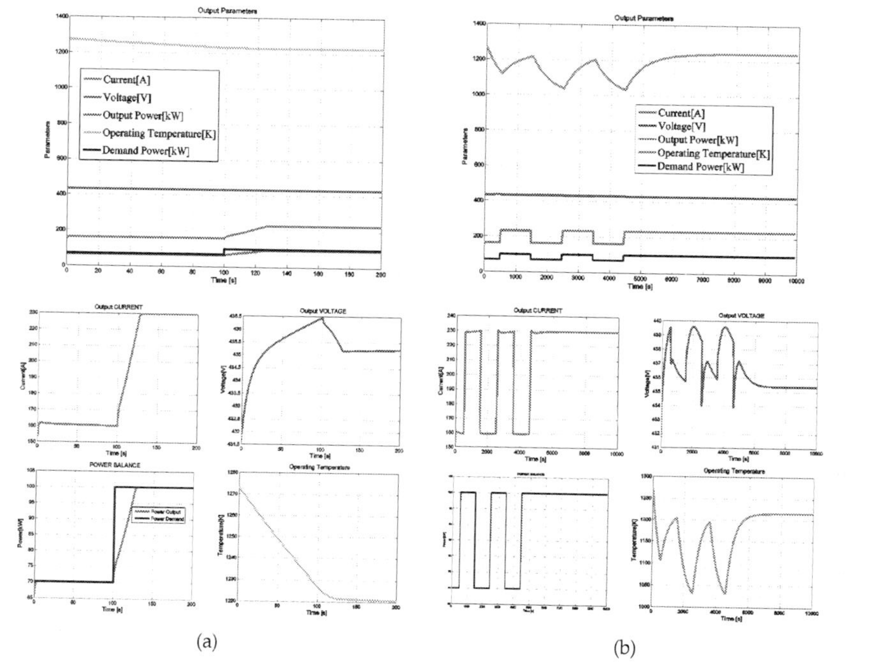

Fig. 13. a. Dynamics model in small timescale - 200[s]; b. Dynamics model in large timescale - 10000[s]

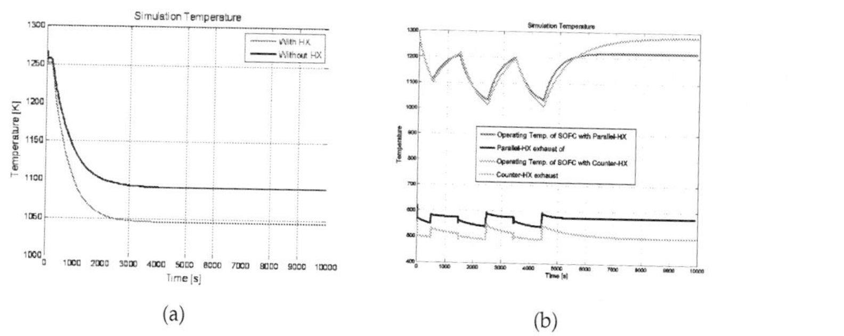

Fig. 14. a. Heat temperature comparison between SOFC with and without HX; b. HX operating temperature consideration between Counter and parallel Configuration

5.3 Operating temperature control by excess air

This section mentioned the study of temperature control in the SOFC operation. Rapid changes in heat in the fuel cell will lead to the deterioration of the material for the cell, SOFC it is important to properly maintain the temperature inside the stack. The operating temperature control method by using the excess air (O_2+N_2) into the fuel cell is shown in Figure 15(b).

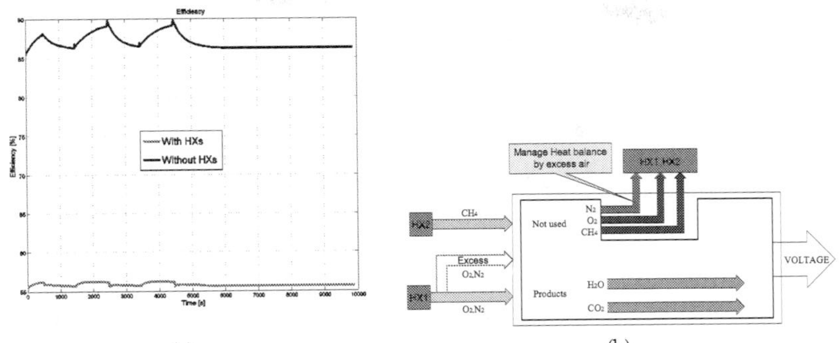

(a) (b)

Fig. 15. a. Efficiency comparison result; b. Concept of control temperature

Excess air is sent to take sensible heat of hot air, stabilize the temperature. The amount of excess air used is determined by the actual temperature which is adjusted by the feedback control.

The excess air bases on the required air for methane reactions, $q^{in}_{O_2}$. The amended air supply target is controlled by detecting the difference between the operating temperature and initial temperature 1273 [K]. The correction coefficient is determined by manufactures. This correction expression is shown in Equation 69. Figure 15(b) shows the model to control input air for maintaining operating temperature by using Equation 69.

$$q^{in}_{O_2-ex} = q^{in}_{O_2}[1 + K_{air}(\frac{T - 1273}{1273})] \tag{69}$$

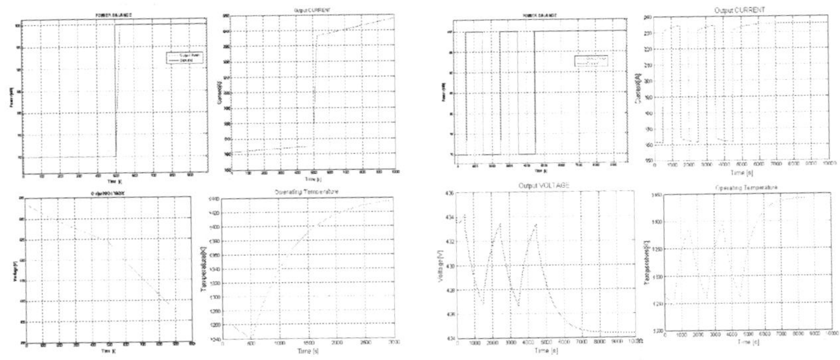

(a) (b)

Fig. 16. a. Without control of temperature - 1000[s]; b. Without control of temperature - 10000[s]

Following is the evaluation for using above excess air method. The output power at the various stages makes change in output voltage, current, operating temperature without excess

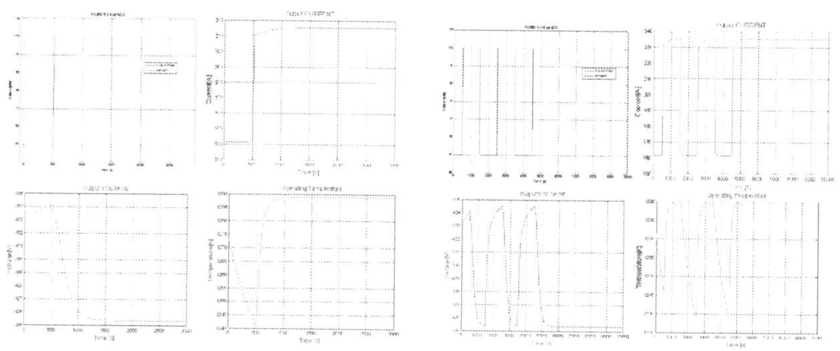

(a) (b)

Fig. 17. a. Consideration control of temperature - 1000[s]; b. Consideration control of temperature - 10000[s]

air in Figure 16(a), 16(b). In Figure 55 simulation time is 1000 [s] with changing demand for one time, and in Figure 56, 10000 [s] with changing demand for several times are shown. These pictures pay attention to rapidly changing temperature and the SOFC stack material degradation will be suffered which reduces life time of SOFC stack.

The extra injection of air fed to control the temperature is shown in Figure 17(a), 17(b). Figure 17(a) expresses that the operating temperatures variation range is smaller over time. This temperature variation reducing can be expected to prevent the short fuel cell stack life because of the rapid changes in operating temperature. From this result, one idea emerges that if we fed SOFC stack with as much air flow rate as it needs to make the variation range smallest. This is actually impractical because of compressor configuration as well as the electrode pressure.

6. Conclusion

A dynamic model of SOFC power unit was developed in Simulink. The load change was subjected to a step change in the reference real power from 70 to 100 [kW]. The characteristics of the fuel cell (voltage, current and power) have a slower gradual change at the instant of step changes. Some goals of this chapter include:

1. Calculate heat balance inside SOFC power unit that effects on the operating temperature and therefore to the output voltage. The N_2 gas is also involved into consideration.
2. This model is applied for internal reforming that uses natural gas (CH_4) as a direct fuel.
3. Add the heat exchangers into SOFC power system and calculate the gas flows temperature attaching to heat balance to compact SOFC model. Evaluate the different heat characteristic of two popular configurations of HX, and therefore increasing system energy efficiency.
4. Evaluate the load following ability of SOFC power unit by using feedback control the fuel and air flow which respond to the load change.
5. Control the operating temperature by excess air.

7. References

A.C. Burt, R.S. Gemmen, A. S. (2004). A numerical study of cell-to-cell variations in a sofc stack, *Journal of Power Sources* Vol.126(76): 76–87.

Achenbach & Elmar (1995). Response of a solid oxide fuel cell to load change, *Journal of Power Sources* Vol.57: 105–109.

Ali Volkan Akkaya, B. S. & Erdem, H. H. (2009). Thermodynamic model for exegetic performance of a tubular sofc module, *Renewable Energy* Vol.1: 1–8.

Caisheng Wang, M. H. N. (2007). A physically based dynamic model for solid oxide fuel cells, *IEEE Transaction on Energy Conversion* Vol.22(4): 887–897.

CaishengWang, M.HashemNehrir, S. R. S. (2005). Dynamic models and model validation for pem fuel cells using electrical circuits, *IEEE Transaction on Energy Conversion* Vol.20(2): 442–451.

D. Sanchez, R. Chacartegui A. Mun, T. (2008). On the effect of methane internal reforming modeling in solid oxide fuel cells, *Journal of Hydrogen Energy* Vol.33: 1834–1844.

David L. Damm, A. G. F. (2005). Radiation heat transfer in sofc materials and components, *Journal of Power Sources* Vol.143: 158–165.

Graham M. Goldin, Huayang Zhu, R. D. B. S. A. B. (2009). Multidimensional flow, thermal, and chemical behavior in solid-oxide fuel cell button cells, *Journal of Power Sources* Vol.187: 123–135.

J. Padulles, G.W. Ault, J. M. (2000). A numerical study of cell-to-cell variations in a sofc stack, *Journal of Power Sources* Vol.86: 495–500.

James Larminie, A. D. (2003). *Fuel Cell Systems Explained, 2nd ed.*, Wiley, England.

Kourosh Sedghisigarchi, A. F. (2004). Dynamic and transient analysis of power distribution systems with fuel cells-part i:fuel-cell dynamic mode, *IEEE Transaction on Energy Conversion* Vol.19(2): 423–428.

Li, P.-W. & Chyu, M. K. (2003). Simulation of the chemicalsimulation of the chemical/electrochemical reactions and heat/mass transfer for a tubular sofc in a stack, *Journal of Power Sources* Vol.124: 487–498.

M. Uzunoglu, M. (2006). Dynamic modeling, design, and simulation of a combined pem fuel cell and ultracapacitor system for stand-alone residential applications, *IEEE Transaction on Energy Conversion* Vol.21(3): 767.

Mitsunori Iwata, Takeshi Hikosaka, M. M. T. I. K. I. K. O. Y. E. Y. S. S. N. (2000). Performance analysis of planar-type unit sofc considering current and temperature distributions, *Solid State Ionics* Vol.132: 297–308.

M.Y. El-Sharkh, A. Rahman, M. A. P. B. A. S. T. T. (2004). A dynamic model for a stand-alone pem fuel cell power plant for residential applications, *Journal of Power Sources* Vol.138: 199–204.

N. Lu, Q. Li, X. S. M. K. (2006). The modeling of a standalone solid-oxide fuel cell auxiliary power unit, *Journal of Power Sources* Vol.161: 938–948.

of Energy, U. D. (2004). *Fuel Cell Handbook, 7th ed.*, EG and G Technical Services, Inc., Morgantown, West Virginia, USA.

P. Piroonlerkgul, W. Kiatkittipong, A. A. A. S. W. W. N. L. A. A. S. A. (2009). Integration of solid oxide fuel cell and palladium membranereactor: Technical and economic analysis, *International journal of hydrogen energy* Vol.34(9): 3894–3907.

P.R. Pathapati, X. Xue, J. T. (2005). A new dynamic model for predicting transient phenomena in a pem fuel cell system, *Journal of Renewable Energy* Vol.30: 1–22.

S. Campanari, P. I. (2004). Definition and sensitivity analysis of a finite volume sofc model for a tubular cell geometry, *Journal of Power Sources* Vol.132: 113–126.

S.H. Chan, K.A. Khor, Z. X. (2001). A complete polarization model of a solid oxide fuel cell and its sensitivity to the change of cell component thickness, *Journal of Power Sources* Vol.93: 130–140.

S.H.Chan, C.F.Low, O. (2002). Energy and energy analysis of simple solid-oxide fuel-cell power system, *Journal of Power Sources* Vol.103: 188–200.

Susumu Nagata, Akihiko Momma, T. K. Y. K. (2001). Numerical analysis of output characteristics of tubular sofc with internal reformer, *Journal of Power Sources* Vol.101: 60–71.

Tadashi Gengo, Nagao Hisatome, Y. A. Y. K. T. K. K. K. (2007). Progressing steadily, development of high-efficiency sofc combined cycle system, *Mitsubishi Heavy Industries, Ltd.Technical Review* Vol.44(1): 1–5.

Takanobu Shimada, Akihiko Momma, K. T. T. K. (2009). Numerical analysis of electrical power generation and internal reforming characteristics in seal-less disk-type solid oxide fuel cells, *Journal of Power Sources* Vol.187: 8–18.

Tomoyuki Ota, Michihisa Koyama, C.-j. W. K. Y. H. T. (2003). Object-based modeling of sofc system: dynamic behavior of micro-tube sofc, *Journal of Power Sources* Vol.118: 430–439.

Wang, C. & Nehrir, M. H. (2007). Load transient mitigation for stand-alone fuel cell power generation systems, *IEEE Transaction on Energy Conversion* Vol.22(4): 864–872.

Xiongwen Zhang, Guojun Li, J. L. Z. F. (2007). Numerical study on electric characteristics of solid oxide fuel cells, *Energy Conversion and Management* Vol.48: 977–989.

21

Defining a Gas Turbine Performance Reference Database Model Based on Acceptance Test Results

Norberto Pérez Rodríguez, Erik Rosado Tamariz,
Alfonso Campos Amezcua and Rafael García Illescas
Instituto de Investigaciones Eléctricas (IIE)
Mexico

1. Introduction

With the growing participation of natural gas (NG) in power generation industry, a global worry about the effects of chemical composition changes on the behavior of gas turbines. A comparison of fossil fuels consumption used in power industry is shown in Figure 1 [1]. This figure represents a projection for the years 2003-2013 according to what is expected by the Mexican electric industry.

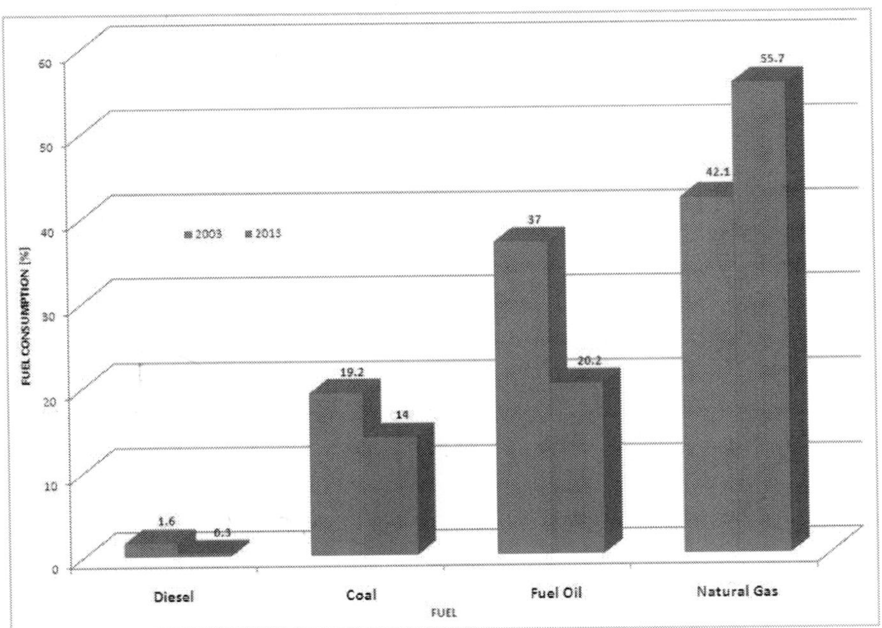

Fig. 1. Fuels Consumption in Mexican Power Industry.

The reduction of the fuel oil consumption observed in the previous figure is due to the process of substitution of this fuel by natural gas. This is because environmental reasons and techniques for incrementing the installed capacity through turbo gas plants and combined cycles based on natural gas.

As a sample of interest of the industry about this topic, it is possible to see that several technical papers related to this problematic, have been published during the last decade. All these works deal with gas quality, the variations on emissions, the increase in the noise levels and the consequences on gas turbine operation [2-10].

As departure point to evaluate the performance of turbogas power plants is analyzes, the increasing demand of electrical power in the world, as well as the necessity count with more modern and efficient electrical power plants orients to the design, installation and commercialization of new power plants which will be developed in harmony with the environment and promote energy saving. Like a behavior model of this phenomenon in Mexico is shown in Figure 2 where is presented the growth perspective to year 2016 of the national power sector in terms of generated gross energy.

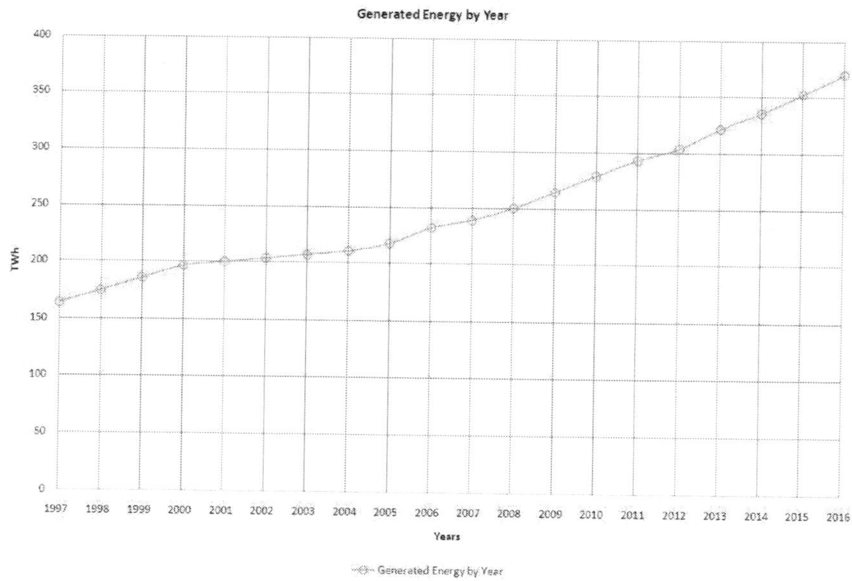

Fig. 2. Generated Gross Energy by Years in Mexico [11].

The growth expectation and modernization in terms of generation of electrical power in the central zone of Mexico gave rules to the Mexican Company of Electric Energy Generation (MCEEG) that operate in this zone to realize the acquisition of electrical power plants of distributed generation on strategic and critics zones around of Mexico city. Therefore MCEEG drove the installation of fourteen 32MW turbo gas electrical power plants impelled by aero derivatives turbines which burn natural gas.

As a result of the technical and administrative necessity of the MCEEG; combined with the effective national regulations in Mexico on terms of energetic production, it is realized collaboration on Mexican Electrical Research Institute (MERI) with the objective to develop this task during the works development until the commercial operation of each unit generation.

The main requirement to obtain the authorization certificate for beginning of the commercial operation of each one of fourteen power plants was the performance evaluation of each unit expressed in terms of average net power capacity, heat consumption rate, maximum noise levels and permissible levels of emissions. The MERI was to realize the behavior evaluation of these parameters through the determination of the average net power capacity, heat consumption rate, maximum noise levels and permissible levels of emissions from data collected during acceptance test of each turbo gas unit which was lead by a company specialized in the subject proposed by the supplier of the commissioning service of power plants and guaranteed by the MCEEG. As an additional product of the work carried out by the MERI, was develop a thermodynamics model based on real parameters obtained in each acceptance test which represents the initial operational state of the units and specifies the guidelines for a future trending monitoring system. The power plant has an aeroderivative turbine installed in simple cycle, operating with a cooling system (chiller) with effective capacity for a specific altitude and operation under any load regimen up to 32 MW.

2. Performance evaluation process

These tests are made on the final individual devices of each plant and integrate approaches to assessing the quality of the product. This is done in order to verify agreement between the guaranties and collected data on their performance under normal conditions of operation of the turbo unit, at different load levels with all its controls in automatic mode.

The goal of the tests is to verify if all devices of the open cycle turbo generator comply with the requirements established before, such as the guaranteed net capacity with the net unit heat consumption, guaranteed low level of noise and guaranteed pollutant emissions.

2.1 Power plant performance

The performance tests of Average Net Power Capacity (P_{NET}) and Heat Consumption Rate (Q_{RATE}) for 32MW turbo gas power plants were carried out in operation condition under base load with the air cooling system at turbine entrance running.

The measurements were developed with a combination of high precision instruments installed isolated and independent of the whole system (not connected to it). The measuring system implemented in the plant itself was also done. The measurements with high precision instruments were developed with instruments installed specifically for the performance test. Measurements of the plant are those made with permanent instrumentation of the plant, including measurements made by the flow meters from the gas station.

The net capacity (P_{NET}) was measured on the high side of the generator step-up transformer with the installed watt-hour meter (Digital-Multi-Meter DMM-B). The energy readings were recorded manually from the screen meter. The readings of energy (kWh) and the precise schedules (time) were used to calculate the average net capacity (P_{NET}) and net unit heat consumption (Q_{RATE}) carrying out the test using the following formulations:

$$P_{NET} = \frac{PM_{END} - PM_{START}}{\Delta t} \quad (1)$$

$$Q_{CONS} = \frac{m_{\tau}(Q_{ATM} + h_{CC} + h_{ATM})}{\Delta t} \quad (2)$$

$$Q_{RATE} = \frac{Q_{CONS}}{P_{NET}} \quad (3)$$

Where: PM_{END} and PM_{START} represented the power measurement at end and beginning the test (kWh), Δt is the time period of the test (h), Q_{CONS} is the heat consumption (kJ/h), m_{τ} is the mass of fuel used during the test period (kg), Q_{ATM} is Reference specific energy of fuel (kJ/kg), h_{CC} and h_{ATM} represented the specific enthalpy of the fuel (combustion chamber) and the reference specific enthalpy of the fuel (kJ/kg).

These tests results of the P_{NET} and Q_{RATE} should be corrected considering the variation between the current environmental conditions at the time of the test and the normal environmental conditions. Those are listed in Table 1.

PARAMETER
Fuel
Equipment Status (Load)
Atmospheric Temperature
Relative Humidity
Chiller Temperature
Altitude
Barometric Pressure
Low Calorific Power (Fuel)

Table 1. Basis Conditions of the Tests.

In order to meet the guaranteed values of performance in each turbo gas unit, it was necessary to adjust the values of (P_{NET}) and (Q_{RATE}) considering the base conditions. This was done by means of the use of correction curves by temperature, barometric pressure, relative humidity and fuel calorific value.

The Corrected Average Net Power Capacity (CP_{NET}) and The Corrected Net Unit Heat Consumption (CQ_{RATE}) are calculated by correcting the P_{NET} and Q_{RATE} values to the guarantee basis conditions; this is done following the correction methodology shown below:

$$CP_{NET} = \frac{P_{NET}}{F_{1P} \cdot F_{2P} \cdot F_{3P} \cdot F_{4P}} \quad (4)$$

$$CQ_{RATE} = \frac{Q_{RATE}}{F_{1HR} \cdot F_{2HR} \cdot F_{3HR} \cdot F_{4HR}} \tag{5}$$

Where: F_{1P}, F_{2P}, F_{3P} and F_{4P} represented the correction factor of measured power versus atmospheric temperature, humidity, barometric pressure and low calorific power (fuel) for corrected net capacity. F_{1HR}, F_{2HR}, F_{3HR} and F_{4HR} represented the correction factor of measured power versus atmospheric temperature, humidity, barometric pressure and low calorific power (fuel) for corrected heat consumption rate.

The correction factors are calculated by linear interpolation between the data points shown in the correction curves (see Figures 3, 4, 5 and 6). The data measured in each run test are averaged before entering the calculation of interpolation.

2.2 Noise level emissions evaluation

In Mexico, noise contamination produced by utilities is regulated by the Secretary of the Environment and Natural Resources ("SEMARNAT" by its initials in Spanish) [12]. This office establishes the policies and applicable regulations. On the other hand, the Mexican Official Standard ("NOM" by its initials in Spanish) [13] is the responsible for establishing maximum limits allowable for noise emissions produced by each industrial or power plant inside or outside the facilities.

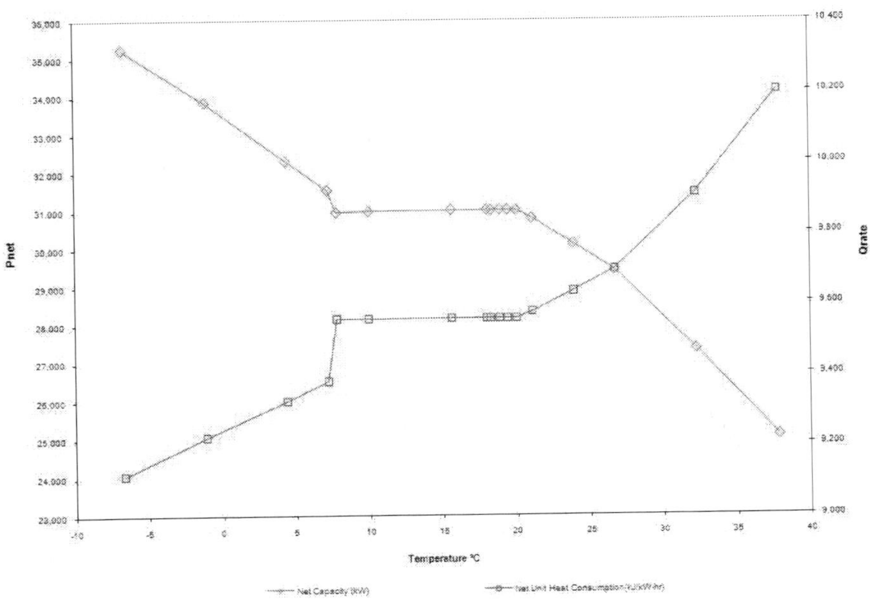

Fig. 3. Correction curve by temperature.

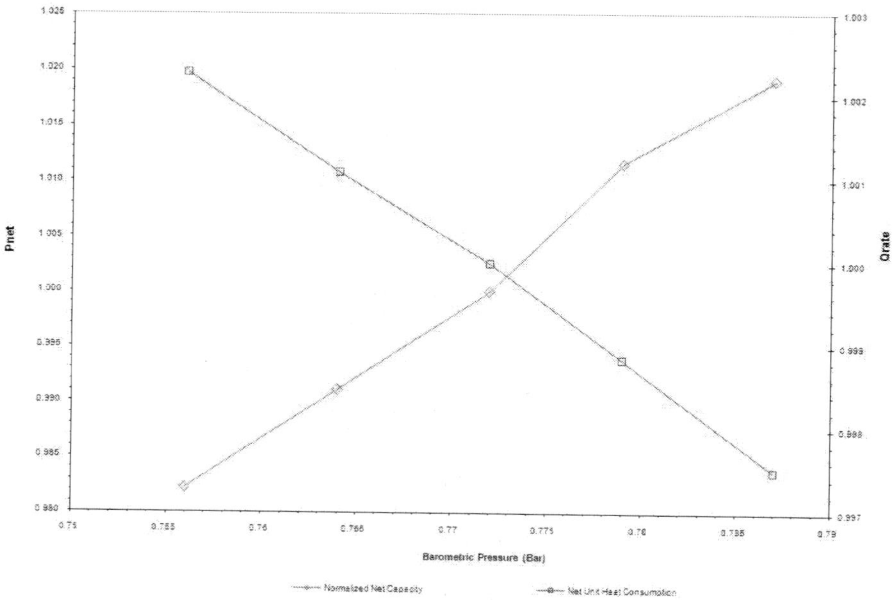

Fig. 4. Correction curve by barometric pressure.

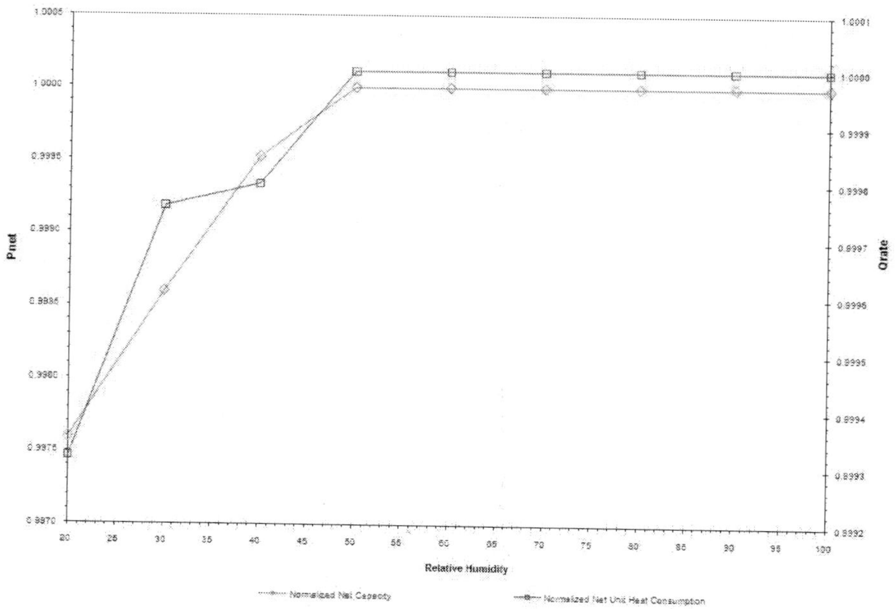

Fig. 5. Correction curve by humidity.

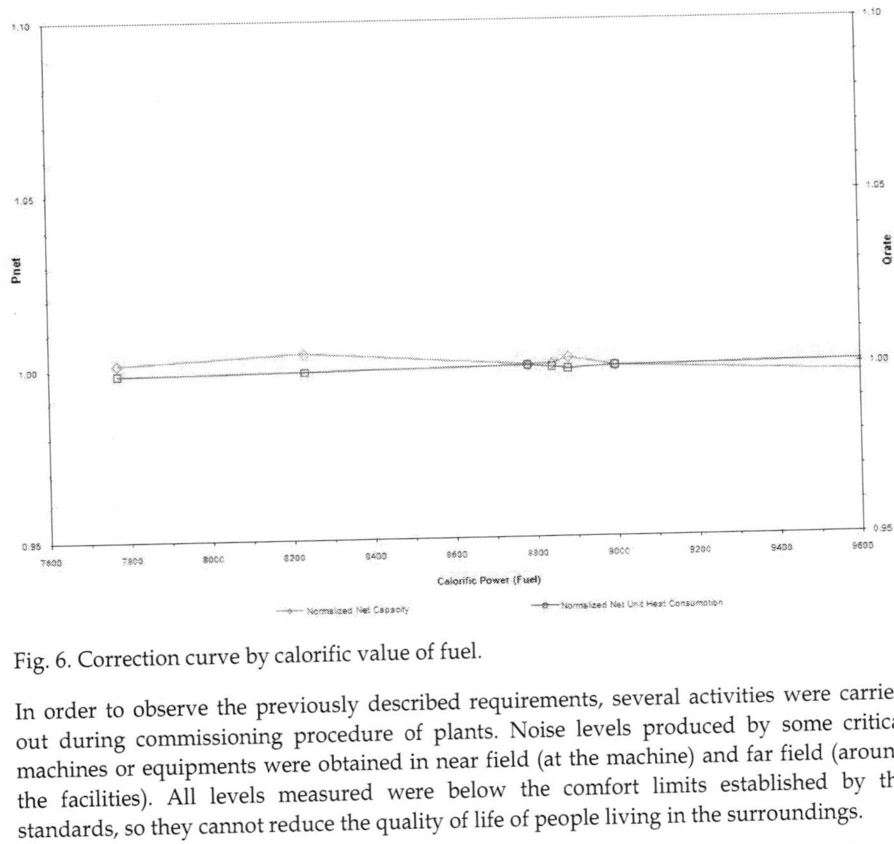

Fig. 6. Correction curve by calorific value of fuel.

In order to observe the previously described requirements, several activities were carried out during commissioning procedure of plants. Noise levels produced by some critical machines or equipments were obtained in near field (at the machine) and far field (around the facilities). All levels measured were below the comfort limits established by the standards, so they cannot reduce the quality of life of people living in the surroundings.

In order to get a correct measuring data processing acquired during far and near field noise evaluation, time intervals should be taken into account as well as minimum and maximum levels for each measuring point.

The equivalent sound level at each measuring point is determined through the mathematical expression (6):

$$N_{EQ} = 10\log\frac{1}{m}\left(\sum\nolimits_{m} 10\frac{N}{10}\right) \tag{6}$$

Then the means must be calculated from the measured values for each point by means of equation (7):

$$N_{50} = \frac{\sum_{i}^{n} N_i}{n} \tag{7}$$

In a similar way, from equation (8) the numerical values of standard deviation can be assessed for each point of the test:

$$\sigma = \frac{\sqrt{\sum (N_i - N_{50})^2}}{n-1} \tag{8}$$

It is also important to find out the noise level values that were present during more than 10% of total recording time through the equation (9):

$$N_{10} = N_{50} + 1.28817\sigma \tag{9}$$

Then corrected numerical values should be obtained from previous measurements:
Correction due to the presence of extreme values:

$$C_e = 0.9023\sigma \tag{10}$$

N_{50} average values from the fixed source and back noise:

$$\Delta_{50} = (N_{50})_{Source}(N_{50})_{Far} \tag{11}$$

Back noise correction:

$$C_f = -(\Delta_{50} + 9) + 3\sqrt{4\Delta_{50} - 3} \tag{12}$$

The numeric noise pressure value Nff from the noise source is obtained from equation (8):

$$N_{ff} = N_{50} + C_e \tag{13}$$

Finally, the last correction from back noise is used to determine the real numerical value of the noise source (equation 13)

$$N'_{ff} = N_{ff} + C_f \tag{14}$$

Where: N_{EQ} is the equivalent sound level (dBA), N is the fluctuant sound level (dBA), m is the total number of measurements, N_{50} is the mean of measured values (dBA), N_i is the sound level during certain time (dBA), n is the number of observations for each measuring point and σ is the standard deviation of measurements.

2.3 Emissions evaluation

Air quality and effects on the environment that produce the greenhouse gases emissions are of major importance for development of any country like Mexico. The nitrogen oxide emissions (NOx) from power generation equipment that uses technologies based on combustion cycles; are strictly regulated in order to develop monitoring and verification process of its generated atmospheric emissions during operation, according to Mexican Official Standards (NOM) which are compatible with the International Standards. For that previous reason, and based on the geographic location of the power plant facilities and according to Mexican Official Standards [14], this site is denominated like a critical zone (C_{ZN}) in Mexico because of high atmospheric pollution levels. This demonstrates the importance of establishing an emissions reference model that the plant generates during operation process.

The reference model was generated from the emission diagnosis develop to the power plant during the performance test at the end of the commissioning procedure. This model will be used as a reference basis for evaluation of emissions averages rates that the unit must handle during the history of operation.

Emissions testing and measurements were accomplished when the unit was operating at base load (100%) and at partial load of 75% with a 15% of O2.

The equipment used during testing consists of a fixed CEMS (Continuous Emission Monitoring System) which was calibrated prior and posterior to tests. This is done by comparing to a certified composition of reference gases found to be within the allowable deviation of +/- 3% [15].

Measurements were recorded manually from PC interface of the emission monitoring. Emissions were stored and endorsed every 110 minutes for both cases (full and partial load) observing a stable behavior of emissions.

After NOx measurements, some statistical calculations with the obtained data were done determining the mean and standard deviation for each load and observing the behavior of such values during load changes.

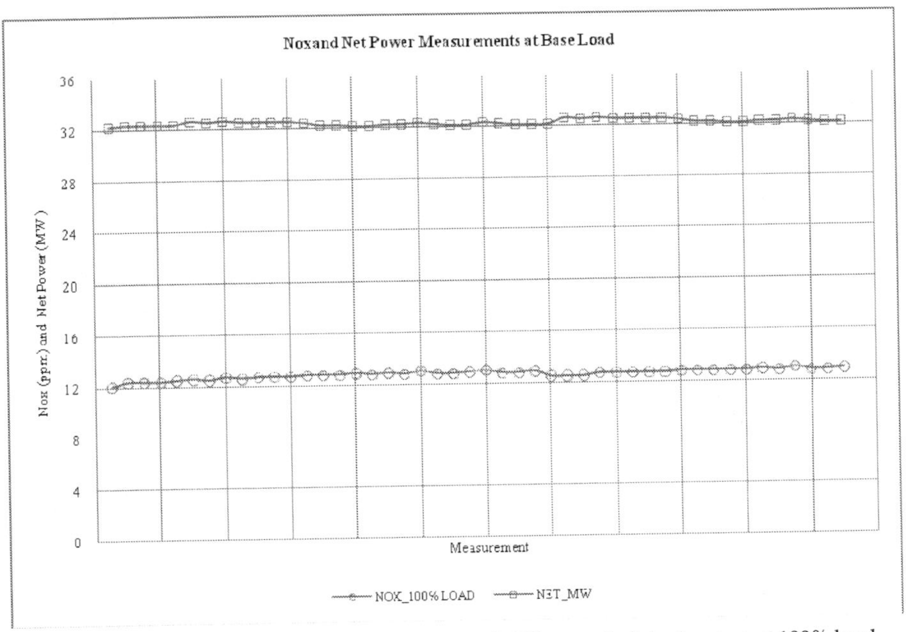

Fig. 7. NOx measurements (ppm) and Net Power (MW) recorded during test at 100% load.

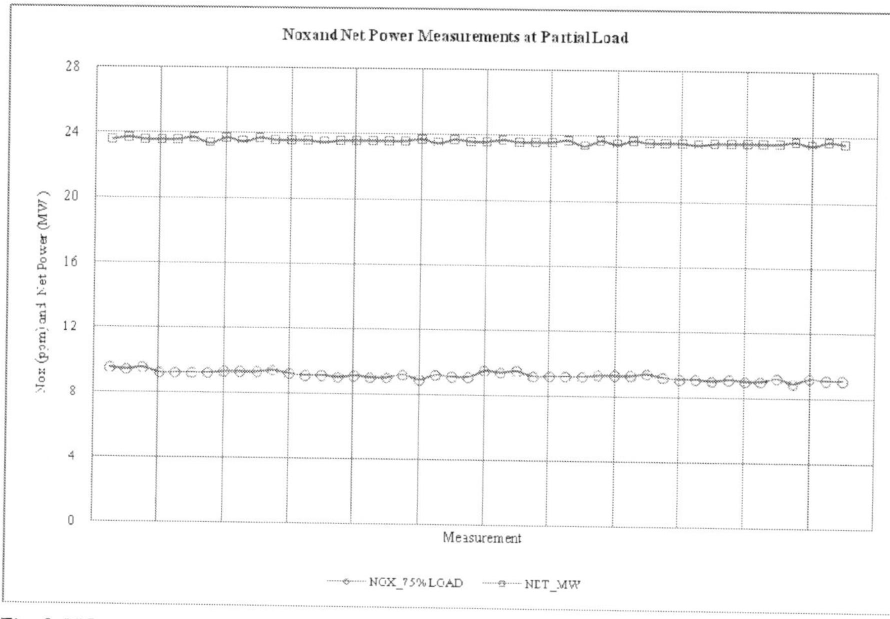

Fig. 8. NOx measurements (ppm) and Net Power (MW) recorded during test at 75% load.

3. Evaluation results

Once the tests, measurements, partial calculations and corrections of the mathematical models have been completed, it is come to make an evaluation of the results obtained with respect to the reference values guaranteed by the manufacturer and with the applicable standards for each parameter.

Firstly the results evaluation on terms of Average Net Power Capacity (P_{NET}) and Net Unit Heat Consumption (Q_{RATE}) at base load and partial load (75 percent) it is realized.

According to the values provided by the manufacturer and based on the average basic operations conditions for each one of the fourteen power plants, the guaranteed reference values of Average Net Power Capacity (P_{NET}) and Net Unit Heat Consumption (Q_{RATE}) are shown in the table 2:

PARAMETER	GUATANTEE VALUE	UNITS
Guaranteed Net Capacity	31,368	kW
Guaranteed Net Unit Heat Consumption	9,545	kJ/kWh

Table 2. Performance guarantees.

A comparison summary on term of Guaranteed Average Net Power Capacity with respect to the obtained results in test for fourteen power plants at base load and partial load are shown in Figures 9 and 10.

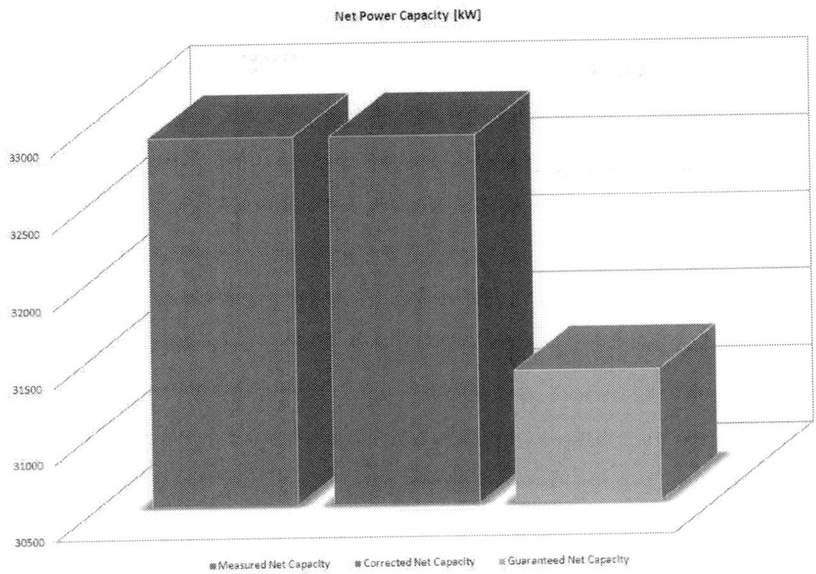

Fig. 9. Average Net Power Capacity comparison test results vs. guaranteed values of the units at base load.

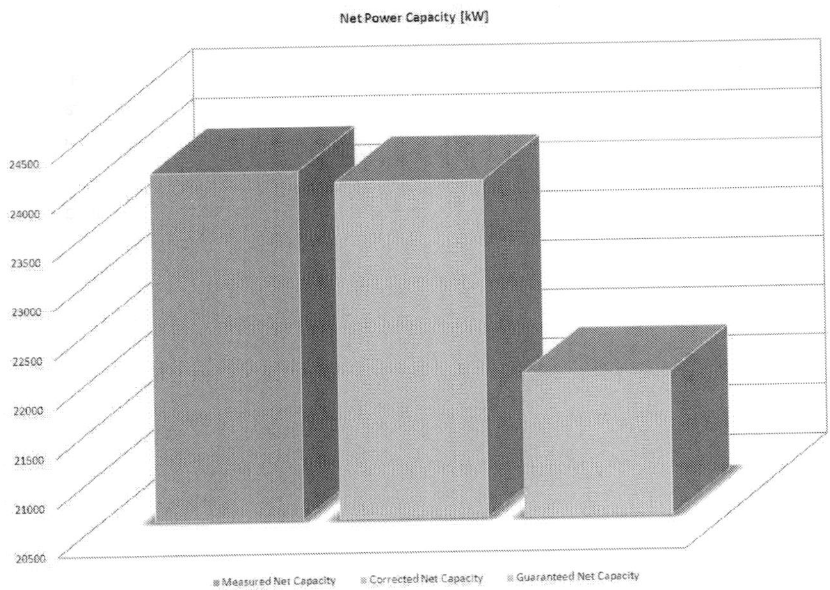

Fig. 10. Average Net Power Capacity comparison test results vs. guaranteed values of the units at partial load (75 percent).

Similar form to a the previous summary now appears a comparison in terms of Guaranteed Net Unit Heat Consumption with respect to the obtained results in test for fourteen power plants at base load and partial load are shown in Figures 11 and 12.

In the other hand the NOx measurements, some statistical calculations with the obtained data were done determining the mean and standard deviation for each load (100% and 75%) and observing the behavior of such values during load changes. Figure 13 shows graphically the recorded deviations.

Figure 14 shows a comparison of reference values established by the national standards [15] an international standards [16], the guaranteed value by the manufacturer [17] and the mean values obtained during the tests.

Once obtained the near and far field noise pressure levels during the test for the fourteen power plants, they were compared to the reference value found in the Mexican standard [13] as shown in Figure 15. From that standard it can be seen that the reference value is higher than the measured one guaranteed by the manufacturer.

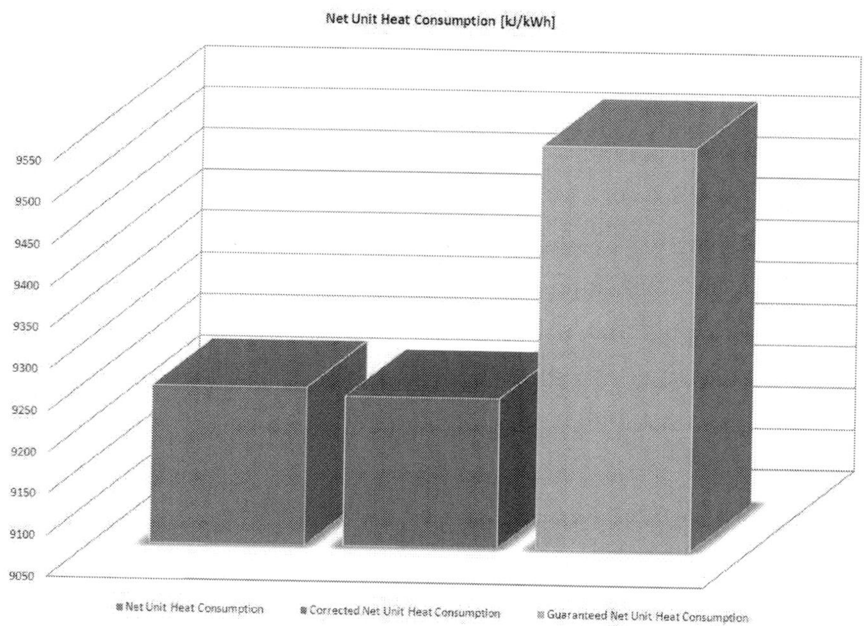

Fig. 11. Net Unit Heat Consumption comparison test results vs. guaranteed values of the units at base load.

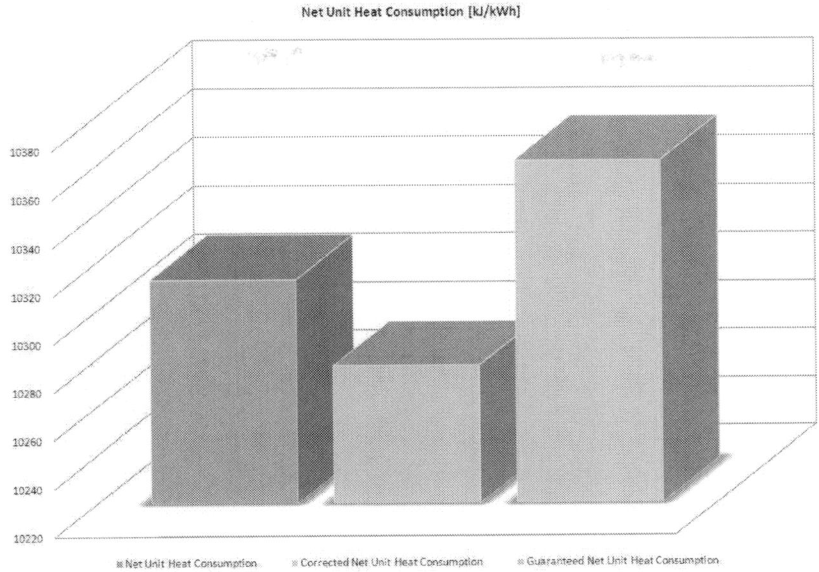

Fig. 12. Net Unit Heat Consumption comparison test results vs. guaranteed values of the units at partial load (75 percent).

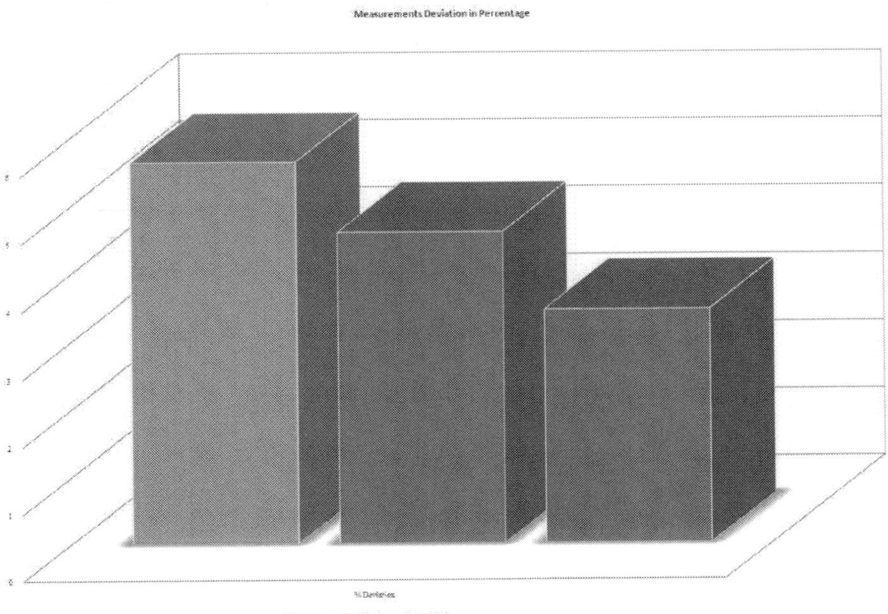

Fig. 13. NOx measurement deviation in percentage obtained during tests.

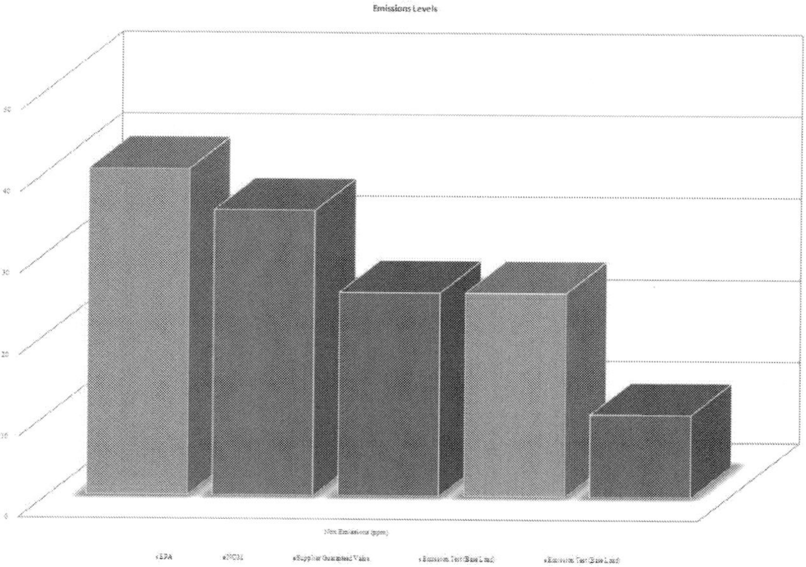

Fig. 14. Comparison of the results obtained during the tests with regard to national and international standards and guaranteed values by the manufacturer.

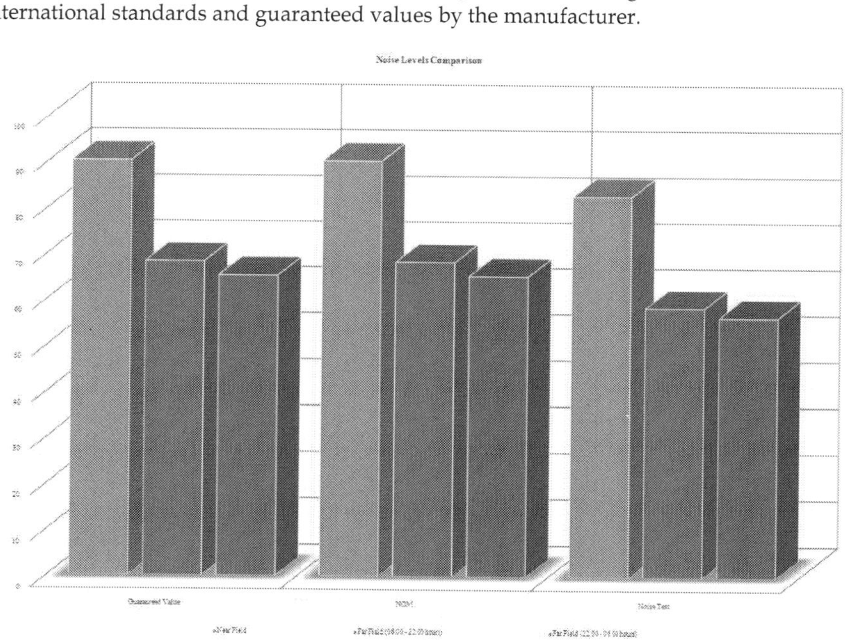

Fig. 15. Comparison of the results obtained during the tests with regard to national and guaranteed values by the manufacturer.

4. Reference model

The reference model based on acceptance test results were generated from the diagnosis develops to the power plants during performance test at the end of the commissioning procedure. This model will be used as a reference basis for the evaluation averages rates that the unit must handle during the history of operation.

In figure 16 is show behavior tendency in terms of average net capacity, developed from the calculated values during the performance test for each one of the fourteen power plants. That is behavior and variation rates that must be have any unit during history operation to maintain it is energetic efficiency.

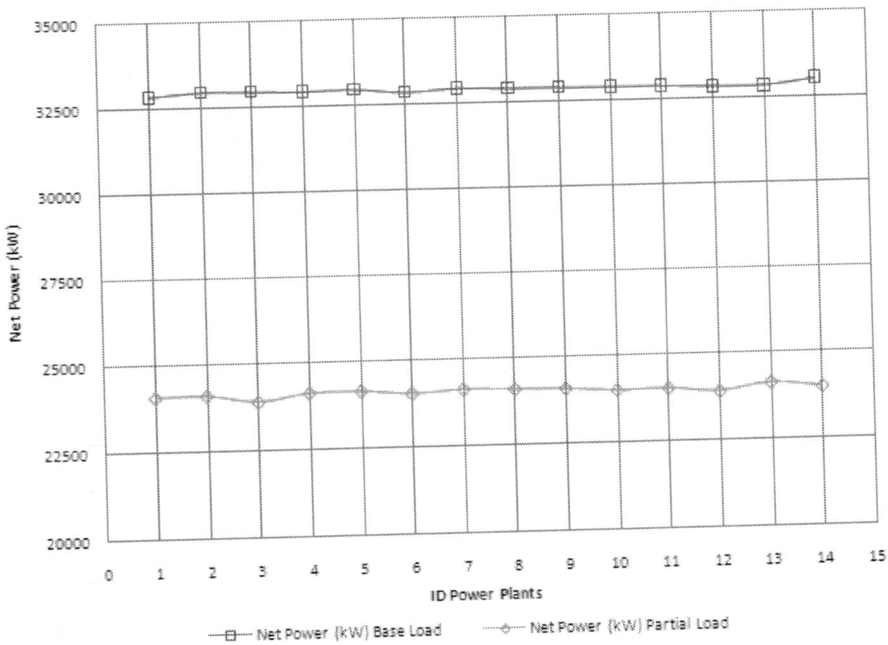

Fig. 16. Reference model of the variation rate in term of average net capacity for a power plant unit at base load or partial load (75 percent).

Furthermore is shown the performance of the unit in terms of emissions to determinate the grade of air pollution exhaust for the power plants. This reference model also was developing from the calculated values during the performance test for each one of the fourteen power plants. It is show in Figure 17.

As a final result of the reference model developed its obtains a Mathematical expressions that consider the average net capacity of the unit based on the emissions levels of the power plant as it show in equation 15:

$$P_{NET} = \alpha NOx^4 - \beta NOx^3 + \phi NOx^2 - \theta NOx + \omega \tag{15}$$

Where: α, β, ϕ, θ and ω are real constans values determined from data colection.

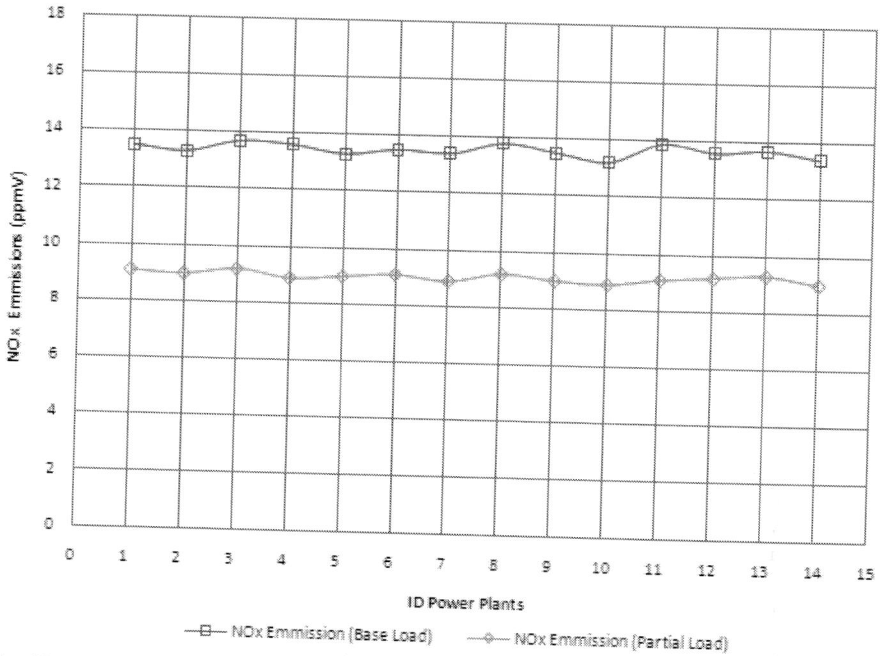

Fig. 17. Reference model of the variation rate in term of emissions for a power plant unit at base load or partial load (75 percent).

5. Conclusions

The goal of this work is to present the fundamentals for determining, through evaluation, a reference model for average net capacity and unit heat consumption as well as pollutant emissions level; furthermore the noise level in fourteen turbo gas power plants.

The model will serve as a reference in terms of Net Power and NOx emissions in the net effects of air quality due to this means of power generation. It will be possible to evaluate, during life history of the unit, the impact through the time due to natural degradation of equipment on the environment. This will be presented as a lower efficiency of the system and higher emissions and noise probably.

The noise levels have been determined for normal operation of unit to establish allowable limits to assure population's health and in the surroundings and personnel. Similarly, the NOx emissions and average net capacity model of reference will be a tool for comparison purposes in future operation.

On the other hand, as a main result of this analysis, it was verified that the unit observes all national and international regulations as well as all requirements previously established between costumer and manufacturer.

Finally, as a result of this study, it is guaranteed to the costumer that this plant will operate correctly satisfying technical requirements, according to unit design and local needs of generation in terms of efficiency and emissions.

6. Acknowledgment

The authors wish to acknowledge the support of the Mexican Federal Electricity Commission (Comisión Federal de Electricidad, CFE) to carry out this study.

7. References

[NOM-081-ECOL-1994 "*Establishment of the maximum permissible limits of noise emissions from fixed sources and its measurement method*", in Spanish

Andrés A. Amell, César A. Bedoya, Bladimir Suárez, Efectos del cambio de composición química del gas natural sobre el comportamiento de turbinas a gas: Una aproximación al caso colombiano, *Revista Energética*, Vol. 35, 2006.

ASME PTC-22-1997., "*Performance Test Code on Gas Turbine*", Performance Test Codes, ASME, 1997

Bengt Gudmundsson, Jenny Larfeldt, Stable dry low emission combustion with fuel flexibility, *7° Fórum de Turbomáquinas Petrobras,* Rio de Janeiro - 24 a 26 de novembro de 2009.

Christian Arnold, Hubert Skiba, Alexander Walke, Gas Market Liberalization in Germany and the Possible Consequences for Gas Turbine Operation, *OMMI* Vol. 2 Issue 3, December 2003.

D J Abbott, The impact of fuel gas composition on gas turbine operation: Future challenges, *British – French Flame Days*, Lille 9 / 10 March 2009.

G. Peureux, S. Carpentier, G. Lartigue, NOx emissions prediction for natural gas engines with fuel quality variations, *Proceedings of the European Combustion Meeting* 2009.

HQ-CTP904 - Commissioning Manual. "*Emissions Test Procedure*", June, 2009.

K S Chana, K J Syed, M I Wedlock, R W Copplestone, M S Cook and G Bulat, Novel unsteady temperature/ heat transfer instrumentation and measurements in the presence of combustor instabilities, *ASME Turbo Expo, 6-9 June 2005*, Reno-Tahoe, Nevada, USA.

Lars O. Nord and Helmer G. Andersen, Influence of variations in the natural gas properties on the combustion process in terms of emissions and pulsations for a heavy-duty gas turbine, *Proceedings of International Joint Power Generation Conference*, June 16-19, 2003, Atlanta, Georgia, USA

NOM-085-ECOL-2003 "*Atmospheric pollution in stationary sources, for fixed sources that use solid fuels fossil, liquid or gaseous or anyone of its combinations that it establishes the maximum permissible levels of emission to the atmosphere of smoke and suspended particles*", in Spanish.

Perspective of investment of the Mexican electrical sector, *COPARMEX-CFE*, June 2007

Report "*Distribution of industrial pollution by source of generation in Mexico*" RECT (Registro de Emisiones y Transferencias de Contaminantes, in Spanish), November 2007.

Robin McMillan, Peter Martin, Richard Noden, Mike Welch, Gas Fuel Flexibility in a Dry Low Emissions Combustion System, *Demag Delaval Industrial Turbomachinery Ltd.*, UK, February 2004.

S. Zaheer Akhtar. Gas Quality: Gas Turbine Boom Places Premium on Ensuring High-Quality Fuel Gas, *Power-Gen* Vol. 105, Issue 10, October 2001.

Secretary of the Environment and Natural Resources ("SEMARNAT" by its initials in Spanish), www.semarnat.gob.mx

Teresa Hansen, Meeting Gas Quality Challenges, *Power-Gen* Vol. 112, Issue 5, May 2008.

U.S. Environmental Protection Agency, http://www.epa.gov/